The 4th International Conference on Advances in Environmental Engineering

The 4th International Conference on Advances in Environmental Engineering

Editors

Adriana Estokova
Natalia Junakova
Tomas Dvorsky
Vojtech Vaclavik
Magdalena Balintova

Basel • Beijing • Wuhan • Barcelona • Belgrade • Novi Sad • Cluj • Manchester

Editors

Adriana Estokova
Technical University of Kosice
Kosice
Slovakia

Natalia Junakova
Technical University of Kosice
Kosice
Slovakia

Tomas Dvorsky
VSB—Technical University of Ostrava
Ostrava
Czech Republic

Vojtech Vaclavik
VSB—Technical University of Ostrava
Ostrava
Czech Republic

Magdalena Balintova
Technical University of Kosice
Kosice
Slovakia

Editorial Office
MDPI
St. Alban-Anlage 66
4052 Basel, Switzerland

This is a reprint of articles from the Proceedings published online in the open access journal *Engineering Proceedings* (ISSN 2673-4591) (available at: https://www.mdpi.com/2673-4591/57/1).

For citation purposes, cite each article independently as indicated on the article page online and as indicated below:

Lastname, A.A.; Lastname, B.B. Article Title. *Journal Name* **Year**, *Volume Number*, Page Range.

ISBN 978-3-7258-0239-5 (Hbk)
ISBN 978-3-7258-0240-1 (PDF)
doi.org/10.3390/books978-3-7258-0240-1

© 2024 by the authors. Articles in this book are Open Access and distributed under the Creative Commons Attribution (CC BY) license. The book as a whole is distributed by MDPI under the terms and conditions of the Creative Commons Attribution-NonCommercial-NoDerivs (CC BY-NC-ND) license.

Contents

Adriana Estokova, Natalia Junakova, Tomas Dvorsky, Vojtech Vaclavik and Magdalena Balintova
Preface: The 4th International Conference on Advances in Environmental Engineering
Reprinted from: *Eng. Proc.* **2023**, *57*, 1, doi:10.3390/engproc2023057001 1

Adriana Estokova, Natalia Junakova, Tomas Dvorsky, Vojtech Vaclavik and Magdalena Balintova
Statement of Peer Review
Reprinted from: *Eng. Proc.* **2023**, *57*, 2, doi:10.3390/engproc2023057002 5

Natalia Junakova, Magdalena Balintova, Jozef Junak and Stefan Demcak
The Impact of Anthropogenic Activity on the Quality of Bottom Sediments in the Watershed of the Delňa Creek
Reprinted from: *Eng. Proc.* **2023**, *57*, 3, doi:10.3390/engproc2023057003 6

Magdalena Balintova, Natalia Junakova and Yelizaveta Chernysh
The Influence of Acidic Mine Waters on Physico-Chemical Processes in the Aquatic Environment
Reprinted from: *Eng. Proc.* **2023**, *57*, 4, doi:10.3390/engproc2023057004 12

Adriana Eštoková, Martina Fabianová and Marek Radačovský
Life Cycle Assessment and Environmental Impacts of Building Materials: Evaluating Transport-Related Factors
Reprinted from: *Eng. Proc.* **2023**, *57*, 5, doi:10.3390/engproc2023057005 19

Eva Kridlova Burdova, Jana Budajova, Peter Mesaros and Silvia Vilcekova
Potential for Reducing the Carbon Footprint of Buildings
Reprinted from: *Eng. Proc.* **2023**, *57*, 6, doi:10.3390/engproc2023057006 27

Marketa Vašinková
pH Variation during Bioaccumulation of Selected Toxic Metals by Newly Isolated Microscopic Fungi from the Ostramo Lagoons
Reprinted from: *Eng. Proc.* **2023**, *57*, 7, doi:10.3390/engproc2023057007 35

Petra Malíková and Jitka Chromíková
The Water Quality of Revitalized Ponds in the Czech Republic Post-Mining Area
Reprinted from: *Eng. Proc.* **2023**, *57*, 8, doi:10.3390/engproc2023057008 43

Radka Matová, Petra Malíková, Silvie Drabinová and Jitka Chromíková
The Evaluation of the Rapid Sand Filter Wash Interval at the Central DWTP in the Czech Republic
Reprinted from: *Eng. Proc.* **2023**, *57*, 9, doi:10.3390/engproc2023057009 53

Sanja Nosalj, Alexandra Šimonovičová, Elena Pieckova, Domenico Pangallo and Bruno Gabel
The Activity of Two Different Solutions against Selected Fungal Phyto-Pathogens
Reprinted from: *Eng. Proc.* **2023**, *57*, 10, doi:10.3390/engproc2023057010 62

Petra Liszoková, Jan Škrkal, Barbara Stalmachová, Věra Záhorová and Helena Pilátová
Occurrence of ^{137}Cs in Soil and Agricultural and Forest Products of the Contaminated Northeastern Part of the Czech Republic
Reprinted from: *Eng. Proc.* **2023**, *57*, 11, doi:10.3390/engproc2023057011 67

Adam Pochyba, Dagmar Samešová, Juraj Poništ and Jozef Salva
Pulp and Paper Mill Sludge Utilization by Biological Methods
Reprinted from: *Eng. Proc.* **2023**, *57*, 12, doi:10.3390/engproc2023057012 77

Karla Placova, Jan Halfar, Katerina Brozova and Silvie Heviankova
Issues of Non-Steroidal Anti-Inflammatory Drugs in Aquatic Environments: A Review Study
Reprinted from: *Eng. Proc.* **2023**, *57*, 13, doi:10.3390/engproc2023057013 84

Radmila Kucerova, Michal Zavoral, Jaroslav Mudrunka, David Takac and Lucie Marcalikova
Analysis of Firewater Samples from Simulated Fires in Illegal Waste Dumps
Reprinted from: *Eng. Proc.* **2023**, *57*, 14, doi:10.3390/engproc2023057014 92

Jiří Kupka, Adéla Brázdová and Tereza Chowaniecová
Environmental Interaction Elements in the Post-Mining Landscape of the Karviná District (Czech Republic)
Reprinted from: *Eng. Proc.* **2023**, *57*, 15, doi:10.3390/engproc2023057015 98

Jan Halfar, Kateřina Brožová, Karla Placová and Miroslav Kyncl
Determining the Presence of Micro-Particles in Drinking Water in the Czech Republic—An Exploratory Study Focusing on Microplastics and Additives
Reprinted from: *Eng. Proc.* **2023**, *57*, 16, doi:10.3390/engproc2023057016 106

Petr Jadviscok and Tereza Gottvaldova
Watercourses and Their Geodetic Mapping for Water Management
Reprinted from: *Eng. Proc.* **2023**, *57*, 17, doi:10.3390/engproc2023057017 113

Jiří Kupka, Adéla Brázdová and Jana Vodová
Problematic Perspectives of Units of Military Fortification in Landscape Management (Teschen Silesia, Czech Republic)
Reprinted from: *Eng. Proc.* **2023**, *57*, 18, doi:10.3390/engproc2023057018 122

Jitka Chromíková, Veronika Rogozná and Tomáš Dvorský
Revitalization of Small Watercourse
Reprinted from: *Eng. Proc.* **2023**, *57*, 19, doi:10.3390/engproc2023057019 132

Zuzana Zemanova, Sarka Krocova and Patrik Sirotiak
Risk Management in the Water Industry
Reprinted from: *Eng. Proc.* **2023**, *57*, 20, doi:10.3390/engproc2023057020 141

Lukáš Kupka and Barbara Stalmachová
Post-Mining Landscape of the Karviná Region and Its Importance for Nature and Landscape Conservation
Reprinted from: *Eng. Proc.* **2023**, *57*, 21, doi:10.3390/engproc2023057021 147

Danka Barloková, Ján Ilavský, Jana Sedláková and Alena Matis
Removal of Humic Substances from Water with Granular Activated Carbons
Reprinted from: *Eng. Proc.* **2023**, *57*, 22, doi:10.3390/engproc2023057022 155

Svitlana Delehan, Hanna Melehanych and Andrii Khorolskyi
The Traditions and Technologies of Ecological Construction in Portugal
Reprinted from: *Eng. Proc.* **2023**, *57*, 23, doi:10.3390/engproc2023057023 164

Lukas Balcarik, Bohdana Simackova, Samaneh Shaghaghi and Lucie Syrova
Effect of Ash from Biomass Combustion on Tailings pH
Reprinted from: *Eng. Proc.* **2023**, *57*, 24, doi:10.3390/engproc2023057024 175

Vladislav Blažek and Jaroslav Závada
The Resistance of an Enamelled Material to Biochemical Leaching
Reprinted from: *Eng. Proc.* **2023**, *57*, 25, doi:10.3390/engproc2023057025 . . . . **184**

Lucie Syrová, Bohdana Šimáčková, Lukáš Balcařík and Samaneh Shaghaghi
Evaluation of the Effect of De-icing Materials on Soil Quality in Selected Areas of the Moravian-Silesian Region
Reprinted from: *Eng. Proc.* **2023**, *57*, 26, doi:10.3390/engproc2023057026 . . . . **190**

Silvie Drabinová, Miroslav Kyncl and Martin Minář
Assessment of the Risks to the Drinking Water Supply System of the Nový Malín Communal Waterworks
Reprinted from: *Eng. Proc.* **2023**, *57*, 27, doi:10.3390/engproc2023057027 . . . . **196**

Lenka Demková, Lenka Bobuľská, Ľuboš Harangozo and Július Árvay
Using Bio-Monitors to Determine the Mercury Air Pollution in a Former Mining Area
Reprinted from: *Eng. Proc.* **2023**, *57*, 28, doi:10.3390/engproc2023057028 . . . . **202**

Vojtěch Václavík, Tomaš Dvorský and Pavlína Richtarová
Separator Systems for Light Liquids
Reprinted from: *Eng. Proc.* **2023**, *57*, 29, doi:10.3390/engproc2023057029 . . . . **209**

Oto Novak
Phytotoxicity Assessment of Wastewater from Industrial Pulp Production
Reprinted from: *Eng. Proc.* **2023**, *57*, 30, doi:10.3390/engproc2023057030 . . . . **217**

Lenka Bobuľská, Lenka Demková and Tomáš Lošák
The Determination of Soil Microbial Biomass Carbon and Adenosine Triphosphate Concentrations at Different Temperatures
Reprinted from: *Eng. Proc.* **2023**, *57*, 31, doi:10.3390/engproc2023057031 . . . . **223**

Pavel Buchta, Vojtěch Václavík and Tomaš Dvorský
The Design for the Reconstruction of Settling Tanks
Reprinted from: *Eng. Proc.* **2023**, *57*, 32, doi:10.3390/engproc2023057032 . . . . **228**

Ján Ilavský and Danka Barloková
The Removal of Selected Pharmaceuticals from Water by Adsorption with Granular Activated Carbons
Reprinted from: *Eng. Proc.* **2023**, *57*, 33, doi:10.3390/engproc2023057033 . . . . **234**

Jaroslav Mudruňka, Kateřina Matunová Kavková, Radmila Kučerová, Lucie Marcaliková, David Takač, Nikola Drahorádová, et al.
Technology of the Biological Treatment of Mine Water at the Kohinoor II Mine
Reprinted from: *Eng. Proc.* **2023**, *57*, 34, doi:10.3390/engproc2023057034 . . . . **243**

Helena Hybská, Martina Mordáčová and Mária Gregušová
Biodegradation of the Personal Care Products
Reprinted from: *Eng. Proc.* **2023**, *57*, 35, doi:10.3390/engproc2023057035 . . . . **250**

Miriama Čambál Hološová, Adriana Eštoková and Miloslav Lupták
Rapid Chloride Permeability Test of Mortar Samples with Various Admixtures
Reprinted from: *Eng. Proc.* **2023**, *57*, 36, doi:10.3390/engproc2023057036 . . . . **257**

Ida Antonie Bogáňová and Petr Hluštík
Ecotoxicity of Wastewater in the Czech Republic
Reprinted from: *Eng. Proc.* **2023**, *57*, 37, doi:10.3390/engproc2023057037 . . . . **264**

Martina Zeleňáková, Tatiana Soľáková, Mladen Milanović, Milan Gocić and Hany F. Abd-Elhamid
Drought Risks Assessment Using Standardized Precipitation Index
Reprinted from: *Eng. Proc.* **2023**, *57*, 38, doi:10.3390/engproc2023057038 273

Veronika Bilkova, Bohdana Simackova, Oto Novak and Lukas Balcarik
Ecotoxicity Assessment of Substrates from a Thermally Active Coal Tailing Dump Using Tests for *Daphnia magna*
Reprinted from: *Eng. Proc.* **2023**, *57*, 39, doi:10.3390/engproc2023057039 277

Lucia Domaracka, Simona Matuskova, Marcela Tausova, Barbara Kowal and Katarina Culkova
A Comparison and Development of Municipal Waste Management in Three Countries, Slovakia, the Czech Republic and Poland, with an Emphasis on the Slovak Republic
Reprinted from: *Eng. Proc.* **2023**, *57*, 40, doi:10.3390/engproc2023057040 283

Martin Valica, Tomáš Lempochner, Linda Machalová, Vanda Adamcová, Patrícia Marková, Lenka Hutárová, et al.
Removal of Cadmium from Aqueous Solution Using Dried Biomass of *Euglena gracilis* var. *bacillaris*
Reprinted from: *Eng. Proc.* **2023**, *57*, 41, doi:10.3390/engproc2023057041 291

Marketa Dreslova
Effects of Vermicompost Application on Plant Growth Stimulation in Technogenic Soils
Reprinted from: *Eng. Proc.* **2023**, *57*, 42, doi:10.3390/engproc2023057042 303

Silvia Vilčeková, Peter Mésároš, Eva Krídlová Burdová and Jana Budajová
End-of-Life Stage Analysis of Building Materials in Relation to Circular Construction
Reprinted from: *Eng. Proc.* **2023**, *57*, 43, doi:10.3390/engproc2023057043 309

Benito Guillermo Mendoza Trujillo, Leonardo Sebastián Cadena Rojas and Andrés Santiago Cisneros Barahona
Identification of the Hydrogeological Potential in Langos-San Andres, by Means of Electrical Resistivity Tomography Interpretation
Reprinted from: *Eng. Proc.* **2024**, *57*, 44, doi:10.3390/engproc2023057044 320

Editorial

Preface: The 4th International Conference on Advances in Environmental Engineering [†]

Adriana Estokova [1], Natalia Junakova [2,*], Tomas Dvorsky [3], Vojtech Vaclavik [3] and Magdalena Balintova [2]

[1] Department of Material Engineering, Institute for Sustainable and Circular Construction, Faculty of Civil Engineering, Technical University of Kosice, Vysokoskolska 4, 04200 Kosice, Slovakia; adriana.estokova@tuke.sk

[2] Department of Environmental Engineering, Institute for Sustainable and Circular Construction, Faculty of Civil Engineering, Technical University of Kosice, Vysokoskolska 4, 04200 Kosice, Slovakia; magdalena.balintova@tuke.sk

[3] Department of Environmental Engineering, Faculty of Mining and Geology, VSB—Technical University of Ostrava, 17. Listopadu 15/2172, 708 00 Ostrava, Czech Republic; tomas.dvorsky@vsb.cz (T.D.); vojtech.vaclavik@vsb.cz (V.V.)

* Correspondence: natalia.junakova@tuke.sk; Tel.: +421-55-602-4266

[†] Presented at the 4th International Conference on Advances in Environmental Engineering, Ostrava, Czech Republic, 20–22 November 2023.

1. Introduction

The 4th International Conference on Advances in Environmental Engineering (AEE 2023) was hosted by the Faculty of Mining and Geology at the VŠB-Technical University of Ostrava in the Czech Republic, in collaboration with the Faculty of Civil Engineering at the Technical University of Kosice in Slovakia. This conference took place from 20 to 22 November 2023, in Ostrava, Czech Republic. It successfully gathered more than 57 participants, both PhD students and representatives from various industries, hailing from five different countries: The Slovak Republic, the Czech Republic, Spain, Poland, and Iraq. This conference builds upon the previous successful editions, where it was possible to bring together experts from both academia and industry to discuss current topics, seek new solutions, and, above all, establish mutual cooperation and create a network of experts in the field of environmental engineering at the regional level. The conference was held under the auspices of the representatives of the faculties: Prof. Ing. Hana Staňková, Ph.D., the Dean of the Faculty of Mining and Geology at the VŠB-Technical University of Ostrava, and Prof. Ing. Dušan Katunský, CSc., the Dean of the Faculty of Civil Engineering at Technical University of Kosice.

From the numerous submissions received, 42 contributions were chosen for inclusion in this *Engineering Proceedings* volume. The organizing committee expresses deep gratitude for the dedicated efforts of the Editorial Board and the reviewers for their invaluable work in preparing this volume. Special thanks are also extended to all the authors for their cooperation and adherence to deadlines.

2. Conference Topics

The conference primarily concentrated on the broad spectrum of environmental engineering, encompassing topics such as:

- Water and wastewater management;
- Water and sediment pollution control and remediation;
- Waste management;
- Soil degradation, conservation, and remediation;
- Landscape management;
- Sustainable building materials;
- Building ecology;

- Environmentally friendly building materials and technologies;
- Life cycle analysis of materials and constructions;
- Air pollution prevention;
- Environmental urbanism and sustainable cities;
- Environmentally sustainable entrepreneurship;
- Solutions for climate change mitigation.

The scientific presentations, which encompassed lectures and posters, covered an extensive array of research domains. This volume offers a comprehensive perspective on the current trends in environmental engineering and addresses the challenges in achieving sustainability.

3. Scientific Committee

3.1. Chairs of the Conference

Professor Adriana Eštoková, the conference chair and a full professor at the Faculty of Civil Engineering of the Technical University of Kosice in Slovakia, specializes in materials and environmental engineering. Her research centers on sustainable construction, particularly the development of building materials derived from waste and secondary materials, their characterization, and evaluation of their environmental safety, including heavy metals, leachability, and radionuclide activity. She also investigates the environmental impact of these materials in buildings, employing life cycle assessment (LCA) and assessing their carbon footprint. Additionally, her work explores material durability in aggressive environments, addressing concerns such as bio-corrosion and chemical corrosion.

Co-chairperson Associate Professor Vojtěch Václavik, from VSB – Technical University of Ostrava in the Czech Republic, is an environmental engineering expert, with a special focus on materials recycling, recovery, and water management. His research activities are mainly focused on the treatment and use of industrial byproducts for the production of building materials. He maintains active collaboration with industry practitioners.

Another co-chairperson, Professor Magdaléna Bálintová, a full professor at the Faculty of Civil Engineering of the Technical University of Kosice in Slovakia, is focused on wastewater treatment, water and sediment quality, remediation methods, removal of heavy metals, and soil protection. She is affiliated with various associations and serves on the editorial boards of multiple journals. Additionally, she actively participates in scientific committees for both international and national conferences. Her contributions extend to editing, guest editing, and reviewing for several reputable scientific journals.

3.2. Scientific Committee Members

The scientific committee comprised distinguished experts in the field of environmental engineering:

1. Peter ANDRÁŠ, Matej Bel University in Banská Bystrica, Slovakia;
2. Magdaléna BÁLINTOVÁ, Technical University of Kosice, Slovakia;
3. Rui Alexandre CASTANHO, University of Johannesburg, Republic of South Africa;
4. Sanja DIMTER, University of Osijek, Croatia;
5. Endre DOMOKOS, University of Pannonia, Hungary;
6. Konstiantyn DYADYURA, Sumy State University, Ukraine;
7. Anatolijus EISINAS, Kaunas University of Technology, Lithuania;
8. Adriana EŠTOKOVÁ, Technical University of Kosice, Slovakia;
9. Bozena GAJDZIK, Silesian University of Technology, Poland;
10. Silvie HEVIÁNKOVÁ, Technical University of Ostrava, Czech Republic;
11. Petr HLUŠTÍK, Brno University of Technology, Czech Republic;
12. Natália JUNÁKOVÁ, Technical University of Kosice, Slovakia;
13. Sebastian KOT, Czestochowa University, Poland;
14. Šárka KROČOVÁ, VSB—Technical University of Ostrava, Czech Republic;
15. František KUDA, VSB—Technical University of Ostrava, Czech Republic;

16. Radmila KUČEROVÁ, VSB—Technical University of Ostrava, Czech Republic;
17. Sergio António Neves LOUSADA, Universidade de Madeira, Portugal;
18. Antonio MESSINEO, University of Enna "Kore", Italy;
19. Daniel Francois MEYER, University of Johannesburg, Republic of South Africa;
20. NGUYEN THI MAI LINH, Ton Duc Thang University, Vietnam;
21. Martin Tchingnabé PALOU, Slovak Academy of Sciences, Slovakia;
22. Sudhakara PANDIAN R, Vellore Institute of Technology, India;
23. Peter PECIAR, Slovak University of Technology in Bratislava, Slovakia;
24. Eva PERTILE, VSB—Technical University of Ostrava, Czech Republic;
25. PHAM ANH DUC, Ton Duc Thang University, Vietnam;
26. Jiří POKORNÝ, VSB—Technical University of Ostrava, Czech Republic;
27. Janusz RAK, Rzeszow University of Technology, Poland;
28. Mindaugas RIMEIKA, Vilnius Gediminas Technical University, Lithuania;
29. Dagmar SAMEŠOVÁ, Technical University of Zvolen, Slovakia;
30. Serkan SAHINKAYA, Nevsehir Haci Bektas Veli University, Turkey;
31. Khrystyna SOBOL', Lviv Polytechnic National University, Ukraine;
32. Tomáš SVĚRÁK, Brno University of Technology, Czech Republic;
33. Miloslav ŠLEZINGR, Mendel University, Czech Republic;
34. Nadežda ŠTEVULOVÁ, Technical University of Kosice, Slovakia;
35. Hana ŠTVERKOVÁ, VSB—Technical University of Ostrava, Czech Republic;
36. Ekaterina TRUSOVA, Belarussian State Technological University, Belarus;
37. Vojtěch VÁCLAVÍK, VSB—Technical University of Ostrava, Czech Republic;
38. Nikolai VATIN, Peter the Great St. Petersburg Polytechnic University, Russia;
39. Hana VOJTKOVÁ, VSB—Technical University of Ostrava, Czech Republic;
40. Silvia VILČEKOVÁ, Technical University of Kosice, Slovakia;
41. Martins VILNITIS, Riga Technical University, Latvia;
42. Tatjana VOLKOV-HUSOVIC, University of Belgrade, Serbia;
43. Wang YU BO, Hubei University of Technology, China;
44. Tadeusz ZABOROWSKI, Poznan Technical University, Poland.

4. Organizing Committee

The organizational guarantee of the conference was assumed by representatives of the institutions that participated in the organization of the conference: Assoc. Prof. Tomáš Dvorský from the VSB—Technical University of Ostrava, Czech Republic, and Assoc. Prof. Natália Junáková from the Technical University of Kosice, Slovakia.

Organizing Committee Members
1. Jan Halfar
2. Kateřina Brožová
3. Jakub Charvát
4. Kateřina Máčalová
5. Miriama Čambál Hološová
6. Jana Budajová

5. Sponsors

A heartfelt thank you goes out to all the partners and sponsors of the AEE 2023 conference (Figure 1).

Figure 1. Logos of the sponsors [1–12].

Conflicts of Interest: The authors declare no conflict of interest.

References

1. AZ GEO. Available online: https://www.azgeo.cz/ (accessed on 7 November 2023).
2. ECOCOAL. Available online: https://www.ecocoal.cz/ (accessed on 7 November 2023).
3. LESOSTAVBY. Available online: https://lesostavby.cz/ (accessed on 7 November 2023).
4. NICOLET CZ, Molecular Spectroscopy. Available online: https://nicoletcz.cz/ (accessed on 7 November 2023).
5. KUBICEK Blowers. Available online: https://kubicekvhs.cz/cs/ (accessed on 7 November 2023).
6. Ostravské Vodárny a Kanalizace a.s. Available online: https://www.ovak.cz/ (accessed on 7 November 2023).
7. Cementáreň Ladce. Available online: https://www.pcla.sk/en/stranka/home (accessed on 7 November 2023).
8. MERCI. Laboratory as It Should Be. Available online: https://www.merci.cz/en/ (accessed on 7 November 2023).
9. Severomoravské Vodovody a Kanalizace Ostrava, a.s. Available online: https://www.smvak.cz/web/guest/home (accessed on 7 November 2023).
10. Czech Chamber of Authorized Engineers and Technicians Active in Construction. Available online: https://www.ckait.cz/ (accessed on 7 November 2023).
11. Český Svaz Stavebních Inženýrů (ČSSI). Available online: http://www.cssi-cr.cz/ (accessed on 7 November 2023).
12. LENZIG. Available online: https://www.lenzing.com/ (accessed on 7 November 2023).

Disclaimer/Publisher's Note: The statements, opinions and data contained in all publications are solely those of the individual author(s) and contributor(s) and not of MDPI and/or the editor(s). MDPI and/or the editor(s) disclaim responsibility for any injury to people or property resulting from any ideas, methods, instructions or products referred to in the content.

Editorial

Statement of Peer Review †

Adriana Estokova [1], Natalia Junakova [2,*], Tomas Dvorsky [3], Vojtech Vaclavik [3] and Magdalena Balintova [2]

[1] Department of Material Engineering, Institute for Sustainable and Circular Construction, Faculty of Civil Engineering, Technical University of Kosice, Vysokoskolska 4, 04200 Kosice, Slovakia; adriana.estokova@tuke.sk

[2] Department of Environmental Engineering, Institute for Sustainable and Circular Construction, Faculty of Civil Engineering, Technical University of Kosice, Vysokoskolska 4, 04200 Kosice, Slovakia; magdalena.balintova@tuke.sk

[3] Department of Environmental Engineering, Faculty of Mining and Geology, VSB—Technical University of Ostrava, 17. Listopadu 15/2172, 708 00 Ostrava, Czech Republic; tomas.dvorsky@vsb.cz (T.D.); vojtech.vaclavik@vsb.cz (V.V.)

* Correspondence: natalia.junakova@tuke.sk; Tel.: +421-55-602-4266

† Presented at the 4th International Conference on Advances in Environmental Engineering, Ostrava, Czech Republic, 20–22 November 2023.

In submitting conference proceedings to *Engineering Proceedings*, the editors certify that all papers published in this volume have been subjected to peer review by the volume editors. The reviews were conducted by expert referees to the professional and scientific standards expected of a proceedings journal. Reviewers who requested to see the manuscript after revision were re-sent the revised manuscript. All reviewers subsequently agreed to the publication of the revised manuscript.

- Type of peer review: double-blind.
- Conference submission management system: web page, e-mail.
- Number of submissions sent for review: 57.
- Number of submissions accepted: 42.
- Acceptance rate (number of submissions accepted/number of submissions received): 73.68.
- Average number of reviews per paper: 2.
- Total number of reviewers involved: 65.
- Any additional information on the review process: The manuscripts were checked for scientific quality by independent reviewers and for technical quality by members of the editorial team.

Conflicts of Interest: The authors declare no conflict of interest.

Disclaimer/Publisher's Note: The statements, opinions and data contained in all publications are solely those of the individual author(s) and contributor(s) and not of MDPI and/or the editor(s). MDPI and/or the editor(s) disclaim responsibility for any injury to people or property resulting from any ideas, methods, instructions or products referred to in the content.

Citation: Estokova, A.; Junakova, N.; Dvorsky, T.; Vaclavik, V.; Balintova, M. Statement of Peer Review. *Eng. Proc.* **2023**, *57*, 2. https://doi.org/10.3390/engproc2023057002

Published: 17 November 2023

Copyright: © 2023 by the authors. Licensee MDPI, Basel, Switzerland. This article is an open access article distributed under the terms and conditions of the Creative Commons Attribution (CC BY) license (https://creativecommons.org/licenses/by/4.0/).

Proceeding Paper

The Impact of Anthropogenic Activity on the Quality of Bottom Sediments in the Watershed of the Delňa Creek [†]

Natalia Junakova *, Magdalena Balintova, Jozef Junak and Stefan Demcak

Department of Environmental Engineering, Faculty of Civil Engineering, Technical University of Košice, Vysokoškolska 4, 042 00 Košice, Slovakia; magdalena.balintova@tuke.sk (M.B.); jozef.junak@tuke.sk (J.J.); stefan.demcak@tuke.sk (S.D.)

* Correspondence: natalia.junakova@tuke.sk
† Presented at the 4th International Conference on Advances in Environmental Engineering, Ostrava, Czech Republic, 20–22 November 2023.

Abstract: This paper is focused on evaluating the quality of bottom sediments and water in the watershed of the Delňa creek, where gold, antimony and mercury were mined in the past. The results showed that the biggest source of pollution was a heap of mining material, where the limit values of Sb, As, Hg and Pb concentrations in the sediments were exceeded. Other sources of pollution in the river basin were the right-hand tributaries. A comparison of the dependencies of the concentrations of potentially toxic metals in the water and sediments shows that while the concentrations of pollutants in the waters react to the current state of water quality in the basin and tributaries (pH, concentration, discharge), the sediments exhibit a stable concentration character.

Keywords: potentially toxic metals; mining activity; sediment quality; water quality

1. Introduction

Contamination of the aquatic environment with potentially toxic metals is receiving considerable attention worldwide. One of the possible sources of environmental pollution with these elements is mining activity, the consequences of which, despite the cessation of activity, persist in the environment for a long time [1]. Due to the influence of mining activity on watersheds, not only water but also sediments are contaminated. Sediments play an important role in the transport of pollutants into the water system, as they can affect water quality due to remobilization processes. Monitoring changes in the chemical composition of river sediments therefore provides important qualitative information on the overall system of water courses and their adjacent areas [2]. Changes in the pollutants' concentration in the sediment, as well as their potential risk to the aquatic environment and health [3], can be detected by monitoring the quality of sediments. Many studies have been carried out in order to determine the impact of mining activity on land quality. Many studies have been carried out with the aim of finding out the immediate or long-term impact of mining activity on the quality of the water ecosystem and its sediments [4–8]. Long-term observation revealed, for example, that increased concentrations of potentially toxic metals due to mining were detected at the beginning of mining. However, after the end of the mining, a decreasing trend in the concentrations of these pollutants in the sediments was observed [9].

2. Material and Methods

This study of the accumulation of selected pollutants in bottom sediments was carried out in the watershed of the Delňa creek, which flows through the Prešov district (Slovakia). It is a left-hand tributary of the Torysa and is 16 km long.

It springs above the Zlatá Baňa village, and above the village of Kokošovce the creek fills the Sigord water reservoir. The reservoir was built for the purpose of flood protection

Citation: Junakova, N.; Balintova, M.; Junak, J.; Demcak, S. The Impact of Anthropogenic Activity on the Quality of Bottom Sediments in the Watershed of the Delňa Creek. *Eng. Proc.* **2023**, *57*, 3. https://doi.org/10.3390/engproc2023057003

Academic Editors: Adriana Estokova, Tomas Dvorsky and Vojtech Vaclavik

Published: 27 November 2023

Copyright: © 2023 by the authors. Licensee MDPI, Basel, Switzerland. This article is an open access article distributed under the terms and conditions of the Creative Commons Attribution (CC BY) license (https://creativecommons.org/licenses/by/4.0/).

and, in the past, this area was a sought-after recreational location; currently the reservoir is mainly used for fishing. The area of the Zlatá Baňa village is known for its past mining activities. In the 18th century, gold mining experiments began to be carried out in the cadastre of the village. Later clay, antimony and opal were mined there. In the 1990s, geological exploration work was carried out on the Zlatá Baňa deposit, which confirmed the presence of sulfidic, mainly Pb-Zn, ores [10].

Bottom sediment and water samples were taken from 11 sampling points in the Delňa creek basin and its right-bank tributaries due to their presumed pollution (Figure 1).

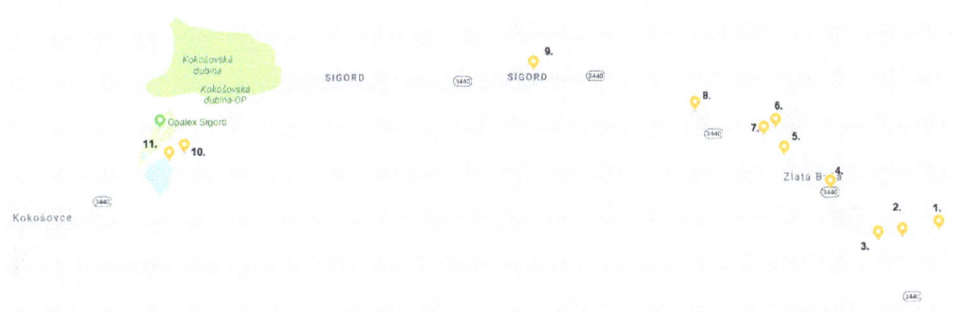

Figure 1. Sampling points in the watershed of the Sigord reservoir.

Sampling point No. 1—a heap of mining material—is located above the built-up area of the Zlatá Baňa village and is detected as the main source of water and sediment pollution in the watershed. The sampling points No. 4, 6 and 8 are located on the right-bank tributaries of the Delňa creek in the built-up area of the village Zlatá Baňa. Other sampling locations (No. 2, 3, 5, 7, 9, 10) are situated on the Delňa creek. Sampling point No. 11 is the inlet to the Sigord reservoir. Sediment samples were taken in plastic bags, water samples were taken in small closable glass bottles. Water and sediment samples were analyzed in laboratory conditions.

Chemical composition of the sediments and water was determined via X-ray fluorescence (XRF) spectrometry using SPECTRO iQ II (Germany). To determine the pH of leachates (prepared in ratio sediment to water of 1:5) the multifunction device MX 300 Xmatepro (METTLER TOLEDO, Columbus, OH, USA) was used.

3. Results and Discussion

3.1. Chemical Composition of the Sediments and Water

To identify the mining activity's impact on the quality of sediments and surface water in the monitored area, the presence of contaminants such as Sb, As, Ni, Cu, Hg, Cr, Al, Pb, Zn, Fe and Mn was determined. The results of the chemical analysis of the monitored contaminants in the sediments and water are shown in Tables 1 and 2. The quality of the bottom sediments was evaluated according to the criteria specified in the Guideline of the Ministry of the Environment of the Slovak Republic No. 549/98-2 for the assessment of risks from polluted streams and reservoirs sediments. The quality of the surface water in the watershed was evaluated according to the Slovak Government Regulation no. 269/2010 Coll., laying down the requirements for the achievement of good water status. Sediments and water concentrations exceeding the maximum allowable values, according to the mentioned regulations, are indicated in bold in Tables 1 and 2.

Table 1. Results of chemical analysis of selected elements in sediments from the Delňa creek watershed.

Sample/Distance from the Source of Contamination	Concentrations in Sediments								(%)		
	(mg/kg)										
	Sb	As	Hg	Cr	Pb	Cu	Zn	Ni	Fe	Al	Mn
S1/0 km (heap)	203.3	980.9	40.9	101.7	997.0	51.2	114.3	31.4	3.2	7.5	<0.0010
S2/0.29 km	21.5	166.4	14.5	163.9	117.9	84.0	1226.0	44.3	4.3	5.1	0.3105
S3/0.65 km	13.2	185.0	4.6	104.1	110.8	76.2	2378.0	45.3	4.5	4.3	0.5259
S4 (tributary)	10.9	7.6	2.9	82.4	<2.0	47.4	75.4	36.0	2.6	5.0	0.1060
S5/1.72 km	13.4	35.5	<2.0	134.1	64.4	68.6	621.9	43.3	3.5	5.9	0.2831
S6 (tributary)	7.2	<1.0	<2.0	180.1	<2.0	37.1	47.4	34.5	2.5	4.6	0.1738
S7/1.92 km	17.7	41.5	<2.0	41.3	5.8	37.1	215.5	31.4	2.7	4.3	0.1022
S8 (tributary)	12.0	<1.0	<2.0	173.8	<2.0	35.9	88.8	35.5	3.4	5.6	0.1017
S9/3.88 km	20.3	181.9	<2.0	176.3	<2.0	20.5	65.9	16.8	3.8	2.3	0.2122
S10/6.86 km	10.1	<1.0	5.0	120.6	<2.0	50.4	167.6	41.8	2.9	5.5	0.1164
S11/6.96 km, inflow to the reservoir	8.7	<1.0	3.4	125.5	6.2	41.7	124.6	43.1	2.3	5.3	0.0404
Limit (Guideline No. 549/98-2)	15.0	55.0	10.0	380.0	530.0	73.0	620.0	44.0	-	-	-

Table 2. Results of chemical analysis of selected elements in surface water from the Delňa creek watershed.

Sample/Distance from Source of Contamination	Concentrations in Water (mg/L)										
	Sb	As	Hg	Cr	Pb	Cu	Zn	Ni	Fe	Al	Mn
W1/0 km (heap)	10.6	<1.0	<2.0	104.9	<2.0	16.2	<1	17.3	89.2	37.8	<5.1
W2/0.29 km	8.4	<1.0	<2.0	126.1	<2.0	12.4	<1	13.4	47.2	27.0	<5.1
W3/0.65 km	<2.4	<1.0	<2.0	124.9	<2.0	18.8	<1	16.5	52.1	14.4	<5.1
W4 (tributary)	7.7	<1.0	<2.0	107.7	<2.0	9.1	<1	15.5	47.3	32.1	<5.1
W5/1.72 km	7.2	<1.0	<2.0	114.4	<2.0	17.8	<1	16.1	51.1	67.3	<5.1
W6 (tributary)	10.1	<1.0	<2.0	110.8	<2.0	16.3	<1	14.0	48.5	16.4	<5.1
W7/1.92 km	7.2	<1.0	<2.0	178.7	<2.0	14.9	<1	14.9	49.2	48.9	<5.1
W8 (tributary)	5.4	<1.0	<2.0	113.6	<2.0	9.0	<1	15.7	51.9	72.5	<5.1
W9/3.88 km	9.9	<1.0	<2.0	125.2	<2.0	10.9	<1	14.9	49.5	37.6	<5.1
W10/6.86 km	4.9	<1.0	<2.0	116.2	<2.0	15.7	<1	14.2	49.4	20.3	<5.1
W11/6.96 km, inflow to the reservoir	8.0	<1.0	<2.0	155.1	<2.0	8.9	<1	8.1	49.4	44.8	<5.1
Limit values (Reg. No. 269/2010)	-	0.0075	0.00007	-	-	-	-	-	2.0	0.2	0.3

The results showed that the heap of mining material was a very significant source of pollution in the studied basin, where the maximum allowable concentrations in sediments and surface water were significantly exceeded for some elements (Sb, As, Hg, Pb, Cu and Zn for sediments, and Fe and Al for surface water).

Due to the influence of pH on the mobilization of contaminants at the sediment/water interface, the pH values of the sediments were determined in the leachates (Table 3).

Table 3. The pH values of sediments.

Sample	pH
S1	3.39
S2	8.72
S3	7.45
S4	7.63
S5	8.23
S6	8.11
S7	8.17
S8	8.51
S9	7.72
S10	7.80
S11	6.77

The lowest pH value was found in S1 sample, which was also assumed to be due to the presence of sulfidic mining material deposited on the heap.

3.2. Accumulation of the Selected Pollutants in the Sediments

Because the increased concentration of potentially toxic metals in waters usually has a direct impact on the concentrations of metals in the sediments [11], in the next part of our study we focused on studying the mobilization of Fe, Al and Sb between the sediment and water. The dependence of the mobilization of these elements on pH values and their distance from the pollution source is shown in Figure 2.

The highest concentrations of iron in the surface water were recorded in sampling point S1—the heap of mining material—which is considered the most significant source of pollution in the studied area (Figure 2). The subsequent decrease in the concentration of Fe in surface water is caused by an increase in the pH value of the water above 3.5 [12,13], which leads to the precipitation of iron cations and their accumulation in the sediment.

The concentrations of aluminum in the water and sediment increased or decreased in direct proportion in all monitored sampling points. An exception was observed in sampling point S10, where there was a decrease in the aluminum concentration in the surface water and a significant increase in the Al in the sediment due to accumulation.

Regarding antimony concentrations, the highest values in the water and sediment were recorded in sampling point 1 (the heap). With the increasing distance from the source of pollution, the concentration of Sb in the sediment decreased. But the right-hand tributaries had the effect of increasing the concentration of antimony in the water due to the presence of dissolved antimony in the tributaries.

A comparison of the dependences of the concentrations of potentially toxic metals in the water and sediments shows that while the concentrations of pollutants in the waters react to the current state of water quality in the basin and tributaries (pH, concentration, discharge), the sediments exhibit a stable concentration character. And with increasing distance from the source of pollution, the accumulation of toxic substances in the sediment occurs to a large extent.

On the other hand, the change in pH due to inflows causes the leaching of heavy metals from the sediments and their subsequent mobility in the watershed. However, it is necessary to add that the concentration of heavy metals is also significantly influenced by their concentration in the tributaries, as was observed in the case of antimony and aluminum.

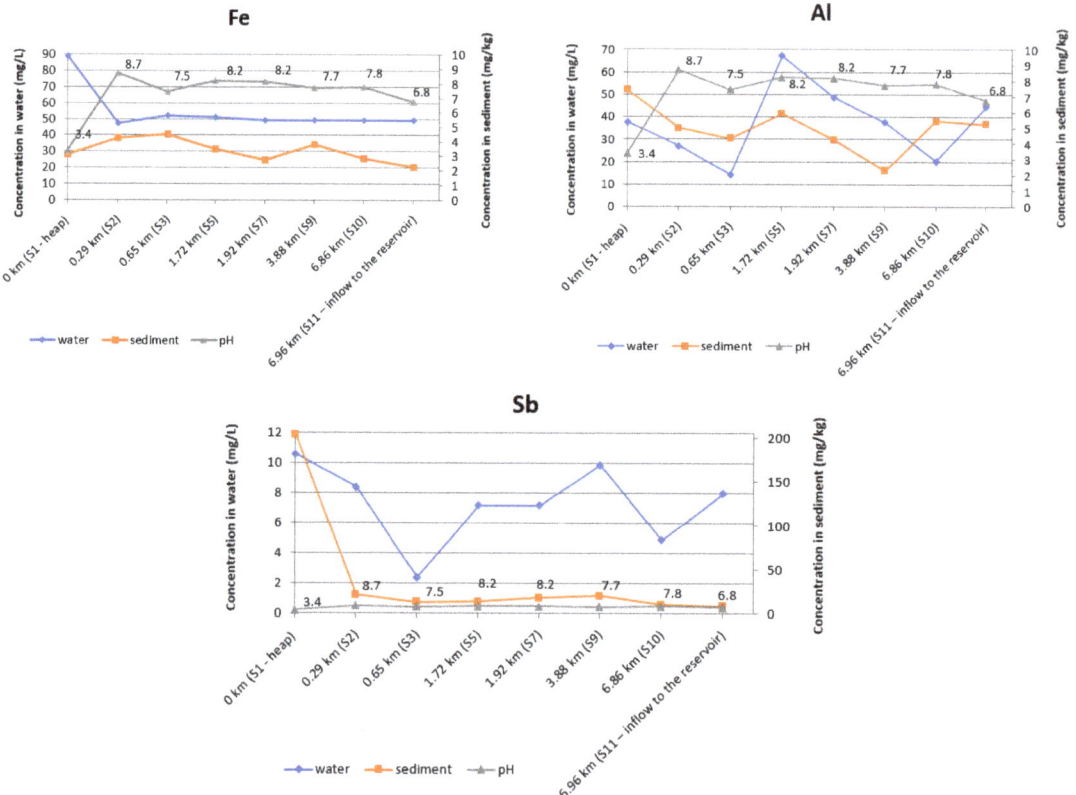

Figure 2. The dependence of the mobilization of Fe, Al and Sb between water and sediment on pH values and their distance from the pollution source.

4. Conclusions

Sediments are a priority source of pollutants in the environment of contaminated areas. They are a reliable indicator of the state of the polluted environment. This work studied the quality of bottom sediments in the Delňa creek watershed, which is affected by mining activity. The results showed that the heap of mining material is a very significant source of pollution in the studied river basin, where the maximum allowable concentrations in the sediments and surface water were significantly exceeded for some potentially toxic elements (Sb, As, Hg, Pb, Cu and Zn for sediments, and Fe and Al for the surface water). In addition to other parameters (discharge, concentration, proportion of clay particles, etc.), the mobilization of contaminants at the interface between the sediment and water is also influenced by the water's pH values, which directly affect the accumulation of heavy metals in sediments, or their release into the surface water. Based on the study of the dependencies between sediment and water in the Delňa creek basins, we can state that increasing the concentration of heavy metals in the water will also increase their concentration in the sediment, but only up to the particular capacity to which the sediment is able to absorb them. However, long-term monitoring is necessary to predict the contamination of water and sediments in the monitored basin.

Author Contributions: Conceptualization, N.J. and M.B.; methodology, N.J. and S.D.; formal analysis, N.J. and J.J.; investigation, N.J., M.B. and J.J.; resources, N.J. and J.J.; data curation, N.J.; writing—original draft preparation, N.J., M.B., J.J. and S.D.; writing—review and editing, N.J. and M.B.; visualization, N.J., M.B., J.J. and S.D.; supervision, N.J. and M.B.; project administration, M.B.; funding acquisition, M.B. All authors have read and agreed to the published version of the manuscript.

Funding: This research was supported by the Slovak Grant Agency for Science (Grant No. 2/0108/23) and by the Slovak Research and Development Agency under the contract No. APVV-20-0140.

Institutional Review Board Statement: Not applicable.

Informed Consent Statement: Not applicable.

Data Availability Statement: The authors declare that all data supporting the results of this research are available in this article.

Conflicts of Interest: The authors declare no conflict of interest.

References

1. Förstner, U.; Wittmann, G. *Metal Pollution in the Aquatic Environment*; Springer: Berlin, Germany, 1983.
2. Junakova, N.; Balintova, M. Predicting Nutrient Loads in Chosen Catchment. *Chem. Eng. Trans.* **2012**, *26*, 591–596.
3. Singovszka, E.; Balintova, M.; Junakova, N. The Impact of Heavy Metals in Water from Abandoned Mine on Human Health. *SN Appl. Sci.* **2020**, *2*, 934. [CrossRef]
4. Wei, W.; Ma, R.; Sun, Z.; Zhou, A.; Bu, J.; Long, X.; Liu, Y. Effects of Mining Activities on the Release of Heavy Metals (HMs) in a Typical Mountain Headwater Region, the Qinghai-Tibet Plateau in China. *Int. J. Environ. Res. Public Health* **2018**, *15*, 1987. [CrossRef] [PubMed]
5. Shang, G.; Wang, X.; Zhu, L.; Liu, S.; Li, H.; Wang, Z.; Wang, B.; Zhang, Z. Heavy Metal Pollution in Xinfengjiang River Sediment and the Response of Fish Species Abundance to Heavy Metal Concentrations. *Int. J. Environ. Res. Public Health* **2022**, *9*, 11087. [CrossRef] [PubMed]
6. Capparelli, M.V.; Cabrera, M.; Rico, A.; Lucas-Solis, O.; Alvear-S, D.; Vasco, S.; Galarza, E.; Shiguango, L.; Pinos-Velez, V.; Pérez-González, A.; et al. An Integrative Approach to Assess the Environmental Impacts of Gold Mining Contamination in the Amazon. *Toxics* **2021**, *9*, 149. [CrossRef] [PubMed]
7. Beck, K.K.; Mariani, M.; Fletcher, M.S.; Schneider, L.; Aquino-López, M.A.; Gadd, P.S.; Heijnis, H.; Saunders, K.M.; Zawadzki, A. The impacts of intensive mining on terrestrial and aquatic ecosystems: A case of sediment pollution and calcium decline in cool temperate Tasmania, Australia. *Environ. Pollut.* **2020**, *265*, 114695. [CrossRef] [PubMed]
8. Smolders, A.J.; Lock, R.A.; Van der Velde, G.; Medina Hoyos, R.I.; Roelofs, J.G. Effects of mining activities on heavy metal concentrations in water, sediment, and macroinvertebrates in different reaches of the Pilcomayo River, South America. *Arch. Environ. Contam. Toxicol.* **2003**, *44*, 314–323. [CrossRef] [PubMed]
9. Lidman, J.; Olid, C.; Bigler, C.; Berglund, Å.M. Effect of past century mining activities on sediment properties and toxicity to freshwater organisms in northern Sweden. *Sci. Total Environ.* **2023**, *872*, 162097. [CrossRef] [PubMed]
10. Košuth, M. Technological characteristics of the Zlatá Baňa ores. In *9. Medzinárodná Banícka Konferencia*; Technical university of Košice: Košice, Slovakia, 1997; pp. 53–57. (In Slovak)
11. Andráš, P.; Buccheri, G.; Turisová, I.; Andráš, P., Jr.; Kupka, J. Heavy metal contamination of the environment in the heap fields of the abandoned Cu deposits of Caporciano (Montecatini Val di Cecina) and Libiola, Italy. *Acta Univ. Matthiae Belii* **2015**, *17*, 34–58. (In Slovak)
12. Xinchao, W.; Roger, C.; Viadero, J.; Karen, M. Recovery of Iron and Aluminum from Acid Mine Drainage by Selective Precipitation. *Environ. Eng. Sci.* **2005**, *22*, 745–755.
13. Balintova, M.; Petrilakova, A. Study of pH influence on selective precipitation of heavy metals from acid mine drainage. *Chem. Eng. Trans.* **2011**, *25*, 345–350.

Disclaimer/Publisher's Note: The statements, opinions and data contained in all publications are solely those of the individual author(s) and contributor(s) and not of MDPI and/or the editor(s). MDPI and/or the editor(s) disclaim responsibility for any injury to people or property resulting from any ideas, methods, instructions or products referred to in the content.

Proceeding Paper

The Influence of Acidic Mine Waters on Physico-Chemical Processes in the Aquatic Environment †

Magdalena Balintova [1,*], Natalia Junakova [1] and Yelizaveta Chernysh [2,3,4]

1. Institute for Sustainable and Circular Construction, Faculty of Civil Engineering, Technical University of Kosice, Vysokoskolska 4, 04200 Kosice, Slovakia; natalia.junakova@tuke.sk
2. Department of Ecology and Environmental Protection Technologies, Sumy State University, 2 Rymskogo-Korsakova St., 40007 Sumy, Ukraine; e.chernish@ssu.edu.ua or chernysh@ftz.czu.cz
3. Faculty of Tropical Agrisciences, Czech University of Life Sciences Prague, Kamýcká 129, 16500 Prague, Czech Republic
4. Department of Water Supply and Wastewater Treatment, T. G. Masaryk Water Research Institute, Podbabská 2582/30, 16000 Prague, Czech Republic
* Correspondence: magdalena.balintova@tuke.sk; Tel.: +421-55-602-4127
† Presented at the 4th International Conference on Advances in Environmental Engineering, Ostrava, Czech Republic, 20–22 November 2023.

Abstract: Acidic mine drainage (AMD) discharged from the abandoned Smolník mine (Pech shaft, Slovakia) contaminates surface water in the Smolník creek due to the decreasing pH and the production of heavy metals. Mixing AMD with surface waters results in an increase in pH, which affects the metal precipitation. Using statistical methods, the effect of pH on the concentration of selected metals (Fe, Mn, Al, Cu and Zn) in the water of the contaminated Smolník creek is described in this work. Polynomial curves were used to identify trends in pH and metal concentration in the surface water. The analysis showed that the second-degree polynomial functions as a candidate for explaining metals' concentration based on the measured surface water's pH with a goodness of model fit, based on a coefficient of determination ranging from 0.4 to 0.7 depending on the determined metal concentration and location.

Keywords: heavy metals; acid mine drainage; pH; polynomial curve

1. Introduction

A specific source of environmental contamination is acid mine drainage (AMD). Its production represents one of the biggest environmental problems already during mining, but especially after mining in polymetallic deposits containing sulfides. When exposed to water and oxygen, most sulfide minerals can oxidize, producing sulfuric acid, metal sulfates that can contaminate surface and groundwater [1]. The source of AMD is primarily the remains of mining activities, e.g., flooded shafts and tunnels, heaps and tailings ponds, representing so-called old mining loads [2,3].

Mining waters are formed during mining, but especially after the end of mining of mineral raw materials in the contact zones of the water and geological environment [4–6]. The amount and composition of minerals in the deposit have a significant influence on the pH of mine water [7,8]. Acid mine drainage (AMD) with pH values below 4.5 occur mainly in sulphide deposits. Their formation is also influenced by iron and the sulfur oxidizing bacteria of the genus Acidithiobacillus. They are autochthonous microorganisms that occur in ore and coal mines where pyrite is found [9]. Metal cations dissolved in AMD are transported to surface waters. Dissolved Fe^{2+} ions are oxidized on the surface by oxygen from the air to Fe^{3+}. This chemical reaction is accompanied by the formation of ocher precipitates (e.g., goethite, jarosite and schwertmanite), which absorb some of the metals on their surface [10]. Heavy metals and sulfates present in AMD contaminate groundwater and surface water, which has a negative impact not only on aquatic organisms, but also on soil contamination and the food chain [11].

There are several mines with AMD generation in the Slovak Republic. The Smolník deposit is one of the historically best-known and richest Cu-Fe ore deposits in Slovakia. After the end of mining in 1990, the mine was flooded, and in 1994, AMD penetrated the surface water, which had a negative impact on aquatic organisms. Because it is a partially open geochemical system into which rainwater and surface water flow, the formation of AMD in the Smolník area cannot be stopped and there is no chance of improving the situation [12,13]. Therefore, the abandoned mining area of Smolník in Slovakia currently belongs to the old environmental burden. The oxidation of pyrite and formation of free sulfuric acid causes water acidification and the dissolution of heavy metals from metal ores. This AMD acidifies and contaminates the waters of the Smolník creek, which transfers pollution to the Hnilec basin [12,14–16]. Increasing the pH of water is associated with the precipitation of metals in the form of hydroxides. Precipitated metals are subsequently accumulated in sediments and can be released again into the aquatic environment upon changes in hydrobiological and physicochemical conditions, such as pH, redox potential, and salinity [11,17–19]. The influence of physicochemical conditions on the accumulation of metals in bottom sediments is also the subject of many investigations [20–25]. They are mainly aimed at studying the influence of pH, redox potential or salinity on the behavior of metals in the water environment.

Polynomial curve fitting is a valuable statistical technique employed in the analysis of chemical results. This equation can help to model the relationship between variables, estimate unknown values, and identify trends or patterns within the chemical data [26].

The aim of this study is to use statistical methods to analyze the effect of acidic mine drainage from the Pech shaft (Smolník mine) on the quality of surface water in the Smolník creek, as well as the effect of pH on the concentration of selected metals in the water.

2. Materials and Methods

Two sampling sites along the Smolník creek were selected for the study of surface water quality (1—approx. 200 m from the Pech shaft, 2—tributary to Hnilec (approx. 9 km)). AMD quality from the Pech shaft was also monitored. The samples were taken in the years 2006–2021. Surface water samples were filtered into a 100 mL plastic container and acidified with 2 mL of ultrapure HNO_3 (67%).

The physical and chemical parameters of the water were determined by a METTLER TOLEDO multifunctional device in situ and the chemical analysis of the water by the AAS method (SpectrAA-30, Varian, Australia).

The results of the measurements and analyses were further evaluated by statistical methods. For the determination of the dependency between concentrations of individual metals and the resulting water pH we have used polynomial fitting. Multiple degrees of polynomials were tested with the coefficient of determination (r^2) being the main criterion for curve selection. Mean squared error (MSE) was used as a secondary, supportive, polynomial selection indicator.

3. Results and Discussion

The average values of metal concentrations and pH values of the surface water (sampling points 1 and 2 from the Smolník creek and acidic mine water from the Pech shaft) during the years 2006–2021 are shown in Table 1. The results are compared with the limit values according to Regulation of the Government of the Slovak Republic no. 269/2010 Coll. Table 1 shows that AMD discharge from the Pech shaft has a permanent negative effect on the water quality in the Smolník creek (samples 1 and 2). Of the monitored metals, the concentrations of elements Fe, Mn, Al, Cu and Zn exceed the limit values. The increase in the concentration of metals is accompanied by a decrease in the pH of the surface water. At the same time, chemical analysis showed that all monitored parameters are exceeded in the AMD.

Table 1. The results of chemical analysis of water in 2006–2021—Smolník creek and Pech shaft.

Parameter		Pech Shaft	Sampling Site 1	2	Limits
Fe		322.70 ± 87.60	12.7 ± 8.9	5.46 ± 5.64	2.00
Mn		25.30 ± 6.30	1.17 ± 0.74	0.91 ± 0.62	0.30
Al	[mg/L]	65.1 ± 19.4	1.17 ± 1.60	0.35 ± 0.72	0.20
Cu		1.51 ± 0.74	0.097 ± 0.116	0.043 ± 0.64	0.02
Zn		7.23 ± 2.27	0.349 ± 0.250	0.254 ± 0.213	0.10
pH		4.0 ± 0.1	5.8 ± 1.1	6.1 ± 1.1	6.0–8.5

As can be seen from Table 1, the concentrations of iron, manganese, aluminum, copper and zinc were exceeded in the Smolník creek. It is known from the literature and our previous studies [11,13,18,20,24] that the behavior of metals in aqueous solutions is influenced by pH, which affects their precipitation and deposition in sediments. Knowledge about the precipitation intervals of selected metals was used to study the effect of pH on the concentration of Al, Cu, Zn, Mn, Fe in the surface water of the Smolník creek.

The statistical analysis was performed for the metal concentrations in water samples taken at sampling points 1 and 2 in the Smolník creek (Table 1). In Figures 1–5 are graphically shown dependences of the concentration of the evaluated metals depending on the pH of the surface water.

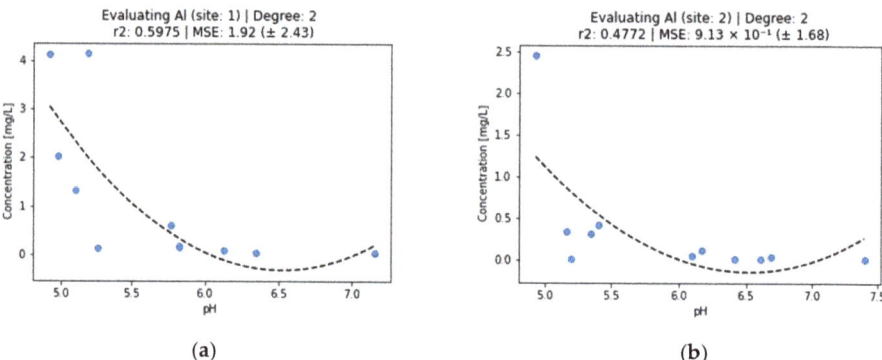

Figure 1. Effect of pH on aluminum concentration in surface water (**a**) sampling point 1, (**b**) sampling point 2.

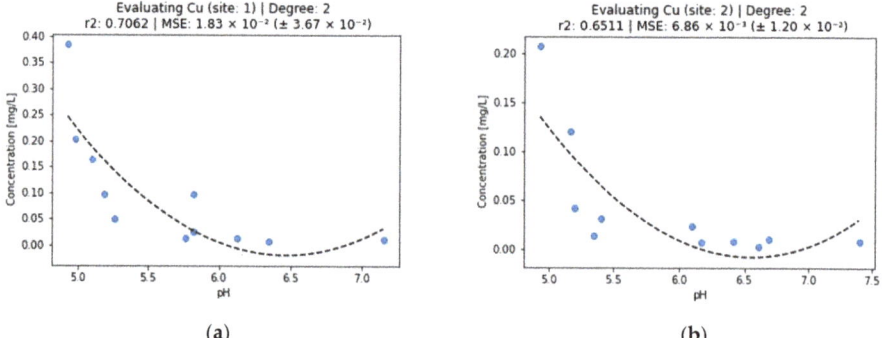

Figure 2. Effect of pH on copper concentration in surface water (**a**) sampling point 1, (**b**) sampling point 2.

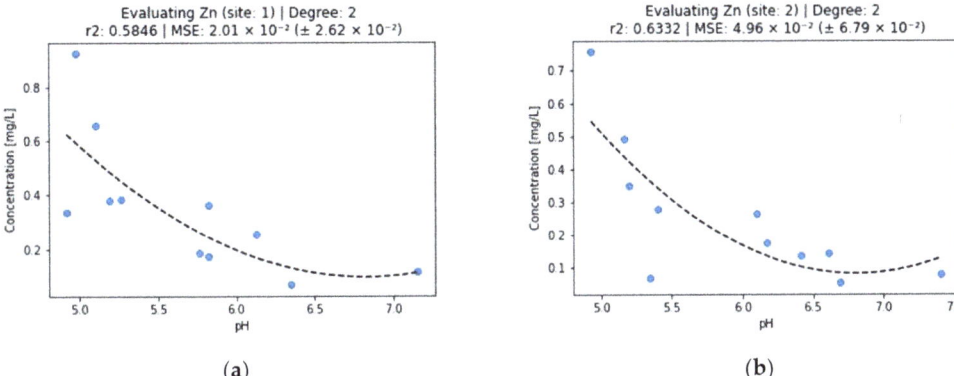

Figure 3. Effect of pH on zinc concentration in surface water (**a**) sampling point 1, (**b**) sampling point 2.

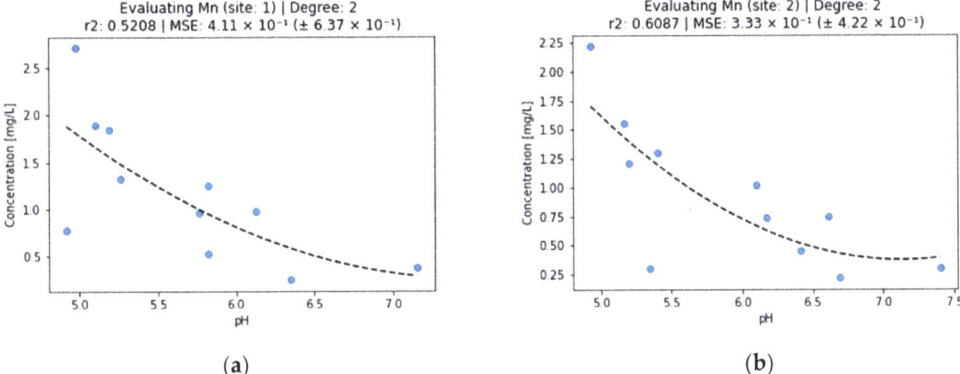

Figure 4. Effect of pH on manganese concentration in surface water (**a**) sampling point 1, (**b**) sampling point 2.

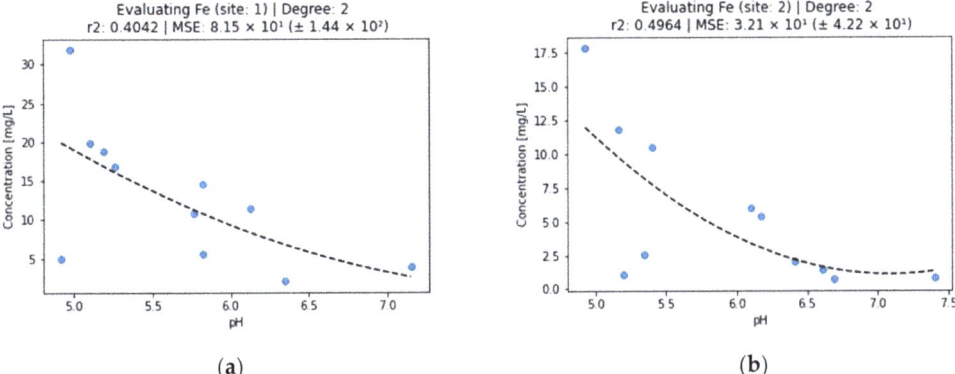

Figure 5. Effect of pH on iron concentration in surface water (**a**) sampling point 1, (**b**) sampling point 2.

In order to statistically evaluate all measurements from both measurement sites for multiple degrees of polynomials, we fitted the respective polynomial curves with a scikit-learn toolbox library available for the programming language Python. The computational evaluation allowed us to test in multiple iterations the polynomial curves of multiple degrees, originally from degree 2 to 10. As a threshold for the best curve candidates' selection, r2 was set to a minimum of 0.5; only for Al and Fe did we need to lower the threshold to 0.45 in the case of Al and to 0.40 for Fe. This was eliminated for all studied chemical elements polynomials of a degree higher than 3. Afterwards, we evaluated r2 in combination with a mean squared error (MSE) to fine-tune the obtained results. The best fitted curves, all with a second-degree polynomial function, are characterized in Figures 1–5.

Figure 1 shows that aluminum concentration decreases with increasing pH. This is in agreement with the literature where aluminum hydroxide precipitates at pH > 5.0 but redissolves at pH > 9.0 [27,28]. The decrease in Al concentration in Figure 1b is also related to the distance of the sampling point from the source of contamination and the longer time of interaction of the contaminant with the water. Looking at the second-degree polynomial curves for Al, we can observe that pH explains the concentration better for site 1—with r2 (0.5975) being higher than for the second site (0.4772).

The precipitation of copper and thus the reduction in its concentration in the water (Figure 2) was carried out in accordance with the literature, according to which copper precipitates at pH > 4 and completely precipitates at pH 6 [19,20]. As can be seen in Figure 2, the concentration of Cu at pH 6 was lower than 0.05 mg/L. Concentrations of copper in both sites have the best explainability by pH from all analyzed metals. Similarly to aluminum, site 1 has a stronger relation between pH and concentration than site 2.

According to [19,27], zinc precipitates in the pH range of 5.5–7. This was reflected in a decrease in concentration below 0.1 mg/L at a pH above 6 (Figure 3). Zinc has the second best explainability of a relationship between water pH and the concentration of the metal, right after copper. In this case, site 2 concentrations are better explained by pH (r2 = 0.6332) than the concentrations of the first site (r2 = 0.5846) by using second-degree polynomials.

Manganese is a common pollutant in mine waters worldwide [29]. Although the hydroxide precipitation of metal cations is usually an effective method for their elimination from aqueous solutions (eg Fe, Cu, Zn, Ni), it is not effective in reducing Mn concentrations below 1 mg/L. In the work [30], various procedures for removing Mn from wastewater from different sources and compositions are presented, while confirming the highest efficiency at pH 8.5 and higher. This statement is not completely valid for Mn in surface waters. Figure 4 shows the decrease in Mn concentration below 0.5 mg/L at pH around 7. Similarly to zinc, the manganese concentrations can also be better explained by the pH for the second site. In this case, the r2 coefficient is between 0.5 and 0.6.

Iron should occur in AMD mainly as an Fe^{2+} cation, which precipitates at pH < 8.5 [27]. The fact that iron precipitated throughout the studied interval was caused by the oxidation of Fe^{2+} to Fe^{3+} by atmospheric oxygen and the precipitation of $Fe(OH)_3$, which starts at pH 3.5. The values of the determination coefficients $r^2 < 0.5$ for the dependence of Fe concentration on pH (Figure 5) confirm a different course of Fe precipitation due to the simultaneous oxidation of Fe^{2+} to Fe^{3+} [18,20]. This simultaneous oxidation explains the worst goodness of fit for the second-degree polynomial curves we use in this study. The relationship was the weakest of all studied metals.

4. Conclusions

The location of the Smolník mine is included among the old environmental burdens due to AMD production and surface water contamination in the Smolník creek. This fact was confirmed by exceeding the limit values of pH and the monitored concentrations of heavy metals in the surface water according to Slovak legislation. Fluctuations in the pH value also affect the concentration of heavy metals (e.g., Fe, Cu, Zn, Al, Mn) in the Smolník creek polluted by acid mine drainage, which was confirmed by the presented results. Statistical analysis showed that pH has a significant effect on the concentration of metals in surface water. It is important to note that this relationship cannot be properly interpreted without a deeper understanding of the chemical properties of the studied metals. The weakest link observed in iron can be explained by the simultaneous oxidation of Fe^{2+} to Fe^{3+}. On the other hand, the best explainability of metal concentration by pH was achieved for copper, using a second-degree polynomial.

Author Contributions: Conceptualization, M.B. and N.J.; methodology, M.B.; software, Y.C.; validation, M.B., N.J. and Y.C.; formal analysis, M.B.; investigation, N.J.; data curation, M.B.; writing M.B. All authors have read and agreed to the published version of the manuscript.

Funding: This research was funded by the Scientific Grant Agency of the Ministry of Education, Science, Research and Sport of the Slovak Republic and the Slovak Academy of Sciences, project VEGA Grant No. 2/0108/23 and APVV-20-0140.

Institutional Review Board Statement: Not applicable.

Informed Consent Statement: Not applicable.

Data Availability Statement: The data presented in this study are available on request from the corresponding author.

Conflicts of Interest: The authors declare no conflict of interest. The funders had no role in the design of the study, in the collection, analyses, or interpretation of data, in the writing of the manuscript, or in the decision to publish the results.

References

1. Akcil, A.; Koldas, K. Acid mine drainage (AMD): Causes, treatment and case studies. *J. Clean. Prod.* **2006**, *14*, 1139–1146. [CrossRef]
2. Lintnerová, O.; Šottník, P.; Šoltés, S. Abandoned Smolnik mine (Slovakia) a catchment area affected by mining activities. *Estonian J. Earth Sci.* **2008**, *57*, 104–110. [CrossRef]
3. Dvořáček, J.; Malíková, P.; Sousedíková, R.; Heviánková, S.; Rys, P.; Osičková, I. Water production as an option for utilizing closed underground mines. *J. S. Afr. Inst. Min. Metall.* **2022**, *122*, 571–577.
4. Skousen, J.G.; Ziemkiewicz, P.F.; McDonald, L.M. Acid mine drainage formation, control and treatment: Approaches and strategies. *Extr. Ind. Soc.* **2019**, *6*, 241–249. [CrossRef]
5. Yuan, J.; Ding, Z.; Bi, Y.; Li, J.; Wen, S.; Bai, S. Resource Utilization of Acid Mine Drainage (AMD): A Review. *Water* **2022**, *14*, 2385. [CrossRef]
6. Nishimoto, N.; Yamamoto, Y.; Yamagata, S.; Igarashi, T.; Tomiyama, S. Acid Mine Drainage Sources and Impact on Groundwater at the Osarizawa Mine, Japan. *Minerals* **2021**, *11*, 998. [CrossRef]
7. Valová, B.; Kotalová, I.; Heviánková, S. Determination of Risk Elements in Mine Waste Dump Soil Sample Using Sequential BCR Extraction. *Inzynieria Miner.* **2020**, *2020*, 217–220.
8. Archundia, D.; Prado-Pano, B.; González-Méndez, B.; Loredo-Portales, R.; Molina-Freaner, F. Water resources affected by potentially toxic elements in an area under current and historical mining in northwestern Mexico. *Environ. Monit. Assess.* **2021**, *193*, 236. [CrossRef]
9. Johnson, D.B.; Hallberg, K.B. Acid mine drainage remediation options: A review. *Sci. Total Environ.* **2005**, *338*, 3–14. [CrossRef]
10. Andras, P.; Adam, M.; Chovan, M.; Slesarova, A. Environmental hazards of the bacterial leaching of ore minerals from waste at the Pezinok deposit (Malé Karpaty MTS., Slovakia). *Carpathian J. Earth Environ. Sci.* **2008**, *3*, 7–22.
11. Hakansson, K.; Karlsson, S.; Allard, B. Effects of pH on the accumulation and redistribution of metals in polluted stream bed sediment. *Sci. Total Environ.* **1989**, *87/88*, 43–57. [CrossRef]
12. Luptakova, A.; Kusnierova, M. Bioremediation of Acid Mine Drainage by SRB. *Hydrometallurgy* **2005**, *77*, 97–102. [CrossRef]
13. Singovszka, E.; Balintova, M.; Holub, M. Heavy metal contamination and its indexing approach for sediment in Smolnik creek (Slovakia). *Clean Technol. Environ. Policy* **2016**, *18*, 305–313. [CrossRef]
14. Bajtoš, P. Mine waters in the Slovak part of the Western Carpathians: Distribution, classification and related environmental issues. *Slovak Geol. Mag.* **2016**, *16*, 139–158.
15. Kupka, D.; Pállová, Z.; Horňáková, A.; Achimovičová, M.; Kavečanský, V. Effluent water quality and the ochre deposit characteristics of the abandoned Smolnik mine, East Slovakia. *Acta Montan. Slovaca* **2012**, *17*, 56–64.
16. Lintnerová, O.; Šottník, P.; Šoltés, S. Stream sediment and soil pollution in the Smolnik mining area (Slovakia). *Slovak Geol. Mag.* **2003**, *9*, 201–203.
17. Calmano, W.; Hong, J.; Forstner, U. Binding and mobilization of heavy metals in contaminated sediments affected by pH and redox potential. *Water Sci. Technol.* **1993**, *28*, 223–235. [CrossRef]
18. Balintova, M.; Singovszka, E.; Vodicka, R.; Purcz, P. Statistical Evaluation of Dependence Between pH, Metal Contaminants, and Flow Rate in the AMD-Affected Smolnik Creek. *Mine Water Environ.* **2016**, *35*, 10–17. [CrossRef]
19. Kruopiene, J. Distribution of Heavy Metals in Sediments of the Nemunas River (Lithuania). *Polish J. Environ. Stud.* **2007**, *16*, 715–722.

20. Balintova, M.; Petrilakova, A. Study of pH Influence on Selective Precipitation of Heavy Metals from Acid Mine Drainage. *Chem. Eng. Trans.* **2011**, *25*, 345–350.
21. Funes, A.; De Vicente, J.; Cruz-Pizarro, L.; De Vicente, I. The influence of pH on manganese removal by magnetic microparticles in solution. *Water Res.* **2014**, *53*, 110–122. [CrossRef] [PubMed]
22. Feng, D.; Aldrich, C.; Tan, H. Treatment of acid mine water by use of heavy metal precipitation and ion exchange. *Miner. Eng.* **2000**, *13*, 623–642. [CrossRef]
23. Balintova, M.; Petrilakova, A.; Singovszka, E. Study of metal ion sorption from acidic solutions. *Theor. Found. Chem. Eng.* **2012**, *46*, 727–731. [CrossRef]
24. Balintova, M.; Petrilakova, A.; Singovszka, E. Study of metals distribution between water and sediment in the Smolnik Creek (Slovakia) contaminated by acid mine drainage. *Chem. Eng. Trans.* **2012**, *28*, 73–78.
25. Luptakova, A.; Balintova, M.; Jencarova, J.; Macingova, E.; Prascakova, M. Metals recovery from acid mine damage. *Nova Biotechnol.* **2010**, *22*, 1111–1118.
26. Pedregosa, F.; Varoquaux, G.; Gramfort, A.; Michel, V.; Thirion, B.; Grisel, O.; Blondel, M.; Prettenhofer, P.; Weiss, R.; Dubourg, V.; et al. Scikit-learn: Machine Learning in Python. *J. Mach. Learn. Res.* **2011**, *12*, 2825–2830.
27. Xinchao, W.; Roger, C.; Viadero, J.; Karen, M. Recovery of Iron and Aluminum from Acid Mine Drainage by Selective Precipitation. *Environ. Eng. Sci.* **2005**, *22*, 745–755.
28. Jenke, D.R.; Diebold, F.E. Recovery of valuable metals from acid mine drainage by selective titration. *Waters Res.* **1983**, *17*, 1585–1590. [CrossRef]
29. Kulkarni, S.J. A Review on Studies and Research on Manganese Removal. *Int. J. Sci. Res.* **2016**, *1*, 45–48.
30. Patil, D.S.; Chavan, S.M.; Oubagaranadin, J.U.K. A review of technologies for manganese removal from wastewaters. *J. Environ. Chem. Eng.* **2016**, *4*, 468–487. [CrossRef]

Disclaimer/Publisher's Note: The statements, opinions and data contained in all publications are solely those of the individual author(s) and contributor(s) and not of MDPI and/or the editor(s). MDPI and/or the editor(s) disclaim responsibility for any injury to people or property resulting from any ideas, methods, instructions or products referred to in the content.

Proceeding Paper

Life Cycle Assessment and Environmental Impacts of Building Materials: Evaluating Transport-Related Factors [†]

Adriana Eštoková *, Martina Fabianová and Marek Radačovský

Institute for Sustainable and Circular Construction, Faculty of Civil Engineering, Technical University of Kosice, Vysokoskolska 4, 042 00 Kosice, Slovakia; martina.wolfova25@gmail.com (M.F.); marek.radacovsky@student.tuke.sk (M.R.)
* Correspondence: adriana.estokova@tuke.sk
[†] Presented at the 4th International Conference on Advances in Environmental Engineering, Ostrava, Czech Republic, 20–22 November 2023.

Abstract: The construction industry plays a significant role in resource consumption and environmental degradation, making it crucial to analyze the sustainability aspects of construction materials and their transportation processes. This paper focuses on conducting a life cycle assessment (LCA) analysis of building materials, specifically considering the environmental impacts associated with their transportation to construction sites. By incorporating the transport phase into the assessment, a more holistic understanding of the environmental implications of construction materials can be achieved. The study aims to quantify the environmental burdens of both material production and transportation, providing valuable insights for sustainable decision making in the construction industry. The analysis revealed that transport of building materials for the studied house by diesel lorry, covering a distance of 150 km, contributed 16% to climate change and a significant 53.5% to abiotic resource depletion. Additionally, it had a 15–18% impact on acidification and photo-oxidant formation.

Keywords: LCA; construction materials; climate change; transportation; environmental impact

1. Introduction

The construction sector is a key contributor to environmental degradation due to its extensive use of materials and energy, as well as associated transportation activities [1,2]. Construction materials, such as concrete, steel, and timber, are often sourced from remote locations, necessitating long-distance transportation [3]. This transportation phase introduces additional environmental burdens, including greenhouse gas emissions, air pollution, and habitat disruption. The construction sector's environmental impact extends beyond the production and use of building materials, transportation of these materials to construction sites plays a significant role as well [4]. Hence, it is essential to comprehensively evaluate the environmental impacts of both construction materials and their transportation to construction sites [5].

To evaluate the environmental impacts of construction materials, a life cycle assessment (LCA) approach is widely used in construction sector today [6–8]. This approach encompasses the entire life cycle of materials, from extraction and manufacturing to transportation and disposal [9]. Primary data can be collected from various sources, including construction material suppliers, transportation companies, and relevant industry databases [10]. The collected data are analyzed using appropriate LCA methodologies to quantify the environmental impacts associated with different materials and transportation modes. Many LCA studies have been employed recently evaluating construction materials and technologies and whole buildings as well [11–14]. However, previous LCA studies often neglect the transport phase, resulting in an incomplete assessment of the overall environmental performance of building materials. To accomplish the goal, a rigorous LCA

methodology is employed, encompassing the various stages of the building material life cycle, from extraction and manufacturing to transport and eventual disposal.

This paper addresses this gap by including transportation-related factors in the LCA analysis, aiming to provide an extended evaluation to the production phase of building materials. Through the integration of transport-related data into the LCA framework, the environmental impacts associated with material transportation were quantified.

2. Materials and Methods

The LCA analysis focused on evaluating the environmental impact of the materials used in a masonry family house during its product phase (A1–A3) and construction phase (A4), following the guidelines of the EN 15804 standard [15]. The selected house for the analysis is a single-story, L-shaped detached family house without a basement. It features a gable roof, while the garage has a flat roof design. The total built-up area of the house is 231.2 m^2, with a usable floor area of 178.5 m^2. The maximum height of the building is from ± 0.000 to $+6.373$ m. The interior layout comprises a day section with an entrance hall, toilet, corridor, bathroom, utility room, kitchen with a pantry, and a living room with a dining area. The night section includes two rooms and a bedroom with a wardrobe. The house also includes an integrated garage.

The LCA procedure, based on ISO 14040 [16], consisted of the following 4 steps: (i.) definition of goal and scope, (ii.) analysis of life cycle inventory (LCI), (iii.) life cycle impact assessment (LCIA), and (iv.) interpretation.

2.1. LCA Goal and Scope

The objective of this LCA analysis was to quantify the environmental impact associated with the materials used in the selected family house. In addition to the standard evaluation of the production phase of materials, the assessment was expanded to include the transportation phase, from the material suppliers to the construction site. The analysis followed the guidelines set by the EN 15804 standard, which defines four life cycle phases for LCA assessments in building construction: the product phase (A1–A3), construction process phase (A4–A5), use phase (B1–B7), and end-of-life phase (C1–C4). For this study, modules A1–A3 and module A4 were evaluated.

The functional unit chosen for this study was a single building representing a total of 448 t of materials. To determine the projected lifespan of the building, various factors such as construction type, assemblies, and local climatic conditions were taken into account. The estimated duration of long-term elements was set at 50 years, while short-term construction elements were assumed to have a lifespan of 25 years [17].

2.2. Inventory (LCI)

To assess the environmental impact of the family house's materials, a detailed analysis of the construction material masses was performed during the inventory step. The foundation structures comprised concrete foundation strips, foundation footings, and reinforced concrete blocks. Beneath the base concrete, an embankment made of aggregate was used. The vertical structures were constructed using pre-cast concrete blocks, asphalt strips, and reinforced concrete crowns. For the horizontal structures, prefabricated monolithic reinforced concrete, steel beams, steel profiles, and extruded polystyrene for insulation were utilized. The roof structure consisted of a wooden gable roof with a ceramic covering, along with steel fasteners and asphalt insulating tape. Regarding thermal insulation, the house employed various materials of different thicknesses, including expanded polystyrene (EPS) boards, extruded polystyrene (XPS), and mineral wool. It should be noted that the evaluation did not consider doors and windows.

By analyzing the masses and types of materials used in the family house's construction, we gain insights into its potential environmental impact, which is crucial for making informed decisions in sustainable building practices.

2.3. Impact Assessment (LCIA)

The environmental impact assessment of the materials used in individual structures was conducted using the ReciPe method in SimaPro software, version 9.3.0.3 [18]. For the life cycle impact analysis in this study, the primary mid-point impact categories were employed to characterize environmental effects [19]. These categories include climate change, ozone depletion, photochemical ozone formation, acidification, eutrophication, depletion of abiotic resources, and water consumption.

To gather data for the materials' production phase (A1–A3), information was extracted from the Ecoinvent database. The data considered typical manufacturing processes of the materials, taking into account the associated energy consumption. The electricity mix used in Slovakia was also taken into consideration during this phase. In the assessment of transportation, various transport distances ranging from 5 to 150 km were modeled. The mean mode of transport for construction materials was considered to be lorries with emission standard EURO 5. Both gasoline and diesel fuels were compared in this context.

2.4. Interpretation

The LCA study's findings are presented in two parts. Firstly, the environmental impacts of materials integrated into the analyzed structures within the A1–A3 modules are provided, excluding considerations of material transport. Secondly, the impacts of construction material transportation to the building site are presented separately for module A4.

By presenting the results in this manner, we can gain a clear understanding of the environmental burdens associated with the materials themselves (A1–A3) and the additional impacts arising from their transportation to the construction site (A4). This comprehensive analysis allows for a more accurate assessment of the overall environmental impact of the building, facilitating informed decision-making and promoting sustainable construction practices.

3. Results

3.1. Impacts of Product Phase of Building (A1–A3 Modules)

The percentage share of the construction materials in individual structures on the overall environmental impacts during their production are illustrated in Figure 1. The environmental impacts are expressed through specific environmental indicators for each environmental category as follows: climate change by global warming potential (GWP), ozone depletion by depletion potential of the stratospheric ozone layer (ODP), photochemical ozone formation by formation potential of tropospheric ozone (POCP), acidification by acidification potential (AP), eutrophication by eutrophication potential (EP), depletion of abiotic resources by abiotic depletion potential (ADP), and water use by water (user) deprivation potential, deprivation-weighted water consumption (WDP).

3.1.1. Climate Change

The highest values of global warming potential (GWP) were observed in the foundation structures (25%) and insulation (29%) of the building which correlate to information in [20]. The vertical structures accounted for 19% of the total GWP, followed by horizontal structures (16%), and roof structures (11%). Load-bearing aerated concrete masonry stands out as the primary emitter of carbon dioxide (CO_2) into the air, contributing to 16% of the total GWP. Masonry is followed by foundation strips (13%) and mineral wool (10%) in terms of their carbon emissions. Ceramic roofing (7%) and cement screed (4%) exhibit lower GWP values but still significantly impact the overall structure. Conversely, materials like gravel, ŽB column, vapor-impermeable foil, and geotextile have a negligible carbon footprint, making them the least burdensome materials in the construction.

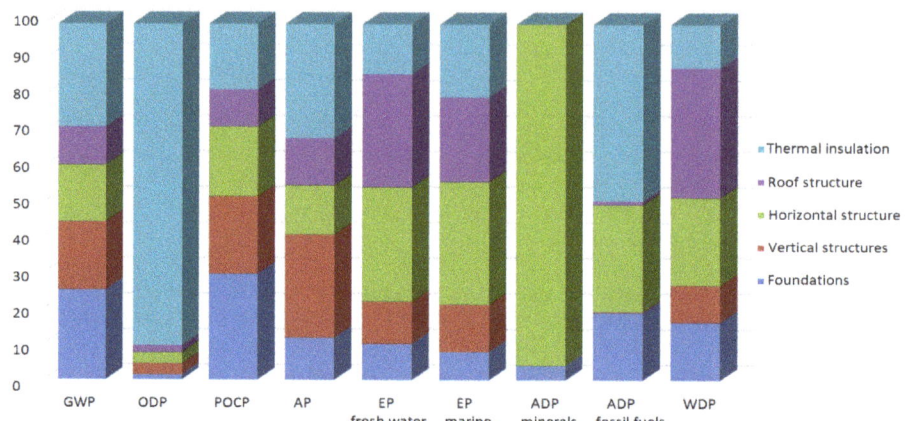

Figure 1. The contribution of materials in structures to the overall environmental impacts (%).

3.1.2. Ozone Depletion

Thermal insulation materials exert the most substantial influence on ozone depletion potential (ODP), constituting a significant portion (91%) of the total ODP. In comparison, foundations contribute only 1%, while vertical structures, horizontal structures, and roof structures contribute 3%, 3%, and 2%, respectively. Combined, these constructions represent a negligible fraction (9%) of the total ODP. Among the thermal insulation materials, XPS insulation boards account for 45% of the ODP, closely followed by EPS insulation boards at 44%. In contrast, concrete structures, screeds, construction timber, and plasterboard have an almost negligible impact on ODP. Additionally, materials such as geotextile, gravel, construction films, and lintels have minimal effects on ODP.

3.1.3. Photochemical Ozone Formation

The environmental impact on photochemical ozone creation potential (POCP) is most significant for foundation materials, contributing to 30% of the total impact. Vertical structures (22%) and horizontal structures (20%) also exert substantial ozone-forming effects. In contrast, insulation (18%) and roof structures (10%) have a relatively lower impact on POCP. Among the construction materials, load-bearing aerated concrete masonry exhibits the highest values (18%) in terms of POCP. Foundation belts show a similar significant impact at 14.5%. Other materials with notable contributions include mineral wool (10%), ceramic roofing (6.5%), and reinforced concrete wreath (4%) in the ozone formation. Conversely, several materials have little to no impact on POCP, including gravel, building foil, vapor-permeable and vapor-impermeable foil, geotextiles, and formwork blocks.

3.1.4. Acidification

Vertical structures (29%) and insulation (32%) account for the largest share of the acidification potential value in the construction. Other construction components have a relatively smaller influence on acidification; foundations contribute 12%, horizontal structures contribute 14%, and roof structures contribute 13%. Among the materials, aerated concrete masonry (24%) and mineral wool (22%) are the primary contributors to the high acidification potential values. Other materials with notable impact in construction include foundation strips (6%), cement screed (4%), and construction lumber (7%). In contrast, gravel, lintels, geotextiles, and foils have minimal effects on acidification potential.

3.1.5. Eutrofication

Horizontal constructions (32%) and roof constructions (32%) are prominent contributors to eutrophication potential (EP). Foundations (10%), vertical structures (12%), and insulation (14%) show relatively lower impacts on eutrophication. Among the materials, cement screed has the most significant share of the total EP (17%), closely followed by ceramic roofing with a similarly substantial influence (16%). Aerated concrete blocks (12%) and foundation strips (8%) also contribute significantly to eutrophication. Conversely, insulating materials, foils, and plasterboard exhibit minimal impacts on eutrophication potential.

3.1.6. Abiotic Resource Depletion

Regarding the depletion of abiotic resources, specifically minerals and metals, horizontal structures account for a significant share (96%), while foundations have minimal impact (4%). Among these materials, steel profiles make up over half of the total depletion (58%). Anhydrite screeds (28%) and plasterboard (9%) exhibit smaller values in this category. However, in the case of fossil resource depletion, the impact is more evenly distributed. Insulation is the most burdensome structure (50%), closely followed by horizontal structures (30%) and foundations (19%). Vertical constructions have a negligible influence, as do roof structures (1%). Mineral wool contributes the most to fossil resource depletion, representing 50% of the total amount. Among horizontal structures, anhydrite screed has the largest share in resource depletion at 8%, while foundation boards represent the most substantial portion of foundations with 18%. Materials such as formwork blocks, cement screed, XPS, and EPS insulation boards show no depletion values.

3.1.7. Water Use

In terms of water use, horizontal constructions (25%) and roof constructions (37%) have the most significant share, followed by foundations (16%), insulation (12%), and vertical structures (10%). The materials contributing the most to water use include wooden elements of the roof (21.5%) and ceramic roofing (15%). Foundation strips also have a notable impact, accounting for 14.5% of water use, similar to the roofing. Load-bearing aerated concrete masonry (9%) and beam ceilings (8%) exhibit lower values of water use in comparison.

3.2. Impacts of Transportation (A4 Module)

This study focused on assessing the environmental impacts of transportation, considering variations in transportation distances and fuel types. Specifically, the impact of material transport on the overall environmental impact of the studied house was evaluated, focusing on the environmental categories with the most significant transportation contribution. Additionally, a new category was introduced to account for the emissions of particulate matter during transportation.

As expected, the environmental impacts show an increasing trend with the transportation distance (Figure 2). The highest increase in the impact of transport occurred in the depletion of abiotic resources, with a maximum increase of up to 2.3 times compared to the minimum distance (5 km). Similarly, transportation had a significant impact on the photochemical formation of ozone, showing an increase of 1.8 times, and on seawater eutrophication, with a recorded increase of 1.6 times. Smaller differences were observed in emissions of solid particles, with an increase of 1.6 times, and acidification, with a difference of 1.3 times. This analysis demonstrates the importance of considering transportation distance in the assessment of environmental impacts associated with material transport. Longer transportation distances can significantly exacerbate certain environmental impacts, highlighting the need for efficient logistics and sustainable transportation practices to minimize the overall environmental footprint of construction activities.

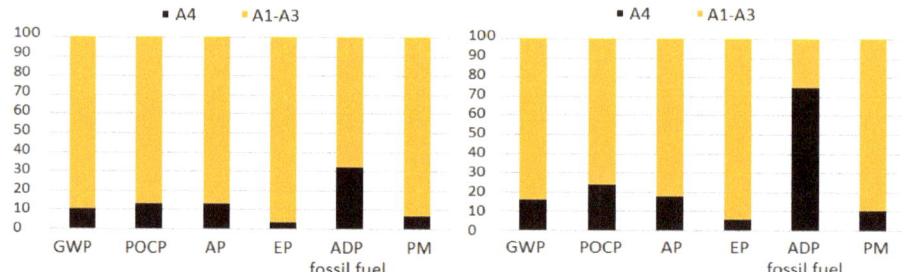

Figure 2. The contribution of transportation of construction materials by diesel lorry (A4 module) in % for distances of 5 km (**left**) and 150 km (**right**).

In the overall environmental impact of a family house, transportation plays a substantial role, particularly in the depletion of abiotic resources, constituting 53.48% of the total impact. Transportation also significantly influences the photochemical formation of ozone (18.35%) and acidification (15.44%). On the other hand, the lowest share is attributed to transport-related emissions of solid particles (8.5%) and eutrophication (4.77%).

Of particular interest is the contribution of transportation of construction materials to climate change. Figure 3 illustrates the percentage contribution of diesel lorries to the overall global warming potentials (GWPs).

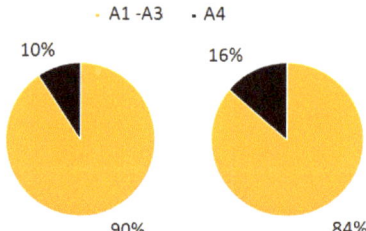

Figure 3. The percentage contribution of diesel lorry to the overall GWP for distances of 5 km (**left**) and 150 km (**right**).

According to Figure 3, the carbon footprint of transport from a distance of 5 km represents 10% of the total global warming potential (GWP), while at a distance of 150 km, this share increases to 16%. The difference in the increase in the carbon footprint of transport between the maximum and minimum distances is 1.5 times. On average, transport accounts for about 13.17% of the total GWP which is slightly higher than reported in [2].

Furthermore, the study also investigated the impact of different fuel types on climate change. Figure 4 compares the effect of fuel type, specifically gasoline versus diesel, on GWP values.

Figure 4 provides compelling evidence that gasoline cars exhibit higher values compared to diesel cars as the distance increases. At a distance of 5 km, the gasoline car has only 0.24% higher values than the diesel car. However, at a distance of 150 km, this difference increases significantly to 3.8%. If this upward trend were to continue, we can extrapolate that at 300 km, the gasoline car would have values 4.17% higher than the diesel car, and at 500 km, the difference would be 4.55%.

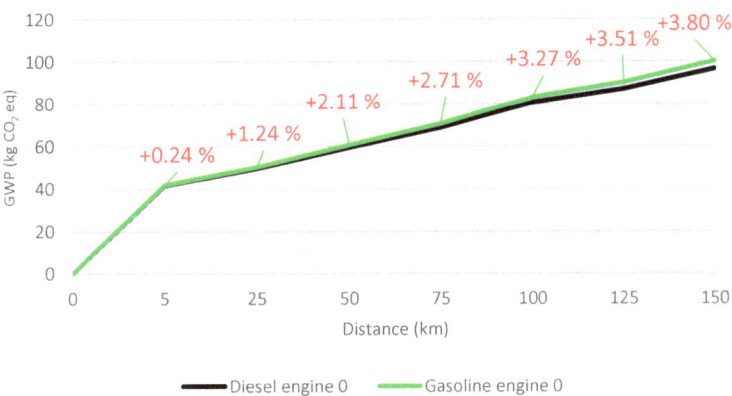

Figure 4. Comparison of the GWP due to transportation for gasoline and diesel car.

4. Conclusions

By incorporating the transport phase into the LCA analysis of building materials, this study provides a better understanding of the environmental impacts associated with construction materials. Based on the presented data and analysis the following conclusions could be drawn.

The materials used in construction have varying degrees of impact on different environmental categories. The analysis of the house construction process (A4 module) revealed significant contributions from the transport of building materials. Specifically, at a transport distance of 150 km, this transportation accounted for 16% of the overall impact on climate change. Surprisingly, transport's role in depleting abiotic resources was even more substantial, amounting to a striking 53.5%. Furthermore, the environmental impact of transport extended to other categories as well. The contributions to acidification and the formation of photo-oxidants ranged from 15% to 18%, signifying their relevance in the overall assessment. These findings emphasize the importance of considering transport-related emissions and resource usage when addressing environmental concerns in the construction industry.

Based on the findings, mitigation strategies to address the environmental impacts of building materials and their transportation should be proposed. These strategies may include optimizing transport routes, promoting regional sourcing of materials, adopting more sustainable transportation modes, and exploring alternative materials with lower environmental footprints.

Author Contributions: Conceptualization, A.E. and M.F.; methodology, M.F.; investigation, M.R. and M.F.; writing, A.E. All authors have read and agreed to the published version of the manuscript.

Funding: This research was funded by the Scientific Grant Agency of the Ministry of Education, Science, Research and Sport of the Slovak Republic and the Slovak Academy of Sciences, project VEGA Grant Nos. 1/0230/21 and 2/0108/23 and APVV-20-0140.

Institutional Review Board Statement: Not applicable.

Informed Consent Statement: Not applicable.

Data Availability Statement: The data presented in this study are available on request from the corresponding author.

Conflicts of Interest: The authors declare no conflict of interest. The funders had no role in the design of the study, in the collection, analyses, or interpretation of data, in the writing of the manuscript, or in the decision to publish the results.

References

1. Ding, Z.; Zhu, M.; Tam, V.W.; Yi, G.; Tran, C.N. A system dynamics-based environmental benefit assessment model of construction waste reduction management at the design and construction stages. *J. Clean. Prod.* **2018**, *176*, 676–692. [CrossRef]
2. Sezer, A.A.; Fredriksson, A. Environmental impact of construction transport and the effects of building certification schemes. *Resour. Conserv. Recycl.* **2021**, *172*, 105688. [CrossRef]
3. Xu, J.; Deng, Y.; Shi, Y.; Huang, Y. A bi-level optimization approach for sustainable development and carbon emissions reduction towards construction materials industry: A case study from China. *Sustain. Cities Soc.* **2020**, *53*, 101828. [CrossRef]
4. Chang, B.; Kendall, A. Life cycle greenhouse gas assessment of infrastructure construction for California's high-speed rail system. *Transp. Res. Part D Transp. Environ.* **2011**, *16*, 429–434. [CrossRef]
5. Eriksson, V. Transport Efficiency: Analysing the Transport Service Triad. Ph.D. Thesis, Chalmers Tekniska Hogskola, Gothenburg, Sweden, 2019.
6. Vasishta, T.; Mehany, M.H.; Killingsworth, J. Comparative life cycle assesment (LCA) and life cycle cost analysis (LCCA) of precast and cast–in–place buildings in United States. *J. Build. Eng.* **2023**, *67*, 105921. [CrossRef]
7. Barbhuiya, S.; Das, B.B. Life Cycle Assessment of Construction Materials: Methodologies, Applications and Future Directions for Sustainable Decision-making. *Case Stud. Constr. Mater.* **2023**, *19*, e02326. [CrossRef]
8. Ramón-Álvarez, I.; Batuecas, E.; Sánchez-Delgado, S.; Torres-Carrasco, M. Mechanical performance after high-temperature exposure and Life Cycle Assessment (LCA) according to unit of stored energy of alternative mortars to Portland cement. *Constr. Build. Mater.* **2023**, *365*, 130082. [CrossRef]
9. Rebitzer, G.; Ekvall, T.; Frischknecht, R.; Hunkeler, D.; Norris, G.; Rydberg, T.; Schmidt, W.P.; Suh, S.; Weidema, B.P.; Pennington, D.W. Life cycle assessment: Part 1: Framework, goal and scope definition, inventory analysis, and applications. *Environ. Int.* **2004**, *30*, 701–720. [CrossRef] [PubMed]
10. Monteiro, N.B.R.; Neto, J.M.M.; da Silva, E.A. Environmental assessment in concrete industries. *J. Clean. Prod.* **2021**, *327*, 129516. [CrossRef]
11. Tinoco, M.P.; de Mendonça, É.M.; Fernandez, L.I.C.; Caldas, L.R.; Reales, O.A.M.; Toledo Filho, R.D. Life cycle assessment (LCA) and environmental sustainability of cementitious materials for 3D concrete printing: A systematic literature review. *J. Build. Eng.* **2022**, *52*, 104456. [CrossRef]
12. De Souza, D.M.; Lafontaine, M.; Charron-Doucet, F.; Chappert, B.; Kicak, K.; Duarte, F.; Lima, L. Comparative life cycle assessment of ceramic brick, concrete brick and cast-in-place reinforced concrete exterior walls. *J. Clean. Prod.* **2016**, *137*, 70–82. [CrossRef]
13. Rodrigo-Bravo, A.; Cuenca-Romero, L.A.; Calderon, V.; Rodriguez, A.; Gutiérrez-González, S. Comparative Life Cycle Assessment (LCA) between standard gypsum ceiling tile and polyurethane gypsum ceiling tile. *Energy Build.* **2022**, *259*, 111867. [CrossRef]
14. Balasbaneh, A.T.; Sher, W.; Yeoh, D.; Koushfar, K. LCA & LCC analysis of hybrid glued laminated Timber–Concrete composite floor slab system. *J. Build. Eng.* **2022**, *49*, 104005. [CrossRef]
15. EN 15804; 2012 + A2 + AC: 2021—Sustainability of Construction Works—Environmental Product Declarations—Core Rules for the Product Category of Construction Products. European Committee for Standardization (CEN): Brussels, Belgium, 2019.
16. ISO 14040; 2006—Environmental Management—Life Cycle Assessment—Principles and Framework. European Committee for Standardization (CEN): Brussels, Belgium, 2006.
17. Sartori, I.; Hestnes, A.G. Energy use in the life cycle of conventional and low-energy buildings: A review article. *Energy Build.* **2007**, *39*, 249–257. [CrossRef]
18. SimaPro Database Manual Methods Library. Pré Sustainability B.V. Available online: https://simapro.com/wp-content/uploads/2021/07/DatabaseManualMethods920.pdf (accessed on 1 August 2023).
19. Huijbregts, M.A.; Steinmann, Z.J.; Elshout, P.M.; Stam, G.; Verones, F.; Vieira, M.; Zijp, M.; Hollander, A.; Van Zelm, R. ReCiPe2016: A harmonised life cycle impact assessment method at midpoint and endpoint level. *Int. J. Life Cycle Assess.* **2017**, *22*, 138–147. [CrossRef]
20. Ondova, M.; Estokova, A. Environmental impact assessment of building foundation in masonry family houses related to the total used building materials. *Environ. Prog. Sustain. Energy* **2016**, *35*, 1113–1120. [CrossRef]

Disclaimer/Publisher's Note: The statements, opinions and data contained in all publications are solely those of the individual author(s) and contributor(s) and not of MDPI and/or the editor(s). MDPI and/or the editor(s) disclaim responsibility for any injury to people or property resulting from any ideas, methods, instructions or products referred to in the content.

Proceeding Paper

Potential for Reducing the Carbon Footprint of Buildings [†]

Eva Kridlova Burdova [1], Jana Budajova [1], Peter Mesaros [2] and Silvia Vilcekova [1,*]

[1] Institute of Sustainable and Circular Construction, Faculty of Civil Engineering, Technical University of Kosice, 04200 Kosice, Slovakia; eva.kridlova.burdova@tuke.sk (E.K.B.); jana.budajova@tuke.sk (J.B.)
[2] Institute of Technology, Economics and Management in Construction, Faculty of Civil Engineering, Technical University of Kosice, 04200 Kosice, Slovakia; peter.mesaros@tuke.sk
* Correspondence: silvia.vilcekova@tuke.sk; Tel.: +421-55-602-4260
[†] Presented at the 4th International Conference on Advances in Environmental Engineering, Ostrava, Czech Republic, 20–22 November 2023.

Abstract: The construction sector produces more than 35% of the total amount of waste in Europe and for around 36% of emissions. This paper deals with the life cycle analysis of three alternatives of a residential building in terms of carbon footprint. At the same time, the analysis focuses on the end-of-life phase and its significance for the reduction of the carbon footprint. From the results obtained, it can be concluded that variant three has the lowest CO_{2eq}/m^2 emission levels compared to those of the other two variants. Based on the overall investigation of the end-of-life phase, this study found that the reuse of brick material contributed the most to the reduction of the overall emissions.

Keywords: building; life cycle assessment; carbon footprint; end of life

1. Introduction

Globalization and growth in this century is accompanied by a huge amount of building materials used for housing and have expanded the need for aggregates in concrete-based materials. The residential wastes obtained via natural resource extraction include concrete, stone, block concrete, solid and tiles, hollow fired clay bricks, mortar, mineral plaster, asphalt, sand and glass [1]. In the European Union, the built environment produces more than 25% of all waste, highlighting the demand to support a circular economy. To illustrate the level of circularity, the characteristic indicators mainly target the amount of primary materials and the amount of non-renewable waste and the lifetime of the product [2]. The construction sector, which accounts for 37% of global greenhouse gas emissions and 36% of global energy resources consumption, is transitioning to a low-carbon and low-energy model [3]. The construction of residential buildings requires the extraction of a large amount of natural resources. Buildings are capital-demanding and require important physical, economic and social financing [4]. One study [5] examines the potential strategies to reduce the embodied energy and greenhouse gas emissions through the flexible reuse of non-residential buildings for a residential objective compared to the new construction of residential buildings. In a renovated building, compared to the construction of a new apartment building, approximately 56% of the embodied energy, 34–48% of the CO_2 equivalent emissions and 72% of the weight of the materials can be saved [5]. The circular economy is a new economic model that aims to overcome today's linear "take, make, dispose" model, delinking global economic development from limited resource consumption [6]. Bio-based circular building materials are materials obtained in whole or in part with a renewable biological origin or from the by-products and biological waste of plants and/or animal biomass that can be used as raw materials for building materials and decorative objects in construction in their original forms. The literature shows that the use of these materials can represent a consistent solution for mitigating the climate impacts of the construction sector, according to the circular economy model [7]. To assess the potential impact of the materials, products or systems, the LCA method has

Citation: Burdova, E.K.; Budajova, J.; Mesaros, P.; Vilcekova, S. Potential for Reducing the Carbon Footprint of Buildings. *Eng. Proc.* **2023**, *57*, 6. https://doi.org/10.3390/engproc2023057006

Academic Editors: Adriana Estokova, Natalia Junakova, Tomas Dvorsky, Vojtech Vaclavik and Magdalena Balintova

Published: 29 November 2023

Copyright: © 2023 by the authors. Licensee MDPI, Basel, Switzerland. This article is an open access article distributed under the terms and conditions of the Creative Commons Attribution (CC BY) license (https://creativecommons.org/licenses/by/4.0/).

been to approach their environmental, economic or social consequences to affect decision making. This has led the construction industry on an eco-efficient path that allows you to do more with less [8]. This study investigated the potential of producing geopolymer bricks from industrial waste, including ferrosilicon and aluminum slag. Design Builder was used to establish that the proposed manufactured brick samples and results, showing an energy reduction of 5.70–14.90% and carbon dioxide reduction of 0.47–7.67%, which were compared to those of ordinary brick. For the manufactured brick samples, the return on investment was determined to be 8.76–15.79 years [9]. Residential and service sector buildings contribute significantly to climate change through energy consumption in these buildings and indirectly through construction activities and the production and disposal of building materials [10]. The decarbonization of the construction industry plays an essential role for achieving the climate change mitigation goals. The results of this study show that the annual greenhouse gas emissions would be decreased by around 40% under the reference scenario, while the annual greenhouse gas emissions can be reduced by around 90% using the ambitious variant, in which all the decarbonization strategies are implemented simultaneously [11].

The aim of this paper is to evaluate three residential building alternatives in terms of GHG emissions and end-of-life waste management.

2. Materials and Methods

An analytical method of environmental management called a life cycle assessment, LCA, is used in the research. LCA is a method of comparing the environmental impacts of products, goods or services during their life cycle. It acknowledges emissions to all components of the environment during production, use and disposal of the product. The processes of extraction of raw materials, production of materials and energy, and ancillary processes or sub-processes are also included [12,13]. The LCA methodology consists of four phases [14]:

- The definition of the objectives and scope. This is used to define how much of the product life cycle will be included in the assessment and what the assessment will be used for. It describes the criteria used to compare the systems and the time reprezentativeness chosen.
- Inventory analysis. This includes a description of the material and energy flows within the product system, and in particular, its interaction with the environment, the raw materials consumed and the emissions into the environment. It describes all the significant processes and ancillary flows of energy and materials.
- Impact assessment. The results of the indicators of all impact categories are calculated here; the relative significance of each impact category is assessed via normalization and, where appropriate, via weighting. The impact assessment tends to result in a tabular summary of all the impacts.
- Interpretation. Includes a critical review, the sensitivity of data and the presentation of results.

2.1. Goal and Scope

The original residential building (Figure 1) is located in the village Raslavice in north-eastern Slovakia. The building is designed as a detached three-story building without a basement, with a gable roof with a slight slope. Residential units are located on each floor. The gross floor area (GFA) is 418.1 m^2. A condensing boiler is used for hot water preparation and heating. The energy consumption total for heating is 34.44 kWh/m^2. Variant B1 is constructed on strip foundations reinforced with perimeter wreaths, with monolithic ceilings and a staircase. The external walls are made of aerated concrete blocks. In variant B1, the external wall materials have been changed, and two additional alternatives have been created. The external walls of variant B2 are made of sand-lime blocks. The external walls of variant B3 are made of hollow bricks.

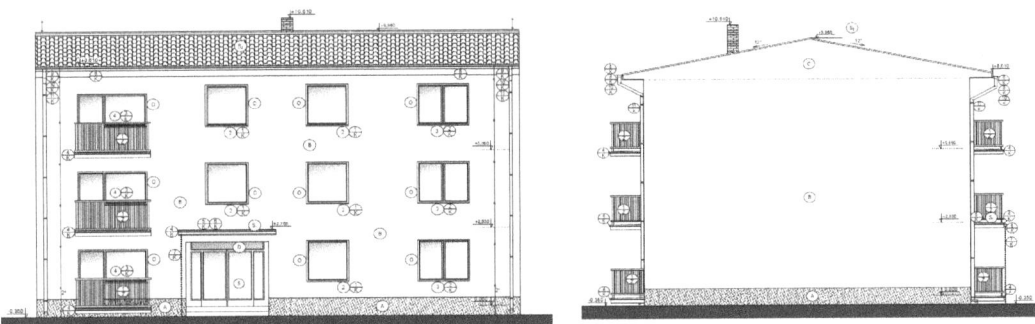

Figure 1. View of the building.

2.2. Life Cycle Inventory

The designed materials were manually entered into the software. This method of input eliminated the possible misclassification of materials within the software database that can occur during automatic uploading. The proposed construction materials were entered into the software, as designed. That is, the materials with the same parameters and the same manufacturer were entered. For the LCA, One Click LCA software, Levels life cycle assessment tool was used, which works in accordance with the ISO standards. For the materials that were not included in the OneClick database, materials with comparable parameters were selected. The windows were not in the database, so similar plastic windows were selected. Waterproof materials with the required parameters from the manufacturer mentioned in the project were replaced by waterproof materials from another manufacturer. LCA also included operational energy consumption and operational water consumption. For the proposed materials, the transport distance between the production plant and the point of use was considered in the analysis, so that the environmental impact of transport was also considered during analysis.

The building was designed and reviewed to understand its potential quantified environmental impacts. The materials of the original B1 building are shown in Table 1. The building's emissions were calculated per functional unit m^2 of GFA for a period of 60 years.

Table 1. Materials of building B1.

Material	B1	Unit
Structural timber, spruce,	3.37	t
Rock wool insulation boards, glass wool, expanded polystyrene	2415.07	m^2
Paint	1474.47	m^2
Aggregate	56.31	m^3
Reinforcement/steel	10.72	t
Waterproofing membrane	874.83	m^2
Ready-mix concrete	142.25	m^3
Gypsum plasterboard	139.29	m^2
Windows	55.67	m^2

Table 2 shows the end-of-life phase scenarios for the building alternatives B1, B2 and B3. Scenario 1 represents the market scenario that is the most typical in Slovakia. The purpose of Scenario 2 was to set the end-of-life phase so that the impacts are further reduced through reuse. Scenario 3 represents the landfilling of the waste.

Table 2. End-of-life scenarios.

Material	Scenario 1	Scenario 2	Scenario 3
Aerated concrete block	Crushed to aggregate	Reuse as material	Landfilling
Sand-lime block	Crushed to aggregate	Reuse as material	Landfilling
Hollow brick	Crushed to aggregate	Reuse as material	Landfilling

2.3. Life Cycle Impact Assessment (LCIA)

The LCA method has a fixed structure and was carried out according to the international standards ISO 14040 and EN 15978, STN EN 15804+A2+AC and PCR (Product Category Rules) as a supplement to STN EN 15804+A2+AC, which is an additional set of rules for specific product categories. Figure 2 shows the system boundary for LCA. Commercially available databases of processes as well as material and energy flows were used for the efficient processing of this LCA study. The LCA method can be defined as the assemblage and assessment of inputs, outputs and potential environmental impacts of a product system during its life cycle [15].

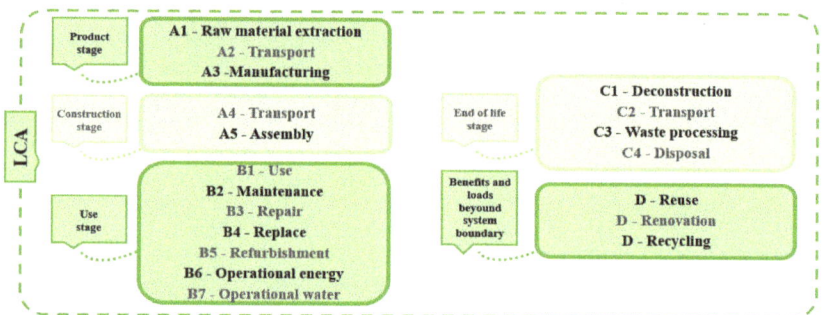

Figure 2. System boundary for LCA.

The LCA methodology allows different categories of environmental impacts to be assessed. According to STN EN 15804+A2+AC, the global warming potential—total (GWP—total) is the sum of three subcategories of climate change, consisting of GWP—fossil, GWP—biogenic and GWP—LULUC) [16]. The temporal changes in aboveground biomass, along with the impact on biodiversity caused by LULUC due to forestry activities, were factored into the calculation. A separate impact category is biogenic carbon storage (CO_2 bio), which has a beneficial impact on the environment. Biological CO_2 storage represents the carbon sequestered in growing vegetation when plants remove carbon dioxide from the air during photosynthesis, which is then converted into oxygen. Sequestration is expressed in kilograms of CO_2 bio-equivalents.

3. Results and Discussion

Table 3 shows the results of the investigated impact categories for the three variants of the building.

Table 3. Results of impact categories.

Impact Category	Unit	B1	B2	B3
GWP—total	(kg CO_{2eq})	8.60×10^5	8.92×10^5	8.35×10^5
GWP—fossil	(kg CO_{2eq})	8.58×10^5	8.90×10^5	8.33×10^5
GWP—bio	(kg CO2e bio)	0	0	0
GWP—LULUC	(kg CO2e)	1.65×10^3	1.67×10^3	1.64×10^3

According to the results for the GWP—total, variant B2 achieves the worst result, 8.92×10^5 kg CO_{2eq}, and B3 achieves the best result, 8.35×10^5 kg CO_{2eq}. Figure 3 shows the results for the GWP—total expressed in kg CO_{2eq} emissions per gross floor area (GFA). The GWP—total values expressed per m² of GFA are 2.06×10^3, 2.13×10^3 and 2.00×10^3 kg CO_{2eq}/m^2 for B1, B2 and B3, respectively. Variant B3 has the lowest emission levels, and variant B2 has the highest. The building materials (A1–A3) produce the largest share of CO_{2eq} emissions at 33.71% (2.90×10^5 kg CO_{2eq}) material replacement and refurbishment (B4-B5) produce 28.88% (2.48×10^5 kg CO_{2eq}) and operational energy (B6) produce 22.93% (1.97×10^5 kg CO_{2eq}). The main contributors to emissions are sand-lime bricks (21.2%), aerated concrete blocks (13.1%) and ready-mix concrete (6.3%).

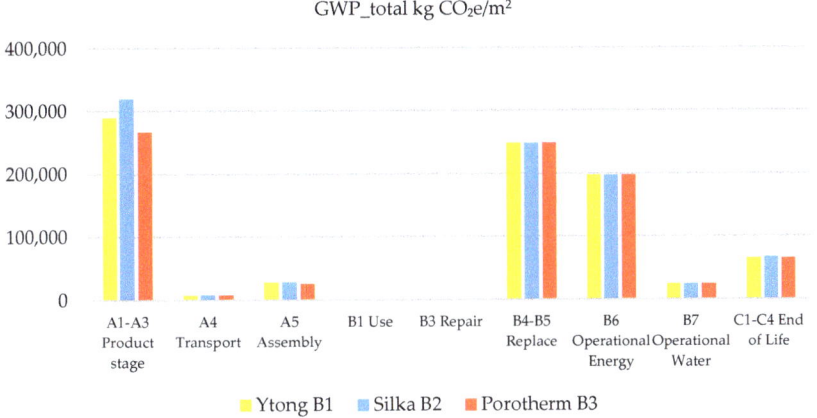

Figure 3. GWP—total in kg CO_{2eq}/m^2.

According to the results for GWP—fossil (Figure 4) and GWP—LULUC (Figure 5), variant B2 achieves the worst result (8.90×10^5, 1.67×10^3 kg CO_{2eq}) and B3 achieves the best result (8.33×10^5, 1.64×10^3 kg CO_{2eq}). Figures 4 and 5 show the results for GWP—fossil and GWP—LULUC expressed in kg CO_{2e} emissions per gross floor area (GFA). The results for GWP—fossil are 2.05×10^3, 2.13×10^3 and 1.99×10^3 kg CO_{2eq}/m^2 for B1, B2 and B3, respectively. The results for GWP—LULUC are 3.95, 4.0 and 3.92 kg CO_{2ee}/m^2 for B1, B2 and B3, respectively.

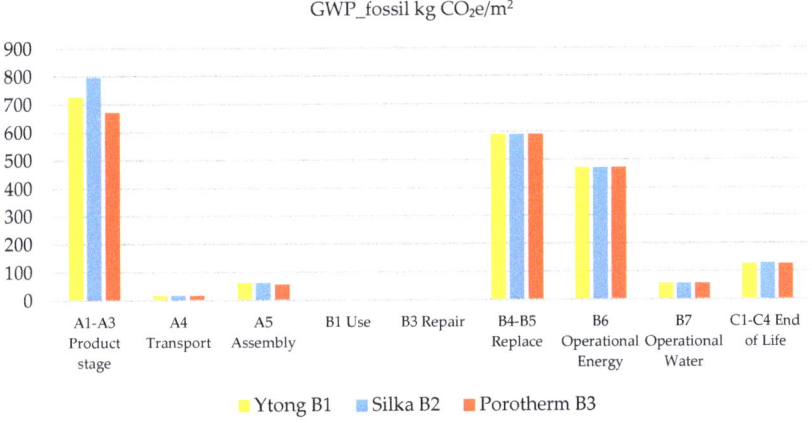

Figure 4. GWP—fossil in kg CO_{2eq}/m^2.

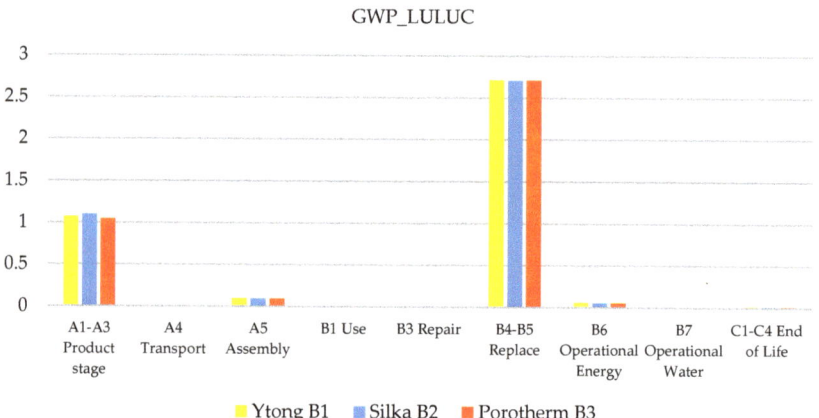

Figure 5. GWP—LULUC in kg CO_{2eq}/m^2.

End of life (EoL) is the last phase of the product life cycle (Figure 6). The product will be recycled, incinerated, used as a new product or deposited in landfill. When materials are incinerated or recycled, they can become useful. Incinerating waste releases usable energy and heat. And by recycling materials, "new" materials are created. In the EoL phase, we, thus, obtain the so-called secondary products. Secondary products have their own life cycle. In this way, the LCA results of the original product can be improved [17].

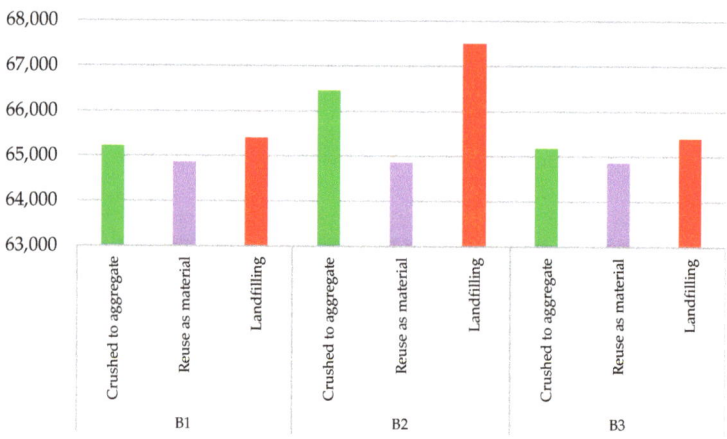

Figure 6. End of life (EOL).

In the original design, the envelope wall materials used for the three variants of building were aerated concrete block (B1), sand-lime block (B2) and hollow brick (B3). The end-of-life phase C1–C4 was changed from the original design. For variants B1, B2 and B3, the last phase of the life cycle for the aerated concrete blocks, lime-sand blocks and hollow bricks was changed from brick/stone crushed into an aggregate to reused as a material and being landfilled. This further reduced the overall emissions by 0.54%, 2.39% and 0.48% in B1, B2 and B3, respectively.

Through the change of EoL of the waste to being reused as a material, the total emissions were further reduced by 0.54%; 2.39% and 0.48% in variants B1, B2 and B3, respectively. Replacing it with landfill, the total emissions increased by 0.021%; 0.129% and 0.027% in B1, B2 and B3, respectively. Only the main material was changed in the variants, but if we also modified the end-of-life phase with other materials, the total emissions would be reduced even more. In a case study from Italy [18], the EoL phase was varied in a similar three-story building. The study suggests the key role of recycling different waste streams, particularly, the reinforcing steel fraction. The positive role of recycling construction plastics also appears interesting, particularly in terms of the impact on the non-renewable energy potential.

From the comparison of the three residential building alternatives, it can be concluded that variant B2 with the lime-sand walls was the best in almost all the categories of impacts. On the contrary, building B3 with the classic hollow bricks was assessed as the best. The greenhouse gas emissions were reduced by 6.39% by using a different external wall material. During the life cycle assessment, the environmental influences relating to the end-of-life phase of the residential building were also examined, with a focus on waste management in phases C1-C4. In buildings B1, B2 and B3, different EoL scenarios were compared. This study found that the re-use of brick waste as brick recyclate contributes the most to the reduction of greenhouse gas emissions.

4. Conclusions

This LCA study examines and quantifies the environmental influences associated with the end-of-life phase of a residential building, with a specific focus on demolition waste management. The analysis compared the environmental characteristics of alternative scenarios characterized by distinct criteria for the demolition of the mentioned building, the management of the demolition waste and the evaluation of the prevention of the burden of recycled materials. The results point to the important role of recycling and material reuse, which represent the most-eliminated impacts in the key categories, as well as in other studies [18,19], where they investigated the importance of the sustainable management of demolition waste. The results of the study are in line with the objectives set out in the European Union's Circular Economy Action Plan. This study quantifies the benefits of using an appropriate selective demolition technique that has the potential to increase both the quality and quantity of residues sent for resource recovery and safe disposal. In addition, the data obtained from the study can be used to establish criteria that can advance the overall environmental performance during the end-of-life phase of the residential building analyzed.

Author Contributions: Conceptualization, S.V. and J.B.; methodology, S.V.; software, J.B.; validation, E.K.B. and P.M.; formal analysis, E.K.B.; investigation, J.B.; resources, E.K.B.; data curation, S.V.; writing—original draft preparation, J.B. and E.K.B.; writing—review and editing, S.V. and P.M.; visualization, J.B.; supervision, P.M.; project administration, S.V.; funding acquisition, S.V. All authors have read and agreed to the published version of the manuscript.

Funding: This research was funded by VEGA, grant number 1/0512/20 and APVV-22-0576.

Institutional Review Board Statement: Not applicable for studies not involving humans or animals.

Informed Consent Statement: Not applicable.

Data Availability Statement: Data is contained within the article.

Conflicts of Interest: The authors declare no conflict of interest.

References

1. Tazi, N.; Idir, R.; Fraj, A.B. Towards achieving circularity in residential building materials: Potential stock, locks and opportunities. *J. Clean. Prod.* **2021**, *281*, 124489. [CrossRef]
2. Cottafava, D.; Ritzen, M. Circularity indicator for residential buildings: Addressing the gap between embodied impacts and design aspects. *Resources, Conserv. Recycl.* **2021**, *164*, 105120. [CrossRef]
3. Izaola, B.; Akizu-Gardoki, O.; Oregi, X. Setting baselines of the embodied, operational and whole life carbon emissions of the average Spanish residential building. *Sustain. Prod. Consum.* **2023**, *40*, 252–264. [CrossRef]
4. Wuyts, W.; Miatto, A.; Sedlitzky, R.; Tanikawa, H. Extending or ending the life of residential buildings in Japan: A social circular economy approach to the problem of short-lived constructions. *J. Clean. Prod.* **2019**, *231*, 660–670. [CrossRef]

5. Grusel, A.P.; Shehabi, A.; Horvath, A. What are the energy and greenhouse gas benefits of repurposing non-residential buildings into apartments? *Resour. Conserv. Recycl.* **2023**, *198*, 107143. [CrossRef]
6. Businge, C.N.; Mazzoleni, M. Impact of circular economy on the decarbonization of the Italian residential sector. *J. Clean. Prod.* **2023**, *408*, 136949. [CrossRef]
7. Le, D.L.; Salomone, R.; Nguyen, Q.T. Circular bio-based building materials: A literature review of case studies and sustainability assessment methods. *Build. Environ.* **2023**, *24*, 110774. [CrossRef]
8. Antwi-Afari, P.; Ng, S.T.; Chen, J. Determining the optimal partition system of a modular building from a circular economy perspective: A multicriteria decision-making process. *Renew. Sustain. Energy Rev.* **2023**, *185*, 113601. [CrossRef]
9. Tarek, D.; Ahmed, M.M.; Hussein, H.S.; Zeyad, A.M.; Al-Enizi, A.M.; Yousef, A.; Ragab, A. Building envelope optimization using geopolymer bricks to improve the energy efficiency of residential buildings in hot arid regions. *Case Stud. Constr. Mater.* **2022**, *17*, e01657. [CrossRef]
10. Deetman, S.; Marinova, S.; Van Der Voet, E.; Van Vuuren, D.P.; Edelenbosch, O.; Heijungs, R. Modelling global material stocks and flows for residential and service sector buildings towards 2050. *J. Clean. Prod.* **2020**, *245*, 118658. [CrossRef]
11. Yang, X.; Hu, M.; Zhang, C.; Steubing, B. Key strategies for decarbonizing the residential building stock: Results from a spatiotemporal model for Leiden, the Netherlands. *Resour. Conserv. Recycl.* **2022**, *184*, 106388. [CrossRef]
12. Simickova, V. *Product Life Cycle in the Manufacturer's Care for the Environment*, 1st ed.; ASPEK, Association of Industrial Ecology in Slovakia: Bratislava, Slovakia, 2002; p. 56. ISBN 80-88995-05-1.
13. One Click LCA Ltd. *Life Cycle Assessment for Buildings*; One Click LCA Ltd.: Helsinki, Finland, 2021.
14. Ling-Chin, J.; Heidrich, O.; Roskilly, A.P. Life cycle assessment (LCA)–from analyzing methodology development to introducing an LCA framework for marine photovoltaic (PV) systems. *Renew. Sustain. Energy Rev.* **2016**, *59*, 352–378. [CrossRef]
15. Vieira, P.S.; Horvath, A. Assessing the end-of-life impacts of buildings. *Environ. Sci. Technol.* **2008**, *42*, 4663–4669. [CrossRef]
16. EN 15804:2013+A2:2023; Sustainability of Construction Works. Environmental Product Declarations. Core Rules for the Product Category of Construction Products. CEN: Brussels, Belgium, 2023.
17. Nickel, L. End-of-life (EOL) in LCA: Impacts of Waste & Benefits of Recycling. 2 September 2023. Available online: https://ecochain.com/blog/end-of-life-eol-in-lca/ (accessed on 10 January 2023).
18. Vitale, P.; Arena, N.; Di Gregorio, F.; Arena, U. Life cycle assessment of the end-of-life phase of a residential building. *Waste Manag.* **2017**, *60*, 311–321. [CrossRef] [PubMed]
19. Wang, J.; Wei, J.; Liu, Z.; Huang, C.; Du, X. Life cycle assessment of building demolition waste based on building information modeling. *Resour. Conserv. Recycl.* **2022**, *178*, 106095. [CrossRef]

Disclaimer/Publisher's Note: The statements, opinions and data contained in all publications are solely those of the individual author(s) and contributor(s) and not of MDPI and/or the editor(s). MDPI and/or the editor(s) disclaim responsibility for any injury to people or property resulting from any ideas, methods, instructions or products referred to in the content.

Proceeding Paper

pH Variation during Bioaccumulation of Selected Toxic Metals by Newly Isolated Microscopic Fungi from the Ostramo Lagoons [†]

Marketa Vašinková

Department of Environmental Engineering, Faculty of Mining and Geology, VŠB–Technical University of Ostrava, 17. Listopadu 2172/15, 708 00 Ostrava, Czech Republic; marketa.vasinkova.st@vsb.cz

[†] Presented at the 4th International Conference on Advances in Environmental Engineering, Ostrava, Czech Republic, 20–22 November 2023.

Abstract: The emergence of new biotechnologies has increased interest in the study of the impact of environmental contamination on microbial communities, particularly in relation to the potential benefits that stress-adapted microorganisms offer in environmental protection, industrial ecology, and mineral and waste processing. This study aimed to compare the bioaccumulation abilities of recently isolated microscopic fungi belonging to the genera *Aspergillus*, *Phoma*, *Cystobasidium*, *Cladosporium* and *Exophiala*. These fungi were isolated from a site contaminated with both toxic metals and organic pollutants. The study monitored the bioaccumulation of selected toxic metal ions (Cu(II), Zn(II), Ni(II), Cr(III), Pb(II)), as well as pH changes, over a 30-day of biomass growth. The medium containing Pb(II) exhibited a statistically significant pH change during the 30-day accumulation period (Mann-Whitney U Test, $p < 0.05$). These findings provide valuable insights for the potential industrial application of microscopic fungi in bioaccumulation processes.

Keywords: microscopic fungi; bioaccumulation; Cu(II); Zn(II); Ni(II); Cr(III); Pb(II)

1. Introduction

Fungi have garnered significant scientific attention owing to their capacity for biotransformations of toxic metals and their adaptability to varying levels of anthropogenic contamination in the surrounding environment [1–4]. The main sources of metal-resistant species are found in contaminated sites, where indigenous fungal strains isolated from these areas demonstrate remarkable tolerance to elevated heavy metal concentrations [5,6]. The ecological function of fungi is subject to the influence of multiple factors [7], which typically exhibit variability, including soil composition, soil properties, climatic conditions, and the ever-evolving nature of organic and inorganic anthropogenic pollutants due to the introduction of novel chemical compounds and technological advancements. Toxic metals found in soil impact the composition of fungal communities and exert influence on their ecological function, particularly with regards to metabolite production and the subsequent decomposition of organic matter. Changes in fungal community structure may occur as a result of selective pressures exerted on specific species. Research has corroborated the ability of soil-inhabiting fungi to adapt to varying degrees of toxic metal contamination by modifying the abundance and structure of their communities [8]. The relationship between toxic metal contamination and the structure and species composition of fungal microbial communities has been observed in soils contaminated with various toxic metal ions [9,10].

The emergence of novel biotechnological advancements has spurred a growing interest in investigating the effects of toxic metal contamination on microbial communities. This research area is intricately linked to the potential benefits that stress-adapted microorganisms can offer in terms of environmental protection, industrial ecology, and mineral and waste processing sectors. Presently, environments contaminated with multiple pollutants, specifically the co-occurrence of organic matter and toxic metals, pose significant environmental challenges. In such scenarios, the utilization of heterotrophic organisms, including

Citation: Vašinková, M. pH Variation during Bioaccumulation of Selected Toxic Metals by Newly Isolated Microscopic Fungi from the Ostramo Lagoons. *Eng. Proc.* **2023**, *57*, 7. https://doi.org/10.3390/engproc2023057007

Academic Editors: Adriana Estokova, Natalia Junakova, Tomas Dvorsky, Vojtech Vaclavik and Magdalena Balintova

Published: 29 November 2023

Copyright: © 2023 by the author. Licensee MDPI, Basel, Switzerland. This article is an open access article distributed under the terms and conditions of the Creative Commons Attribution (CC BY) license (https://creativecommons.org/licenses/by/4.0/).

fungi, holds particular promise as they possess the ability to concurrently remove both types of pollutants.

1.1. Characteristics of the Ostramo Lagoons Site

Processes reliant on cellular metabolism, collectively known as bioaccumulation, encompass the transportation of metals across the cell wall and their subsequent localization within specific cellular compartments [11,12]. Numerous microorganisms, including filamentous fungi and yeasts, have demonstrated the capability to accumulate toxic metals within their cellular structures [13,14]. Bioaccumulation is categorized as a physiological mechanism for metal immobilization that is dependent upon cellular metabolism. Unlike biosorption, bioaccumulation is a more intricate process involving active transport of substances across cell membranes facilitated by transport proteins [15,16]. This active transport is preceded by a brief and rapid passive phase occurring immediately upon contact of the organism with toxic metal ions, involving ion exchange at the cell surface and physical adsorption [17].

The success of bioaccumulation relies on the internal structural and biochemical properties of cells, genetic and physiological adaptations of the organism, and the different oxidation states of metals [18]. One limiting factor in the utilization of living biomass for bioremediation processes is the toxicity of the metal to the microorganism. Elevated concentrations of toxicants can damage cell membranes and organelles, induce lipid peroxidation, and disrupt enzymatic functions, leading to the generation of reactive oxygen species and cell apoptosis [19]. Consequently, research efforts should also focus on the potential adaptation of respective organisms to stress conditions, which could enhance their efficiency in metal removal processes [20].

1.2. Effect of pH on Bioaccumulation of Toxic Metals

pH is one of the key factors that influences both the availability of metals and their uptake by fungi [21]. pH determines the chemical form in which a metal exists, whether as an ion, complex compound, or precipitate, which subsequently affects the potential for bioaccumulation by microorganisms [22]. Strains of filamentous fungi tolerant to elevated concentrations of toxic metals may exhibit increased production of organic acids. This phenomenon can be described as one of the resistance mechanisms employed by fungi against toxic metals, as organic acids react with metals to form organic compounds [23,24]. The change in pH in the solution during bioaccumulation may thus be a determining factor indicating the type of resistance mechanism in fungi or may be related to the bioaccumulation potential of the strain. Moreover, the availability and reactivity of fungal cell wall functional groups can be influenced by pH, which is widely used during biosorption processes [25,26].

Furthermore, pH exerts a notable influence on the physiology and metabolism of fungi, leading to indirect effects on metal accumulation. Fungal growth, enzyme activity, and various cellular processes are intricately linked to pH levels. Any pH-induced alterations in these factors can influence the ability of fungi to accumulate and tolerate toxic metals.

2. Materials and Methods

2.1. Characteristics of the Ostramo Lagoons Site

The Ostramo Lagoons, located in Ostrava, Czech Republic, were classified as significant ecological burdens. Their origins could be traced back to the deposition of waste generated from refinery production, which began in the late 19th century. Since 1965, the lagoons also served as a repository for waste resulting from the regeneration of used lubricating oils. The primary organic pollutants found in the lagoons included petroleum hydrocarbons, polycyclic aromatic hydrocarbons, polychlorinated biphenyls, and phenols. In addition to organic pollution, the lagoons were found to contain various toxic metals, including Arsenic, Cadmium, Copper, Mercury, Nickel, and Lead. Furthermore, the pH levels within the lagoons ranged from acidic to neutral [27,28].

2.2. Experimental Bioaccumulation of Toxic Metals by Selected Species of Isolated Microscopic Filamentous Fungi

The present study aimed to investigate the bioaccumulation of toxic metals in the biomass of microscopic fungi under controlled laboratory conditions. The following strains of isolated microscopic fungi were used: *Aspergillus niger*, *Aspergillus candidus*, *Aspergillus iizukae*, *Aspergillus westerdijkiae*, *Aspergillus ochraceus*, *Aspergillus clavatus*, *Phoma* sp., *Cystobasidium* sp., *Cladosporium* sp., *Exophiala xenobiotica*. The bioaccumulation experiments were conducted using static incubation methods. The respective metal salts solved in demineralized water were initially added to the cultivation medium to achieve a final concentration of 100 mg L^{-1} of metal. The metal ions investigated included Cu(II) from Cu(NO$_3$)$_2$·3H$_2$O (Sigma Aldrich, Darmstadt, Germany, purity 99.9%), Zn(II) from Zn(NO$_3$)$_2$·6H$_2$O (Sigma Aldrich, purity > 99.0%), Ni(II) from NiCl$_2$·6H$_2$O (Sigma Aldrich, purity > 99.9%), Cr(III) from Cr(NO$_3$)$_3$·9H$_2$O (Sigma Aldrich, purity 99.0%) and Pb(II) from Pb(NO$_3$)$_2$ (Penta Chemicals, Prague, Czech Republic, purity 99.0%).

The cultivation media used were Sabouraud Dextrose Broth (HiMedia Laboratories, Mumbai, India) for Cu(II), Zn(II), Ni(II), and Cr(III), and Czapek Dox Liquid Medium (HiMedia Laboratories) for Pb(II). The preparation of both media followed the instructions provided by the manufacturer, using demineralized water. The resulting pH of Sabouraud Dextrose Broth was measured to be 5.6 ± 0.2, while Czapek Dox Liquid Medium had a pH of 6.8 ± 0.2. Autoclave sterilization was performed at a temperature of 121 °C and a pressure of 15 lbs to ensure sterility of the media.

Static cultivation of the fungi was carried out in Erlenmeyer flasks for a duration of 30 days. The pH of the media was monitored throughout the cultivation period using standardized pH meter (HANNA, HI 991001). After 30 days of growth, the live biomass was harvested, filtered, dried, and weighed. The medium containing metal ions was filtered using a Gridded MCE sterile filter with a pore size of 0.45 μm (Membrane Solutions, Auburn, WA, USA). The metal content in both the biomass and nutrient medium was determined using atomic absorption spectroscopy (AAS) performed on a VARIAN AA 280FS instrument (Agilent, Santa Clara, CA, USA).

The efficiency of bioaccumulation was quantified as the ratio of the amount of metal accumulated by the biomass after 30 days of cultivation to the initial amount of metal present in the nutrient medium prior to the introduction of microscopic fungi according to the following equation:

$$\mu = \frac{m_{acc}}{m_{in}} \cdot 100, \quad (1)$$

where μ is expressed as a percentage and represents the efficiency at which a metal is accumulated within the total amount of dried biomass. The parameter m_{acc} denotes the amount of metal accumulated, while m_{in} represents the initial amount of metal present in the medium at the beginning of the incubation. For instance, in this particular scenario, the metal concentration in the medium is 100 mg L^{-1} in a 0.5 L volume, resulting in an initial metal amount of 50 mg.

3. Results

The bioaccumulation efficiency shown in Table 1 has been determined by evaluating the obtained data, which quantifies the total amount of toxic metal that has accumulated in the dried biomass, expressed as a percentage according to Equation (1).

Table 1. Calculation of bioaccumulation efficiency: total amount of toxic metal accumulated in dried biomass (%).

Fungi	Cu(II)	Zn(II)	Ni(II)	Cr(III)	Pb(II)
A. niger	20.15	7.91	5.89	6.82	10.78
A. candidus	9.52	92.52	3.98	7.67	9.74
A. iizukae	5.96	29.54	14.26	21.15	1.43
A. westerdijkiae	21.07	47.66	7.13	7.66	4.47
A. ochraceus	57.42	56.88	6.68	37.73	2.24
A. clavatus	33.35	30.84	13.74	7.69	4.85
Phoma sp.	86.43	32.91	18.29	18.67	7.06
Cystobasidium sp.	NA *	NA *	2.46	NA	6.00
Cladosporium sp.	54.68	4.45	13.35	12.03	19.11
Exophiala xenobiotica	NA *	NA *	5.69	NA *	5.23

* Not applicable.

The results regarding bioaccumulation efficiency revealed that *A. ochraceus* exhibited the highest efficiency among the tested metal ions. It achieved 57.42% efficiency during Cu(II) bioaccumulation, 56.88% efficiency during Zn(II) bioaccumulation, and 37.73% efficiency during Cr(III) bioaccumulation. This strain demonstrates a high tolerance to the toxic effects of these transition metal ions and efficiently accumulates them within its cells, utilizing its metabolic potential.

The pH was measured at weekly intervals, starting immediately after the addition of a spore suspension into the medium, and continuing on days 3, 10, 17, 24, and 30. Figure 1 illustrates the pH course in the media containing specific metal ions and the microscopic fungus. The pH values were plotted against time, providing insights into the changing acidity or alkalinity levels. Each graph showcases the pH trends for each specific metal ion and the microscopic fungus under investigation. Additionally, the boxplot in Figure 1f represents the distribution of pH changes in media containing particular metal ions, each plot contains the values of pH change for fungi 1–10.

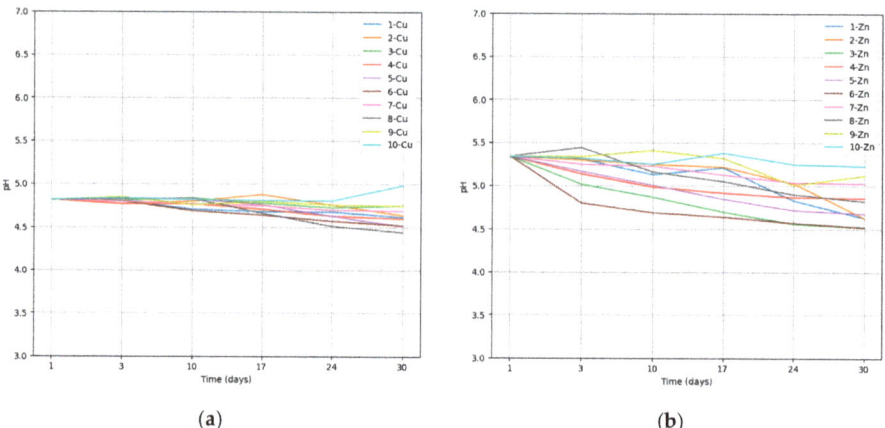

Figure 1. Cont.

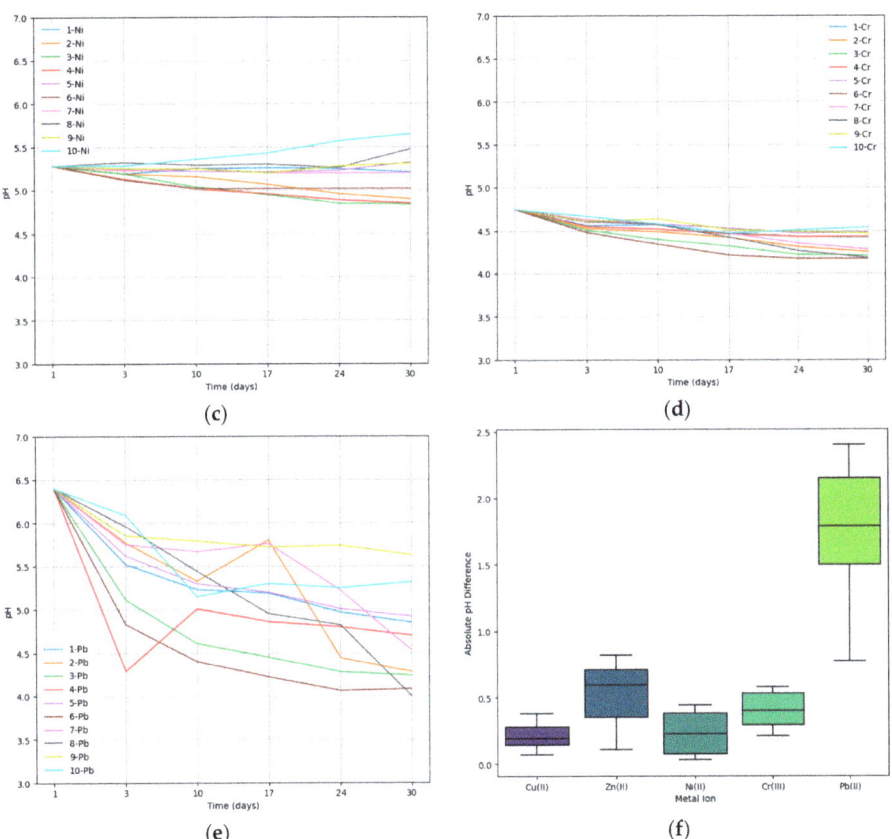

Figure 1. The course of pH in the bioaccumulation process. 1—*A. niger*, 2—*A. candidus*, 3—*A. iizukae*, 4—*A. westerdijkiae*, 5—*A. ochraceus*, 6—*A. clavatus*, 7—*Phoma* sp., 8—*Cystobasidium* sp., 9—*Cladosporium* sp., 10—*Exophiala xenobiotica* (**a**) Change in pH in Cu-containing media during bioaccumulation.; (**b**) change in pH in Zn-containing media during bioaccumulation.; (**c**) change in pH in Ni-containing media during bioaccumulation; (**d**) change in pH in Cr-containing media during bioaccumulation; (**e**) change in pH in Pb-containing media during bioaccumulation; (**f**) pH variation in metal ion-enriched media between day 1 and day 30.

4. Discussion

When describing the bioaccumulation of metals present in a nutrient medium through growing biomass, it is appropriate to consider the ways in which growing filamentous microscopic fungi can interact with metals. One possible scenario involves the uptake of metal cations into the cell through membrane transporters and their compartmentalization or biotransformation inside the cell, or conversely, efflux from the cell to the outside [29]. Another mode of interaction involves the production of organic acids, which have the ability to react with metals, resulting in the formation of insoluble compounds that can be present in the nutrient medium and on the surface of fungal mycelia. Certain species of microscopic fungi, such as those belonging to the genus *Aspergillus*, have the ability to produce extracellular polymeric substances (EPS) that can act as surface-active agents during the removal of toxic metals [30]. These genetically determined mechanisms directly influence the process of metal biotransformation, including the bioaccumulation potential of individual strains.

The average efficiency of bioaccumulation varied depending on the metal being accumulated, with the following preference order: Zn(II) > Cu(II) > Cr(III) > Ni(II) > Pb(II). These findings indicate that fungi possess an enhanced ability to accumulate metal ions, especially those that serve as essential micronutrients like Zn(II) an Cu(II). This heightened accumulation can be attributed to specific mechanisms that facilitate fungi in the efficient uptake and subsequent accumulation of these metals, which play crucial roles in various critical cellular processes [31,32].

To gain further insights into the bioaccumulation process of toxic metals by microscopic fungi, the pH of the enriched medium, which can be influenced by the production of secondary metabolites by the fungi, was measured. Fungi act as chelating agents and detoxify the effects of some toxic metal ions. Metal chelation through oxalic acid form oxalate crystals which can be in form of precipitates in growth medium or can be entrapped in fungal hyphae [24]. Furthermore, a correlation has been demonstrated between the higher production of organic acids by fungi and the degree of adaptation of a specific fungal strain to the presence of toxic metals in the environment [33]. The observed variations in pH within the metal ion-enriched medium underscore the dynamic nature of the bioaccumulation processes. Changes in pH can affect the ionization and speciation of metal ions, thereby affecting their solubility and bioavailability, as well as their ability to bind to specific sites in the fungal cell wall, which is widely exploited during bioremediation [34]. Consequently, these variations influence the interactions between microscopic fungi and the metals [35,36].

The medium containing Pb(II) experienced the largest average change in pH over the 30-day accumulation period. On average, the pH changed by 1.75, which indicates that the presence of Pb(II) had a statistically significant impact on the acidity or alkalinity of the medium (Mann-Whitney U test, $p < 0.05$). The low average bioaccumulation efficiency for Pb(II) among tested fungal strains (7.01%) is consistent with the observation that low pH values are not effective in removing Pb(II) from aqueous solutions and wastewater [37]. In the case of other metal ions, no significant change was observed. The lowest average change in pH was recorded for Cu(II) and Ni(II). In the case of Ni(II), even in four cases the pH of the medium increased by 0.04, 0.19, 0.03 a 0.37, respectively. Ni(II) was also accumulated with the second lowest average efficiency of the metal ions tested (9.15%), indicating the low ability of tested fungal strains to produce organic acids in the presence of Ni(II) and also to accumulate them inside their cells or in the cell wall.

5. Conclusions

The results of the study highlight that the bioaccumulation potential of microscopic fungi is influenced by both the type of metal ion and the particular strain of autochthonous fungi. The interaction between the growing fungal biomass and the metal ions within the medium encompasses various processes, including bioaccumulation, biosorption, ion exchange, and precipitation, among others. Alterations in the medium's pH during this process provide insights into the fungus's environmental interactions. These pH shifts indicate whether the fungus is producing more or fewer organic acids, aiding in the understanding of the underlying biochemical processes within the organism's resistance mechanisms. Bioaccumulation experiments for 30 days revealed a statistically significant effect of Pb(II) on the resulting pH of the culture medium.

Funding: Research was funded by The Project for Specific University Research (SGS) No. SP2023/4 by the Faculty of Mining and Geology of VŠB—Technical University of Ostrava.

Institutional Review Board Statement: Not applicable.

Informed Consent Statement: Not applicable.

Data Availability Statement: Data are contained within the article.

Conflicts of Interest: The author declares no conflict of interest.

References

1. Farkas, B.; Vojtková, H.; Bujdoš, M.; Kolenčík, M.; Šebesta, M.; Matulová, M.; Duborská, E.; Danko, M.; Kim, H.; Kučová, K.; et al. Fungal mobilization of selenium in the presence of hausmannite and ferric oxyhydroxides. *J. Fungi* 2021, *7*, 810. [CrossRef] [PubMed]
2. Duborská, E.; Szabó, K.; Bujdoš, M.; Vojtková, H.; Littera, P.; Dobročka, E.; Kim, H.; Urík, M. Assessment of *Aspergillus niger* strain's suitability for arsenate-contaminated water treatment and adsorbent recycling via bioextraction in a laboratory-scale experiment. *Microorganisms* 2020, *8*, 1668. [CrossRef]
3. Urík, M.; Polák, F.; Bujdoš, M.; Miglierini, M.B.; Milová-Žiaková, B.; Farkas, B.; Goneková, Z.; Vojtková, H.; Matúš, P. Antimony leaching from antimony-bearing ferric oxyhydroxides by filamentous fungi and biotransformation of ferric substrate. *Sci. Total Environ.* 2019, *664*, 683–689. [CrossRef] [PubMed]
4. Šimonovičová, A.; Ferianc, P.; Vojtková, H.; Pangallo, D.; Hanajík, P.; Kraková, L.; Feketeová, Z.; Čerňanský, S.; Okenicová, L.; Žemberyová, M.; et al. Alkaline Technosol contaminated by former mining activity and its culturable autochthonous microbiota. *Chemosphere* 2017, *171*, 89–96. [CrossRef] [PubMed]
5. Gadd, G.M.; Sayer, J.A. *Environmental Microbe-Metal Interactions*; Lovley, D.R., Ed.; ASM Press: Washington, DC, USA, 2000; p. 237.
6. Šimonovičová, A.; Vojtková, H.; Nosalj, S.; Piecková, E.; Švehláková, H.; Kraková, L.; Drahovská, H.; Stalmachová, B.; Kučová, K.; Pangallo, D. *Aspergillus niger* environmental isolates and their specific diversity through metabolite profiling. *Front. Microbiol.* 2021, *12*, 658010. [CrossRef] [PubMed]
7. Nosalj, S.; Šimonovičová, A.; Vojtková, H. Enzyme production by soilborne fungal strains of *Aspergillus niger* isolated from different localities affected by mining. In Proceedings of the IOP Conference Series Earth and Environmental Science, Surakarta, Indonesia, 24–25 August 2021; Volume 900, p. 012027. [CrossRef]
8. Zhao, X.; Huang, J.; Lu, J.; Sun, Y. Study on the influence of soil microbial community on the long-term heavy metal pollution of different land use types and depth layers in mine. *Ecotoxicol. Environ. Saf.* 2019, *170*, 218–226. [CrossRef] [PubMed]
9. Rajapaksha, R. Heavy metal tolerance of culturable bacteria and fungi in a long-term cultivated tropical ultisol. *Eur. J. Soil Biol.* 2011, *47*, 9–15. [CrossRef]
10. Zafar, S.; Aqil, F.; Ahmad, I. Metal tolerance and biosorption potential of filamentous fungi isolated from metal contaminated agricultural soil. *Bioresour. Technol.* 2007, *98*, 2557–2561. [CrossRef]
11. Blackwell, K.; Tobin, J. Cadmium accumulation and its effects on intracellular ion pools in a brewing strain of *Saccharomyces cerevisiae*. *J. Ind. Microbiol. Biotechnol.* 1999, *23*, 204–208. [CrossRef]
12. Gadd, G.M. Biosorption: Critical review of scientific rationale, environmental importance and significance for pollution treatment. *J. Chem. Technol. Biotechnol. Int. Res. Process Environ. Clean Technol.* 2009, *84*, 13–28. [CrossRef]
13. Šebesta, M.; Vojtková, H.; Cyprichová, V.; Ingle, A.P.; Urík, M.; Kolenčík, M. Mycosynthesis of metal-containing nanoparticles—Fungal metal resistance and mechanisms of synthesis. *Int. J. Mol. Sci.* 2022, *23*, 14084. [CrossRef] [PubMed]
14. Šebesta, M.; Vojtková, H.; Cyprichová, V.; Ingle, A.P.; Urík, M.; Kolenčík, M. Mycosynthesis of metal-containing nanoparticles—Synthesis by *Ascomycetes* and *Basidiomycetes* and their application. *Int. J. Mol. Sci.* 2023, *24*, 304. [CrossRef] [PubMed]
15. Bibbins-Martínez, M.; Juárez-Hernández, J.; López-Domínguez, J.; Nava-Galicia, S.; Martínez-Tozcano, L.; Juárez-Atonal', R.; Cortés-Espinosa, D.; Díaz-Godinez, G. Potential application of fungal biosorption and/or bioaccumulation for the bioremediation of wastewater contamination: A review. *J. Environ. Biol.* 2023, *44*, 135–145. [CrossRef]
16. Argüello, J.M.; Raimunda, D.; González-Guerrero, M. Metal transport across biomembranes: Emerging models for a distinct chemistry. *J. Biol. Chem.* 2012, *287*, 13510–13517. [CrossRef] [PubMed]
17. Dursun, A.; Uslu, G.; Cuci, Y.; Aksu, Z. Bioaccumulation of copper (II), lead (II) and chromium (VI) by growing *Aspergillus niger*. *Process Biochem.* 2003, *38*, 1647–1651. [CrossRef]
18. Timková, I.; Sedláková-Kaduková, J.; Pristaš, P. Biosorption and bioaccumulation abilities of actinomycetes/streptomycetes isolated from metal contaminated sites. *Separations* 2018, *5*, 54. [CrossRef]
19. Igiri, B.E.; Okoduwa, S.I.; Idoko, G.O.; Akabuogu, E.P.; Adeyi, A.O.; Ejiogu, I.K. Toxicity and bioremediation of heavy metals contaminated ecosystem from tannery wastewater: A review. *J. Toxicol.* 2018, *2018*, 2568038. [CrossRef]
20. Valix, M.; Loon, L. Adaptive tolerance behaviour of fungi in heavy metals. *Miner. Eng.* 2003, *16*, 193–198. [CrossRef]
21. Ayele, A.; Haile, S.; Alemu, D.; Kamaraj, M. Comparative utilization of dead and live fungal biomass for the removal of heavy metal: A concise review. *Sci. World J.* 2021, *2021*, 5588111. [CrossRef]
22. Sintorini, M.; Widyatmoko, H.; Sinaga, E.; Aliyah, N. Effect of pH on metal mobility in the soil. In Proceedings of the IOP Conference Series: Earth and Environmental Science, Surakarta, Indonesia, 24–25 August 2021; Volume 737, p. 012071. [CrossRef]
23. White, C.; Sayer, J.; Gadd, G. Microbial solubilization and immobilization of toxic metals: Key biogeochemical processes for treatment of contamination. *FEMS Microbiol. Rev.* 1997, *20*, 503–516. [CrossRef]
24. Fomina, M.; Hillier, S.; Charnock, J.; Melville, K.; Alexander, I.J.; Gadd, G. Role of oxalic acid overexcretion in transformations of toxic metal minerals by *Beauveria caledonica*. *Appl. Environ. Microbiol.* 2005, *71*, 371–381. [CrossRef] [PubMed]
25. Michalak, I.; Chojnacka, K.; Witek-Krowiak, A. State of the art for the biosorption process—A review. *Appl. Biochem. Biotechnol.* 2013, *170*, 1389–1416. [CrossRef] [PubMed]
26. Filote, C.; Rosca, M.; Hlihor, R.M.; Cozma, P.; Simion, I.M.; Apostol, M.; Gavrilescu, M. Sustainable application of biosorption and bioaccumulation of persistent pollutants in wastewater treatment: Current practice. *Processes* 2021, *9*, 1696. [CrossRef]

27. Vojtková, H. New strains of copper-resistant pseudomonas bacteria isolated from anthropogenically polluted soils. In Proceedings of the International Multidisciplinary Scientific GeoConference Surveying Geology and Mining Ecology Management, SGEM, Albena, Bulgaria, 17–26 June 2014; Volume 1, pp. 451–457. [CrossRef]
28. Vojtková, H. Biodiversity of Pseudomonas bacterial strains isolated from Ostrava Lagoons, Czech Republic. In Proceedings of the 15th International Multidisciplinary Scientific Geoconference SGEM 2015, Albena, Bulgaria, 18–24 June 2015; pp. 291–296. [CrossRef]
29. Xu, X.; Hao, R.; Xu, H.; Lu, A. Removal mechanism of Pb (II) by *Penicillium polonicum*: Immobilization, adsorption, and bioaccumulation. *Sci. Rep.* **2020**, *10*, 9079. [CrossRef] [PubMed]
30. Paria, K.; Pyne, S.; Chakraborty, S.K. Optimization of heavy metal (lead) remedial activities of fungi *Aspergillus penicillioides* (F12) through extra cellular polymeric substances. *Chemosphere* **2022**, *286*, 131874. [CrossRef]
31. Antsotegi-Uskola, M.; Markina-Iñarrairaegui, A.; Ugalde, U. New insights into copper homeostasis in filamentous fungi. *Int. Microbiol.* **2020**, *23*, 65–73. [CrossRef]
32. Toledo, H.; Sánchez, C.I.; Marín, L.; Amich, J.; Calera, J.A. Regulation of zinc homeostatic genes by environmental pH in the filamentous fungus *Aspergillus fumigatus*. *Environ. Microbiol.* **2022**, *24*, 643–666. [CrossRef]
33. Ge, W.; Zamri, D.; Mineyama, H.; Valix, M. Bioaccumulation of heavy metals on adapted *Aspergillus foetidus*. *Adsorption* **2011**, *17*, 901–910. [CrossRef]
34. Ramya, D.; Kiruba, N.J.M.; Thatheyus, A.J. Biosorption of heavy metals using fungal biosorbents—A review. *Fungi Bio-Prospect. Sustain. Agric. Environ. Nano-Technol.* **2021**, *2*, 331–352. [CrossRef]
35. Gadd, G.M. Interactions of fungi with toxic metals. In *The genus Aspergillus: From Taxonomy and Genetics to Industrial Application*; Springer: Boston, MA, USA, 1994; pp. 361–374.
36. Robinson, J.R.; Isikhuemhen, O.S.; Anike, F.N. Fungal–metal interactions: A review of toxicity and homeostasis. *J. Fungi* **2021**, *7*, 225. [CrossRef]
37. Karimi, H. Effect of pH and initial pb (II) concentration on the lead removal efficiency from industrial wastewater using $Ca(OH)_2$. *Int. J. Water Wastewater Treat* **2017**, *3*, 1–4. [CrossRef]

Disclaimer/Publisher's Note: The statements, opinions and data contained in all publications are solely those of the individual author(s) and contributor(s) and not of MDPI and/or the editor(s). MDPI and/or the editor(s) disclaim responsibility for any injury to people or property resulting from any ideas, methods, instructions or products referred to in the content.

Proceeding Paper

The Water Quality of Revitalized Ponds in the Czech Republic Post-Mining Area [†]

Petra Malíková * and Jitka Chromíková

Department of Environmental Engineering, Faculty of Mining and Geology, VSB—Technical University of Ostrava, 17. Listopadu 2172/15, 708 00 Ostrava, Czech Republic; jitka.chromikova@vsb.cz
* Correspondence: petra.malikova@vsb.cz
† Presented at the 4th International Conference on Advances in Environmental Engineering, Ostrava, Czech Republic, 20–22 November 2023.

Abstract: This study assesses the water quality of the Sušanecké Ponds in the Czech Republic post-mining area. Four monitoring profiles were chosen: the Sušanka River inflow into the Sušanecké Ponds, two ponds and the outflow of the Sušanecké Ponds to the Sušanka River. The sampling took place in a 14-day interval from March to October 2022. The monitored parameters were temperature, O_2, pH, electrical conductivity, turbidity, nitrate nitrogen, ammoniacal nitrogen, total phosphorus, chlorophyll-a, COD_{Cr}, BOD_5 and metals—Fe, Mn, Cu, Zn, Ni, Pb, Co, Cd, Cr. The results were evaluated in accordance with the Czech standard ČSN 75 7221 and Government Regulation No. 401/2015 Coll. Based on the evaluation data, it was found that the area of the Sušanecké Ponds does not meet the limits of the government regulation for three out of the twenty parameters. According to the standard, Sušanecké Ponds are mainly classified as highly polluted waters.

Keywords: water quality; surface water; post-mining area; Ostrava-Karviná District (OKD); Moravian-Silesian Region

1. Introduction

The environmental quality of post-mining cities is slowly improving. On the one hand, these cities are pursuing sustainable development, taking into account the role of blue-green infrastructure. On the other hand, they must perform certain economic and social functions in the places of settlement [1]. The emphasis is on the adequacy of water sources for recreational purposes [2,3], especially in dense urban areas. In particular, these activities relate to water quality issues (water bathing, water sports and fishing) [4]. In addition, post-mining cities face depopulation due to their negative image [5]. Consequently, all actions co-existing with the mitigation of this attitude are welcome. Many cities are thus trying to restore the original function of existing water bodies that were used in the past, focusing on restoring their natural after-use potential in terms of water quality. Finally, efforts are also being made to green densely built areas and improve climate resilience [6,7].

The Ostrava-Karviná District (OKD) is only a small part of the vast Upper Silesia coal basin, most of which is located in neighboring Poland (more than 90% of the area). With regard to the black coal mining industry, the region is one of the most important in Europe. Approximately 90% of all the black coal production in the former Czechoslovakia was extracted from OKD until the late of 1980s, and three quarters of the total was extracted from Karviná region. In the early 1990s, the extraction of black coal in the Ostrava region and other regions of the Czech Republic was reduced after the transformation of the Czech mining industry [8].

The aim of this study was to evaluate the water quality of Sušanecké Ponds in the OKD post-mining city.

Citation: Malíková, P.; Chromíková, J. The Water Quality of Revitalized Ponds in the Czech Republic Post-Mining Area. *Eng. Proc.* 2023, 57, 8. https://doi.org/10.3390/engproc2023057008

Academic Editors: Adriana Estokova, Natalia Junakova, Tomas Dvorsky, Vojtech Vaclavik and Magdalena Balintova

Published: 29 November 2023

Copyright: © 2023 by the authors. Licensee MDPI, Basel, Switzerland. This article is an open access article distributed under the terms and conditions of the Creative Commons Attribution (CC BY) license (https://creativecommons.org/licenses/by/4.0/).

2. Materials and Methods

2.1. Study Area

The study area is located in the Havířov City, Horní Suchá district, in the Moravian-Silesian Region of the Czech Republic (Figure 1B). This area is in the OKD post-mining district. The climate in this region is moderately warm, with an average annual temperature of 19 °C in summer and an average annual rainfall of 650 mm [9]. The earliest references to the Sušanecké Ponds date from 1764 to 1768 and 1780 to 1783, documented from military maps [10–12]. Previously, the Sušanecké Ponds were composed of ten water bodies fed by the Lučina River, the Sušanka River and the Dolnosušský Stream. Between 1836 and 1852, the number of water areas dropped from ten to five, and from 1869 to 1885, the number of water areas was further reduced from ten to four by dividing the ponds into large units. Subsequently, the water area was again divided, with an increase to seven.

Until the late 1950s, seven ponds were used for fishing and recreational purposes and by the inhabitants of Horní Suchá and its surroundings. The ponds were also the source of water for the Dukla Mine. After they were no longer used as a source of water for the Dukla Mine, in 1959, the Sušanecké Ponds became part of the sludge lagoons in the wastewater treatment system; then the sewage from the surrounding villages was also discharged there. The mine water from the Plant František, Mine Odra, Plant 9th May, Mine Darkov and, in an emergency, also from mines Dukla and Lazy were all pumped into the ponds. The actual construction of the treatment system took place between 1959 and 1966, when the sedimentation tanks were completed [9]. With the decline of mining and after the closure of the wastewater treatment plant, remediation and reclamation work began in 1995. Coal sludge and flotation tailings from the tanks began to be removed in 2000. This process lasted until 2016. As part of the reclamation work, some parts of the tanks were filled in and the remaining areas were flooded with water. Since 2009, dice sneak urchin (*Natrix tessellate*) has been registered in the area as a specially protected species [13]. It is a critically endangered species in the Czech Republic and its number is being monitored. Since then, all work on the reservoirs is carried out in the period when the viper hibernates. The Dulka Mine was closed in June 2015. Currently, there are six water ponds with a total area of 215,872 m^2 (Figure 1A) [9].

2.2. Sampling

The selection of the monitored profiles was determined on the basis of availability and stability throughout the period. A total of four monitoring profiles were selected (see Figure 1A). The first sampling point (SP) was the Sušanka River, more specifically the tributary of the Sušanecké Ponds. The second and third sampling points were two of six ponds. The last sampling point was the Sušanka River, which flows from the last pond. Consequently, the profiles observed in the Sušanka River can show whether the pond cascade has a positive effect on the self-cleaning processes. Water sampling was chosen to meet the minimum number of 12 samples from each sampling point to assess surface water quality in accordance with the standard ČSN 75 7221 [14] between March and October 2022. The sampling was carried out at intervals of 14 days, every time at 7 a.m. If the sampling date fell on a rainy day or after the rain, the sampling was performed at an alternative date in order to avoid results distortion (turbidity, dilution).

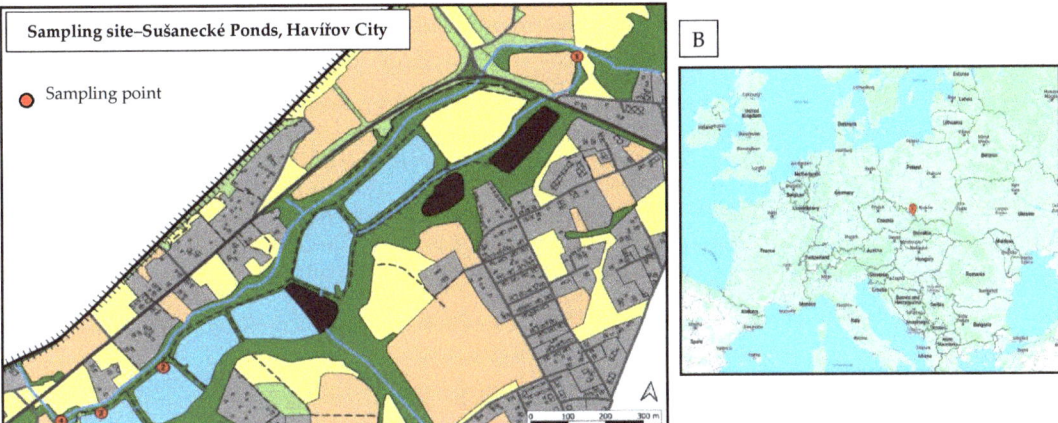

Figure 1. The simplified sketch map of the Sušanecké Ponds (**A**—sampling points, **B**—localization in Czech Republic) (adopted by [15]).

Water sampling and filling of the sample bottles were carried out in accordance with the standards ČSN EN ISO 5667-1, ČSN ISO 5667-4, ČSN EN ISO 5667-6, and ČSN EN ISO 5667-6 change A11 in [16–19]. For sampling, a PP 600 mL beaker connected to an aluminum telescopic sampling rod by a snap-in joint was used. Before the actual sampling, the sampler beaker and the PP and glass sampling bottles were rinsed several times with sampled water in the sample site. For the chlorophyll-a indicator, the samples were collected in dark glass sample bottles with a volume of 0.5 L. For other monitoring indicators, the samples were filled into a PP sample bottle with a volume of 2 L. All sample bottles were properly labelled with the number of samples and date of collection. The handling and storage of samples was carried out in accordance with the standard ČSN EN ISO 5667-3 in [20]. The samples were placed in coolers and transported to the laboratories of the Department of Environmental Engineering at the Faculty of Mining and Geology of the VSB-Technical University of Ostrava.

2.3. Measurements

The 48 water samples were characterized by 20 physicochemical and biological parameters, including temperature, water reaction (pH), dissolved oxygen (DO), electrical conductivity, turbidity, 5-day biochemical oxygen demand (BOD), chemical oxygen demand by dichromate (COD), total phosphorus (P_{TOT}), nitrate nitrogen, ammonia nitrogen, chlorophyll-a and metals (Fe, Mn, Cu, Zn, Ni, Pb, Co, Cd, Cr). Temperature, pH, electrical conductivity and DO values were measured at the sample site. The measurements were performed using multiple WTW pH/Cond multimeter (WTW, Weilheim, Germany) and HQ 30d oximeter (HACH Company, Loveland, CO, USA). Rest analyses were carried out using standard methods in the laboratories of the Department of Environmental Engineering. The UV-VIS spectrometer DR 3900 (HACH Company, Loveland, CO, USA) was used to measure turbidity, P_{TOT}, nitrate nitrogen, ammonia nitrogen and chlorophyll-a. Metals were performed using the AAS VARIAN AA280FS (Agilent, Santa Clara, CA, USA).

2.4. Data Analysis

The resulting values of the monitoring parameters were assessed according to the standard ČSN 75 7221 Classification of surface water quality in [14], Government Regulation No. 401/2015 Coll. on indicators and values of permissible pollution of surface water and wastewater, the requirements for permits for the discharge of wastewater into surface water and sewers and on sensitive areas, as amended, and the Environmental Quality Standards, which are annexed to the No. 401/2015 Coll. for specific pollutants for surface water bodies and the values of permissible pollution of surface water used for water supply purposes. The annual average (AA-EQS) and maximum allowable concentrations (MAX-EQS) were then used for the assessment in [21]. The samples were set up and processed in MS Excel.

The classification of water quality in accordance with ČSN 75 7221 was performed by comparing the calculated characteristic values of the indicator with the corresponding limit valve sets. The characteristic value of the water quality indicator is the value with a probability not exceeding 90%. The characteristic value of chlorophyll-a is the maximum value of a specified number of measured values for the growing season from March to October as prescribed in the standard. Water quality is classified separately for each indicator. The resulting class of the monitoring profile must be determined by the most negative classification found for each selected indicator. In Table 1, a list of monitoring indicators and the intervals of the individual classes of monitored indicators are provided in accordance with ČSN EN 75 7221.

Table 1. List of water quality indicators for the classification of Sušanecké Ponds (adapted from [14]).

Quality Indicator	Unit	Quality Classes				
		I	II	III	IV	V
Electrical conductivity	mS/m	<40	<70	<110	<160	≥160
DO	mg/L	>8.5	>7.5	>6	>4	≥4
$N-NO_3^-$	mg/L	<3	<6	<10	<13	≥13
$N-NH_4^+$	mg/L	<0.3	<0.7	<2	<4	≥4
P_{TOT}	mg/L	<0.05	<0.15	<0.4	<1	≥1
BOD	mg/L	<2	<4	<8	<15	≥15
COD	mg/L	<15	<25	<45	<60	≥60
Chlorophyll-a	µg/L	<10	<25	<50	<100	≥100
Mn	µg/L	<100	<300	<500	<800	≥800
Zn	µg/L	<15	<50	<100	<200	≥200
Cu_{TOT}	µg/L	<5	<15	<30	<60	≥60
Fe	µg/L	<500	<1000	<2000	<3000	≥3000
Cr	µg/L	<5	<15	<35	<70	≥70
Cd_{TOT}	µg/L	<0.1	<0.5	<1	<2	≥2
Co	µg/L	<1.5	<3	<6	<12	≥12
Ni_{TOT}	µg/L	<3	<6	<12	<40	≥40
Pb_{TOT}	µg/L	<3	<8	<15	<30	≥30

3. Results and Discussion

3.1. Water Quality Assessment in Accordance with GR No. 401/2015 Coll.

The Government Regulation (GR) No. 401/2015 Coll. does not include limit values of permitted pollutants for electrolytic conductivity, turbidity and chlorophyll-a. The trends of monitored water parameters of the Sušanecké Ponds SPs (SP1–SP4) in the monitoring period from March to October 2022 are shown in Figures 2–4.

Figure 2A shows how the temperature changed throughout the measurement process. At the beginning of the March measurement period, the temperatures rose gradually. In early August, the highest temperatures in the SP2, SP3 and SP4 samples were recorded, with the annual average (AA) temperature not exceeding 29 °C. In contrast, the lowest temperature (4.4 °C) was measured at the beginning of the SP4 growth season.

As shown in Figure 2B, the measured values were within the permissible pH value limit based on the GR, so it can be concluded that the Sušanecké Ponds, its inflow and outflow, meet the government-mandated parameters.

Figure 2C shows the DO pattern during the growing season. The DO values decreased during the measurement period. However, fluctuation was observed during the period from June to July. According to the GR, the annual average DO value should be more than 9 mg/L. All SPs, except for SP2, with an average value of 8.91 mg/L, meet the GR.

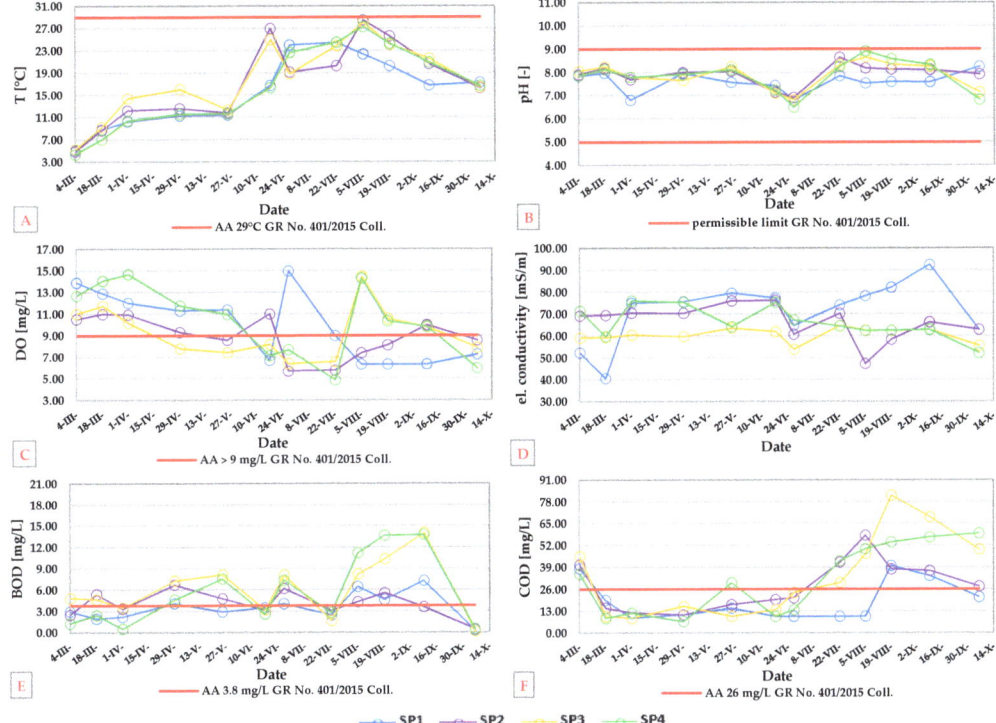

Figure 2. Water parameter trends at the Sušanecké Ponds sample points during the monitoring period from March to October 2022—part I (**A**—temperature trend, **B**—pH trend, **C**—DO trend, **D**—el. conductivity trend, **E**—BOD trend, **F**—COD trend).

The electrical conductivity values (in Figure 2D) remained constant during the study period. The lowest value was 40.6 mS/m in March, and the highest value was 92.6 mS/m in September.

During the monitoring period, most samples showed variations in the BOD values measured (Figure 2E). However, in early August, when the values of the SP3 and SP4 samples gradually reached their maximum values, the biggest change was observed. The next major change occurred in early October, when all values dropped to less than 0.5 mg/L. The GR established an average permitted BOD (five day) value of 3.8 mg/L. After adjusting, the average value was found to be greater than the values of three samples, namely SP2, SP3 and SP4. As a result, the Sušanecké Ponds and its outflow did not comply with the parameters prescribed by the government.

The course of the COD values measured is shown in Figure 2F. COD showed a decrease in the early spring. From May, there was a change in values as they gradually increased. The AA COD value according to the GR was 26 mg/L. After the average, it was found that the three SPs, namely SP2, SP3 and SP4, exceeded this value. Therefore, the Sušanecké Ponds and outflow did not meet the government mandated parameters.

The development of nutrients is shown in Figure 3A–C. The increase in values may be related to the addition of spring fertilizers in agricultural areas and to additional feeding by fishermen during the fishing season. The AA value of ammonia nitrogen (0.23 mg/L) was exceeded at all SPs, so the Sušanecké Ponds, inflow and outflow did not meet the GR parameters. Neither the storage nor the self-cleaning capacity regarding nutrients was confirmed; everything that flows into the ponds flows out. There is evidence of phosphorus release in the ponds, which could be influenced by the anoxic conditions at the bottom of the sediments. This is also reflected in the development of chlorophyll-a.

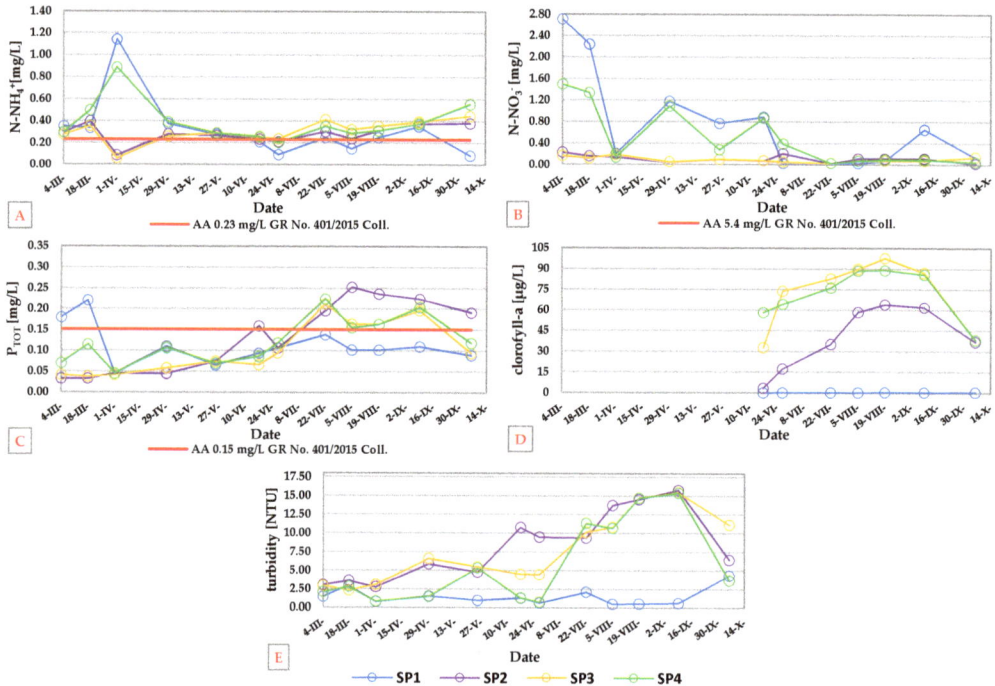

Figure 3. Water parameter trends at the Sušanecké Ponds sample points during the monitoring period from March to October 2022—part II (**A**—N-NH$_4^+$ trend, **B**—N-NO$_3^-$, **C**—P$_{TOT}$ trend, **D**—chlorophyll-a trend, **E**—turbidity trend).

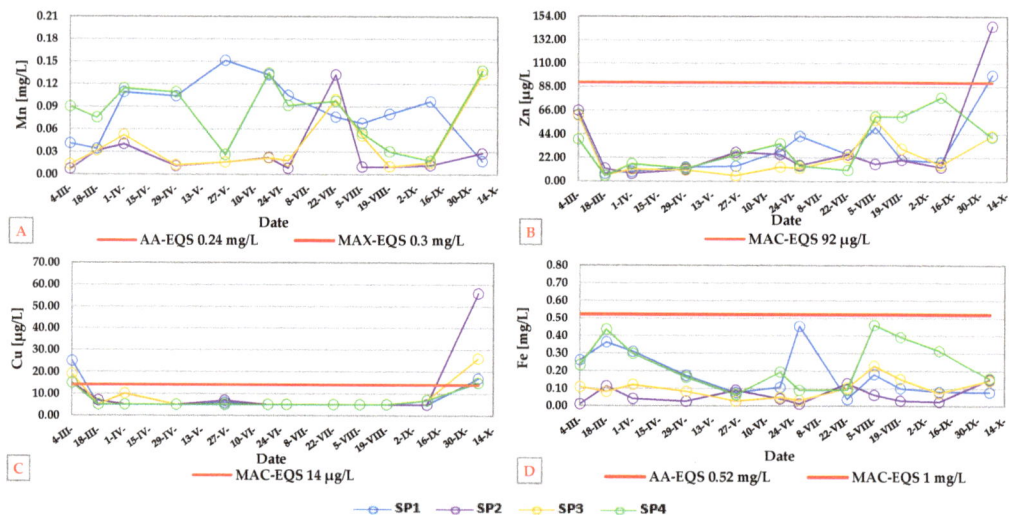

Figure 4. Water parameter trends at the Sušanecké Ponds sample points during the monitoring period from March to October 2022—part III (**A**—Mn trend, **B**—Zn trend, **C**—Cu trend, **D**—Fe trend).

Figure 3E shows an increase trend of turbidity values from SP2 to SP4. For SP1, the turbidity value was approximately 2 NTU over the entire monitoring period, but this value only increased to 4.34 NTU in October. In September, SP2 achieved the highest turbidity value, 15.75 NTU. Conversely, at the end of June, the lowest value was 0.7 NTU in SP1.

Similarly, the development of metal concentrations is shown in Figure 4. The AA values according to EQS for Mn, Zn, Cu and Fe were not exceeded. Metals—Cd, Cr, Co and Pb—were not evaluated because their concentrations were below the detection limit. Specifically, the concentration of Cd_{TOT} was less than 1 µg/L, and the others were less than 10 µg/L. In case further monitoring is required, the ICP-MS method would be more appropriate.

3.2. Water Quality Assessment in Accordance with the Standard ČSN 75 7221

As mentioned in Section 3.1, metals Cd, Cr, Co and Pb were not evaluated because they were below the detection limit. The classification of the individual SPs into water quality classes is shown in Tables 2 and 3. Because the chlorophyll-a value from March to May was not measurable, only the values from June were included.

Table 2. Classification of SP1 and SP2 into water quality classes during the monitoring period.

Quality Indicator	Unit	Number of Determinations	Arithmetic Mean	Median	C_{90}	Quality Class	Total Classification
SP1—Tributary of the Sušanka River							
Electrical conductivity	mS/m	12	71.36	75.45	80.79	III	
DO	mg/L	12	9.87	10.15	6.35	III	
$N-NO_3^-$	mg/L	12	0.74	0.42	1.65	I	
$N-NH_4^+$	mg/L	12	0.32	0.26	0.34	III	
P_{TOT}	mg/L	12	0.11	0.10	0.14	II	
BOD	mg/L	12	3.56	3.21	5.45	III	III
COD	mg/L	12	19.04	13.00	29.03	III	
Chlorophyll-a	µg/L	7	0.26	0.29	0.59 *	I	
Mn	µg/L	12	84.58	88.50	119.58	II	
Zn	µg/L	12	32.25	22.5	55.52	III	
Cu_{TOT}	µg/L	12	7.75	5.00	5.00	II	
Fe	µg/L	12	184.58	139.50	332.38	I	
SP2—Pond No. 1							
Electrical conductivity	mS/m	12	66.55	69.50	70.54	III	
DO	mg/L	12	8.91	8.95	7.03	III	
$N-NO_3^-$	mg/L	12	0.10	0.11	0.17	I	
$N-NH_4^+$	mg/L	12	0.28	0.28	0.37	II	
P_{TOT}	mg/L	12	0.13	0.13	0.22	III	
BOD	mg/L	12	4.09	3.95	5.76	III	IV
COD	mg/L	12	28.27	24.32	41.46	III	
Chlorophyll-a	µg/L	7	39.62	37.02	64.02 *	IV	
Mn	µg/L	12	27.25	14.00	35.14	I	
Zn	µg/L	12	31.83	18.00	44.94	II	
Cu_{TOT}	µg/L	12	10.41	5.00	10.68	II	
Fe	µg/L	12	60.83	41.5	119.90	I	

Note: (*) In case of the chlorophyll-a indicator, the maximum value of all measured values is taken.

The water quality of SP1, a Sušanka River tributary, was classified as water quality class III—polluted water, based on the values of the parameters shown in Table 2. This quality class is characterized by the following: water affected by human activity, and water quality indicators do not correspond with values suitable for a balanced and sustainable ecosystem. The water quality of SP2, Pond No. 1, was classified as water quality class IV—highly polluted water, based on the parameters listed in Table 2, when the chlorophyll-a parameter was the worst. This quality class is described as follows: water is heavily influenced by human activity, and water quality indicators indicate values that are not suitable for a balanced and sustainable ecosystem. The water quality of the remaining two SPs (SP3—Pond No. 2 and SP4—outflow to Sušanka River) were also classified as water quality class IV based on the indicators BOD, COD and chlorophyll-a listed in Table 3.

Table 3. Classification of SP3 and SP4 into water quality classes during the monitoring period.

Quality Indicator	Unit	Number of Determinations	Arithmetic Mean	Median	C_{90}	Quality Class	Total Classification
SP3—Pond No. 2							
Electrical conductivity	mS/m	12	60.63	61.35	63.26	II	
DO	mg/L	12	9.34	9.00	7.04	III	
N-NO_3^-	mg/L	12	0.08	0.06	0.14	I	
N-NH_4^+	mg/L	12	0.31	0.30	0.39	II	
P_{TOT}	mg/L	12	0.10	0.08	0.17	III	
BOD	mg/L	12	6.17	6.07	9.19	IV	IV
COD	mg/L	12	33.86	27.00	58.76	IV	
Chlorophyll-a	µg/L	7	71.52	82.63	97.42 *	IV	
Mn	µg/L	12	39.25	18.50	73.62	I	
Zn	µg/L	12	23.58	14.00	48.90	II	
Cu_{TOT}	µg/L	12	8.50	5.00	14.14	II	
Fe	µg/L	12	98.16	90.00	145.68	I	
SP4—Sušanka River outflow							
Electrical conductivity	mS/m	12	66.27	64.51	75.44	III	
DO	mg/L	12	10.36	10.65	6.90	III	
N-NO_3^-	mg/L	12	0.48	0.20	1.19	I	
N-NH_4^+	mg/L	12	0.39	0.33	0.51	III	
P_{TOT}	mg/L	12	0.48	0.20	1.19	I	
BOD	mg/L	12	5.68	3.70	12.32	IV	IV
COD	mg/L	12	31.70	32.67	55.84	IV	
Chlorophyll-a	µg/L	7	71.28	76.11	88.71 *	IV	
Mn	µg/L	12	81.41	90.50	123.20	II	
Zn	µg/L	12	32.83	29.50	60.00	III	
Cu_{TOT}	µg/L	12	6.83	5.00	5.92	II	
Fe	µg/L	12	239.41	211.50	410.70	I	

Note: (*) In case of the chlorophyll-a indicator, the maximum value of all measured values is taken.

4. Conclusions

According to the evaluation of the data based on the standard ČSN 75 7221 and Government Regulation No. 401/2015 Coll. it was found that the area of Sušanecké Ponds does not meet the GR limit values for three of the parameters, namely for ammonia nitrogen, BOD and COD. According to the standard, the area is largely classified as the water quality class IV. This fact can be attributed to the earlier use of Sušanecké Ponds as waste reservoirs, which served as "wastewater treatment plants", expanded agricultural activity in the area, abundant population and constantly growing industrial activity. It is obvious that the quality of the water in the ponds and the Sušanka River will be significantly affected by these factors. In conclusion, it can be stated that the water quality indicators of the Sušanecké Ponds reach the values of heavily polluted water. The self-cleaning effect of the Sušanecké Ponds has not been proven; the water quality has deteriorated.

Currently, the City of Havířov is considering the revitalization of the surroundings of Sušanecké Ponds and their use as a water element in a possible urban forest park. Therefore, in future steps it would be advisable to measure flow and runoff conditions.

Author Contributions: P.M. and J.C. were involved in all aspects of the manuscript such as conceptualization, methodology, software, validation, formal analysis, investigation, resources, data curation, writing—original preparation, writing—review and editing, visualization, supervision, and project administration. All authors have read and agreed to the published version of the manuscript.

Funding: This research received no external funding.

Institutional Review Board Statement: Not applicable.

Informed Consent Statement: Not applicable.

Data Availability Statement: Data sets are contained within the article. Further data and materials requests should be addressed to petra.malikova@vsb.cz (P.M.).

Acknowledgments: The authors would like to thank the Department of Environmental Engineering, Faculty of Mining and Geology for technical and material support connected with laboratory analyses. Furthermore, the authors would like to thank the internal materials of the Havířov City Magistrate's Office and Bardoňovská Sára for the cooperation with the creation of Figure 1A.

Conflicts of Interest: The authors declare no conflict of interest. The funders had no role in the design of the study; in the collection, analyses, or interpretation of data; in the writing of the manuscript; or in the decision to publish the results.

References

1. Harfst, J. Utilizing the past: Valorizing post-mining potential in Central Europe. *Extr. Ind. Soc.* **2015**, *2*, 217–224. [CrossRef]
2. Kantor-Pietraga, I.; Krzysztofik, R.; Solarski, M. Planning Recreation around Water Bodies in Two Hard Coal Post-Mining Areas in Southern Poland. *Sustainability* **2023**, *15*, 10607. [CrossRef]
3. Nelson, E.; Rogers, M.; Wood, S.A.; Chung, J.; Keeler, B. Data-driven predictions of summertime visits to lakes across 17 US states. *Ecosphere* **2023**, *14*, e4457. [CrossRef]
4. Keeler, B.L.; Wood, S.A.; Polasky, S.; Kling, C.; Filstrup, C.T.; Downing, J.A. Recreational demand for clean water: Evidence from geotagged photographs by visitors to lakes. *Front. Ecol. Environ.* **2015**, *13*, 76–81. [CrossRef] [PubMed]
5. Sokołowski, J.; Frankowski, J.; Mazurkiewicz, J.; Lewandowski, P. Hard coal phase-out and the labour market transition pathways: The case of Poland. *Environ. Innov. Soc. Transit.* **2022**, *43*, 80–98. [CrossRef]
6. Wagner, I.; Krauze, K.; Zalewski, M. Blue aspects of green infrastructure. *Sustain. Dev. Appl.* **2013**, *4*, 145–155.
7. Bertram, C.; Meyerhoff, J.; Rehdanz, K.; Wüstemann, H. Differences in the recreational value of urban parks between weekdays and weekends: A discrete choice analysis. *Landsc. Urban Plan.* **2017**, *159*, 5–14. [CrossRef]
8. Havrlant, J.; Krtička, L. Reclamation of devastated landscape in the Karviná region (Czech Republic). *Environ. Socio-Econ. Stud.* **2014**, *2*, 1–12. [CrossRef]
9. Anonymous. Havířov City, Czech Republic. Internal Documentation. Unpublished, 2023.
10. *Ist Military Survey—Silesia, Map Sheet No.12*. Austrian State Archive/Military Archive, Vienna, Geoinformatics Laboratory at J.E.Purkyně. Available online: http://oldmaps.geolab.cz/map_viewer.pl?lang=en&map_root=1vm&map_region=sl&map_list=s012 (accessed on 18 August 2023).
11. *IInd Military Survey—Moravia, Map Sheet No. O_5_X*. Austrian State Archive/Military Archive, Vienna, Geoinformatics Laboratory at J.E.Purkyně. Available online: http://oldmaps.geolab.cz/map_viewer.pl?lang=en&map_root=2vm&map_region=mo&map_list=O_5_X (accessed on 18 August 2023).
12. *IIIrd Military Survey—1:25,000, Map Sheet No. 4061_3*. Austrian State Archive/Military Archive, Vienna, Geoinformatics Laboratory at J.E.Purkyně. Available online: http://oldmaps.geolab.cz/map_viewer.pl?lang=en&map_root=3vm&map_region=25&map_list=4061_3 (accessed on 18 August 2023).
13. Vlček, P.; Zavadil, V.; Jablonski, D.; Mebert, K. Dice Snake (*Natrix tessellate*) in the Baltic Sea Drainage Basin (Karvinsko District in Silesia, Czech Republic). *Merensiella* **2011**, *18*, 177–187.
14. ČSN 75 7221; Water Quality—Classification of Surface Water Quality. Czech Standard Institute: Prague, Czech Republic, 2017.
15. Bardoňová, S. *Sušanecké Ponds: Water Sampling Sites in the Area of the Sušanecké Ponds [1:50,000]*; Characteristics of the site and description of the Sušanecké Ponds and their proposed use; Department of Urban Development: Havířov City, Czech Republic; Magistrate of the Havířov City, Czech Republic, 2021.
16. ČSN EN ISO 5667-1; Water Quality—Sampling—Part 1: Guidance on the Design of Sampling Programs and Sampling Techniques. Czech Standard Institute: Prague, Czech Republic, 2023.
17. ČSN ISO 5667-4; Water Quality—Sampling—Part 4: Guidance on Sampling from Lakes, Natural and Man-Made. Czech Standard Institute: Prague, Czech Republic, 2018.
18. ČSN EN ISO 5667-6; Water Quality—Sampling—Part 6: Guidance on Sampling of Rivers and Streams. Czech Standard Institute: Prague, Czech Republic, 2017.
19. ČSN EN ISO 5667-6; Change A11 Water Quality—Sampling—Part 6: Guidance on Sampling of Rivers and Streams. Czech Standard Institute: Prague, Czech Republic, 2020.

20. ČSN EN ISO 5667-3; Water Quality—Sampling—Part 3: Preservation and Handling of Water Samples. Czech Standard Institute: Prague, Czech Republic, 2019.
21. Czech Republic. The Decree No 401/2015 Coll., on Indicators and Values of Permissible Pollution of Surface Water and Wastewater, the Details of Permits for the Discharge of Wastewater into Surface Water and Sewers and on Sensitive Areas, as Amended. Available online: https://www.psp.cz/sqw/sbirka.sqw?cz=401&r=2015 (accessed on 15 May 2023). (In Czech).

Disclaimer/Publisher's Note: The statements, opinions and data contained in all publications are solely those of the individual author(s) and contributor(s) and not of MDPI and/or the editor(s). MDPI and/or the editor(s) disclaim responsibility for any injury to people or property resulting from any ideas, methods, instructions or products referred to in the content.

Proceeding Paper

The Evaluation of the Rapid Sand Filter Wash Interval at the Central DWTP in the Czech Republic [†]

Radka Matová, Petra Malíková *, Silvie Drabinová and Jitka Chromíková

Faculty of Mining and Geology, Department of Environmental Engineering, VSB—Technical University of Ostrava, 17 Listopadu 2172/15, 708 00 Ostrava, Czech Republic; radka.matova@vsb.cz (R.M.); silvie.drabinova@vsb.cz (S.D.); jitka.chromikova@vsb.cz (J.C.)

* Correspondence: petra.malikova@vsb.cz
[†] Presented at the 4th International Conference on Advances in Environmental Engineering, Ostrava, Czech Republic, 20–22 November 2023.

Abstract: This paper evaluates the washing interval of the rapid sand filter at the central drinking water treatment plant Nová Ves—Frýdlant nad Ostravicí in the Czech Republic (DWTP Nová Ves). The aim was to conduct automated flow cytometry measurements (FCM) and find the link between FCM and turbidity. The monitor parameters were the length of the wash cycle in hours, the flow rate of the filter and the production, the pumping of the recirculating wash water, and the physico-chemical and microbial analysis of the water samples. The focus of this paper is the detailed characteristics of the filtration mode evaluated during the summer and winter periods. During the measurements, it was confirmed that turbidity replicated the FCM data measured by the FC BactoSense instrument. Turbidity can be identified as one of the key features that can be related to the measurements made. Turbidity and the cell count itself are influenced, among other things, by the pumping of the return water, whereby an increase in the cell count can be observed after the pumping has stopped but gradually stabilizes at the values measured before pumping.

Keywords: rapid sand filter (RSF); drinking water treatment plant (DWTP); flow cytometry (FCM); turbidity; filter washing; conventional water treatment; drinking water

1. Introduction

The purpose of drinking water treatment plants (DWTPs) is to provide safe water of good sanitary and chemical quality to the public [1,2]. The key processes used in drinking water treatment plants to remove dissolved and solid contaminants from source water include coagulation/flocculation, sedimentation, oxidation, and filtration systems [3]. Membrane filtration and granular bed filtration are used as filtration systems. Depending on the origin of raw water, different configurations are used to remove microbiological or chemical pollutants. Surface water usually requires multiple treatment steps [1] compared to groundwater, which often requires simpler treatment. The final step is then the sanitation. The quality of drinking water is determined by the legislation and also ensures the protection of the water supply (distribution) system, especially against corrosion and fouling.

Current climate change may be a major challenge for water supplies, especially in countries such as the Czech Republic, where all rivers originate, and surface water quality may be unstable [4,5].

With regard to the development and optimization of filtration technologies, it should be noted that water filtration is still a very conservative area and that some important advances are being made. These developments, although based on basic concepts, make the entire process more efficient and have a positive impact on the economy. The concept has remained unchanged for decades, but only variations developed and adapted to different

conditions, such as different raw water quality and conceptual changes in process lines, allowing for the processes to be optimized according to different requirements [6–9].

The theoretically permitted filter cycle time considers the possible growth of organic matter in the filter bed. For each filter, the theoretical filtration cycle length should be determined. High-quality raw water filters can achieve a seven-day filtering time. However, it should be considered that a relatively long time period may result in an increase in organic and microorganic matter in the filter bed and may lead to staleness of the filter bed. The unwanted result is both the smell and taste of the filtered water, as well as the formation of slime compounds that affect the difficulty of filter washing. For this reason, the operation is usually guaranteed by shorter filtration cycles, higher filter washing frequency, and higher washing water intake associated with this process [6]. In particle tracking methods, it was shown that turbidity measurements themselves are insufficient to properly map the filter process. The measurement of the number of particles is about 20 to 25 times more sensitive than measuring the residual turbidity of treated water [10]. Therefore, new methods are proposed, such as particle counters and flow cytometers (FCMs) [11,12], which may be used to monitor water filtration processes more accurately.

FCM is a method that has become increasingly popular in water management practices in developed countries [12–14]. It is used to accurately quantify the total microbial recovery in water, not only the total number of cells per mL, but also classifying cells into small, large, alive, and dead cells (using simultaneous light scattering and fluorescence measurements combined with dye stains for counting and analyzing). This also gives a clear graphical representation of the cell distribution (the so-called "fingerprint"). In addition to laboratory FCM, there is now a complete automation online FCM that can be easily attached to a sample from any point of treatment technology or any point of the water network [12].

The aim of this study was to measure FCM with other monitoring parameters and measurement under the 13N filter of the DWTP Nová Ves and to find the link between FCM and turbidity. On the basis of measured data, evaluation and proposal options focus on optimizing and assessing the frequency of rapid filter washing at the operating conditions.

2. Materials and Methods

2.1. Drinking Water Treatment Plant

For the evaluation of the rapid sand filter washing cycle using FCM, the central DWTP Nová Ves—Frýdlant nad Ostravicí in the Czech Republic (DWTP Nová Ves), owned and operated by the SmVaK Ostrava, a.s. company, was selected. DWTP Nová Ves is a part of the Beskydy group water supply system and has been operating since the 1970s (it has undergone several partial reconstructions over the years). Its current used capacity is 800 L/s, but it can reach the maximum output of 2200 L/s. The valley water supply reservoir Šance in the Beskydy Mountains is the raw water source for the DWTP Nová Ves, from which water is transported by gravity to the DWTP. DWTP Nová Ves supplies drinking water to almost 60 towns and villages in the Czech Republic in the regions of Frýdek-Místek, Karviná, Nový Jičín, part of Ostrava, and the border region of Poland (the Jastrebie—Zdroj district) [15].

In terms of raw water properties, the water is very soft, with a pH neutral over the year and a low level of biological and microbiological pollution. The basic parameters monitored reach average values of COD_{Mn} 2.4 mg/L, color 9.2 mg/L Pt, and turbidity 4.0 NTU. Raw surface water is fed to 20 sand rapid filters (European type WABAG) with a total area of 1940 m^2 with a filling of quartz sand grain size from 1.0 to 1.2 mm within the framework of one-stage coagulation filtration. The current filtration cycle, based on operational experience, typically lasts 40–96 h, at which point the filters regenerate/wash at the end of the cycle. The length of the filtration cycle depends on the quality of the surface water, the total daily water production, the pumping of recirculation wash water, the dosage of chemicals, etc. Regeneration of the filters is always performed in five filter units, at a frequency of twice a day, with five filters regenerated in the morning and five in the afternoon. The filter washing process combines air and water washing—2.5 min of air washing, 4.5 min of air and water washing at 300 L/s, and 2.5 min of water washing at 500 L/s. For oxidation of organic matter, chlorine dioxide is dosed into the raw water (before the sand filters), and aluminum sulfate (20% solution) is dosed as a coagulant before the filters. To adjust the pH, lime hydrate is dosed in the form of lime water, and hygiene is ensured by chlorine water and chlorine dioxide. The sludge from the filter-washing effluent with lime is settled in two horizontal sedimentation tanks and subsequently treated by centrifuge draining. In times of turbid state conditions, when turbidity is increasing due to excessive rainfall or snowmelt, the 1st separation stage is activated. It consists of flocculation with hydromixing and subsequent sedimentation in a lamella build-up, where the coagulant agent is combined with a polymeric organic flocculant [15].

2.2. FCM Measuring and Monitored Parameters

For evaluation of the washing cycle, the rapid sand filter 13N at DWTP Nová Ves was selected. The measurements were divided into two stages to cover different climatic periods (summer, winter). The 1st stage was carried out from 9 August to 23 August 2019, and the 2nd stage was carried out from 27 November to 18 December 2019. The current measured control parameters for the work filtration cycle were wash cycle, flow rate under the filter, average daily flow rate under the filter, the average flow rate per filtrate cycle, water production, times of pumping recirculation wash water (RWW), and turbidity—of raw water, RWW and water after filtration; next monitored parameters were physico-chemical and microbiology indicators (the list of monitored indicators is in Table S1). The newly proposed tested control parameters were parameters measured by FCM (the list of FC-measured results is in Table S1).

The 1st stage was the manual measurement of input data on the rented flow cytometer (FC) SIGRIST BactoSence (bNovate Technologies, Zürich, Switzerland), with the LDC—Live/Dead Count cartridge to determine the number of damaged/dead and intact/alive cells, followed by mounting under the 13N filter to monitor microbiological recovery during the filtration cycle in automatic mode, with the simultaneous monitoring of selected parameters during the measurement period. The time interval of the analyses and sampling was chosen to be 4 h, so 6 analyses were performed per day. Along with the FCM measurements, turbidity was automatically measured using the online photometer Aquascat 2 WTM A (Sigrist-Photometer AG, Ennetbürgen, Switzerland). At this stage, laboratory cultivation analyses—Heterotrophic Plate Count method (HPC) at 22 °C and 36 °C were carried out simultaneously; the flow rate under the filter 13N was monitored due to the pumping of the RWW, and finally, the instrument was tested under operating conditions.

The aim of the 2nd stage was to carry out more detailed measurements in the winter season, based on the trend observed after the 1st stage of measurements, with extended filtering cycles. The aim of this measurement was to obtain the necessary data to determine the link between FCM and turbidity, depending on the results of the previous measurements, and to ensure trouble-free operation of the instrument under operating conditions. As the optimal interval of measurements, a time interval of 2 h was chosen from 27 November to 28 November 2019, followed by measurements at an interval of 4 h, i.e., 6 analyses were performed per day. The following parameters were monitored during the measurements: filter 13N wash cycle length in hours; flow rates under filter 13N and production in L/s; and RWW pumping times and turbidity. The turbidity monitored was for raw water turbidity, RWW turbidity, and turbidity under the 13N filter, measured with the same above-mentioned photometer in the NTU unit.

Physico-chemical and microbial analysis of the water samples was conducted in the laboratory of the DWTP (SmVaK Ostrava, a.s., Ostrava, Czech Republic). All parameters were measured via the standard methods. All data sets were normalized using means and standard deviations of the variables using Microsoft Excel 2010.

3. Results and Discussion

3.1. The First Stage of Measurement

The average values of the number of live and dead, small and large cells per mL of water, and the percentage of live and large cells are presented in Table 1. The RWW sample had to be diluted 10 times due to exceeding the FCM detection limit. This verified that no correlation between cell count and organisms could be found, and a comparison of the measurements was made (see Table S2). The graphical representation of the so-called "fingerprint" in manual mode of RWW 10-times diluted sample by FC can be seen in Figure 1 (additional photos are in Figure S1). The quality of raw water is presented in Table S3.

Table 1. The initial data measured manually with the FC at DWTP Nová Ves.

Sample	Date and Time of Analysis	TCC [1] Cells/mL	ICC [2] Cells/mL	ICP [3] %	HNAP [4] %
raw water	9 August 2019 12:08	1,596,150	1,166,375	73.1	90.9
RWW [5] 10 times diluted	9 August 2019 11:38	894,500	489,877	54.8	90.9
RWW [5] recalculation	9 August 2019 11:38	8,945,000	4,898,770	54.8	90.9
13N filter	9 August 2019 13:17	633,398	577,288	91.1	88.7
inflow DWR [6]	9 August 2019 10:55	616,171	19,575	3.1	39.1
outflow DWR [6]	9 August 2019 10:16	542,900	14,733	2.7	38.8

[1] Total Cell Count (live and dead) per mL. [2] Intact Cell Count (intact or live) per mL. [3] Intact Cell Percentage ICP = ICC/TCC. [4] High Nucleic Acid Percentage HNAP = HNAC/ICC. [5] Recirculation Wash Water. [6] Drinking Water Reservoir.

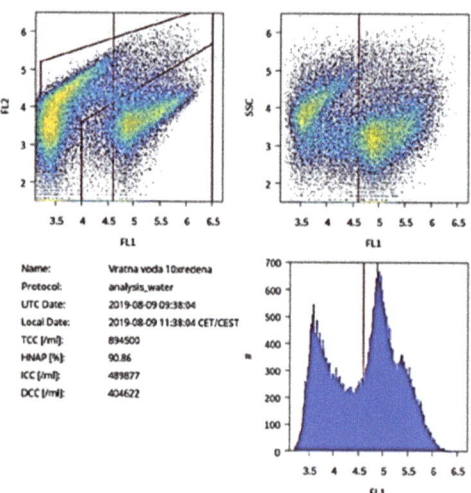

Figure 1. Data measured with FC Bactosence—"fingerprint" of WWR 10 times diluted sample (DCC—Death Cell Count).

The results of FCM prove that even high-quality drinking water contains from 100 to 1000 times more cells than can be cultivated by classical laboratory methods [16], i.e., the limits of cell counts cannot be set, according to the experience of Switzerland, which has "Flow Cytometry Analysis of Water Samples" included by the Swiss Federal Office of Public Health in Swiss regulations [17]. The limits on the number of cells depend on several factors, such as the source of water, the method of treatment, the method of sanitation, the place of measurement, the residence time of the water in the network, etc. In the analyses, it is necessary to collect the actual empirical cell count data and determine the limit quantity based on these data. The trends of FCM and turbidity after the rapid sand filtration are presented in the figures below.

Based on the measurements in the automatic mode, the measured data were processed graphically to show the length of the wash cycles, interleaved on the right axis in Figure 2, with the length of the filtration cycles reported in hours. The next parameter monitored was turbidity, whereby the FCM measurements on the right axis interleaved with turbidity in Figure 3, a relationship that can be seen between the turbidity increase and the microbial recovery increase. Due to planned downtime and an unplanned measurement outage, there was a noticeable disruption to the entire measurement, but the graph shows that by extending the filtration cycle, the filters were incorporated, cell counts stabilized, and gradually decreased. This can be observed in the measurement period from 16 August to 22 August 2019, even though there was a planned shutdown during the measurement period. Turbidity is measured in a very small range from 0 to 0.7 NTU. This parameter approximately follows the trend of the FC measurements and can be identified as one of the key parameters that can be related to the measurements made. Turbidity and the cell count itself are influenced by, among other things, the RWW pumping, where the cell count increase can be observed after the pumping is stopped in the next measurement but gradually declines to the values measured before pumping. Of course, a slight increase in turbidity can also be explained by the higher turbidity of the RWW. This parameter and its evolution were also monitored in the second stage of measurements. The prolongation of the filtration cycle could bring along, as a negative effect, a reduction in the flow rates under the filter due to clogging and fouling. Therefore, during the measurements, the flow rates under the filter were recorded to verify whether the extension of the washing cycle would result in a reduction in the capacity and flow rates through the filter. Based on these, it can be said that the filter cycle extending will not significantly affect the flow rate. Comparing the measurements from 14 August to 16 August 2019, the average flow rate is 36.48 L/s. Extending the filtration cycle from 16 August to 22 August 2019, the average flow rate is 35.88 L/s.

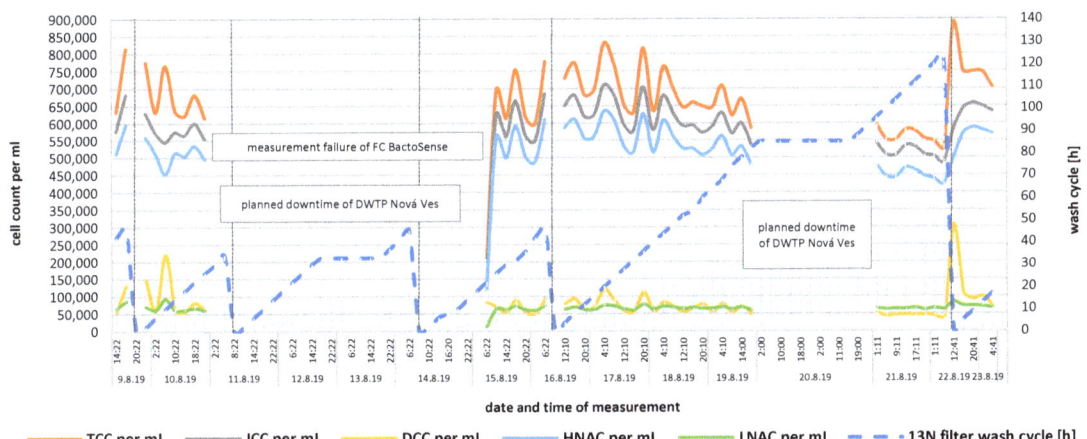

Figure 2. The 1st-stage measurement from 9 August to 23 August 2019: FCM and wash cycle.

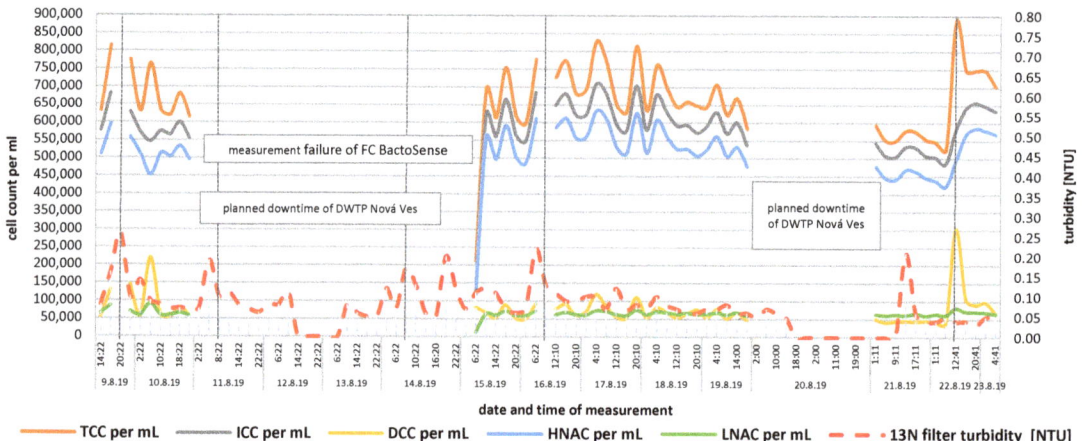

Figure 3. The 1st-stage measurement from 9 August to 23 August 2019: FCM and turbidity.

3.2. The Second Stage of Measurement

The evolution of Figure 4 evidently confirms the results of the first measurement; during the filter regeneration, there is a sharp increase in microbiological recovery, and there is a gradual decrease in the measured values over time. This is evident in the case of extended cycles, where, compared to shorter cycles, the time between cycles is so short that there is no flickering and stabilization of the values, and the microbiological increase is again observed during washing. In this context, it is necessary to take RWW, which enters the filtration process and influences the evolution of the measured values, into consideration. According to Figures 4–6, the pumping of the RWW is followed by the microbiological recovery increase, which turns into the values measured before the RWW pumping in the following hours.

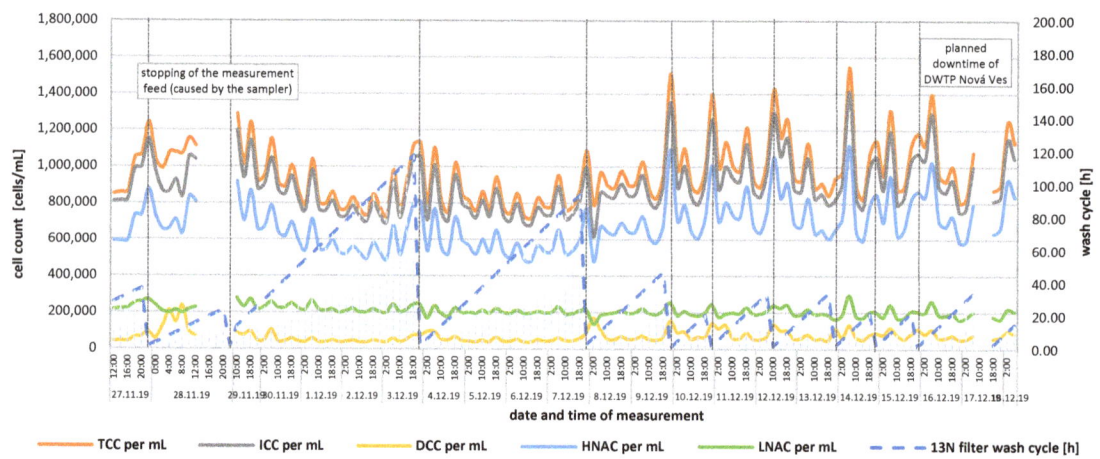

Figure 4. The 2nd stage of measurement from 27 November to 18 December 2019: FCM and wash cycle.

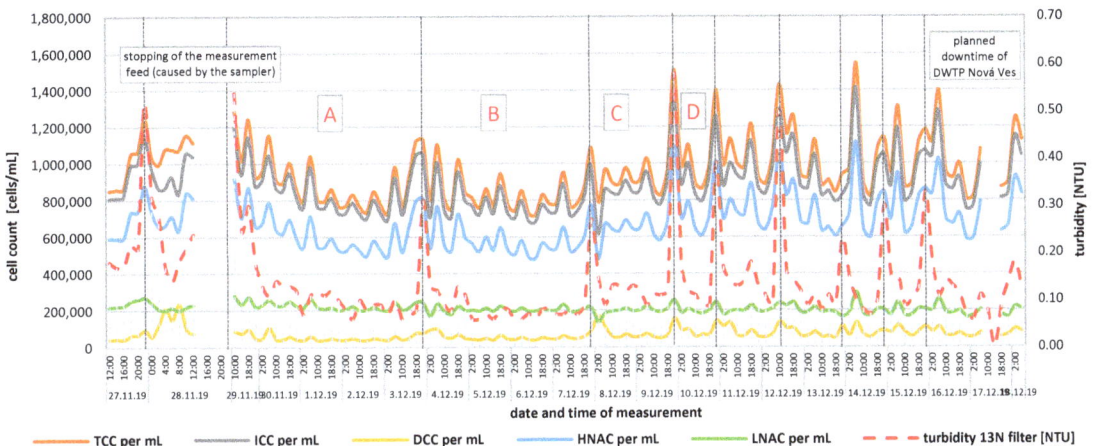

Figure 5. The 2nd stage of measurement from 27 November to 18 December 2019: FCM and turbidity (detailed **A–D** cycles are in Figure 6).

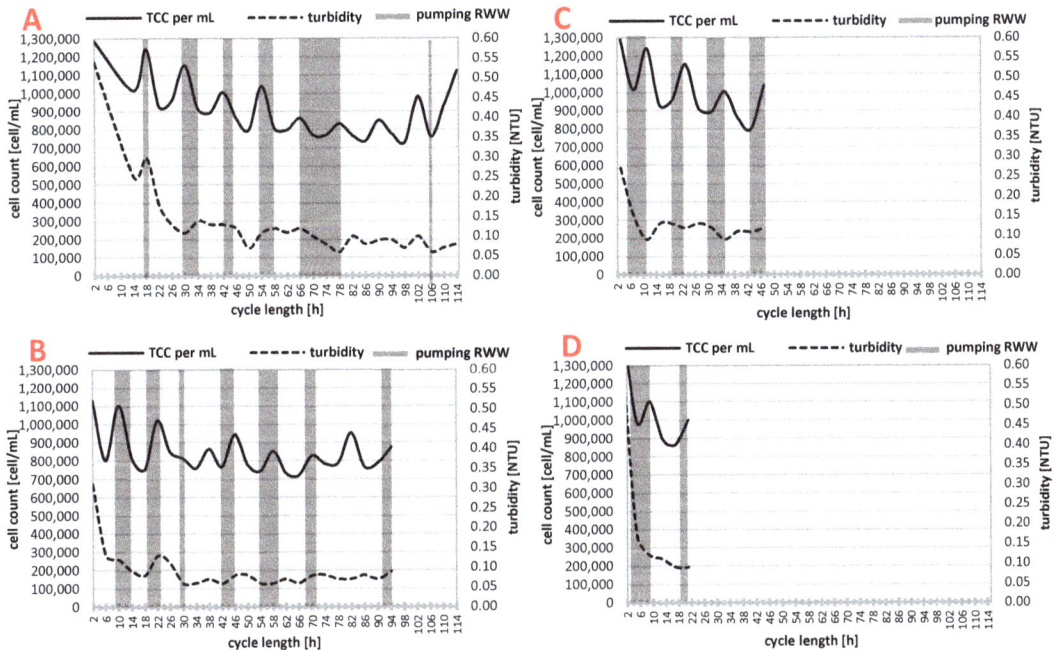

Figure 6. The 2nd stage of measurement from 27 November to 18 December 2019: detailed cycle length—TCC and turbidity affected by the RWW pumping (detailed **A–D** cycles marked in Figure 5).

In this context, the RWW pumping is not determinant for the filtration technology optimization, and while maintaining the method of pumping the RWW, an extension of the filtration cycles is possible. The measurements confirmed the evolution of turbidity, which was very similar to the measured values by the FC in the first measurement, so as already reported, the turbidity increase is directly related to the increase in microbiological recovery. This increase occurs again while the filters are being washed and the RWW is being pumped, which shows even higher turbidity values measured in the range of 3.7

to 60.2 NTU. For comparison, values in the range from 0.06 to 0.54 NTU are measured on the turbidity meter below the filter. In addition, the raw water turbidity was recorded as a parameter entering the technological process. Based on the increase in the turbidity of raw water, it is possible to consider its aggravated quality, which, especially for surface sources, is influenced by several factors. According to the above, the turbidity parameter can be considered to be the main one. The quality of raw water is presented in Table S3.

4. Conclusions

Water filtration is a relatively conservative area in terms of development and optimization. However, it is also an area in which important developments are constantly being made that can make the whole process more efficient. As part of the evaluation of the filtration cycle, the following parameters were monitored under operating conditions: length of the wash cycle in hours; the flow rate of the filter and the production; the pumping of the recirculating wash water; the physico-chemical and microbial analysis of the water samples; and turbidity and flow cytometry (FCM). The filtration mode was evaluated during the summer and winter periods.

The measurements performed and its evaluation show that the link between turbidity and FCM is possible. Similar trends have been observed, and there is evidence that FCM is more responsive, which would be more advantageous for the WTP operators, who could react faster if FCs were fitted. The only disadvantages to implementing the FCM in operation are the purchase price and ensuring the operating conditions for its flawless operation.

Supplementary Materials: The following supporting information can be downloaded at https://www.mdpi.com/article/10.3390/engproc2023057009/s1, Figure S1: The DWTP Nová Ves: A—13N filter, B—FC BactoSence fitting under the 13N filter, C—Detail of the connection to the water feed in automatic mode; Table S1: The list of monitored physico-chemical and microbiology indicators in water samples; Table S2: The 1st stage—microbiology analysis: laboratory and FCM; Table S3: Laboratory analysis of raw water DWTP Nová Ves.

Author Contributions: Conceptualization, R.M. and P.M.; methodology, R.M.; validation R.M., P.M. and J.C.; resources, S.D.; data curation, R.M. and S.D.; writing—original draft preparation R.M. and P.M.; writing—review and editing, R.M., P.M., S.D. and J.C.; visualization S.D.; supervision, R.M. and P.M.; project administration, R.M. All authors have read and agreed to the published version of the manuscript.

Funding: This research received no external funding.

Institutional Review Board Statement: Not applicable.

Informed Consent Statement: Not applicable.

Data Availability Statement: Data supporting reported results can be found in Supplementary Materials.

Acknowledgments: The authors would like to thank the company SmVak Ostrava, a.s., for allowing us to perform semi-operational verification at the central DWTP Nová Ves, for technical and material support (rental of the flow cytometer—SIGRIST Bactosense and laboratory analyses).

Conflicts of Interest: The authors declare no conflict of interest. The funders had no role in the design of this study, in the collection, analyses, or interpretation of data, in the writing of the manuscript, or in the decision to publish the results.

References

1. World Health Organization. *Guidelines for Drinking Water Quality: Fourth Edition Incorporating the First and Second Addenda*, 4th ed. + 1st add + 2nd add; WHO: Geneva, Switzerland, 2022; Available online: https://apps.who.int/iris/handle/10665/352532 (accessed on 9 June 2023).
2. Rosario-Ortiz, F.; Rose, J.; Speight, V.; Von Gunten, U.; Schnoor, J. How do you like your tap water? *Science* **2016**, *351*, 912–914. [CrossRef] [PubMed]
3. Gerba, C.P.; Pepper, I.L. Chapter 24—Drinking water treatment. In *Environmental and Pollution Science*, 3rd ed.; Brusseau, M.L., Pepper, I.L., Gerba, C.P., Eds.; Academic Press: Cambridge, MA, USA, 2019; pp. 435–454.

4. Kyncl, M.; Heviánková, S.; Nguien, T.L.C. Study of supply of drinking water in dry seasons in the Czech Republic. *IOP Conf. Ser. Earth Environ. Sci.* **2017**, *92*, 012036. [CrossRef]
5. Kročová, Š.; Kavan, Š. Cooperation in the Czech Republic border area on water management sustainability. *Land Use Policy* **2019**, *86*, 351–356. [CrossRef]
6. Hendricks, D.W. *Water Treatment Unit Processes: Physical and Chemical*, 1st ed.; CRC Press: Boca Raton, FL, USA, 2006. [CrossRef]
7. Cescon, A.; Jiang, J.-Q. Filtration Process and Alternative Filter Media Material in Water Treatment. *Water* **2020**, *12*, 3377. [CrossRef]
8. van der Hoek, J.P.; Bertelkamp, C.; Verliefde, A.R.D.; Singhal, N. Drinking water treatment technologies in Europe: State of the art—Challenges—Research needs. *J. Water Supply Res. Technol.* **2014**, *63*, 124–130. [CrossRef]
9. Amirtharajah, A. Optimum backwashing of filters with air scour: A review. *Water Sci. Technol.* **1993**, *27*, 195–211. [CrossRef]
10. Miska, V.; van der Graaf, J.H.J.M.; de Koning, J. Improvement of monitoring of tertiary filtration with particle counting. *Water Supply* **2006**, *6*, 1–9. [CrossRef]
11. Fujioka, T.; Ueyama, T.; Mingliang, F.; Leddy, M. Online assessment of sand filter performance for bacterial removal in a full-scale drinking water treatment plant. *Chemosphere* **2019**, *229*, 509–514. [CrossRef] [PubMed]
12. Safford, H.R.; Bischel, H.N. Flow cytometry applications in water treatment, distribution, and reuse: A review. *Water Res.* **2019**, *151*, 110–133. [CrossRef]
13. Prest, E.I.; El-Chakhtoura, J.; Hammes, F.; Saikaly, P.E.; van Loosdrecht, M.C.M.; Vrouwenvelder, J.S. Combining flow cytometry and 16S rRNA gene pyrosequencing: A promising approach for drinking water monitoring and characterization. *Water Res.* **2014**, *63*, 179–189. [CrossRef] [PubMed]
14. Props, R.; Rubbens, P.; Besmer, M.; Buysschaert, B.; Sigrist, J.; Weilenmann, H.; Waegeman, W.; Boon, N.; Hammes, F. Detection of microbial disturbances in a drinking water microbial community through continuous acquisition and advanced analysis of flow cytometry data. *Water Res.* **2018**, *145*, 73–82. [CrossRef] [PubMed]
15. SmVak, a.s (Ostrava, Czech Republic). Internal documentation. Unpublished. 2019.
16. Pišan, J. BactoSense, 2019: Presentation of flow cytometry method (FCM). In Proceedings of the Laboratory Committee Meeting SOVAK DWTP Podolí, Průhonice, Czech Republic, 2–3 October 2019; Available online: https://www.technoprocur.cz/data/files/SIGRIST%2520BactoSense/Prezentace_BactoSense.pdf (accessed on 9 June 2023).
17. The Federal Council. Swiss Federal Institute of Aquatic Science and Technology. Available online: https://www.admin.ch/gov/en/start/documentation/media-releases.msg-id-47549.html (accessed on 13 June 2023).

Disclaimer/Publisher's Note: The statements, opinions and data contained in all publications are solely those of the individual author(s) and contributor(s) and not of MDPI and/or the editor(s). MDPI and/or the editor(s) disclaim responsibility for any injury to people or property resulting from any ideas, methods, instructions or products referred to in the content.

Proceeding Paper

The Activity of Two Different Solutions against Selected Fungal Phyto-Pathogens [†]

Sanja Nosalj [1,*], Alexandra Šimonovičová [1], Elena Pieckova [2], Domenico Pangallo [3] and Bruno Gabel [4]

[1] Department of Pedology, Faculty of Natural Sciences, Comenius University, Ilkovičova 6, 842 15 Bratislava, Slovakia; alexandra.simonovicova@uniba.sk
[2] Department of Microbiology, Slovak Medical University, Limbová 12, 833 03 Bratislava, Slovakia; elena.pieckova@szu.sk
[3] Institute of Molecular Biology SAS s.r.i., Dúbravská Cesta 21, 841 04 Bratislava, Slovakia; domenico.pangallo@savba.sk
[4] Vitigroup a. s., Dvořákovo nábrežie 10, 811 02 Bratislava, Slovakia; bruno@gabel.sk
* Correspondence: nosalova15@uniba.sk
[†] Presented at the 4th International Conference on Advances in Environmental Engineering, Ostrava, Czech Republic, 20–22 November 2023.

Abstract: Two different solutions differing in their composition were tested against the following fungal phyto-pathogens: *Cladosporium cladosporioides*, *Alternaria infectoria*, *Botrytis cinerea*, *Monilia fructigena*, *Aspergillus clavatus* and *Penicillium digitatum*. The two different solutions were a fatty acid-based DPB solution and the development batch of a substance labeled M-decanocide. The tested species of fungal phyto-pathogens, namely *Penicillium digitatum*, *Aspergillus clavatus* and *Cladosporium cladosporioides*, grew rapidly and sporulated heavily on the agar plates with the product M-decanocide applied. In *Aspergillus clavatus* and *Cladosporium cladosporioides* cases, growth-free zones were formed around the cuts when the tested solution was applied, which confirmed its positive antifungal effect. The fungicidal effect of the tested DPB solution was not confirmed for the selected fungal phyto-pathogens.

Keywords: fatty acids; M-decanocide; fungal pathogens; fungicidal effect

1. Introduction

The protection of plants, especially those of agricultural importance, is crucial, because various diseases significantly reduce their quality and, consequently, their nutritional value. Some examples of such diseases are diseases caused by animal pests, such as potato tapeworm, potato aphid, snails, bugs and the so-called teddy bear, along with various rodents and so on. Plant diseases caused by viruses (so-called viroses) include the cucumber mosaic virus, plum blight, tomato bronze virus, etc. Diseases caused by bacteria (so-called bacterioses) damage, e.g., potatoes (bacterial ringworm of potatoes, brown rot of potatoes), stone fruits (bacterial spotting of stone fruits), as well as tomatoes, strawberries and other crops important for human nutrition.

Fungal diseases caused by microscopic fungi (so-called mycoses) are also very widespread. Examples include potato blight (caused by *Phytophtora infestans*), botrytis on vines and strawberries (*Botrytis cinerea*), and moniliosis (species of the genus *Monilia*) occurring on various stone fruits (apples, plums, pears, cherries, apricots, etc.). The so-called blacks, which are dark-to-black coatings, e.g., on beet and on other host plants, are caused by species of the genera *Cladosporium* and *Alternaria*. Several fusaria (species of the genus *Fusarium*), as well as species of the genus *Aspergillus* and *Penicillium*, parasitize on grain. Many professional publications are reviewing pests, plant diseases and their protection [1–4].

There is a wide range of chemical preparations at disposal, which are intended not only for spraying already-damaged plants, but especially for their preventive protection against possible disease. More and more emphasis is placed on ecological protection, i.e., the use of substances on an ecological basis, within the framework of not harming the environment.

The aim of the work presented was to test the activity of two different solutions against selected phyto-fungal pathogens, the composition of which corresponds to the current "eco-friendly" requirements placed for phytosanitary agents.

2. Materials and Methods

As part of the mentioned experiment, we used the species *Cladosporium cladosporioides*, *Alternaria infectoria, Botrytis cinerea* and *Monilia fructigena* from the fungal pathogens of plants, which come from the collection of microorganisms at the Institute of Molecular Biology SAV s.r.i. in Bratislava, and the species *Aspergillus clavatus* and *Penicillium digitatum* isolated from grain, which come from the collection of microfungi at the Slovak Medical University in Bratislava.

The mentioned species were cultured on slant SDA (Sabouraud Dextrose Agar, Himedia, Mumbai, India) for 5 to 7 days, respectively, until visible sporulation. Twenty mL of sterile distilled water was then added after thorough shaking, and the entire volume of the spore suspension was poured into 330 mL of the prepared nutrient MEA (malt extract agar, Himedia, Mumbai, India) medium at a temperature of about 40 °C. Such an inoculated medium with each tested fungal pathogen was portioned in Petri dishes (diameter 9 cm, height of the medium after solidification 0.5 cm). We made 4 incisions for each agar plate (Figure 1) with a 10 mm corkscrew in the solid medium, into which we then applied 2–3 drops of the tested solution. The corkscrew was sterilized by a flame after every use. We also used the application of sterile Whatman filter paper discs with a diameter of 10 mm on the surface of the inoculated solid medium, onto which we dripped the tested solution until the paper was soaked (3 drops). After incubation of the inoculated systems up to 7 days at 25 °C, the presence of growth-free zones around incisions/discs with an effective antifungal mixture was observed.

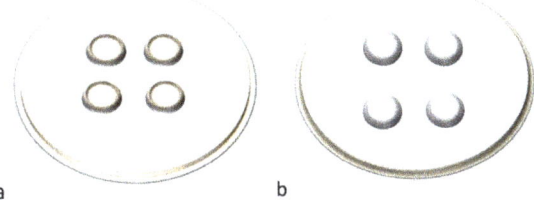

Figure 1. Scheme of Petri dishes with the cutouts (**a**) and filter paper discs (**b**) for testing the solutions.

The first tested solution was an unspecified mixture of fatty acids labeled DPB. In the experiment, we used 4 concentrations of the initial product, namely 1:10, 1:100, 1:500 and 1:1000. We used the second solution with the working designation M-decanocide only at the concentration of 1:10. Each pathogen was tested on 15 Petri dishes (3 × control sample without the test solution, 3 × tested pathogen with the solution diluted at 1:10, 3 × tested pathogen with the solution of 1:100, 3 × tested pathogen with the solution of 1:500 and 3 × tested pathogen with the solution of 1:1000).

When testing the second solution (M-decanocide), we cultured each species in 9 Petri dishes (3 × control sample without the tested solution, 3 × tested pathogen with the solution applied to sections in the agar medium, 3 × tested pathogen with the solution applied to sterile filter paper discs).

Both used solutions are under development and in the process of patenting; therefore, there is no official producer at this moment.

3. Results and Discussion

The DPB (mixture of fatty acids) solution showed a bioactive effect on the tested species of agriculturally important pests (apple borer and marbled borer). The bioactivity of these substances is not based on the classic mechanism of insecticidal products, i.e., the phytosanitary concept of pest eradication in the target agrocenosis, but on the modification of communication signals in the process of pest reproduction. Disruption of communication signals within the insect population is a method relatively well known under the name "mating disruption". The fact that in the last 30 years, in the case of some special crops (vines, fruit orchards), this method has been used on a large scale and repeatedly, has led to selection pressure on the genetic diversity at the population level of insects. As a result, the population of pest communities in the process of fertilization within "different channels" is higher than those "having been tuned" by man 30 years ago.

Therefore, it became necessary to look for other alternatives to the "non-chemical combat", and DPB and M-decanocide products are our answer to the mentioned requirement. Due to the fact that these are products under development, it is not possible to provide more details about their chemical composition and method of preparation.

It would be more than welcome and at the same time financially advantageous if the above-mentioned solutions, in addition to being insecticidal, also prove fungicidal effects. The species *Cladosporium cladosporioides* and *Alternaria infectoria* grew very slowly in the presence of DPB, and individual colonies appeared gradually on the culture medium, but also around the cutouts where the tested solution was applied. In addition, in close proximity to the cutouts, we noticed a relatively dense growth of the *Aspergillus niger* species, probably caused by air contamination. Despite the fact that we did not test the species *Aspergillus niger*, its growth around the cutouts was more or less identical compared to the growth of the other tested species, and more or less directly proportional to the concentration of the test solution used; we always recorded the best growth at the highest concentration of 1:10 (Figure 2).

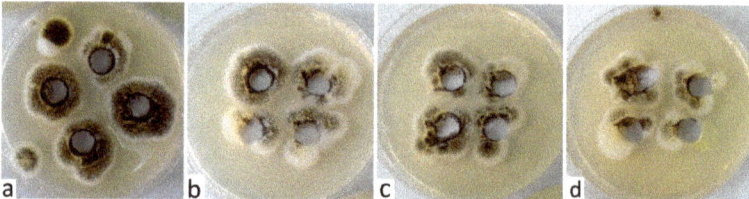

Figure 2. Growth of *Aspergillus niger* around the sections where the test solution (DPB) was applied in concentrations of 1:10 (**a**), 1:100 (**b**), 1:500 (**c**) and 1:1000 (**d**).

We also recorded the growth of mycelium of the species *Penicillium digitatum* and *Aspergillus clavatus*, not only in the control samples, but also in all Petri dishes where the test solution was applied at all four concentrations. The species *Botrytis cinerea* was characterized by relatively fast growth and intense sporulation. However, the inhibitory effect of the applied solution was not recorded in this case either. The growth of the species *Monilia fructigena*, on the other hand, was very weak and slow, both on the control plates and the plates with the tested solution.

Based on the results of the experiments so far, it is clear that no inhibition zones were formed around the cutouts in the nutrient medium with conidia of the fungal pathogen, which would mean the antifungal effect of the tested solution. Since the growth of the fungal pathogen appeared around the incisions, it indicates the un-efficacy of the tested solution as a fungicide.

The tested species of fungal phyto-pathogens, namely *Penicillium digitatum*, *Aspergillus clavatus* and *Cladosporium cladosporioides*, grew rapidly and sporulated heavily on agar plates with the product M-decanocide applied. In *Aspergillus clavatus* and *Cladosporium cladosporioides* cases, growth-free zones were formed around the cuts where we applied the tested solution, which confirms its positive, antifungal effect (Figure 3).

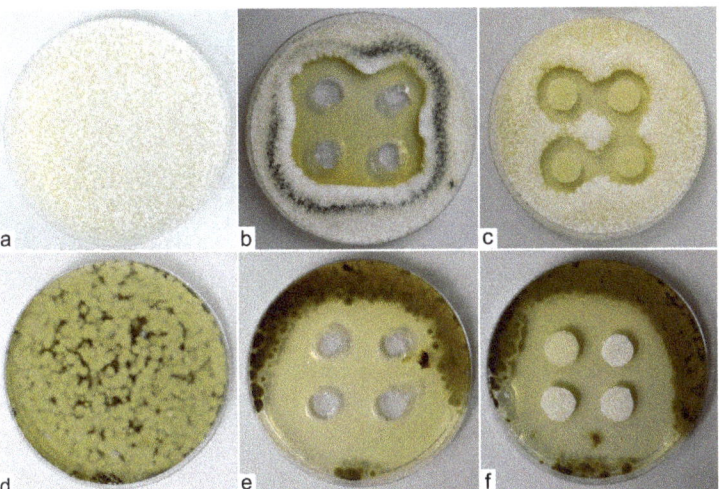

Figure 3. Zones around the applied solution (M-decanocide) for the species *Aspergillus clavatus* (control sample—(**a**), to wells in the agar medium—(**b**), on sterile filter paper targets with a diameter of 1 cm—(**c**)) and *Cladosporium cladosporioides* (control sample—(**d**), to wells in agar medium—(**e**), on sterile filter paper targets with a diameter of 1 cm—(**f**)).

The species *Monilia fructigena*, *Alternaria infectoria* and *Botrytis cinerea* grew very slowly and sporulated poorly on the agar plates, thus, the efficacy of the used solution against them could not be clearly confirmed.

The occurrence of phytopathogenic fungi, which belong to different species such as *Alternaria*, *Aspergillus*, *Cladosporium*, *Fusarium*, *Penicillium* and others, represents a high risk to agricultural production but also human health due to the production of mycotoxins [5,6]. The species *Alternaria infectoria* is known for wheat infection, while *Botrytis cinerea* has a wide host range, affecting various crops, vegetables and fruits [7,8]. The species *Cladosporium cladospoioides* often cause the rotting of grapes and strawberries [9]. *Monilia fructigena* is a widespread phytopatogen that causes brown rot in fruits such as apples, cheries, peaches, apricot and others [10]. *Penicillium digitatum* often spreads on citrus fruits and thus causes losses in harvest due to infection [11]. *Aspergillus clavatus* is occasionaly pathogenic and can cause a disease in humans and other mammals [12]. It often occurs in soils with cultivated potatoes, cotton, sugar canes, etc., but can be also isolated from grapes or various crops, and thus represents a high risk due to the production of mycootxins [13,14].

Many pathogenic fungi are resistant to common fungicides, so there is a need for the development of new, more efficient solutions.

4. Conclusions

The fungicidal effect of the tested DPB solution was not confirmed for the selected fungal phyto-pathogens *Cladosporium cladosporioides*, *Alternaria infectoria*, *Penicillium digitatum*, *Aspergillus clavatus*, *Botrytis cinerea* and *Monilia fructigena*. The inhibitory effect of the M-decanocide solution was confirmed for the species *Aspergillus clavatus* and *Cladosporium cladosporioides*.

Author Contributions: Conceptualization, A.Š. and B.G.; methodology, A.Š. and E.P.; writing—original draft preparation, A.Š.; writing—review and editing, B.G., S.N. and E.P.; fungi preparation, D.P.; visualization, S.N. All authors have read and agreed to the published version of the manuscript.

Funding: This research received no external funding.

Institutional Review Board Statement: Not applicable.

Informed Consent Statement: Not applicable.

Data Availability Statement: All data supporting the findings in this manuscript are available upon request.

Conflicts of Interest: The authors declare no conflict of interest.

References

1. Lohrer, T. *Pests and Plant Disease–Pictorial Atlas*, 1st ed.; Esence: Prague, Czech Republic, 2021; p. 384. (In Czech)
2. Matlák, J. *Year-Round Protection of Garden Crops*, 1st ed.; Juraj Matlák: Bratislava, Slovakia, 2022; p. 102. (In Slovak)
3. Matlák, J. *Atlas of Diseases and Pests of Fruit Crops and Vines*, 1st ed.; Juraj Matlák: Bratislava, Slovakia, 2023; p. 68. (In Slovak)
4. Matušíková, S.; Cagáň, Ľ. *Protection of Fruit Plants against Diseases*, 1st ed.; Plat4MBooks: Bratislava, Slovakia, 2019; p. 264. (In Slovak)
5. Tournas, V.H.; Katsoudas, E. Mould and yeast flora in fresh berries, grapes and citrus fruits. *Int. J. Food Microbiol.* **2005**, *105*, 11–17. [CrossRef] [PubMed]
6. Tancinová, D.; Barboráková, Z.; Mašková, Z.; Mrvová, M.; Medo, J.; Golian, M.; Štefániková, J.; Árvay, J. In vitro antifungal activity of essential oils (family lamiaceae) against cladosporium sp. Strains–postharvest pathogens of fruits. *J. Microbiol. Biotechnol. Food Sci.* **2023**, *13*, 1–6. [CrossRef]
7. Oviendo, M.S.; Sturm, M.E.; Reynoso, M.M.; Chulze, S.N.; Ramirez, M.L. Toxigenic profile and AFLP variability of *Alternaria alternata* and *Alternaria infectoria* occurring on wheat. *Medical. Braz. J. Microbiol.* **2013**, *44*, 447–455. [CrossRef] [PubMed]
8. Filinger, S.; Elad, Y. *Botrytis—The Fungus, the Pathogen and Its Management in Agricultural Systems*; Springer: Cham, Switzerland, 2016; p. 486.
9. Latorre, B.A.; Briceño, E.X.; Torres, R. Increase in *Cladosporium* spp. populations and rot of wine grapes associated with leaf removal. *Crop Prot.* **2011**, *1*, 52–56. [CrossRef]
10. Liu, Z.; Bai, H.; Yang, H.; Tang, S.; Wei, M.; Huang, X.; Li, Y. Biological characteristics of *Monilia fructigena* as pathogen of brown rot in sweet cherry. *Int. J. Fruit Sci.* **2012**, *29*, 423–427.
11. Papoutsis, K.; Mathioudakis, M.M.; Hasperué, J.H.; Ziogas, V. Non-chemical treatments for preventing the postharvest fungal rotting of citrus caused by *Penicillium digitatum* (green mold) and *Penicillium italicum* (blue mold). *Trends Food Sci. Technol.* **2019**, *89*, 479–491. [CrossRef]
12. Razzaghi-Abyaneh, M.; Rai, M. Introductory Chapter: The Genus Aspergillus-Pathogenicity, Mycotoxin Production and Industrial Applications. In *The Genus Aspergillus-Pathogenicity, Mycotoxin Production and Industrial Applications*; IntechOpen: London, UK, 2022; pp. 1–4. [CrossRef]
13. Felšöciová, S.; Tančinová, D.; Rybárik, Ľ.; Mašková, Z. Mycobiota of slovak wine grapes with emphasis on *Aspergillus* and *Penicillium* species in the south Slovak wine region. *Potravinarstvo* **2017**, *11*, 496–502. [CrossRef] [PubMed]
14. Domsch, K.H.; Anderson, T.H.; Gams, W. *Compendium of Soil Fungi*, 1st ed.; Academic Press: London, UK, 1980; pp. 86–88.

Disclaimer/Publisher's Note: The statements, opinions and data contained in all publications are solely those of the individual author(s) and contributor(s) and not of MDPI and/or the editor(s). MDPI and/or the editor(s) disclaim responsibility for any injury to people or property resulting from any ideas, methods, instructions or products referred to in the content.

Proceeding Paper

Occurrence of ^{137}Cs in Soil and Agricultural and Forest Products of the Contaminated Northeastern Part of the Czech Republic [†]

Petra Liszoková [1,*], Jan Škrkal [2], Barbara Stalmachová [1], Věra Záhorová [2] and Helena Pilátová [2]

[1] Faculty of Mining and Geology, VSB—Technical University of Ostrava, 17. Listopadu 2172/15, 708 00 Ostrava, Czech Republic; barbara.stalmachova@vsb.cz
[2] National Radiation Protection Institute (SÚRO, v. v. i.), Bartoškova 28, 140 00 Prague, Czech Republic; jan.skrkal@suro.cz (J.Š.); vera.zahorova@suro.cz (V.Z.); helena.pilatova@suro.cz (H.P.)
* Correspondence: petra.liszokova.staneckova@gmail.com
[†] Presented at the 4th International Conference on Advances in Environmental Engineering, Ostrava, Czech Republic, 20–22 November 2023.

Abstract: In the more contaminated northeastern region of the Czech Republic (Moravian-Silesian Region) in 2018 and 2019, soil samples and selected agricultural and forestry products were collected. The contamination of the environment was caused by radioactive cesium ^{137}Cs from the nuclear Chernobyl disaster, and the activity concentration of ^{137}Cs was determined in all soil samples taken, ranging from 0.177 Bq kg^{-1} dry matter (dm) to up to 299 Bq kg^{-1} dm, with an arithmetic mean of 38.4 Bq kg^{-1} dm. The activity concentrations of ^{137}Cs of agricultural and forestry products ranged from <0.02 Bq kg^{-1} to 1390 Bq kg^{-1} dm, and the transfer factors calculated based on these varied from 0.011 to 31 with an arithmetic mean of 3.4, with the highest values found in forest ecosystem products. The transfer factors and annual committed effective dose was calculated. It was statistically proven that the level of soil contamination with ^{137}Cs is related to the altitude and intensity of the precipitation in April and May of 1986, after the Chernobyl NPP accident.

Keywords: ^{137}Cs; radioactive contamination; soil samples; agricultural products; forestry products; transfer factor

1. Introduction

Although 37 years have passed since the accident at the Chernobyl nuclear power plant (NPP), its legacy is still relevant, both for the continuous monitoring of the consequences and the preparedness for a possible similar situation, the risk of which has now significantly increased with the ongoing war and threat to the Ukrainian NPPs.

The Czech Republic (CR) belongs to the countries that were more significantly affected by the accident in Chernobyl. The Czech Republic (CR) is one of the countries that were more substantially affected by the Chernobyl accident. The mean value of the Chernobyl ground deposition of ^{137}Cs in the Czech Republic was 5 kBq m^{-2}, but, individually, the area exceeded 100 kBq m^{-2} [1–3]. The content of ^{137}Cs in soil immediately after an accident depends on local fallout conditions, e.g., the amount of precipitation, the prevailing direction of atmospheric flow, and the properties and ways of land use [4,5]. The impact of precipitation caused the passage of contaminated air masses across the Czech Republic from the northeast to southwest to release a large amount of ^{137}Cs in the areas of South Bohemia (southwest of the Czech Republic) and northern Moravia (northwest of the Czech Republic). This results in locally higher ground deposition levels, especially in some components of natural ecosystems.

As a result of the higher content of ^{137}Cs, there is a problematic situation with wild boars in the affected areas, where in the past period, animals significantly exceeding the limit values for meat consumption were captured; for example, a sample of wild boar with an activity of 4.9 kBq kg^{-1} was measured in the Olomouc region [6].

The transfer of Chernobyl cesium into plants depends not only on the content of ^{137}Cs in the root layer of the soil [7], but also on the nature of the radionuclide interaction with the soil and with the plants into which cesium passes through the soil solution. Also important are the chemical form of ^{137}Cs, the type of soil [8,9], the physico-chemical and biological properties of the soil [10], and the location of the soil (hill, hillside, slope, valley) [11]. In addition, ^{137}Cs is significantly better bound in organic soil matter with a higher mineral content, as when it is in soils with a high organic content, Cs is mobilized in soil solution [12,13]. It was calculated that there was an approximately tenfold increase in the transfer of Cs to plants when the organic matter content increased from 5 to 50% [14]. Clay minerals effectively reduce the migration of Cs in the soil [15], resulting in a lower uptake by plants [10]. It was stated that Cs is highly absorbed by clay particles and organic matter and is virtually non-exchangeable for other ions [16–18].

Cesium transfer can also be influenced by the type of plant and specific plant variety [19], as well as by the density of sowing [20]. The transfer of Cs is also influenced by other factors such as plant structure, metabolism, habitat and plant nutrition, climate, weather, and, last but not least, the character and form of agricultural activity [21], or even the activity of animals [22].

To evaluate the risk of the radionuclide transfer of radionuclide from soil to living organisms, a parameter called the transfer factor (TF) is used, which expresses the ratio of radionuclide activity concentration in the target matrix to activity concentration in the soil. Transfer factors (TF) expressing the degree of the radionuclide transfer of radionuclide from the soil to plant are normally at the level of 0.001 to 1 for agricultural crops, with values higher than 1 being achieved by plants on sandy and organic soils [22,23].

In order to evaluate the level of contamination in the North Moravian region, we carried out a survey of agricultural and intact soils, from which soil samples were taken and the content of ^{137}Cs was analyzed. Agricultural, wild plants, and mushrooms were also sampled at selected locations, where, in addition to the activity concentration of ^{137}Cs, the TF was also determined.

2. Materials and Methods

2.1. Sampling

The sampling of soils, agricultural and forest products, and permanent grass cover took place in the summer months of 2018 and 2019 under clear weather in the northeastern part of the Czech Republic. The samples were collected in the number of 1 to 3 samples on an area of about 100 m^2 at 73 sites, as shown in Figure 1, using a shovel and a spade from an area of 20 cm × 20 cm to a depth of about 20 cm. The total number of soil samples was 176. Sampling was carried out on an area of 4.328 km^2. The altitudes of the individual sites ranged from 199 m to 827 m above sea level, with an average of 372 m above sea level. The native weight of each sample was weighed immediately after collection. The average native weight of all samples was around 14.5 kg, depending on the water content in the sample. Subsequently, the sample was thoroughly mixed, dried at a room temperature range of 19 °C to 20 °C for 14 days, reweighed, mixed, and crushed. Then, each soil sample was sieved through a 3 mm mesh pedological sieve and reweighed. After sieving, the soil was weighed again. A representative portion with an average weight of approx. 380 g was taken from each sample prepared in this way to measure activity concentration. For all soil samples, cadastral classification was also determined. The altitude and geographical coordinates were determined using the map background of the altimetry analysis. The cadastral classification of soils was determined through the State Administration of Land Surveying and Cadastre, an online cadastral map [24].

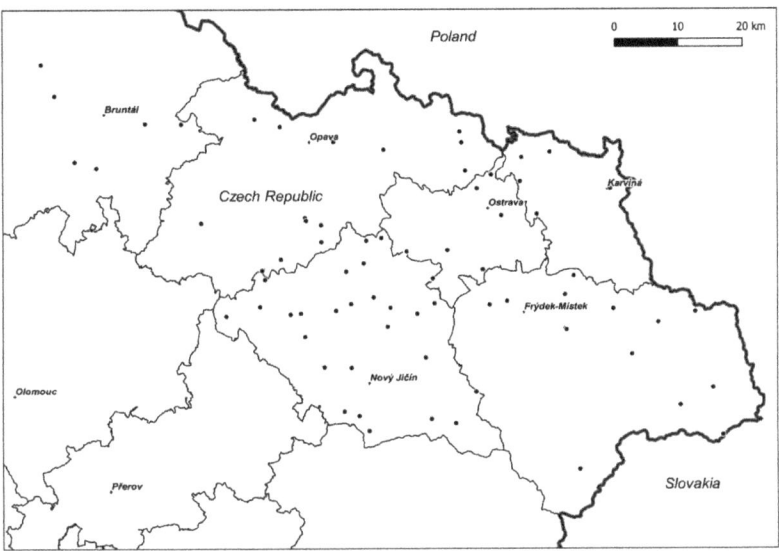

Figure 1. Map of soil sampling locations.

Forest and agricultural products were collected during the harvest period (September–October). Using a knife and a small shovel, the following items were sampled: cereals (*Triticum*), legumes (*Glycine max*), root crops (*Beta vulgaris, Solanum tuberosum, Apium graveolens*), annual fodder crops (*Zea mays*), oilseeds (*Brassica napus, Papaver somniferum*), grasses (*Festuca pratensis*), vegetables and root plants (*Brassica oleracea* var. *Capitata, Armoracia rusticana*), fruit (*Maleventum*), angiosperms (*Rosa canina, Sorbus aucuparia*), gymnosperms (*Pinus sylvestris*), fungi (*Lactarius lignyotus, Leccinum holopus, Neoboletus luridiformis, Amanita rubescens*) and spore-bearing plants (*Bryopsida*). In most cases, the sample consisted separately of the plant part intended for consumption and the rest of the plant body. In other cases, the plant part and the plant body were measured together within one sample. The samples of forest and agricultural products were first dried at room temperature and then in a biomass dryer at 52 °C with a drying time of 720 min. The collected forest and agricultural products were ground into pieces of max. 10 cm in length using a knife and cleaver. For the remaining biomass samples, only fruits, the plant parts of spore plants, and grasses were taken and measured.

2.2. Measurement of ^{137}Cs Activity

The ^{137}Cs activity in all dried samples (see Section 2.1) was measured at the National Radiation Protection Institute (SÚRO, v. v. i., Czech Republic) by semiconductor gamma spectrometry on HPGe detectors with efficiencies of 10–150%. The measurement time was chosen to achieve the uncertainty of ^{137}Cs determination lower than 10%, but no more than 1 week; if the activity was below the detection threshold, a minimum significant activity (MSA) was determined [25]. Both plant and soil samples were measured in 200 mL cylindrical plastic vessels and 500 mL Marinelli beakers on the detector. In the case of a larger sample, the cylindrical containers were around the detector. The energy and efficiency calibrations were completed using gel standards. Since the density of the vegetable and soil samples mostly varied from gel density, a correction for self-absorption had to be made. The measurement method is accredited. The residual moisture and dry matter were determined in a small portion of each sample at 105 °C. The ^{137}Cs activity in all measured samples are related to the sampling date and recalculated for dry matter.

2.3. Data Processing

The maps were created in the QGIS [26] and SAGA GIS [27] programs. The vector polygon layers of the administrative division of the Czech Republic come from the Register of Territorial Identification, Addresses and Real Estate (RÚIAN) (Czech Office for Surveying, Mapping and Cadastre—ČÚZK) [24]. The point data were transferred to the surface map by the Multilevel B-Spline interpolation method in the SAGA GIS program [27].

The transfer factors (TF) were calculated as the ratio of the activity concentration of ^{137}Cs in the dry matter of agricultural and forestry products to the activity concentration of ^{137}Cs in the dry matter of the soil. The IAEA (2010) [28] refers to TF as a concentration ratio.

Using the Excel office program, statistical characteristics were determined: simple arithmetic and geometric means (AM, GM), arithmetic and geometric standard deviations (SD, GSD), the minimum and maximum for soil activity concentration, and transfer factors, which were further divided into groups. Only data with a specified activity of ^{137}Cs higher than MSA were included in the calculation.

3. ^{137}Cs Activity Concentration in Soils

A total of 176 soil samples were taken from 73 localities and 6 districts: Bruntál, Opava, Nový Jičín, Ostrava, Karviná and Frýdek-Místek. Table 1 shows the statistical characteristics of ^{137}Cs activity concentration in all soil samples and in the groups of soils divided by cadastral classification.

Table 1. Statistical characteristics of ^{137}Cs activities in soils divided by cadastral classification.

	All	pg	Arable Land	Other Land	Forest
Number of samples	176	29	52	42	53
AM [Bq kg^{-1} dm]	38.4	21.2	18.1	19.2	84.4
SD [Bq kg^{-1} dm]	57	21	12	19.41	84
AM$_P$ [Bq m^{-2}]	11,213	6190	5285	5198	24,645
GM [Bq kg^{-1} dm]	16.8	14.5	13.3	11.0	37.8
GSD	3.97	2.44	2.52	3.18	5.61
GM/AM ratio	0.44	0.69	0.74	0.57	0.45
minimum [Bq kg^{-1} dm]	0.177	2.50	0.438	0.52	0.177
maximum [Bq kg^{-1} dm]	299	106	52.2	75	299
max/min ratio	1690	42	119	144	1690

Notes: pg is permanent grassland, AM$_P$ [Bq m^{-2}]—the value was calculated using the mean soil density in the Czech Republic, which is 1460 kg/m^3, and a soil abstraction depth of 20 cm, according to the relationship AM$_P$ = AM·292.

Overall, the activity concentration of ^{137}Cs lay in a wide three-order range, with the lowest activity concentration being 0.177 Bq kg^{-1} dry matter (dm) measured in the forest soil of the locality Studénka nad Odrou, and the highest ^{137}Cs activity being 299 Bq kg^{-1} dm, which was measured in the forest soil of the Spálov site. The overall arithmetic mean was equal to 38.4 Bq kg^{-1} dm. The results of the calculations for cadastral classification show that the highest mean activity concentration value of ^{137}Cs was, as expected, in the forest land system (AM 84.4 Bq kg^{-1} dm). The mean activity concentration of ^{137}Cs was about 4 times higher in forest soils than in permanent grasslands and arable and other soils. The arithmetic means for permanent grassland (21.2 Bq kg^{-1} dm), arable land (18.1 Bq kg^{-1} dm), and other land (19.2 Bq kg^{-1} dm) are very close, with the arable and other lands corresponding to systems disrupted by human activity.

Since no area activity was determined for the removed soils, to compare with the total soil deposition density of ^{137}Cs in CR, we calculated the mean area activity of AP$_P$ from the arithmetic mean of the activity concentration of AP, the mean density of soils in the Czech Republic, and the depth of sampling. The results are presented in Table 1. The arithmetic mean of 5 kBq m^{-2} [1] in the post-Chernobyl soil survey was obtained in the CR mainly on unshaded and untouched soils, so this value corresponds most closely to permanent grasslands with AP$_P$ 6.19 kBq m^{-2}. Given that 32 years have passed since 1986 to 2018, the decrease in ^{137}Cs should be, due to physical transformation, at least 50%. In terms of other environmental factors, the mean activity in the CR should now be 2.5 kBq m^{-2} or lower. The level of soil contamination in the region we monitored was therefore about 2.5 times higher than in relation to the whole Czech Republic.

The higher value of ^{137}Cs cannot be attributed to the nuclear accident in Fukushima, which affected the CR only minimally. Nor can it be attributed to the global fallout from nuclear weapons tests from 1945 to 1963 [29], which was approximately 5 kBq m^{-2} relatively homogeneously dispersed at our latitudes (40–50° N and 50–60° N) (calculated from UNSCEAR [30]). Figure 1 shows a map of sampling points covering six districts. Activity concentrations were put on a map [Figure 2]. In the largest part of the territory, the activities were less than 40 Bq kg^{-1} dm. The areas of higher contamination are in the Opava region and in the northwest of the Bruntál region, with activity concentrations exceeding 160 Bq kg^{-1} dm, which corresponds to a surface contamination of 46 kBq m^{-2} and higher. A place with higher activity was also found in the east of the Frýdek-Místek region.

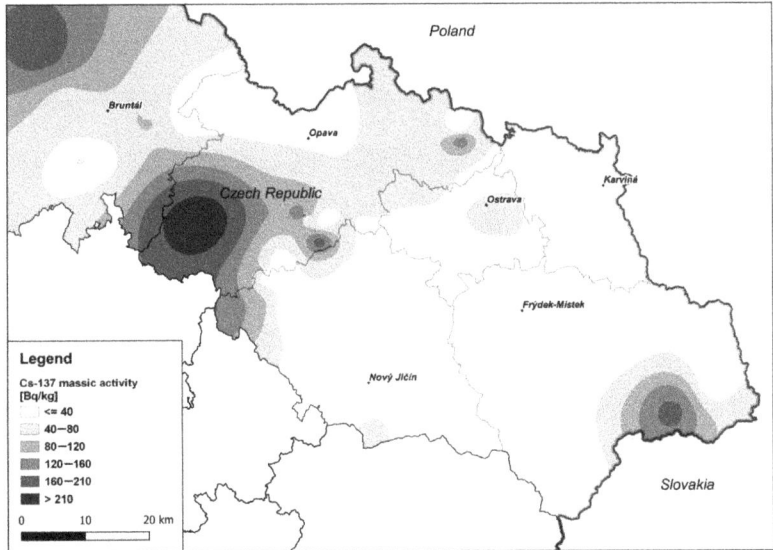

Figure 2. Map of ^{137}Cs activity concentration in soil.

The higher contamination of the Opava and Bruntál regions corresponds approximately to the map of Chernobyl contamination [2], on which the surface activity locally reached 20–40 kBq m^{-2} in these regions.

On the map [2], the region of Ostrava is also more contaminated, locally and with activity reaching the interval of 40–100 kBq m^{-2}. The differences must be attributed to the insufficient sampling density, not only ours but also the Chernobyl one (776 samples for the entire CR), as well as the inaccuracy associated with interpolation.

Due to the higher activity of ^{137}Cs in heavily sloping soils, all values were put into a graph depending on altitude. Figure 3 shows a statistically confirmed upward trend fitted with a linear slope ($p < 0.000$).

The fact that areas with higher altitudes were more contaminated is probably related to the higher frequency of precipitation in these areas in April and May 1986. The altitudes were divided into groups according to altitudes of 100–199 m above sea level (ASL), 200–299 m ASL, 300–399 m ASL, 400–499 m ASL, 500–599 m ASL, 600–699 m ASL, 700–799 m ASL, and 800–899 m ASL. For each interval, the arithmetic mean of the precipitation totals in April and May of 1986 was calculated, and related to half of the relevant altitude interval. The result was a statistically confirmed upward trend ($p < 0.000$).

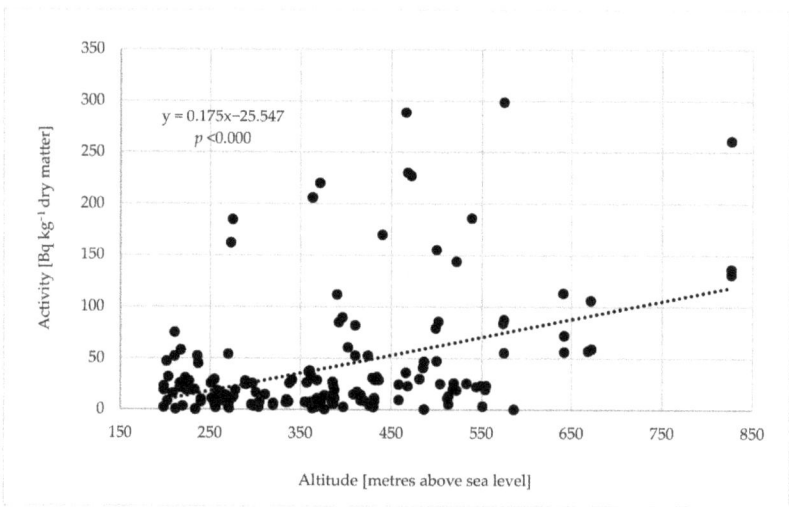

Figure 3. Dependence of ^{137}Cs activity concentration in all soils on the altitude.

The dependence of ^{137}Cs activity on the precipitation totals was also tested. In the vicinity of the soil sampling points, there are 39 weather stations, with available precipitation totals for April and May 1986 [31]. To calculate the values of the precipitation totals, the precipitation for the months of April and May in 1986 was averaged, and the nearest weather station to each sampling location was assigned to construct graphs of the dependence of ^{137}Cs activities in the soil at the altitude of the sampling point. The upward trend for areas with higher precipitation totals has been statistically confirmed ($p < 0.045$).

4. Results of Measurements of Agricultural and Forestry Products

A total of 28 biomass samples were measured. Three plant species were divided into two specimens: the part intended for consumption and the rest of the plant body. One sample contained two types of mushrooms mixed together. In 7 samples of agricultural products and plants of permanent grassland (sugar beet root, wheat grains, potato tubers from two sites, rose hip, soybean, and horseradish), it was not possible to measure the activity of ^{137}Cs, and therefore MSA (designation "<") was determined. The lowest activity concentration of ^{137}Cs < 0.02 Bq kg^{-1} dm was detected in a sample of sugar beet root; the highest ^{137}Cs activity 1390 Bq kg^{-1} dm was measured in a mixture of blacksmith's boletus and rosacea toadstool. The arithmetic mean of the activities of ^{137}Cs in plant samples was 89.4 Bq kg^{-1} dm. The arithmetic mean in the mushroom samples was 821.5 Bq kg^{-1} dm.

5. Transfer to Agricultural and Forestry Products

A total of 23 biomass-soil pairs were available for the calculation of transfer factors. In Tables 2 and 3, the TFs are sorted by cadastral division. Most biomass samples (14) were taken from arable lands, 2 from permanent grasslands, and 7 from forests.

On arable lands, the mean value of TF was equal to 0.12. The TF range agrees with the typical range of 0.001 to 1 for agricultural crops [23]. The determined values of TF [Table 2] are generally approximately at the level of the mean values presented in [30]. Higher values were found in wheat stalks (0.69), poppy seeds (0.11), and rapeseed stalks (0.1), while low values were found in sugar beet root (<0.0046) and fescue (0.012).

Table 2. Transfer factors (TF).

Title	Latin Name	Activity Concentrations Soil ^{137}Cs [Bq kg^{-1} dm]	TF	Cadastra Classification of Land
sugar beet—plant body	*Beta vulgaris*	25.6	0.031	acreage
sugar beet—root	*Beta vulgaris*	25.6	<0.0046	acreage
sugar beet—leaf	*Beta vulgaris*	8.7	0.077	acreage
wheat—stem	*Triticum*	16.0	0.036	acreage
wheat—grain	*Triticum*	16.0	<0.081	acreage
wheat—stem	*Triticum*	16.0	0.69	acreage
soybeans—beans	*Glycine max*	25.3	<0.024	acreage
soybean legume—stem	*Glycine max*	25.3	0.032	acreage
meadow fescue/hay—stem	*Festuca pratensis*	17.5	0.012	acreage
White cabbage—plant body	*Brassica oleracea* var. *Capitata*	29.5	0.041	acreage
rapeseed—stem	*Brassica napus*	45.0	0.1	acreage
poppy—poppy	*Papaver somniferum*	22.3	0.13	acreage
potato eggplant—plant body	*Solanum tuberosum*	7.50	<0.16	acreage
apple—whole fruit	*Maleventum*	52.2	0.011	acreage
wine rose—fruit	*Rosa canina*	14.7	<0.054	permanent grassland
horseradish—plant body	*Armoracia Rusticana*	9.4	<0.12	permanent grassland
moss	*Bryopsida*	170	0.11	forest land
moss	*Bryopsida*	30.1	0.17	forest land
moss	*Bryopsida*	60.0	2	forest land
Rowan—brood	*Sorbus aucuparia*	26.5	0.52	forest land
black grouse—whole fruit	*Lactarius lignyotus*	47.3	5.3	forest land
Boletus blacksmith + toadstool rosacea—whole fruit	*Neoboletus luridiformis + Amanita rubescens*	79.2	18	forest land
pine cones—whole fruit	*Pinus sylvestris*	2.7	31	forest land

Note: < sign is followed by the value of the transfer factor (TF) calculated from the minimum significant activity by weight of dry matter of forest and agricultural products.

Table 3. Statistical characteristics of transfer factors divided by cadastral focus.

	All	Arable Land	Forest Land
number of samples	17	10	7
AM	3.4	0.12	8.2
SD	8.4	0.20	12
GM	0.22	0.051	1.8
GSD	11.4	3.39	9.21
GM/AM ratio	0.064	0.44	0.22
minimum	0.011	0.011	0.11
maximum	31	0.69	31
max/min ratio	2990	65	296

The transfer factor for rapeseed of 0.10 came out approximately the same as the arithmetic mean of the transfer factors for the same plant of 0.13 [32]. The transfer to fruit trees is an order of magnitude lower, and the value (0.011) of the fruit of the apple tree is about twice as high as the mean value (0.0058) [31].

Tyson et al., 1999 [33], showed that ^{134}Cs is unevenly distributed in plants. Its higher concentration was found in young shoots and growth tissues. Radionuclides often transfer in higher concentrations to stems and leaves than to the generative parts of plants [31]. There are also differences in the values of TF found in our results.

The influence of the change in soil composition on the transfer is evident when comparing arable soils with forest soils, which are often characterized by a thick layer of overlying humus. While the total AM transfer factors for all samples were equal to 3.4, the TF for forest land was very high (AM 8.2), and the TF in arable land samples was an order of magnitude lower (AM 0.12). Also, the maximum-to-minimum ratio is higher for forest lands, 4.5 times higher compared to arable lands. The transfer to plants mainly reduces the content of clay minerals, and vice versa, it increases the amount of organic matter. In highly organic and peaty soils, cesium is weakly adsorbed on organic surfaces and can be easily desorbed into soil solution [12,13,34]. In addition, in forest soil, TF increases at a lower pH. A higher transfer to blueberries that prefer acidic soils is known. IAEA [35] described a trend in barley and cabbage towards a decrease in TF ^{137}Cs with an increasing pH. TF increases in the transfer of Cs from humus to spruce bark with a decrease in pH (H_2O) and pH (KCl) [36].

The lowest values of TF in forest soils were within the range of 0.11–2. Interestingly, the TF in pine cones was really high (31); it was even higher than the TF in mushrooms (5.3; 18), to which ^{137}Cs usually accumulate highly. This may be due to the longer reach of the pine roots and, accordingly, their ability to draw nutrients from a more distant place with higher soil activity.

6. Committed Effective Dose

As mentioned in the introduction, the monitored region was one of the most contaminated in the Czech Republic by the post-Chernobyl fallout. To evaluate the exposure risk, from the observed concentration activities, we can approximately estimate the committed effective dose of adults from plant agricultural products and compare it with the mean value of the dose for the whole Czech Republic. In 2015, the geometric mean of the total annual intake of ^{137}Cs of food was 110 Bq, which, when multiplied by a conversion coefficient of 0.013 µSv/Bq, corresponds to an effective dose of 1.4 µSv [36]. Plant agricultural products (vegetables, fruits, cereals, and potatoes) accounted for 4.6% and mushrooms for 76% of the total intake of ^{137}Cs. Thus, the corresponding mean effective dose in 2015 was 0.065 µSv for plant agricultural products and 1.1 µSv for fungi. The effective dose of agricultural crops was calculated from the activity concentrations above the MSA of white cabbage, sugar beet, apple, poppy, beetroot leaves, wheat stalks, leguminous soybean, and oilseed rape for our samples. The geometric mean of ^{137}Cs activity on a fresh basis is equal to 0.37 Bq/kg for our samples grown on agricultural soils. Using the annual consumption of fresh plant agricultural products in the Czech Republic of 181 kg, the annual mean effective dose of adults is 0.9 µSv. For our two mushroom samples, with a mean annual consumption of 2.8 kg and a mean fresh activity concentration of 62.3 Bq/kg, the effective dose is 2.3 µSv.

It can therefore be seen that the mean committed effective dose from ^{137}Cs from plant agricultural products for our area of interest is significantly higher, about 14 times the national mean. The effective dose from mushrooms is approximately twice as high. Of course, there is a time gap, and nationwide doses will be slightly lower today, but due to the currently very slow decrease in the activity of post-Chernobyl ^{137}Cs in agricultural ecosystems, the difference between the doses in 2015 and 2018–2019 is only slight. The observed higher values of the effective dose for agricultural products in our region are consistent with the higher soil contamination discussed above, although the ratio of the effective dose of our region to the rest of the Czech Republic is higher, which may be due to the relatively small number of samples and the associated lack of representativeness.

7. Conclusions

The research analyzed 176 soil samples from six districts in the northeast of the Czech Republic: Bruntál, Opava, Nový Jičín, Ostrava, Karviná, and Frýdek-Místek. For ^{137}Cs activity concentration, 28 biomass samples were also measured.

The average activity concentration of ^{137}Cs in the soil was 38.4 Bq kg^{-1}, while the highest value was found in the soils of forest ecosystems and the lowest in arable soils. The average activity in grass soils was 21.2 Bq kg^{-1}, which corresponds to a surface activity value of 6.19 kBq m^{-2}, which is at least 2.5 times higher than the surface activity of soil in the Czech Republic. The connection between the increase in the activity concentration of the soil with the increasing height above sea level and the intensity of precipitation totals confirms that mountain areas with lower agricultural production are the most contaminated.

The activity concentrations of ^{137}Cs in agricultural and forest products ranged from <0.021 Bq kg^{-1} to 1392 Bq kg^{-1} dry matter, which corresponded to the value of the transfer factor <0.0046 to 31, with the highest transfer to products of the forest ecosystem. Transfer factors were influenced by the plant type, part, and soil composition. The calculated annual effective dose from agricultural products was 0.9 µSv, i.e., approx. 14 times higher than the average for the Czech Republic, which agrees with the higher contamination of the northeast of the Czech Republic.

The results of our experiments therefore confirmed a higher occurrence of ^{137}Cs both in the soil and in plant products. Considering that the detected value is not representative due to the limited number of analyzed samples, it is necessary to further investigate similar areas not only in the Czech Republic, but also elsewhere in the world, and thus protect against the consumption of both contaminated agricultural products and products of forest ecosystems.

Author Contributions: Conceptualization, P.L. and J.Š.; methodology, J.Š.; software, H.P.; validation, P.L. and B.S.; formal analysis, P.L.; investigation, V.Z.; resources, B.S.; data curation, H.P.; writing—original draft preparation, P.L. and H.P.; writing—review and editing, P.L. and J.Š.; visualization, V.Z.; supervision, V.Z. and J.Š.; funding acquisition, B.S. All authors have read and agreed to the published version of the manuscript.

Funding: This research was funded by VSB—Technical University of Ostrava, Czech Republic, grant number SGS SP 2018/18, and also by the Ministry of the Interior of the Czech Republic project number MV VI20192022153—Optimization of procedures for the implementation of crop production in the area affected by a nuclear accident.

Institutional Review Board Statement: Not applicable.

Informed Consent Statement: Not applicable.

Data Availability Statement: Data are contained within the article.

Conflicts of Interest: The authors declare no conflict of interest.

References

1. Hůlka, J.; Thomas, J. Přehled expozice obyvatelstva přírodnímu záření. *Bezpečnost Jad. Energ.* **2007**, *15*, 65–67.
2. De Cort, M.; Dubois, G.; Fridman, S.D.; Germenchuk, M.G.; Izrael, Y.A.; Janssen, A.; Jones, A.R.; Kelly, G.N.; Kvasnikova, E.V.; Matveenko, I.I.; et al. *Atlas of Cesium Deposition on Europe after the Chernobyl Accident*; Office for Official Publications of the European Communities: Brussels, Belgium; Luxemburg, 1998.
3. Meusburger, K.; Evrad, O.; Alewell, C.; Borelli, P.; Cinelli, G. Plutonium aided reconstruction of caesium atmospheric fallout in European topsoils. *Sci. Rep.* **2020**, *10*, 11858. [CrossRef] [PubMed]
4. Handl, J.; Sachre, R.; Jakob, D.; Michel, R.; Evangelista, H.; Gonçalves, A.C.; de Freitas, A.C. Accumulation of ^{137}Cs in Brazilian soils and its transfer to plants under different climatic conditions. *J. Environ. Radioact.* **2008**, *99*, 271–278. [CrossRef] [PubMed]
5. Ivanov, M.M.; Konoplev, A.V.; Walling, D.E.; Konstantinov, E.A.; Gurinov, A.L. Using reservoir sediment deposits to determine the longer-term fate of chernobyl-derived ^{137}Cs fallout in the fluvial system. *Environ. Pollut.* **2021**, *274*, 116588. [CrossRef]
6. Skrkal, J.; Rulik, P.; Fantinova, K.; Mihalik, J.; Timkova, J. Radiocaesium levels in game in the Czech Republic. *J. Environ. Radioact.* **2015**, *139*, 18–23. [CrossRef]
7. Grabovskij, V.A.; Dzendzeljuk, O.S. Features of 137Cs contamination of some plants in the Ukrainian Western Polesie. In Proceedings of the II International Scientific and Practical Conference—Chernobyl Readings, Gomel, Belarus, 24–25 April 2008.
8. Jarošík, J. *Přestup Radioaktivity v Systému Atmosféra—Půda—Rostlina, Studie 5/1988*; Ústřední informační středisko pro jaderný program, Praha 5: Zbraslav, Czech Republic, 1988.
9. Titov, I.E.; Krechetnikov, V.V.; Mikailova, R.A.; Panov, A.V. Geoinformation decision support system for remediation of the 137Cs contaminated agricultural lands after the Chernobyl NPP accident. *Nucl. Eng. Technol.* **2022**, *54*, 2244–2252. [CrossRef]
10. Nimis, P.L. Radiocesium in plants of forest ecosystem. *Stud. Geobot.* **1996**, *15*, 3–49.
11. Miller, M.P.; Singer, M.J.; Nielsen, D.R. Spatial variability of wheat yield and soil properties on complex hills. *Soil Sci. Soc. Am. J.* **1988**, *52*, 1133–1141. [CrossRef]

12. Kruyts, N.; Titeux, H.; Delvaux, B. Mobility of radiocesium in three distinct forest floors. *Sci. Total Environ.* **2004**, *319*, 241–252. [CrossRef]
13. Cremers, A.; Elsen, A.; Valcke, E.; Wauters, J.; Sandalls, F.; Gaudern, S. The sensitivity of upland soils to radiocaesium contamination. In *Transfer of Radionuclides in Natural and Semi-Natural Environments*; Desmet, G., Nassimbeni, P., Belli, M., Eds.; Elsevier Applied Science: London, UK, 1990; pp. 238–248.
14. van Bergeijk, K.E.; Noordijk, H.; Lembrechts, J.; Frissel, M.J. Influence of pH, soil type and soil organic matter content on soil-to-plant transfer of radiocaesium and strontium as analyzed by a nonparametric method. *J. Environ. Radioact.* **1992**, *15*, 265–276. [CrossRef]
15. Liu, W.; Li, Y.; Yu, H.; Saggar, S.; Gong, D. Distribution of ^{137}Cs and ^{60}Co in plough layer of farmland: Evidenced from a lysimeter experiment using undisturbed soil columns. *Pedosphere* **2021**, *31*, 180–190. [CrossRef]
16. Rosen, K.; Vinichuk, M.; Öborn, I.; Gutierrez-Villanueva, J.L. Migration of radionuclides in undisturbed soil. In Proceedings of the NSFS Conference, Reykjavik, Iceland, 22–25 August 2011.
17. Ritchie, J.C.; McHenry, J.R. Application of Radioactive Fallout Cesium-137 for Measuring Soil Erosion and Sediment Accumulation Rates and Patterns: A Review. *J. Environ. Qual.* **1990**, *19*, 215–233. [CrossRef]
18. Meena, S.; Senthilvalavan, P.; Vadivel, A. Chemically induced phytoextraction of caesium-137. In Proceedings of the 19th World Congress of Soil Science, Brisbane, Australia, 1–6 August 2010.
19. Penrose, B.; Beresford, N.A.; Broadley, M.R.; Crout, N.M.J. Inter-varietal variation in caesium and strontium uptake by plants: A meta-analysis. *J. Environ. Radioact.* **2015**, *139*, 103–117. [CrossRef] [PubMed]
20. Lacko, T.V. Cultivation of non-traditional plants from the vetch family in the area of radioactive contamination. In Proceedings of the II International Scientific and Practical Conference—Chernobyl Readings, Gomel, Belarus, 24–25 April 2008.
21. Tölgyessy, J.; Harangozó, M. *Rádioekológia*; Univerzita Mateja Bella, Fakulta prírodných vied: Banská Bystrica, Slovakia, 2000; p. 131.
22. Auerswald, K.; Sippel, R.; Kainz, M.; Demmel, M.; Scheinost, A.; Sinowski, W.; Maindl, F.X. The crop response to soil variability in an agroecosystem. *Adv. GeoEcol.* **1997**, *30*, 39–53.
23. Zhu, Y.-G.; Smolders, E. Plant uptake of radiocaesium: A review of mechanisms, regulation and application. *J. Exp. Bot.* **2000**, *51*, 1635–1645. [CrossRef] [PubMed]
24. State Administration of Land Surveying and Cadastre. Available online: https://www.cuzk.cz/ (accessed on 5 January 2023).
25. Currie, L.A. Limits for qualitative detection and quantitative determination. *Anal. Chem.* **1968**, *40*, 586–593. [CrossRef]
26. QGIS. A Free and Open Source Geographic Information System. Available online: https://www.qgis.org (accessed on 10 January 2023).
27. SAGA GIS. System for Automated Geoscientific Analyses. Available online: https://saga-gis.sourceforge.io/ (accessed on 17 January 2023).
28. IAEA. *Handbook of Parameter Values for the Prediction of Radionuclide Transfer in Terrestrial and Freshwater Environments*; Technical Reports Serie No. 472; IAEA: Vienna, Austria, 2010.
29. Kalan, S.K.; Forkapić, S.; Marković, B.S.; Gavrilov, B.M.; Bikit-Schroeder, K. Deposition of 137Cs and precipitation distribution in Vojvodina, Northern Serbia after the Chernobyl accident. *Chemosphere* **2021**, *264*, 128471. [CrossRef]
30. UNSCEAR. *Annex C, Exposures to the Public from Man-Made Sources of Radiation*; Sales No. E.00.IX.3; United Nations Sci Committee on the Effects of Atomic Radiation, United Nations Publications: New York, NY, USA, 2000; p. 212.
31. Czech Hydrometeorological Institute. Available online: https://www.chmi.cz/ (accessed on 29 December 2022).
32. Procházka, J.; Škrkal, J.; Rulík, P.; Křováková, K.; Šímová, I. Determining the transfer factors for estimates of the radioactive contamination of agricultural crops. *Radiat. Prot. Dosim.* **2022**, *198*, 747–753. [CrossRef]
33. Tyson, M.J.; Sheffield, E.; Callaghan, T.V. Uptake, transport and seasonal recycling of ^{134}Cs applied experimentally to bracken (*Pteridium aquilinum* L. Kuhn). *J. Environ. Radioact.* **1999**, *46*, 1–14. [CrossRef]
34. Valcke, E.; Cremers, A. Sorption-desorption dynamics of radiocaesium in organice matter soils. *Sci. Total Environ.* **1994**, *157*, 275–283. [CrossRef]
35. IAEA. TECDOC 1497: Classification of soil systems on the basis of transfer factors of radionuclides from soil to reference plants. In Proceedings of the Coordination Meeting, Chania, Greece, 22–26 September 2003.
36. Škrkal, J.; Pilátová, H.; Rulík, P.; Suchara, I.; Sucharová, J.; Holá, M. Behaviour of ^{137}Cs in forest humus detected across the territory of the Czech Republic. *Sci. Total Environ.* **2017**, *593–594*, 155–164. [CrossRef] [PubMed]

Disclaimer/Publisher's Note: The statements, opinions and data contained in all publications are solely those of the individual author(s) and contributor(s) and not of MDPI and/or the editor(s). MDPI and/or the editor(s) disclaim responsibility for any injury to people or property resulting from any ideas, methods, instructions or products referred to in the content.

Proceeding Paper

Pulp and Paper Mill Sludge Utilization by Biological Methods [†]

Adam Pochyba, Dagmar Samešová *, Juraj Poništ and Jozef Salva

Department of Environmental Engineering, Faculty of Ecology and Environmental Sciences, Technical University in Zvolen, 960 01 Zvolen, Slovakia; pochybaadam@gmail.com (A.P.); jurajponist1111@gmail.com (J.P.); jsallva@gmail.com (J.S.)
* Correspondence: dsamesova@gmail.com; Tel.: +421-45-5206330
[†] Presented at the 4th International Conference on Advances in Environmental Engineering, Ostrava, Czech Republic, 20–22 November 2023.

Abstract: The aim of our contribution is to evaluate the possibilities for the biochemical utilization of paper wastes. We tested aerobic and anaerobic degradation of paper waste sludges from a plant that processes recycled paper. Testing included the assessment of phytotoxicity. We can conclude that the concentration of paper sludges did not have toxic effects on the bacterial consortium of the anaerobic or aerobic conditions. However, the leachate of paper sludges and water from sludge dewatering had a slightly negative effect on the germination of cress (*Lepidium sativum* L.) and lettuce (*Lepidium sativum* L.) seeds where the germination ranged from 83.3% to 100% but the mass yield was higher.

Keywords: pulp; paper; sludge; biogas; biodegradation; phytotoxicity

1. Introduction

The aim of our contribution is to evaluate the possibilities for the biochemical utilization of paper waste. The global production of waste from paper and paperboard was approximately 417.3 million metric tons in 2021 [1]. The production of paper and paperboard is resource-intensive, and the generation of pollutants is harmful to the environment [2]. The cellulose paper industry in Europe generates approximately 11 million tons of waste annually [3].

Waste materials or by-products generated in paper mills are typically categorized as follows [4]:

1. Residues (including coarse, heavy, and fine residues): These are generated during the use of recycled fiber pulp (RCF) and may contain fibrous pieces, plastics, metals, sand, and glass, respectively;
2. Deinking sludge: Produced during the removal of printing ink from RCF, typically comprising short fibers/fine particles, inorganic fillers, and ink particles;
3. Primary sludge: Generated in the mechanical water cleaning process, primarily consisting of short fibers/fine particles and fillers;
4. Secondary sludge: Formed during the treatment of process water using biological methods;
5. Process water (often referred to as wastewater): It is a significant component of paper production and is generally treated on-site to remove impurities [5].

Paper mill sludge generated from wastewater treatment constitutes the largest portion of waste in terms of volume. It contains the remnants of both organic and inorganic substances that were not utilized in the paper manufacturing process. These substances may include wood matter, sulfuric acid, hydrochloric acid, phosphoric acid, and various types of inorganic salts [6–8].

The disposal of paper waste has historically been addressed primarily through landfilling or incineration [9–11]. However, current preferences strongly favor recycling methods and the utilization of paper waste. Environmentally acceptable processes also include

biological treatment methods, which enable the transformation of waste into useful products such as compost or biogas [12]. Composting is a microbial process that breaks down organic material into a product that can enrich soil. Anaerobic digestion is a process in which organic waste is decomposed in the absence of oxygen, resulting in energetically useful methane and CO_2 [13].

2. Material and Methods

Samples were collected from a paper mill that processes recovered paper (recycling paper) and consisted of dewatered paper mill sludge with a dry matter content of 56.7% and water from the dewatering process. From the paper mill sludge, a water extract was prepared and used in the tests. For anaerobic degradation, sewage sludge from the anaerobic treatment of wastewater treatment plant were used as an inoculum. An inoculum obtained by leaching from the soil was used for the aerobic test. To evaluate the phytotoxic properties for the potential use of the sludge in soil or compost, modified tests on higher plant seeds were conducted.

2.1. Preparation of the Leachate

The preparation of the leachate followed the standard STN EN 12457-4, using a ratio of 1 L of distilled water per 100 g of sludge dry matter. The leaching process occurred for 24 h on an orbital shaker at laboratory temperature, and the extract was pre-filtered before testing [14].

2.2. Biodegradability Tests (Anaerobic and Aerobic Biodegradability of Organic Substances in Paper Mill Sludge)

The biodegradability tests were conducted using the OxiTop® OC 110 instrument (WTW, Weilheim, Germany), which operates based on the principle of respirometric measurement. The working volume of the instrument bottles was 510 cm^3. The tests were conducted according to the following standards: OECD 311, OECD 301 D, ISO/DIS 14851:2016 E, and OECD 302 B. These standards were adapted to our specific requirements and conditions [15–18].

The anaerobic test was carried out with 3 different leachate concentrations A, B and C (Table 1).

Table 1. Content and volume of sample concentrations (anaerobic biodegradability).

Ratio of Ingredients in Mixtures	A	B	C
Sludge (mL) *	1.0	1.2	1.6
Leachate (mL)	50	60	70
Biomedium (mL) **	125	125	125

* Sludge from a wastewater treatment plant (anaerobic stage). ** Composition of the biomedium: K_2HPO_4 KCl. $MgSO_4$, $7H_2O$, $NaNO_3$, 1000 mL distilled water.

The aerobic test was also performed in 3 concentrations of leachate A, B and C (Table 2).

Table 2. Content and volume of sample concentrations (aerobic biodegradability).

Ratio of Ingredients in Mixtures	A	B	C
Leachate (mL)	50	60	70
Inoculum (mL) *	5	5	5
Biomedium (mL) **	125	125	125

* Inoculum obtained by leaching from soil. ** Composition of the biomedium: K_2HPO_4.KCl. $MgSO_4$, .$7H_2O$, $NaNO_3$, 1000 mL distilled water.

We repeated all experiments three times, average values are shown in the tables and graphs.

2.3. Phytotoxicity Test with Lepidium sativum L.

The cress test, or phytotoxicity test with *Lepidium sativum* L., is a modified method for evaluating the intensity of the decomposition of organic materials and the maturity of the resulting compost for use in composting practices. It involves assessing the phytotoxicity of the sample extract using the germination index (GI) of a sensitive plant species, such as garden cress (*Lepidium sativum* L.) [19].

2.4. Phytotoxicity Test on Leaf Lettuce (Lactuca sativa L.)

To determine the toxic effects of pollutants on leaf lettuce (*Lactuca sativa* L.), we followed a modified ISO 18763:2016 procedure. We mixed the soil with different concentrations of the leachate (3 concentrations of leachate and water from sludge dewatering) of paper sludge to determine the optimum concentration for seed germination and growth [17].

2.5. Calculation of Biodegradation

$$Biodegradation\ (\%) = 100 \times \frac{BOD - BOD_{Blank}}{ThOD}$$

where:
BOD is the biochemical oxygen demand of the test substance (mg/L),
BOD blank is the biochemical oxygen demand of the biotic control (mg/L),
ThOD is the theoretical oxygen demand required when the target compound is completely oxidized (mg/L) [20].

3. Results and Discussion

3.1. Anaerobic Biodegradability of Organic Substances in Paper Sludge

The test lasted 28 days, with three concentrations of the leachate. At regular intervals, the chemical oxygen demand (COD_{Cr}), pH, and conductivity were determined in the samples.

The COD_{Cr} value decreased during the test, indicating a gradual decrease in the organic pollution of the sample and an increased amount of organic pollution. The measured pressure (Figure 1), affected by the gas produced by the microorganisms, from the samples over the 28 days increased in direct proportion to the increasing amount of paper sludge leachate that was added, with approximately the same pattern of development. From these results, we can conclude that the increased concentration of sludge did not have toxic or inhibitory effects on the bacterial consortium of the anaerobic sewage sludge and supplied the necessary nutrients for the development and growth of the microorganisms. The positive effect of the addition of paper sludge during anaerobic fermentation has been demonstrated in the works of many authors, e.g., Chynoweth, 1993 [21], Zhu, 2021 [22].

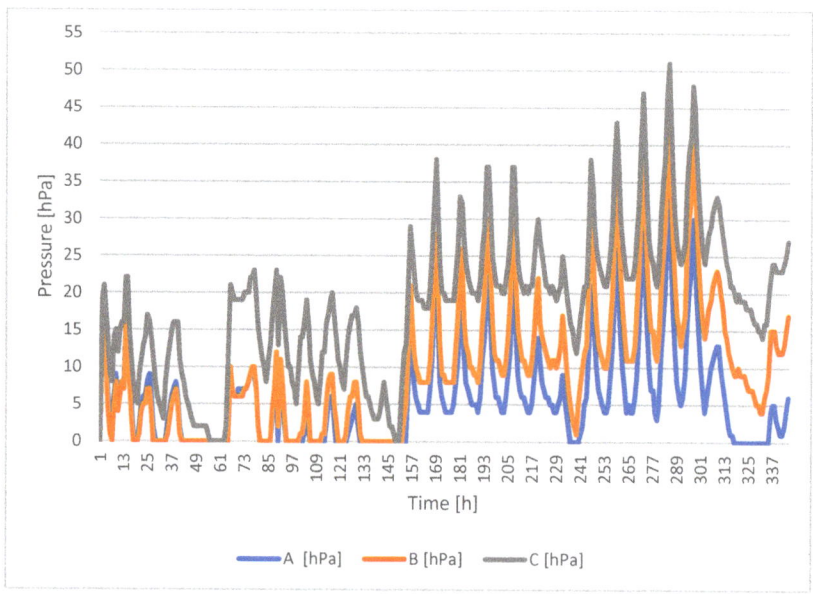

Figure 1. Pressure evolution during the test.

3.2. Aerobic Biodegradability of Organic Substances in Paper Sludge

The test was performed in six repetitions and with three concentration levels. The pH values of all leachate concentrations were close to neutral pH, indicating the stability of the reaction and minimal contamination. The COD_{Cr} values at each sample concentration decreased from the first to the last day, indicating that the samples were undergoing an oxidative process in which oxygen was consumed (Table 3).

Table 3. Sample properties for aerobic biodegradability.

Sample	COD_{Cr} Day 1	COD_{Cr} Day 5
A	847.80	561.90
B	960.84	481.11
C	1073.88	581.97
Blank	489.84	501.70

As the leachate concentration of the paper sludge increased, the biological oxygen demand (BOD) increased in direct proportion (Figure 2). The conditions for the formation of the microbial consortium improved with the increasing concentration of the substances contained in the leachate; we assume enrichment with nutrients and trace elements.

The percentage of biodegradation increased in direct proportion to the increasing concentration of the paper sludge leachate (Figure 3). This indicates that a greater amount of the test substance has been degraded and undergone biological transformation. A higher percentage of biodegradation indicates a higher capacity of the substance to be degraded by microorganisms.

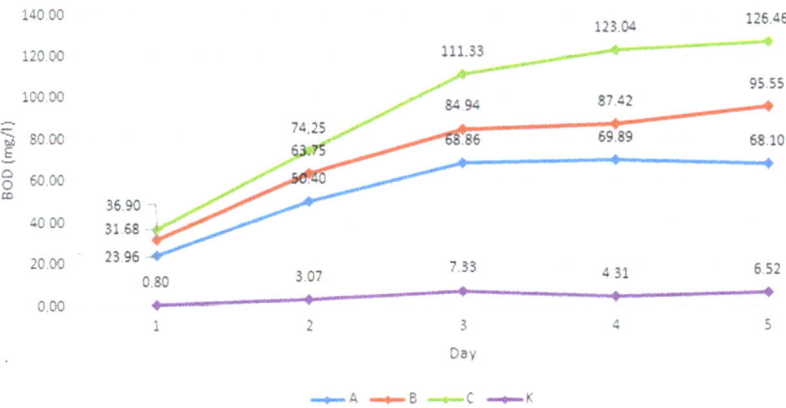

Figure 2. Evolution of average BOD values during the test (mg/L).

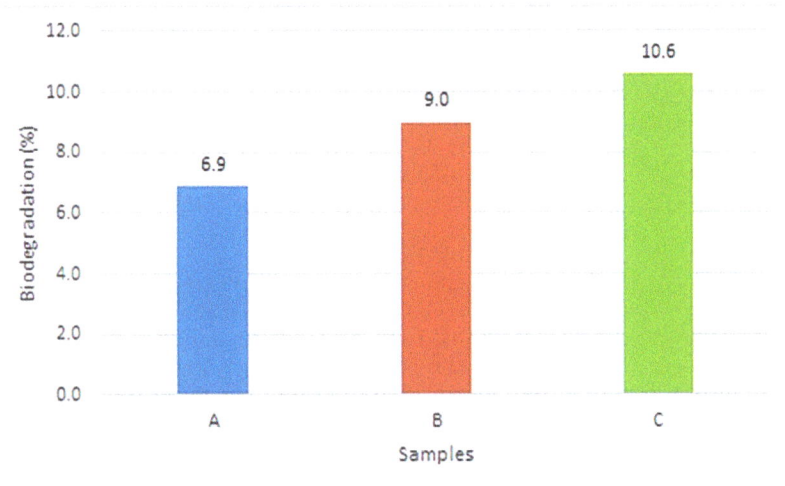

Figure 3. Biodegradation of the samples.

3.3. Phytotoxicity Test with *Lepidium sativum* L.

The GI of *Lepidium sativum* L. shown in Table 4 indicates that the leachate and water from drainage had inhibitory effects. The samples contained substances that adversely affected seed germination. Some of these substances may be present in the callus due to the use of biocides, dyes, adhesives, and other chemicals in the papermaking process. Therefore, it is possible that the presence of these chemicals in the wastepaper sludge caused an inhibitory effect on the germination of *Lepidium sativum* L. seeds.

Table 4. Results of the phytotoxicity test with *Lepidium sativum* L.

Sample	Germination (%)	Germination (%)	Average Root Length (mm)	(GI) (%)
Blind experiment	100	100	6.74	100
Water from sludge dewatering	92.50	92.50	4.09	56.12
Sludge leachate	88.89	88.89	4.81	63.40

3.4. Leaf Lettuce (Lactuca sativa L.) Growth Inhibition Test

As the concentration of the leachate from the paper sludge increased, the ability of the plants to germinate increased. From data shown in Table 5, it is clear that the optimum nutrient concentration for both growth and germination is at concentration C. This concentration of the sludge leachate did not have an inhibitory effect on the plants; we can assume that it had a more beneficial effect on the plants than the blind experiment since the plants had a greater dry weight compared with the blind experiment. Seeds watered with leachate concentration C had 100% germination. Their weight was higher than the weight of the plants in the blind experiment, the highest of the whole test. The dry matter percentage was one hundredth less than that of the blind. The concentration of the leachate from sludge C did not have an inhibitory effect on the plants, and it can be said that it had a better effect than the blind experiment, since these plants have a higher weight and dry matter percentage.

Table 5. Evaluation of inhibition test with *Lactuca sativa* L.

Sample	Germination (%)	Wet Weight (g)	Dry Weight (g)	Dry Matter (%)
Blank	100	0.29	0.27	4.88
A	83.3	0.26	0.25	5.30
B	86.6	0.26	0.25	4.82
C	100	0.33	0.31	5.03
Water from sludge dewatering	90	0.25	0.23	6.72

Paper sludge leachate contains nutrients that stimulate the biomass production of the test organism. However, the leachate tends to have an inhibitory effect on germination, but tends to stimulate the growth and biomass production of *Lactuca sativa* L.

4. Conclusions

The results of the anaerobic biodegradability tests showed that the paper sludge had no toxic or inhibitory effects on the bacterial consortium of the anaerobic sewage sludge and supplied the necessary nutrients for the development and growth of the microorganisms.

Aerobic biodegradability tests showed that increasing the sludge concentration improved the decomposition of organic substances, with pH and COD_{Cr} being positive indicators. The phytotoxicity test showed only a partial negative effect on the seed germination of *Lepidium sativum* L., indicating a possible limitation of seed growth in sludge. These results highlight the need for further tests before sludge is used in agriculture conditions. The results of *Lactuca sativa* L. growth inhibition showed that the leachate can possibly have a beneficial effect on plant growth, however, it can inhibit seed germination. It is important to take this factor into account when using sludge as a fertilizer. These findings suggest the need for further research to investigate the use of sludge as a fertilizer for different plant species and their response to this material.

Author Contributions: Conceptualization, D.S. and A.P.; methodology, D.S.; validation, A.P., D.S., J.S. and J.P.; formal analysis, J.S.; investigation, A.P., D.S., J.S. and J.P.; resources, D.S.; data curation, J.P.; writing—original draft preparation, A.P.; writing—review and editing, D.S., J.S. and J.P.; visualization, J.S.; supervision, D.S.; project administration, D.S.; funding acquisition, D.S. All authors have read and agreed to the published version of the manuscript.

Funding: This research was funded by Technical University in Zvolen under the project of the VEGA No. 1/0524/23 "Assessment of biodegradation in the terms of the energy potential of waste".

Institutional Review Board Statement: Not applicable.

Informed Consent Statement: Not applicable.

Data Availability Statement: All data used is part of the article, no third party consent is required.

Conflicts of Interest: The authors declare no conflict of interest.

References

1. Statista Research Department: Global Production of Paper and Paperboard 1961–2021, 29 August 2023, Available: Global Paper and Paperboard Production 2021 | Statista. Available online: https://www.statista.com/statistics/240598/production-of-paper-and-cardboard-in-selected-countries (accessed on 21 September 2023).
2. Mandeep; Kumar Gupta, G.; Shukla, P. Insights into the Resources Generation from Pulp and Paper Industry Wastes: Challenges, Perspectives and Innovations. *Bioresour. Technol.* **2020**, *297*, 122496. [CrossRef]
3. European Commission. *New Market Niches for the Pulp and Paper Industry Waste Based on Circular Economy Approaches*; European Commission: Brussels, Belgium, 2021. [CrossRef]
4. Gopal, P.M.; Sivaram, N.M.; Barik, D. Paper Industry Wastes and Energy Generation from Wastes. In *Energy from Toxic Organic Waste for Heat and Power Generation*; Woodhead Publishing: Cambridge, UK, 2019; pp. 83–97. [CrossRef]
5. Casco, M.E.; Moreno, V.; Duarte, M.; Sapag, K.; Cuña, A. Valorization of Primary Sludge and Biosludge from the Pulp Mill Industry in Uruguay through Hydrothermal Carbonization. *Waste Biomass Valorization* **2023**, *14*, 3893–3907. [CrossRef]
6. Institute for Prospective Technological Studies (Joint Research Centre); Roudier, S.; Kourti, I.; Sancho, D.; Gonzalo, M.; Suhr, M.; Santonja, G.; Klein, G. Best Available Techniques (BAT) Reference Document for the Production of Pulp, Paper and Board. Available online: https://op.europa.eu/en/publication-detail/-/publication/fcaab1a5-d287-40af-b21c-115e529685fc/language-en (accessed on 21 September 2023).
7. Kumar, A.; Singh, A.K.; Bilal, M.; Prasad, S.; Rameshwari, K.R.T.; Chandra, R. Paper and Pulp Mill Wastewater: Characterization, Microbial-Mediated Degradation, and Challenges. In *Nanotechnology in Paper and Wood Engineering*; Elsevier: Amsterdam, The Netherlands, 2022; pp. 371–387. [CrossRef]
8. Lindholm-Lehto, P.C.; Knuutinen, J.S.; Ahkola, H.S.; Herve, S.H. Refractory Organic Pollutants and Toxicity in Pulp and Paper Mill Wastewaters. *Environ. Sci. Pollut. Res.* **2015**, *22*, 6473–6499. [CrossRef] [PubMed]
9. Show, K.-Y.; Guo, X. *Industrial Waste*; InTech: London, UK, 2012; ISBN 9535102532.
10. De Azevedo, A.R.G.; Alexandre, J.; Pessanha, L.S.; Manhães, R.d.; de Brito, J.; Marvila, M.T. Characterizing the Paper Industry Sludge for Environmentally-Safe Disposal. *Waste Manag.* **2019**, *95*, 43–52. [CrossRef]
11. Xu, J.; Liao, Y.; Yu, Z.; Cai, Z.; Ma, X.; Dai, M.; Fang, S. Co-Combustion of Paper Sludge in a 750 t/D Waste Incinerator and Effect of Sludge Moisture Content: A Simulation Study. *Fuel* **2018**, *217*, 617–625. [CrossRef]
12. Priadi, C.; Wulandari, D.; Rahmatika, I.; Moersidik, S.S. Biogas Production in the Anaerobic Digestion of Paper Sludge. *APCBEE Procedia* **2014**, *9*, 65–69. [CrossRef]
13. Hagelqvist, A.; Granström, K. Co-Digestion of Manure with Grass Silage and Pulp and Paper Mill Sludge Using Nutrient Additions. *Environ. Technol.* **2016**, *37*, 2113–2123. [CrossRef]
14. STN EN 12457-4, 2006 Waste Characterization. Extraction. Verification Test for Leaching Granular Waste Materials and Sludge. Available online: https://eshop.normservis.sk/norma/stnen-12457-4-1.4.2006.html (accessed on 21 September 2023).
15. OECD. Test No. 311: Anaerobic Biodegradability of Organic Compounds in Digested Sludge: By Measurement of Gas Production. Available online: https://www.oecd-ilibrary.org/environment/test-no-311-anaerobic-biodegradability-of-organic-compounds-in-digested-sludge-by-measurement-of-gas-production_9789264016842-en (accessed on 21 September 2023).
16. OECD. Test No. 301: Ready Biodegradability. Available online: https://www.oecd-ilibrary.org/environment/test-no-301-ready-biodegradability_9789264070349-en (accessed on 21 September 2023).
17. ISO-18763-2016. Available online: https://cdn.standards.iteh.ai/samples/63317/309b04cbc836409292017175f5039b36/ISO-18763-2016.pdf (accessed on 21 September 2023).
18. OECD. Test No. 302B: Inherent Biodegradability: Zahn-Wellens/EVPA Test. Available online: https://www.oecd-ilibrary.org/environment/test-no-302b-inherent-biodegradability-zahn-wellens-evpa-test_9789264070387-en (accessed on 21 September 2023).
19. Plíva, P.; Banout, J.; Harant, J.; Jelínek, A.; Kollárová, M.; Roy, A.; Tomanova, D. *Zakládání, Průběh a Řízení Kompostovacího Procesu*; Výzkumný ústav zemědělské techniky: Praha, Czech Republic, 2006; ISBN 80-86884-11-2.
20. Rossetti, I.; Conte, F.; Ramis, G. Kinetic Modelling of Biodegradability Data of Commercial Polymers Obtained under Aerobic Composting Conditions. *Eng—Adv. Eng.* **2021**, *2*, 54–68. [CrossRef]
21. Chynoweth, D.P.; Turick, C.E.; Owens, J.M.; Jerger, D.E.; Peck, M.W. Biochemical Methane Potential of Biomass and Waste Feedstocks. *Biomass Bioenergy* **1993**, *5*, 95–111. [CrossRef]
22. Zhu, A.; Qin, Y.; Wu, J.; Ye, M.; Li, Y.-Y. Characterization of Biogas Production and Microbial Community in Thermophilic Anaerobic Co-Digestion of Sewage Sludge and Paper Waste. *Bioresour. Technol.* **2021**, *337*, 125371. [CrossRef]

Disclaimer/Publisher's Note: The statements, opinions and data contained in all publications are solely those of the individual author(s) and contributor(s) and not of MDPI and/or the editor(s). MDPI and/or the editor(s) disclaim responsibility for any injury to people or property resulting from any ideas, methods, instructions or products referred to in the content.

Proceeding Paper

Issues of Non-Steroidal Anti-Inflammatory Drugs in Aquatic Environments: A Review Study [†]

Karla Placova *, Jan Halfar, Katerina Brozova and Silvie Heviankova

Faculty of Mining and Geology, VŠB–Technical University of Ostrava, 17. listopadu 2172/15, 708 00 Ostrava, Czech Republic; jan.halfar@vsb.cz (J.H.); katerina.brozova@vsb.cz (K.B.); silvie.heviankova@vsb.cz (S.H.)
* Correspondence: karla.placova@vsb.cz
[†] Presented at the 4th International Conference on Advances in Environmental Engineering, Ostrava, Czech Republic, 20–22 November 2023.

Abstract: The quality of wastewater greatly affects the aquatic environment. Currently, a significant amount of emerging pollutants are entering wastewater in the form of pharmaceutical contaminants, industrial chemicals, pesticides, toxins, etc. However, conventional wastewater treatment processes at WWTPs are not effective enough for these emerging pollutants. Therefore, emerging pollutants represent a significant source of wastewater pollution and associated pollution of surface water and even drinking water. The main sources of pharmaceutical contaminants are analgesics and anti-inflammatory drugs, and of these, the most common in wastewater are non-steroidal anti-inflammatory drugs (NSAIDs). The aim is to provide a comprehensive review of the available information on NSAIDs in the aquatic environment, i.e., their occurrence, effects on the environment, formation of main metabolites, and methods of NSAID removal, with a focus on current trends and possible directions for future research.

Keywords: non-steroidal anti-inflammatory drugs; aquatic environment; metabolites; ibuprofen; diclofenac; naproxen; ketoprofen

1. Introduction

The issue of water quality in the aquatic environment (referring to wastewater, surface waters, and groundwater) is a widely discussed topic worldwide. These water sources commonly contain nitrates, heavy metals (e.g., Pb, Cd, Ni), persistent organic pollutants (such as PAHs, PCBs, C_{10-40}), etc. [1–5]. There is an European Pollutant Release and Transfer Register (E-PRTR), which includes a list of such substances totaling nearly one hundred. This register is established based on the regulations of the European Parliament and the Council (EU) and the implementing decisions of the Commission (EU). Its purpose is to control the quality of the environment [6].

Currently, in addition to the common pollution in the aquatic environment, there are so-called emerging contaminants. Emerging contaminants include personal care products, industrial chemicals, pharmaceutical residues, steroid hormones, endocrine-disrupting chemicals, microplastics, and others [7–9]. These emerging contaminants are not part of the E-PRTR, and their occurrence is not regulated by legislative measures. However, these contaminants have been repeatedly found in water sources. Their impact on the environment is not sufficiently known. Therefore, water quality experts are devoting significant efforts to researching the occurrence, adverse effects, and possible methods of removing emerging contaminants from the aquatic environment [8,9].

Among the frequently monitored emerging contaminants are pharmaceutical residues in the form of pharmaceuticals and personal care products (PPCPs) (e.g., carbamazepine, gabapentin, triclosan, furosemide, warfarin), nonsteroidal anti-inflammatory drugs (NSAIDs) (e.g., ibuprofen, diclofenac, naproxen, ketoprofen, aspirin), medications affecting the central

nervous system (e.g., paracetamol, tramadol, metoprolol, oxazepam, diazepam, caffeine), macrolide antibiotics (e.g., erythromycin, clarithromycin, azithromycin), and hormones (e.g., estrone, 17-β-estradiol, 17-α-ethinylestradiol) [10]. At the same time, NSAIDs are among the most frequently occurring contaminants in the aquatic environment [11,12]. This is mainly due to their high consumption, which ranks them among the most consumed medications [13]. The aim of this overview is to provide a comprehensive summary of available information on the issue of NSAIDs (non-steroidal anti-inflammatory drugs) in the aquatic environment, including their occurrence, environmental effects, formation of major metabolites, and methods of NSAID removal, focusing on current trends and potential directions for future research.

2. Selection and Characterization of NSAIDs

NSAIDs, with their effects, belong to analgesics (pain relievers), antipyretics (reduce fever), and anti-inflammatory agents. They are used for colds, flu, and arthritis; to alleviate headaches, toothaches, and muscle pain; during menstruation; and for sprains or strains. They act on the central nervous system similarly to steroids, but without their adverse effects. NSAIDs are typically available over the counter or prescribed by a doctor [11–15]. In the environment, they have been found not only in the aquatic environment but also in soil, sediments, and even in drinking water. Their typical concentrations range from ng/L to µg/L. Despite these low concentrations, they can have toxic effects on living organisms based on their high bioactivity [14,15].

Among the representatives of NSAIDs most commonly found in the environment, there are diclofenac, ibuprofen, naproxen, and ketoprofen [12,13,15–17]. Diclofenac (DICL) is a widely used analgesic for both human and veterinary purposes worldwide. It is used to alleviate all kinds of pain, and its consumption amounts to several thousand tons per year. Diclofenac can accumulate in various environmental components due to its low biodegradability. In addition to being found in water bodies, it has been detected in soil, sediments, and even in drinking water [11]. In 2015, DICL was included in the list of monitored substances based on the Commission Implementing Decision (EU), but it was subsequently removed from the list in 2018 [18]. However, studies, e.g., [12,13,19], have reported its presence in the aquatic environment at concentrations higher than 0.1 µg/L (the proposed threshold by the European Framework Directive for Water) [20]. Therefore, it is still important to gather information not only about its occurrence but also about its potential effects on the environment. Ketoprofen (KET) is utilized for its analgesic and antipyretic effects, while ibuprofen (IBU) and naproxen (NAP) are also used for their anti-inflammatory effects (IBU has short-term effects, while NAP has long-term effects) [11]. Similar to DICL, their global annual consumption is estimated to be several thousand tons. These medications can enter the aquatic environment through industrial and domestic wastewater. These NSAIDs can negatively impact water quality and have chronic effects on living organisms. Therefore, it is essential to investigate the fate of NSAIDs in the aquatic environment [21].

3. Occurrence of NSAIDs in the Aquatic Environment

NSAIDs are usually over-the-counter drugs, and, therefore, they are easily accessible. These medications belong to the group of drugs with high daily dosages, and after passing through the human body, these NSAIDs become a significant source of pollution in the wastewater system (in their unchanged form or as metabolites). Another contribution to wastewater pollution by NSAIDs can come from residues from their production, hospital waste, or agriculture. Through this contamination, wastewater can become contaminated, leading to the pollution of surface waters and, occasionally, underground and drinking water as well. In other words, the production, consumption, disposal, and application of NSAIDs have an impact on overall water pollution [22].

From the data presented in the review study [22], Figure 1 was created by the authors, illustrating the extent of NSAID wastewater pollution worldwide. In South America and

Asia, lower concentrations of NSAIDs (hundreds of ng/L) were observed, while significantly higher concentrations (thousands of ng/L or μg/L) were found in Australia, Africa, North America, and Europe. This varying occurrence of NSAIDs in wastewater can be influenced by population density, the number of hospital facilities connected to the sewage system, the pharmaceutical industry, agriculture, aquaculture, the treatment processes of specific wastewater treatment plants (WWTPs), or relevant legislative regulations [22].

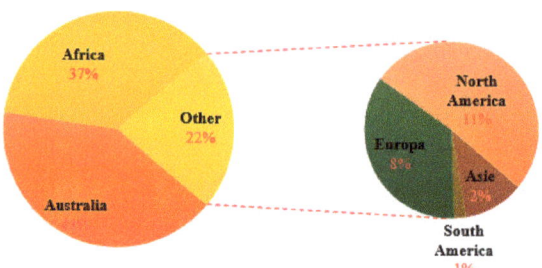

Figure 1. Level of global pollution of wastewater by NSAIDs.

Table 1 presents the results of global studies focusing on the presence of NSAIDs in aquatic environments. Frequently observed concentrations of NSAIDs range from 0.7 to 420 ng/L in Asia [11,23], 9 to 328 ng/L in South America [24,25], 23 to 1830 ng/L in Europe [11,13,18], 5 to 4880 ng/L in North America [23,26], 147 to 9585 ng/L in Africa [11,12,19], and 80 to 11,165 ng/L in Australia [27,28] in aquatic environments. In the case of monitoring lower concentrations, the study focused on surface water, groundwater, or drinking water [11,24,27], whereas inflows of municipal wastewater exhibited higher concentrations of NSAIDs [18,19]. This is also evidenced, for example, by studies on wastewater [23] and drinking water [11] in China or studies on effluent [19] and influent [12] from WWTPs in Algeria, where there is an order of magnitude difference in NSAID concentrations.

Table 1. Occurrence of NSAIDs in aquatic environments worldwide.

	Where/When	Concentration NSAID [ng/L]				Ref.
		IBU	DICL	NAP	KET	
Europe	Spain/2016, 2018	- 179	344 65	96 -	106 357	[13] [11]
	Greece/2022	-	119–620	40	23	[18]
	Finland/2016	1830	-	1687	-	[11]
Asia	China/2005, 2019	420 0.7	- 40	30 5	- 2	[23] [11]
	Korea/2009	414	-	-	-	[23]
Africa	South Africa/2016	3870–8500	-	<1500–5340	1200–9220	[19]
	Algeria/2014 Algeria/2016	340–430 1608	1615–2710 2319	333 9585	1035 -	[12] [19]
	Morocco/2022	274	147	197	198	[11]
North America	Canada/2006	5–8	-	-	-	[23]
	Mexico/2018	231–1106	283–1209	33–4880	-	[26]
South America	Columbia/2017	9–32	-	-	-	[24]
	Brazil/2020	<100–320	<100–328	-	-	[27]

Table 1. Cont.

Where/When		Concentration NSAID [ng/L]				Ref.
		IBU	DICL	NAP	KET	
Australia	Sydney/2020	80–150	-	-	-	[27]
	New Zealand/2019	3149–11,165	142–245	414–7976	-	[28]

4. Effects of NSAIDs on the Environment

NSAIDs occur in trace amounts in environmental compartments (water, soil, sediments). While their primary purpose is the treatment of diseases and the recovery of humans and animals, their presence in the environment can cause adverse effects on non-target organisms [29–31]. It has been demonstrated that NSAID representatives pose a health risk to birds, freshwater fish, mollusks, plants, algae, and bacteria due to their effects [20,29–36]. Another issue is the potential formation of metabolites from the original drug form. For example, it has been proven that IBU and DICL metabolites can occur in higher concentrations in water environments. Such metabolites can be more persistent and pose a more severe risk to non-target organisms [37,38].

The specific effects of IBU include toxicity to phytoplankton [31], algae, and bacteria [36]; kidney and gill damage in freshwater fish [31,35]; effects on bone development, aerobic respiration, and immune functions in freshwater fish [35]; chronic effects on aquatic organisms (reduced sperm motility and hatching) [29]; impact on the immune system; genotoxic effects [33]; and induction of morphological and ultrastructural changes in algal cells [35]. For DICL, the proven negative effects include extreme toxicity to vultures [30]; toxicity to phytoplankton [31], river biofilm communities [29], and broilers [35]; damage to the kidneys, liver, gills, testes, brain, and DNA in fish [20,31]; cardiac anomalies and cardiovascular defects in freshwater fish [34]; chronic effects on fish (reduced hatching, delayed hatching, reduced growth in the egg stage); cytological changes and tissue damage in freshwater fish [29]; and impact on the immune system and genotoxic effects [33]. Similarly to IBU, NAP also acts toxically on phytoplankton [31] and causes damage to the kidneys, liver, and gills in freshwater fish [20,31], additionally affecting the growth and photosynthetic processes of plants [32]. KET, like IBU and DICL, causes cardiac anomalies and cardiovascular defects in freshwater fish, affects the immune system, and has genotoxic effects [33,34,36].

From the studies conducted so far, it can be said that IBU and DICL pose a high eco-toxicological risk, while NAP and KET pose a moderate eco-toxicological risk to non-target organisms [36].

5. Main Metabolites of NSAIDs

During metabolism in living organisms, IBU undergoes oxidation and subsequent conjugation with glucuronic acid. The original form of IBU is transformed into two main metabolites: hydroxy-ibuprofen (OH-IBU) and carboxy-ibuprofen (CA-IBU). Approximately 15% of IBU is excreted unchanged after the metabolic process, while CA-IBU accounts for 43% and OH-IBU for 26% of the total IBU ingested [39]. In the case of DICL, metabolism involves hydroxylation of the methyl group, resulting in the formation of 4'-hydroxydiclofenac (4'OH-DICL). Only 6% of the total amount of DICL ingested is excreted in its original form, while 16% is excreted as 4'OH-DICL [40]. NAP is mainly metabolized into naproxen glucuronide and O-desmethyl-naproxen [11], and KET is transformed into ketoprofen glucuronide through metabolism [41].

6. Methods of Removing NSAIDs from the Aquatic Environment

In the EU, the regular use of around three thousand drugs, including NSAIDs, is estimated. These substances end up in wastewater, and the current technologies for their treatment are not sufficiently effective because the original design of WWTPs did not consider them. As a result, surface waters and, in rare cases, even groundwater and drinking water become contaminated, i.e., the aquatic environment. Therefore, it is necessary to also study the possibilities of an additional treatment stage in WWTPs. This issue has been addressed in studies [13,42–44], where effective methods for treating NSAIDs have been confirmed, such as adsorption processes, membrane separations, advanced oxidation processes, or biodegradation/biotransformation. The adsorption technique is highly versatile, flexible, simple, fast, easily operable, and does not result in sludge formation, unlike membrane separation [45]. The key to adsorption is the selection of a suitable adsorbent, which can be activated carbon [42,43], sewage sludge [13,45], or agricultural/industrial/domestic waste [46]. On the other hand, membrane separation utilizes membranes ((non)porous, with various shapes, made of polymeric, ceramic, or mixed matrices, charged/neutral, etc.). Based on the pore size in membranes, we distinguish reverse osmosis, nanofiltration [13], microfiltration, and ultrafiltration [45]. As the pore size decreases, the efficiency of contaminant elimination increases, but it also leads to higher operational and financial demands [13]. Advanced oxidation processes include chemical [44], electrochemical, or photochemical [13,44,47] methods. However, advanced oxidation processes often result in the formation of intermediate products that are more toxic than the original micropollutants. To improve efficiency, combined techniques are utilized, such as combining advanced oxidation processes with membrane separation or biological processes [48]. The method of biodegradation/biotransformation is based solely on biological processes. In this method, plants assist in the elimination of contaminants [13,42] through the process of microbial degradation, which occurs through adsorption, absorption, and the process of metabolism.

The conducted studies on the elimination of NSAIDs, along with the effectiveness of the selected elimination process, are presented in Table 2. In the case of IBU elimination, high efficiency was achieved through adsorption [43,46], membrane separation (nanofiltration) [13], biodegradation/biotransformation [42], or with the help of combined techniques (adsorption + ozonation [29], adsorption + solid-phase extraction [49], adsorption + membrane separation [44]). Similar results were obtained for DICL and NAP. Successful solutions were also found with the use of a membrane bioreactor, which achieved an elimination efficiency of >90% for IBU, DICL, NAP, and KET [44].

Table 2. Implemented elimination processes of individual NSAIDs and their percentage effectiveness.

PROCESS	IBU	DICL	NAP	KET
ADSORPTION (activated carbon, biochar, or activated sludge)	90–96% [43,46]	89–95% [43]	88–94% [13,42]	-
MEMBRANE SEPARATION	86–99% (nanofiltration) [13]	90–99% (nanofiltration) [45]	95% (reverse osmosis) [13]	-
ADVANCED OXIDATION PROCESSES (AOP)	81% [47]	85–87 [13,47]	83–90% [44]	-
BIODEGRADATION/BIOTRANSFORMATION	90–99% [42]	90–99% [13]	68.6–90% [42]	-
ADSORPTION (activated carbon) + ozonation	>93% [29]	-	-	-
ADSORPTION (magnetic activated carbon) + magnetic solid-phase extraction + UV radiation	-	94% [49]	82% [49]	93% [49]
ADSORPTION and solid-phase microextraction	77–96% [49]	-	-	77–96% [49]
ADSORPTION (metal-organic frameworks) + magnetic solid-phase extraction	84–110% [49]	-	-	-
ADSORPTION (activated sludge) + MEMBRANE SEPARATION, the so-called membrane bioreactor	95% [44]	95% [44]	98% [44]	90% [44]

7. Future Research

This overview is a comprehensive summary of the available information on the issue of NSAIDs in the aquatic environment. It evaluates the occurrence of NSAIDs, their effects on the environment, metabolic byproducts, and potential elimination from wastewater. The main gaps regarding the issue of NSAIDs in the water environment include the following:

- Unspecified requirements for potential tertiary treatment of wastewater from WWTPs;
- Non-existent threshold concentrations of NSAIDs in aquatic environments;
- Lack of information on the occurrence of metabolic byproducts of NSAIDs.

Future research should, therefore, focus on exploring possibilities for improving wastewater treatment using affordable, effective, and sustainable technologies. Additionally, studying the long-term effects of NSAIDs on ecosystems and human health is crucial in order to establish permissible concentrations of NSAIDs in the aquatic environment that do not burden the environment. It is also necessary to investigate the presence of metabolic byproducts from the original form of NSAIDs and their potential adverse effects on the environment.

Author Contributions: K.P. (analysis of the issue, writing—original draft preparation, and final version); J.H. (writing—original draft preparation, resources); K.B. (visualization, writing—original draft preparation); S.H. (supervision). All authors have read and agreed to the published version of the manuscript.

Funding: This research was funded by the Student Grant Competition financed by the VSB-Technical University of Ostrava within the project "Research and detection of micropollutants in drinking and waste water from sewage treatment plants." (no. SP2023/045) and with the financial support of the European Union under the REFRESH—Research Excellence For REgion Sustainability and High-tech Industries project number CZ.10.03.01/00/22_003/0000048 via the Operational Programme Just Transition.

Institutional Review Board Statement: Not applicable.

Informed Consent Statement: Not applicable.

Data Availability Statement: Data are contained within the article.

Conflicts of Interest: The authors declare no conflict of interest.

References

1. Foltová, P.; Heviankova, S.; Chromikova, J.; Kasparkova, A. Increased concentrations of nitrates in groundwater in selected localities of the Moravian–Silesian and Olo-mouc Regions in the Czech Republic. *Inz. Miner.* **2018**, *2018*, 35–38.
2. Fei, Y.; Hu, Y.H. Recent progress in removal of heavy metals from wastewater: A comprehensive review. *Chemosphere* **2023**, *335*, 139077. [CrossRef]
3. Kotalova, I.; Calabkova, K.; Drabinova, S.; Heviankova, S. Contribution to the study of selected heavy metals in urban wastewaters using ICP-MS method. *IOP Conf. Ser. Earth Environ. Sci.* **2020**, *444*, 012028. [CrossRef]
4. Vojtková, H.; Maslanová, I.; Sedlácek, I.; Svanová, P.; Janulková, R. Removal of heavy metals from wastewater by a *Rhodococcus* sp. Bacterial strain. In Proceedings of the 12th International Multidisciplinary Scientific Geoconference & Expo, Albena, Bulgaria, 17–23 June 2012.
5. Rosińska, A. The influence of UV irradiation on PAHs in wastewater. *J. Environ. Manag.* **2021**, *293*, 112760. [CrossRef] [PubMed]
6. European Pollutant Releases and Transfer Register—E-PRTR. Integrovaný Registr Znečišťování Životního Prostředí [Online]. 2021. Available online: www.irz.cz/registry-znecistovani/evropsky-registr-uniku-a-prenosu-znecistujicich-latek-e-prtr (accessed on 30 July 2023).
7. Chen, Y.; Lin, M.; Zhuang, D. Wastewater treatment and emerging contaminants: Bibliometric analysis. *Chemosphere* **2022**, *297*, 133932. [CrossRef] [PubMed]
8. Kumar, R.; Qureshi, M.; Vishwakarma, D.K.; Al-Ansari, N.; Kuriqi, A.; Elbeltagi, A.; Saraswat, A. A review on emerging water contaminants and the application of sustainable removal technologies. *Case Stud. Chem. Environ. Eng.* **2022**, *6*, 100219. [CrossRef]
9. Halfar, J.; Brožová, K.; Čabanová, K.; Heviánková, S.; Kašpárková, A.; Olšovská, E. Disparities in Methods Used to Determine Microplastics in the Aquatic Environment: A Review of Legislation, Sampling Process and Instrumental Analysis. *Int. J. Environ. Res. Public Health* **2021**, *18*, 7608. [CrossRef] [PubMed]
10. Morin-Crini, N.; Lichtfouse, E.; Liu, G.; Balaram, V.; Ribeiro, A.R.L.; Lu, Z.; Stock, F.; Carmona, E.; Teixeira, M.R.; Picos-Corrales, L.A.; et al. Worldwide cases of water pollution by emerging contaminants: A review. *Environ. Chem. Lett.* **2022**, *20*, 2311–2338. [CrossRef]
11. Hawash, H.B.; Moneer, A.A.; Galhoum, A.A.; Elgarahy, A.M.; Mohamed, W.A.; Samy, M.; El-Seedi, H.R.; Gaballah, M.S.; Mubarak, M.F.; Attia, N.F. Occurrence and spatial distribution of pharmaceuticals and personal care products (PPCPs) in the aquatic environment, their characteristics, and adopted legislations. *J. Water Process. Eng.* **2023**, *52*, 103490. [CrossRef]

12. Kermia, A.E.B.; Fouial-Djebbar, D.; Trari, M. Occurrence, fate and removal efficiencies of pharmaceuticals in wastewater treatment plants (WWTPs) discharging in the coastal environment of Algiers. *Comptes Rendus Chim.* **2016**, *19*, 963–970. [CrossRef]
13. Sapingi, M.S.M.; Khasawneh, O.F.S.; Palaniandy, P.; Aziz, H.A. Analytical techniques for the detection of pharmaceuticals in the environment. In *The Treatment of Pharmaceutical Wastewater*; Elsevier: Amsterdam, The Netherlands, 2023; pp. 149–177, ISBN 9780323991605. [CrossRef]
14. Samal, K.; Mahapatra, S.; Ali, H. Pharmaceutical wastewater as Emerging Contaminants (EC): Treatment technologies, impact on environment and human health. *Energy Nexus* **2022**, *6*, 100076. [CrossRef]
15. Bankole, D.T.; Oluyori, A.P.; Inyinbor, A.A. The removal of pharmaceutical pollutants from aqueous solution by Agro-waste. *Arab. J. Chem.* **2023**, *16*, 104699. [CrossRef]
16. Oluwalana, A.E.; Musvuugwa, T.; Sikwila, S.T.; Sefadi, J.S.; Whata, A.; Nindi, M.M.; Chaukura, N. The screening of emerging micropollutants in wastewater in Sol Plaatje Municipality, Northern Cape, South Africa. *Environ. Pollut.* **2022**, *314*, 120275. [CrossRef] [PubMed]
17. Dubey, M.; Vellanki, B.P.; Kazmi, A.A. Fate of emerging contaminants in a sequencing batch reactor and potential of biological activated carbon as tertiary treatment for the removal of persisting contaminants. *J. Environ. Manag.* **2023**, *338*, 117802. [CrossRef] [PubMed]
18. Ofrydopoulou, A.; Nannou, C.; Evgenidou, E.; Christodoulou, A.; Lambropoulou, D. Assessment of a wide array of organic micropollutants of emerging concern in wastewater treatment plants in Greece: Occurrence, removals, mass loading and potential risks. *Sci. Total Environ.* **2022**, *802*, 149860. [CrossRef] [PubMed]
19. Waleng, N.J.; Nomngongo, P.N. Occurrence of pharmaceuticals in the environmental waters: African and Asian perspectives. *Environ. Chem. Ecotoxicol.* **2022**, *4*, 50–66. [CrossRef]
20. Näslund, J.; Asker, N.; Fick, J.; Larsson, D.J.; Norrgren, L. Naproxen affects multiple organs in fish but is still an environmentally better alternative to diclofenac. *Aquat. Toxicol.* **2020**, *227*, 105583. [CrossRef]
21. Letsoalo, M.R.; Sithole, T.; Mufamadi, S.; Mazhandu, Z.; Sillanpaa, M.; Kaushik, A.; Mashifana, T. Efficient detection and treatment of pharmaceutical contaminants to produce clean water for better health and environmental. *J. Clean. Prod.* **2023**, *387*, 135798. [CrossRef]
22. Adeleye, A.S.; Xue, J.; Zhao, Y.; Taylor, A.A.; Zenobio, J.E.; Sun, Y.; Han, Z.; Salawu, O.A.; Zhu, Y. Abundance, fate, and effects of pharmaceuticals and personal care products in aquatic environments. *J. Hazard. Mater.* **2022**, *424*, 127284. [CrossRef]
23. Kumar, M.; Sridharan, S.; Sawarkar, A.D.; Shakeel, A.; Anerao, P.; Mannina, G.; Sharma, P.; Pandey, A. Current research trends on emerging contaminants pharmaceutical and personal care products (PPCPs): A comprehensive review. *Sci. Total Environ.* **2023**, *859*, 160031. [CrossRef]
24. Aristizabal-Ciro, C.; Botero-Coy, A.M.; López, F.J.; Peñuela, G.A. Monitoring pharmaceuticals and personal care products in reservoir water used for drinking water supply. *Environ. Sci. Pollut. Res.* **2017**, *24*, 7335–7347. [CrossRef] [PubMed]
25. Chaves, M.d.J.S.; Barbosa, S.C.; Malinowski, M.d.M.; Volpato, D.; Castro, B.; Franco, T.C.R.d.S.; Primel, E.G. Pharmaceuticals and personal care products in a Brazilian wetland of international importance: Occurrence and environmental risk assessment. *Sci. Total Environ.* **2020**, *734*, 139374. [CrossRef] [PubMed]
26. Rivera-Jaimes, J.A.; Postigo, C.; Melgoza-Aleman, R.M.; Acena, J.; Barcelo, D.; Lopez de Alda, M. Study of pharmaceuticals in surface and wastewater from Cuernavaca, Morelos, Mexico: Occurrence and environmental risk assessment. *Sci. Total Environ.* **2018**, *613–614*, 1263–1274. [CrossRef] [PubMed]
27. McKenzie, T.; Holloway, C.; Dulai, H.; Tucker, J.P.; Sugimoto, R.; Nakajima, T.; Harada, K.; Santos, I.R. Submarine groundwater discharge: A previously undocumented source of contaminants of emerging concern to the coastal ocean (Sydney, Australia). *Mar. Pollut. Bull.* **2020**, *160*, 111519. [CrossRef]
28. Kumar, R.; Sarmah, A.K.; Padhye, L.P. Fate of pharmaceuticals and personal care products in a wastewater treatment plant with parallel secondary wastewater treatment train. *J. Environ. Manag.* **2019**, *233*, 649–659. [CrossRef]
29. Lonappan, L.; Brar, S.K.; Das, R.K.; Verma, M.; Surampalli, R.Y. Diclofenac and its transformation products: Environmental occurrence and toxicity—A review. *Environ. Int.* **2016**, *96*, 127–138. [CrossRef]
30. Okoye, C.O.; Okeke, E.S.; Okoye, K.C.; Echude, D.; Andong, F.A.; Chukwudozie, K.I.; Okoye, H.U.; Ezeonyejiaku, C.D. Occurrence and fate of pharmaceuticals, personal care products (PPCPs) and pesticides in African water systems: A need for timely intervention. *Heliyon* **2022**, *8*, e09143. [CrossRef]
31. Mohan, H.; Rajput, S.S.; Jadhav, E.B.; Sankhla, M.S.; Sonone, S.S.; Jadhav, S.; Kumar, R. Ecotoxicity, Occurrence, and Removal of Pharmaceuticals and Illicit Drugs from Aquatic Systems. *Biointerface Res. Appl. Chem.* **2021**, *11*, 12530–12546. [CrossRef]
32. Zezulka, Š.; Kummerová, M.; Šmeringai, J.; Babula, P.; Tříska, J. Ambiguous changes in photosynthetic parameters of Lemna minor L. after short-term exposure to naproxen and paracetamol: Can the risk be ignored? *Aquat. Toxicol.* **2023**, *259*, 106537. [CrossRef]
33. Mezzelani, M.; Gorbi, S.; Fattorini, D.; D'errico, G.; Consolandi, G.; Milan, M.; Bargelloni, L.; Regoli, F. Long-term exposure of Mytilus galloprovincialis to diclofenac, Ibuprofen and Ketoprofen: Insights into bioavailability, biomarkers and transcriptomic changes. *Chemosphere* **2018**, *198*, 238–248. [CrossRef]
34. Selderslaghs, I.W.; Blust, R.; Witters, H.E. Feasibility study of the zebrafish assay as an alternative method to screen for developmental toxicity and embryotoxicity using a training set of 27 compounds. *Reprod. Toxicol.* **2012**, *33*, 142–154. [CrossRef]

35. Jiang, L.; Li, Y.; Chen, Y.; Yao, B.; Chen, X.; Yu, Y.; Yang, J.; Zhou, Y. Pharmaceuticals and personal care products (PPCPs) in the aquatic environment: Biotoxicity, determination and electrochemical treatment. *J. Clean. Prod.* **2023**, *388*, 135923. [CrossRef]
36. Wojcieszyńska, D.; Guzik, H.; Guzik, U. Non-steroidal anti-inflammatory drugs in the era of the Covid-19 pandemic in the context of the human and the environment. *Sci. Total Environ.* **2022**, *834*, 155317. [CrossRef]
37. La Farré, M.; Pérez, S.; Kantiani, L.; Barceló, D. Fate and toxicity of emerging pollutants, their metabolites and transformation products in the aquatic environment. *TrAC Trends Anal. Chem.* **2008**, *27*, 991–1007. [CrossRef]
38. Leclercq, M.; Mathieu, O.; Gomez, E.; Casellas, C.; Fenet, H.; Hillaire-Buys, D. Presence and Fate of Carbamazepine, Oxcarbazepine, and Seven of Their Metabolites at Wastewater Treatment Plants. *Arch. Environ. Contam. Toxicol.* **2009**, *56*, 408–415. [CrossRef] [PubMed]
39. Tan, S.; Jackson, S.; Swift, C.; Hutt, A. Stereospecific analysis of the major metabolites of ibuprofen in urine by sequential achiral-chiral high-performance liquid chromatography. *J. Chromatogr. B Biomed. Sci. Appl.* **1997**, *701*, 53–63. [CrossRef] [PubMed]
40. Stülten, D.; Zühlke, S.; Lamshöft, M.; Spiteller, M. Occurrence of diclofenac and selected metabolites in sewage effluents. *Sci. Total Environ.* **2008**, *405*, 310–316. [CrossRef]
41. Drover, V.J.; Bottaro, C.S. Determination of pharmaceuticals in drinking water by CD-modified MEKC: Separation optimization using experimental design. *J. Sep. Sci.* **2008**, *31*, 3740–3748. [CrossRef]
42. AL Falahi, O.A.; Abdullah, S.R.S.; Abu Hasan, H.; Othman, A.R.; Ewadh, H.M.; Kurniawan, S.B.; Imron, M.F. Occurrence of pharmaceuticals and personal care products in domestic wastewater, available treatment technologies, and potential treatment using constructed wetland: A review. *Process Saf. Environ. Prot.* **2022**, *168*, 1067–1088. [CrossRef]
43. Liu, T.; Aniagor, C.O.; Ejimofor, M.I.; Menkiti, M.C.; Tang, K.H.D.; Chin, B.L.F.; Chan, Y.H.; Yiin, C.L.; Cheah, K.W.; Chai, Y.H.; et al. Technologies for removing pharmaceuticals and personal care products (PPCPs) from aqueous solutions: Recent advances, performances, challenges and recommendations for improvements. *J. Mol. Liq.* **2023**, *374*, 121144. [CrossRef]
44. Hossein, M.; Asha, R.; Bakari, R.; Islam, N.F.; Jiang, G.; Sarma, H. Exploring eco-friendly approaches for mitigating pharmaceutical and personal care products in aquatic ecosystems: A sustainability assessment. *Chemosphere* **2023**, *316*, 137715. [CrossRef] [PubMed]
45. Osuoha, J.O.; Anyanwu, B.O.; Ejileugha, C. Pharmaceuticals and personal care products as emerging contaminants: Need for combined treatment strategy. *J. Hazard. Mater. Adv.* **2023**, *9*, 100206. [CrossRef]
46. Ayati, A.; Tanhaei, B.; Beiki, H.; Krivoshapkin, P.; Krivoshapkina, E.; Tracey, C. Insight into the adsorptive removal of ibuprofen using porous carbonaceous materials: A review. *Chemosphere* **2023**, *323*, 138241. [CrossRef] [PubMed]
47. Javaid, A.; Latif, S.; Imran, M.; Hussain, N.; Rajoka, M.S.R.; Iqbal, H.M.; Bilal, M. Nanohybrids-assisted photocatalytic removal of pharmaceutical pollutants to abate their toxicological effects—A review. *Chemosphere* **2022**, *291*, 133056. [CrossRef]
48. Alessandretti, I.; Rigueto, C.V.T.; Nazari, M.T.; Rosseto, M.; Dettmer, A. Removal of diclofenac from wastewater: A comprehensive review of detection, characteristics and tertiary treatment techniques. *J. Environ. Chem. Eng.* **2021**, *9*, 106743. [CrossRef]
49. Yang, L.; Wang, S.; Xie, Z.; Xing, R.; Wang, R.; Chen, X.; Hu, S. Deep eutectic solvent-loaded Fe3O4@MIL-101(Cr) with core–shell structure for the magnetic solid phase extraction of non-steroidal anti-inflammatory drugs in environmental water samples. *Microchem. J.* **2023**, *184*, 108150. [CrossRef]

Disclaimer/Publisher's Note: The statements, opinions and data contained in all publications are solely those of the individual author(s) and contributor(s) and not of MDPI and/or the editor(s). MDPI and/or the editor(s) disclaim responsibility for any injury to people or property resulting from any ideas, methods, instructions or products referred to in the content.

Proceeding Paper

Analysis of Firewater Samples from Simulated Fires in Illegal Waste Dumps [†]

Radmila Kucerova *, Michal Zavoral, Jaroslav Mudrunka, David Takac and Lucie Marcalikova

Department of Environmental Engineering, Faculty of Mining and Geology, VSB—Technical University of Ostrava, 708 00 Ostrava, Czech Republic; michal.zavoral@gmail.com (M.Z.); jaroslav.mudrunka@vsb.cz (J.M.); takac.dave@gmail.com (D.T.); lucie.kucerova1@vsb.cz (L.M.)
* Correspondence: radmila.kucerova@vsb.cz
[†] Presented at the 4th International Conference on Advances in Environmental Engineering, Ostrava, Czech Republic, 20–22 November 2023.

Abstract: The aim of the work was to simulate a fire in the illegal waste dumps and to find out whether the used firewater represents a potential danger to the environment after its runoff into the surrounding area. Laboratory analysis confirmed the presence of heavy metals and other chemical compounds (nitrates, sulphates and chlorides). Given the results of the above analysis, we can state that the used firewater contains hazardous substances, which confirms the release of these substances into the water during extinguishing of waste-dump material. This thesis does not aim to criticize the tactical procedures of firefighters in extinguishing such fires, as their main goal is to eliminate waste-dump fires and eliminate the further spread of fire.

Keywords: illegal waste dump; fire; extinguishing; environment; danger

Citation: Kucerova, R.; Zavoral, M.; Mudrunka, J.; Takac, D.; Marcalikova, L. Analysis of Firewater Samples from Simulated Fires in Illegal Waste Dumps. *Eng. Proc.* **2023**, *57*, 14. https://doi.org/10.3390/engproc2023057014

Academic Editors: Adriana Estokova, Natalia Junakova, Tomas Dvorsky, Vojtech Vaclavik and Magdalena Balintova

Published: 1 December 2023

Copyright: © 2023 by the authors. Licensee MDPI, Basel, Switzerland. This article is an open access article distributed under the terms and conditions of the Creative Commons Attribution (CC BY) license (https://creativecommons.org/licenses/by/4.0/).

1. Introduction

There is no current legislation to address the issue of hazardous substances contained in firewater, specifically for water used for extinguishing illegal waste dumps. In real conditions (during an ongoing fire) it is unthinkable to measure the content of hazardous substances in the runoff firewater, but no work has been found that would deal with the issue, and therefore (at least to give you an idea) we have the opportunity to find out something more about what remains in the place of illegal waste dumps after their fire.

It is almost impossible to find a specific number of illegal waste dumps in the Czech Republic, because the Ministry of the Environment does not keep any agenda that would record the statistical number of illegal waste dumps [1]. At least approximate numbers can be obtained through various applications that allow citizens to report illegal waste dumps. Three of them are the applications ZmapujTo, TrashOut or ZlepšemeČesko. For example, in the ZmapujTo application, it is possible to find out that from October 2021 to October 2022, 2674 illegal waste dumps were reported throughout the Czech Republic.

The main feature of an illegal waste dump is the accumulation of waste by an unknown producer. Such waste is characterised by its diversity and collection outside the designed areas [2]. Because of its very existence, the illegal waste dump encourages people to pile up more waste in such places, which causes an increase in the volume of the waste dump and thus increases the risk of possible fire and contamination of the surroundings. The most common locations of illegal waste dumps are remote places of the suburban districts.

The risk of illegal waste dumps burning is relatively high, and fires occur across the country during all seasons. One of the possibilities of an illegal dump fire can undoubtedly be the spontaneous ignition of stored materials. The Fire Rescue Service of the Czech Republic does not distinguish between legal and waste dumps, so there are no statistics available to distinguish the number of fires.

2. Materials and Methods

According to the ZmapujTo.cz portal, ten localities across the Ústí and Labem Region were selected, where illegally stored waste was to be found. Field research revealed that waste is actually present at six of these sites. The structure of the waste was the same in several places, so samples were taken from only four illegal waste dumps. A sufficient amount of waste was collected from each site using the necessary tools.

Sample 1 consisted of PET bottles, a LED bulb, sheet metal packaging from synthetic paint, sheet metal packaging of wood glaze, plastic packaging of fertilizer, Tetrapak packaging and insulating polystyrene.

Sample 2 contained material from a car wreck (seat cover, foam seat filling, plastic bottle with coolant, textiles, printed circuit, plastic box with bulbs and fuses).

Sample 3 consisted of cut parts of the tyre. The composition of a tyre generally consists of rubber, steel reinforcement and a modified fabric.

Sample 4 consisted of dirty fabric, Christmas decorations (wood, plastic, polystyrene), tin food packaging, cigarette wrapper, polystyrene filling material, coffee cup (plastic lid, paper cup with polypropylene) and thermal insulation gold foil.

After treatment, the samples separated or deformed by the tools were placed in transportable sealable plastic boxes. All sample boxes were sealed with insulating tape and numbered (including the lid of the box) according to the location for better clarity in further processing. (Figure 1).

Figure 1. Samples.

All material was carefully dried, shaped into the desired, suitable shape (cut with scissors, cut with a knife or deformed) and prepared for the subsequent simulated fire.

The simulation itself was carried out by burning individual samples in the galvanized containers, which were numerically marked according to the individual samples. Pieces of wood were first set on fire in the containers using paper and a gas lighter (the use of alcohol firelighters was not intended to eliminate the substances contained in these firelighters). After sufficient burning of the wood, the samples were gradually placed in the individual containers to burn them (Figure 2). Each sample was left to burn for 10 min, and then extinguishing was carried out. Using one "D flash", the individual simulated fires were extinguished relatively quickly (the diameter of hose D is 25 mm, it is used for extinguishing forest fires, but also for apartment fires). The firewater corresponded to the commonly used water that the fire protection unit takes from the municipal water mains.

Figure 2. Fire simulation.

After extinguishing successfully, the remnants of the individual burnt samples were allowed to cool in containers. Favourable climatic conditions at an outdoor temperature of 3 °C contributed to the fast cooling. The temperature of individual samples was continuously checked by a thermal camera, and after reaching a temperature of 15 °C, the used firewater was poured into prepared boxes to reduce the amount of burnt waste residues contained in the water (Figure 3).

Figure 3. Samples before filtration.

After pouring it into the boxes, the colouration of the firewater was clearly visible in some of the samples, and in the case of the burnt tyre sample, granular parts were clearly visible, which could not be separated by mere spillage, as they were carried away by water.

For the purposes of this experiment, an accredited laboratory was consulted in advance on how to store the firewater. The laboratory recommended taking a simple sample (single and randomly taken sample) in the PET bottles with a minimum capacity of 1 litre. These PET bottles (sample containers) and their bottle caps were rinsed with unused firewater three times before use. To pour the firewater into the prepared sample containers, it was recommended to use a filter fabric to prevent parts of the incinerated material from entering the samples intended for analysis. In our case, FITPOP I filter fabric was used (filter non-woven fabric with base fabric, 100% polypropylene, heat treated input side, grammage 500 g/m^2, thickness 2 mm, breathability 150 l/m^2/s at 200 Pa, heat resistance 90 °C). The sample container was filled to the brim with used firewater, with the total volume of each sample being 1.5 litres (Figure 4).

Figure 4. Samples after filtration.

The samples prepared in this way were transported to the laboratory for analysis. The time from sampling to submission did not exceed 15 h, with all samples stored at 2–4 °C.

In order to work with the results of the analysis, it was necessary to create "sample 0", i.e., unused firewater from the mains, which was transported to the laboratory together with other samples and subjected to analysis.

An accredited laboratory ALS Czech Republic with a branch in Lovosice was selected for the analysis of firewater samples.

3. Results

Table 1 lists all parameters (marked in red) for which higher values than those allowed were measured [3]. At first glance, it is clear that the limits have been exceeded for all the samples analysed. Out of a total of 116 parameters (4 samples × 29 parameters), the limits were exceeded in a total of 35 cases.

Table 1. Limits exceeded in samples.

Parameter	Sample 1	Sample 2	Sample 3	Sample 4	Units	Limit
Physical parameter						
pH value	7.53	5.62	6.67	6.57	-	5–9
Inorganic parameters						
Ammoniacal nitrogen	0.38	4.51	0.831	3.56	mg/L	0.23
Chlorides	131	254	9.8	27.7	mg/L	150
CHSK$_{Cr}$	137	3050	201	1280	mg/L	26
Sulphates e.g., SO_4^{2-}	37	35.6	68.2	133	mg/L	200
Total nitrogen	4.04	19.3	6.79	15.8	mg/L	6
Total metals/major cations						
Ag	<0.0050	<0.0050	<0.0050	<0.0050	mg/L	0.0035
Al	0.739	0.382	0.152	1.98	mg/L	1
As	<0.010	<0.010	<0.010	<0.010	mg/L	0.011
B	0.016	0.035	0.013	0.101	mg/L	0.3
Ba	0.36	0.281	0.0667	0.129	mg/L	0.18
Be	<0.0002	<0.0002	<0.0002	0.00022	mg/L	0.00005
Ca	99.9	73.3	25.4	74.8	mg/L	190
Cd	<0.0020	0.0173	<0.0020	<0.0020	mg/L	0.00015
Co	<0.0020	0.0894	0.0027	0.0049	mg/L	0.003
Cr	<0.0020	0.0109	<0.0020	0.0138	mg/L	0.018
Cu	0.0316	0.0732	0.034	0.208	mg/L	0.014
Fe	0.368	6.53	0.45	3.58	mg/L	1
Hg	<0.010	<0.010	<0.010	<0.010	mg/L	0.07
Mg	3.97	4.07	4.04	17.5	mg/L	120
Mn	0.036	0.175	0.0344	0.424	mg/L	0.3
Mo	<0.0030	<0.0030	<0.0030	0.0053	mg/L	0.018
Ni	0.0086	0.0498	<0.0050	0.0158	mg/L	0.034
P	<0.050	0.74	<0.050	1.4	mg/L	0.15
Pb	<0.010	0.506	<0.010	<0.010	mg/L	0.014
Sb	0.334	1.14	<0.020	<0.020	mg/L	0.25
Se	<0.030	<0.030	<0.030	<0.030	mg/L	0.002
V	<0.0020	<0.0020	<0.0020	<0.0020	mg/L	0.018
Zn	1.67	92.6	4.38	5.12	mg/L	0.092

4. Discussion

Again, it is necessary to state that the composition of waste dumps (whether legal or illegal) is always diverse, but it is evident that the stated values of parameters increase after a fire in any waste dumps.

During the evaluation of the results of the analysis and the procedure of the experiment, a possible link between the amount of measured zinc values and the use of zinc containers in waste incineration was discovered. The containers used for the experiment were made of sheet steel to which a zinc layer ("galvanized") was applied to prevent corrosion of the material. It is therefore probable that due to the thermal reaction there was a significant release of this element into the resulting samples. For any further experiments, it will certainly be appropriate to use a laboratory environment with adequate equipment to eliminate this phenomenon.

A point for reflection may be the development of extinguishing agents that would "neutralise" hazardous substances typical for waste dumps (identified on the basis of analyses and other research). As already mentioned, cartridges containing wetting agents that reduce the surface tension of water have already been used. Thanks to this, the water effectively wets the surface and at the same time flows better into the depth of the burning material. Therefore, the development of a cartridge on a similar principle and the introduction into the standard equipment of firefighting brigades would be beneficial in the event of such fires to reduce the leakage of hazardous substances into the soil or surface water.

5. Conclusions

For this work, four samples were selected from illegal waste dumps from different locations and of a different composition. After comparing the measured values with the permitted limits, it was found that the greatest burden on the environment is represented by fires in car wrecks, particularly the materials of car interiors. Surprisingly, the least burdensome content was the content of substances after a tyre fire (which is not to say that the substances contained in the water after such a fire are not dangerous).

The analysis found that the main elements released in the fire were potassium, copper, sodium, calcium, iron and zinc. In terms of toxicity, concentrations of antimony, barium, aluminium, cobalt, phosphorus, nickel and lead were measured. Each of these elements, by its characteristics, poses a greater or lesser risk to the environment. Some elements occur naturally in the environment and are dangerous only at higher values. On the contrary, other elements have a negative effect on the environment even at lower concentrations. As far as physical parameters are concerned, all samples experienced a decrease in pH and an enormous increase in chemical oxygen demand (which was predictable due to the fact that the firewater was drinking water and after contamination with foreign material, the chemical consumption increases).

Finally, the recommendations for municipalities in whose territory illegal waste dumps occur could be to assume they may be burnt and be prepared for the subsequent need to decontaminate the environment, which should be treated as an area affected by a leak of a dangerous substance, or to prevent such fires by timely removal of these waste dumps or, in the best case, by frequent monitoring of the places of probable origin.

Author Contributions: Conceptualization, R.K. and M.Z.; methodology, M.Z.; software, D.T.; validation, R.K., J.M. and L.M.; formal analysis, J.M.; investigation, M.Z.; resources, M.Z.; data curation, M.Z.; writing—original draft preparation, M.Z.; writing—review and editing, R.K.; visualization, D.T.; supervision, L.M.; project administration, L.M.; funding acquisition, M.Z. All authors have read and agreed to the published version of the manuscript.

Funding: This research received no external funding.

Institutional Review Board Statement: Not applicable.

Informed Consent Statement: Not applicable.

Data Availability Statement: The data presented in this study could be found at Ref. [4].

Conflicts of Interest: The authors declare no conflict of interest.

References

1. Janouš, V. *Erotic Tools and Mines. Black Landfills Have Increased, People Face High Fines*; Vltava Labe Media: Praha, Czech Republic, 2022; Available online: https://www.denik.cz/ekonomika/cerna-skladka-20220531.html (accessed on 29 November 2022).
2. Tomášková, H. Illegal Landfills. Resistant Water: Ecology in Practice. 2022. Available online: https://www.komunalniekologie.cz/info/nelegalni-skladky (accessed on 11 March 2023).
3. Government Regulation No. 445/2021 Coll.: Government Regulation Amending Government Regulation No. 401/2015 Coll., on Indicators and Values of Permissible Pollution of Surface Water and Wastewater, Requirements for Permits for Discharge of Wastewater into Surface Waters and Sewers, and Sensitive Areas. In Prague: Sagit, 2021, 200/2021, Number 445. Available online: https://www.sagit.cz/info/sb21445 (accessed on 20 June 2023).
4. Zavoral, M. Analysis of Extinguishing Water Samples from Simulated Fires of Illegal Waste Dumps. Bachelor's Thesis, VŠB–Technical University of Ostrava, Ostrava, Czech Republic, 2023. (In Czech).

Disclaimer/Publisher's Note: The statements, opinions and data contained in all publications are solely those of the individual author(s) and contributor(s) and not of MDPI and/or the editor(s). MDPI and/or the editor(s) disclaim responsibility for any injury to people or property resulting from any ideas, methods, instructions or products referred to in the content.

Proceeding Paper

Environmental Interaction Elements in the Post-Mining Landscape of the Karviná District (Czech Republic) [†]

Jiří Kupka [1,*], Adéla Brázdová [2] and Tereza Chowaniecová [1]

[1] Department of Environmental Engineering, Faculty of Mining and Geology, VŠB—Technical University of Ostrava, Poruba, 708 00 Ostrava, Czech Republic; tereza.chowaniecova.st@vsb.cz
[2] Department of Building Materials and Diagnostics of Structures, Faculty of Civil Engineering, VŠB—Technical University of Ostrava, Poruba, 708 00 Ostrava, Czech Republic; adela.brazdova@vsb.cz
* Correspondence: jiri.kupka@vsb.cz
[†] Presented at the 4th International Conference on Advances in Environmental Engineering, Ostrava, Czech Republic, 20–22 November 2023.

Abstract: One of the actual environmental problems of the Karviná District is the loss of 'memory of place' among local inhabitants and the generally accepted preconception that the mining landscape of the Karviná District has 'nothing to offer'. Our article aims to present possible tools and methods that could contribute to solving this problem. One of them may be the elements of environmental interaction.

Keywords: post-mining landscape elements; environmental competence; environmental management; environmental interaction landscape elements

1. Introduction

The northern part of the Karviná District, especially between the cities of Orlová, Karviná and the municipality of Horní Suchá, has the character of a traditional post-mining landscape (P-ML). The Karviná District is located in the Moravian–Silesian Region (Czech Republic). This area is a part of the Upper Silesian Coal Basin, a smaller part of which, located in the Czech Republic, is called the Ostrava–Karviná Coal Basin [1–4]. Deep coal mining has been ongoing here for more than two centuries and the effects have had the greatest impact on the landscape in the Czech Republic [1,5]. Mining has thus imprinted itself indelibly on the landscape and is still perceived by the public in a controversial way [5].

The mining itself not only affected the natural components of the landscape, but also caused sudden changes in the number and composition of the population [6]. To illustrate, the original agricultural village of 'Karvinná', which had 3386 inhabitants in 1870, became a modern industrial town of 22,317 inhabitants in 1930 [6]. The 'original Karvinná' is practically non-existent today (the city centre has been moved and most of the original buildings have disappeared) [7]. The last remnant is only the urban district of Karviná–Doly, where only twenty permanent inhabitants live these days (according to the 2021 census) [6,8]).

The Karviná District still faces several environmental problems, such as air pollution, soil contamination or the impact of deep mining on the hydrological regime [2,5,9,10]. This is also associated with social and economic problems, and not only in connection with the closing of coal mines (since 1990) [3,9,11]. One of the current environmental problems in the Karviná District, which is still only partially reflected, is the loss of 'local memory' among the inhabitants [12]. Our article aims to present the possible tools and methods for environmental management that could contribute to solving this problem.

2. Background

The P-ML can be defined as a landscape in which significant changes in relief caused by subsidence and spoil tips have been induced by the extraction of minerals from the earth's crust, and in which there are landscape elements of which the formation was conditioned by mining activities (mining towers, tailings dams, tailings ponds, etc.) [3,5]. In the case of the P-ML of the Karviná District, the surface effects of deep coal mining were so devastating that entire human settlements were destroyed [7]. Conversely, in areas with high groundwater levels, the P-ML of the Karviná District is subject to waterlogging of the resulting subsidence and the subsequent formation of aquiferous subsidence basins, thus creating subsidence lakes [13]. If left to natural succession, very valuable replacement wetlands and aquatic communities of organisms are gradually formed in their area, which find suitable conditions for reproduction and shelter, and at the same time, there is a rich food supply [3]. The resulting secondary habitats for plants and animals are important for nature and landscape conservation [14].

In the past, deep coal mining in the Karviná District was also connected with a sudden influx of people from the former Austro-Hungarian Empire, including Galicia (a historical and geographic region spanning what is now southeastern Poland and western Ukraine) and Upper Hungary (todays Slovak Republic), and later from other parts of the newly formed Czechoslovakia, i.e., people without deeper ties to the region [15,16]. We can therefore assume that the Karviná District, together with its population, is struggling with the alienation from the landscape that is occurring today in society as a whole. This alienation is caused not only by 'common' changes in the way of life, but also by other problems closely linked to life in the P-ML. These include, above all, the lack of historical continuity, the radical transformation of the landscape and the generally accepted preconception that the PM-L of the Karviná District has nothing to offer and that it lies completely outside the centre of all operations [11,16,17].

According to our philosophy, the local inhabitants' relationship to the PM-L (or to any landscape in general (the concept of a 'small homeland') can be positively influenced by developing their environmental competencies. For the purpose of our paper, it is appropriate to define 'environmental competence' as a set of interrelated knowledge, skills and attitudes that enable an individual to better perceive the landscape of their home, and to realize its important role in their lives. The significant contribution of coal mining to the region or the country in the past, setting trends in the development of mining technologies and, in essence, increasing geodiversity or biodiversity in the landscape (without neglecting the original destructive effects on native biocenoses), cannot be overlooked. The crucial role in the development of environmental competencies in the P-ML of the Karviná District can be seen in the use of elements of the PM-L as key places for changing the perception of the landscape of the Karviná District. Behind this approach is an effort to take it as comprehensive as possible not only to the issue of the P-ML of the Karviná District, but also changing its perception by its inhabitants.

3. Results

Based on the analyses performed, we propose to carry out a thorough passportisation of the P-ML elements to present the possibilities of developing the environmental competencies of the inhabitants of the Karviná District. For this purpose, a simple procedure was proposed as shown in the diagram below (see Figure 1). There are likely several different subjects (persons, organisations, etc.) in the area of interest that approach the P-ML elements individually and work with them at different levels (e.g., the POHO2030 programme led by the Moravian-Silesian Investment and Development Agency (MSID) or The National Heritage Institute). The key then is not only to bring together the results of the individual subjects and eventually make the data available to the wider public (ultimately not only to local inhabitants), but also to find ways to work with these elements further. In addition to an effective passport, we propose to work with the element as with an environmental interaction landscape element (EnILE).

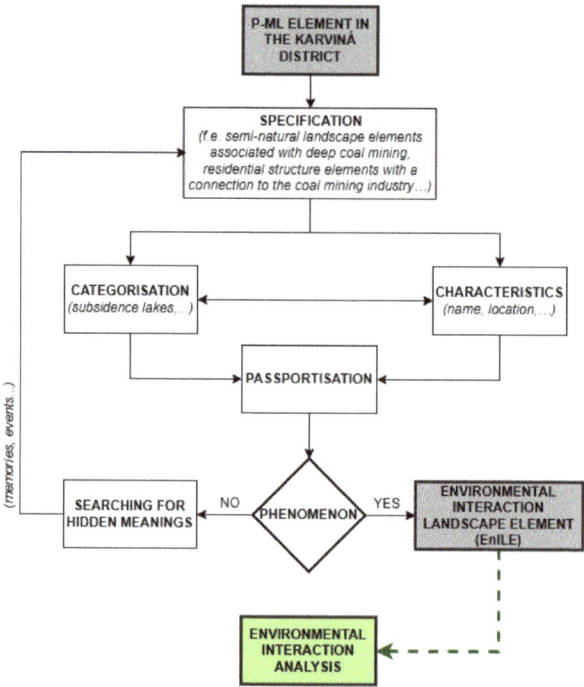

Figure 1. Diagram of the assessment workflow for P-ML elements of the Karviná District in terms of possible environmental interactions.

3.1. Categorisation and Characteristics of Landscape Elements

The diagram presents a simplified procedure for working with P-ML elements in the framework of increasing the environmental competencies of the inhabitants. The first step is the specification of the P-ML elements in terms of the landscape layer with subsequent categorisation. Examples are given in Table 1.

Table 1. An overview of specifications and categorisations of the P-ML elements.

Specifications in Terms of Landscape Layer	Categorisation Examples of P-ML Elements
semi-natural landscape elements associated with deep coal mining	subsidence lakes; reedbeds in wastewaters from coal preparation sites (tailing ponds), etc.
urban structure elements with a connection to the coal mining industry	buildings of mining operations such as mining towers or tailing ponds; mining colonies; religious monuments with a connection to mining; memorials, etc.
landscape elements of an intangible nature and interactions associated with mining	historical events; traditions; associations; projects; literature; mining-related films; subjectively coloured stories, etc.

The characteristics of a landscape element may include, for example, the exact name, location, description, photo documentation, list of relevant information sources or property rights. The structure of the description is adapted to the nature of each category (a subsidence lake requires a completely different characterisation than, for example, a description of a mining tradition). The output of the specification is a passport card as part of the database, which allows for a clear summary of all the information found about the landscape elements in the P-ML.

3.2. Working with Phenomena

For subsequent work with the P-ML elements for the development of environmental competencies, it is important to work with them as with phenomena. In our case, a phenomenon is not only seen in terms of potential or actual uniqueness, but a phenomenon can also mean what appears to a person without distinguishing whether it is a reality or an illusion [18]. For example, a waste dump can be perceived by some as a real 'waste dump' for someone else it can be a potential source of raw materials. For another, it can be an 'ulcer in the landscape' (healed or not), a natural laboratory or a place of which the use is still being explored. At the same time, for some, it is a place for relaxation or a place to experience real adventures and a place of discovery. In our case, the P-ML element under consideration must be found to be a phenomenon in the above sense. If it is not found at 'first sight', then it is necessary to look for the hidden meaning of the element and a more thorough passport. If the phenomenon has been found after a thorough search; then, the P-ML element can be viewed as an EnILE.

3.3. Environmental Interaction Design

For the purpose of the development of environmental competencies of the inhabitants of P-ML the Karviná District, we propose seven basic types of environmental interactions between humans and the landscape. These can also be seen as certain 'needs' of the landscape. The first four interactions have a general validity: observing, discovering, admiring and interpreting. The other three interactions are specific to those types of landscapes in which there has been a significant loss of memory (not necessarily only post-mining landscapes, but landscapes in general, e.g., the landscape of the Nízký Jeseník Mountains, marked by the deportation of the former inhabitants in 1945/46, etc.). In this case, it is mediation, networking and integration.

Applying this view analytically to the landscape elements of P-ML, we propose to speak of them as landscape elements of environmental interaction (EnILE), which can be uniformly processed for these purposes into overview cards (general name including possible motivational name, description of the phenomenon, analysis of the situation, examples of activities related to the development of environmental competences of the inhabitants, etc.).

3.4. Example of Working with Phenomena

In this section, an excerpt of specifications of selected landscape elements in the Karviná District and their phenomena as EnILE are presented to illustrate our approach. For these purposes, three elements of the P-ML were selected: the Church of St. Hedwig of Silesia in Doubrava village, the Kozinec subsidence lake in Doubrava village, and the 'Darkov sea' (or 'Karviná sea').

3.4.1. Church of St. Hedwig of Silesia in Doubrava Village

Characteristic: It was built at the end of the 19th century by the Roman Catholic Church. The land for the construction of the church was donated by baron Richard Mattencloit and designed by Eugen Fridrich Fulda, an important architect and builder from Teschen [19,20].

Phenomenon: The Church of St. Hedwig—the patron saint of Silesia—is an ornament of the square of the picturesque village of Doubrava, which has preserved partly its agricultural character and some elements of the original settlement buildings even in the P-ML of Karviná (see Figure 2A). However, the main thing is only discovered when entering the church. The mural on the Gospel side depicts two miners kneeling in front of St. Procopius, the patron saint of miners, with the mining tower towering in the background (see Figure 2B). This painting shows how mining was an important part of the life of the local people, that they had it projected onto the wall of the church in a spiritual image.

Figure 2. (**A**,**B**) Church of St. Hedwig of Silesia in Doubrava village: general view (**A**) and the mural on the Gospel side depicts two miners kneeling in front of St. Procopius (**B**); (**C**,**D**) Subsidence lake Kozinec in Doubrava village: a general view of its southern part with the artificial island (**C**), which is, among other things, a nesting site for specially protected bird species, such as the Great Crested Grebe (*Podiceps cristatus*) (**D**); (**E**,**F**) 'Darkov sea' (or 'Karviná sea')—overall view from bird's perspective (**E**), this site is used mainly for recreational purposes today and is not only used by young families (**F**) (author T. Chowaniecová).

3.4.2. Subsidence Lake Kozinec in Doubrava Village

Characteristic: It is an example of a semi-nature landscape element created by subsidence and subsequent waterlogging in the Doubrava–Kozinec area as a result of deep coal mining. It is located in the river terrace of the Olza River, and the northern part is bordered by the remains of an alluvial forest (see Figure 2C). Considering that intensive mining in this area started only in the late 1990s, the subsidence lake is relatively young. Its depth reaches up to about 12 metres.

Phenomenon: Several decades ago, the interesting and scientifically important nature reserve 'Loucké rybníky' was legally annulled due to mining activities in Louky nad Olší (todays Karviná–Louky). Today, the area has very little in common with its former state, although it still retains some importance (e.g., a stopover sites for many rare bird species, see Figure 2D). On the other hand, in the P-ML in the Karviná District, a similar process has led to the creation of the Kozinec lake, which is playing an increasingly important role in nature conservation, and it is possible that it may one day become a specially protected area.

3.4.2.1. 'Darkov Sea' (or 'Karviná Sea')

Characteristics: The area in the former village of Darkov has subsided as a result of deep coal mining and seepage of groundwater has created a roughly 32-hectare body of water called the 'Darkov sea' (or 'Karviná sea') (see Figure 2E). The lake is bordered by an asphalt circuit that makes the area accessible to cyclists or in-line skaters, and the well-maintained beaches allow visitors to use the lake for swimming or other water sports. The nearby mining towers of the Darkov Mine and the ČSM Mine, which remind us of the mining history of this place, are also a dominant feature of the site.

Phenomenon: This part of the landscape was traditionally used for fish farming, which mostly gave way to mining in the Czech part of the original Duchy of Teschen (1290–1918) but survived in the part that became part of Poland. If we go deeper into the past, there were originally numerous glacial lakes, some of which have survived into the modern era. The lake 'Darkov sea', which is becoming one of the most popular recreational sites in the P-ML of the Karviná District (see Figure 2F), is thus an integral part of P-ML of the Karviná District not only in its origin but also in its character.

4. Discussion

Within the framework of environmental care, management measures are introduced at various levels to contribute to the quality of life of the inhabitants (not only) in urban agglomerations and, of course, to protect and care for individual components of nature. It can be argued that much greater success is achieved in landscapes where local people are more connected to their history and are proud of their region. In places where people's relationship with their homeland is in decline, awareness-raising activities can take place, for example, in the Moravian–Silesian Region, the POHO2030 project is carrying out public events of various kinds to highlight the value of the P-ML of the Karviná District. The inhabitants who are more connected to their place of residence can therefore automatically assume that they need to behave responsibly towards the landscape or the environment. In the case of inhabitants of the P-ML with its specific problems as described above, a specific approach is needed. The methodological approach outlined shows one possible way forward. Although it is still the subject of our further research, it could become part of such management practices as waste sorting or water conservation. Thus, by analogy in this particular case, in the management aimed at building a responsible relationship between the inhabitants (or 'users') and the specific P-ML.

For a better understanding of this issue, it seems advantageous to introduce the concept of the 'landscape element of environmental interaction' as a focus point in the landscape to build this relationship. For this purpose, a diagram has been designed which is based primarily on the analysis of the P-ML elements of the Karviná District and their characteristics. The result of this step is to be a passport of all P-ML elements of the Karviná District. A subsequent and equally important step is the search for phenomena.

A P-ML element of the Karviná District becomes an EnILE if it is described in terms of the seven basic proposed types of environmental interaction between humans and the landscape, e.g., mediation, which we propose to define as an interaction that helps to (re)build the disturbed relationship of inhabitants to the landscape (offering a balanced approach), or integration, which is based for example on the implementation of landscape elements of environmental interaction in the school curriculum. It is beyond the scope of this article to provide a more detailed specification of environmental interactions and an evaluation of their level in the case of specific P-ML elements of the Karviná District, including the results of guided interviews with 'landscape users'. To illustrate, we present only specific examples of selected EnILE that can become or are becoming 'key places for changing the perception of the landscape' in the relationship of users to P-ML of the Karviná District. A follow-up step, which is closely related to the discussed issue, is the presentation of activities with different groups of local inhabitants, which are closely connected with EnILE in relation to the proposed scheme.

5. Conclusions

Although the P-ML of the Karviná District is often perceived negatively by the public and local inhabitants, and it is true that it has a number of environmental problems, this assessment is often related to a non-objective and one-sided presentation of the region, rather than the fact that this landscape has nothing else to offer except coal. For this reason, one possible solution is to work with the public (users and residents) in a way that is analogous to management in other parts of the environment. Not to be confused with the generally accepted environmental education and public awareness and environmental counselling as important preventive tools of environmental policy, since the P-ML of the Karviná District has its own specifics, which are also essential to consider.

For this purpose, a diagram of the assessment workflow for P-ML elements of the Karviná District is proposed, including their passports in relation to possible environmental interactions, search for phenomena and design of environmental interactions. Key places for changing the perception can be seen in the use of the so-called environmental interaction landscape elements (EnILE) in environmental management. We are working on this issue in our research activities and believe that this can contribute to changing the view of the Karviná District not only as a post-mining landscape (PM-L), but also a landscape with a loss of memory in general.

Author Contributions: Conceptualization, J.K. and T.C.; methodology, J.K. and A.B.; investigation, J.K., T.C. and A.B.; resources, J.K., T.C. and A.B.; data curation, J.K., T.C. and A.B.; writing—original draft preparation, J.K., T.C. and A.B.; writing—review and editing, J.K., T.C. and A.B.; visualization, A.B. and T.C.; supervision, J.K. All authors have read and agreed to the published version of the manuscript.

Funding: This research received no external funding, and the APC was funded by the Faculty of Mining and Geology, Department of Environmental Engineering.

Institutional Review Board Statement: Not applicable.

Informed Consent Statement: Not applicable.

Data Availability Statement: Data are contained within the article.

Conflicts of Interest: The authors declare no conflict of interest.

References

1. Kolář, T.; Čermák, P.; Oulehle, F.; Trnka, M.; Štěpánek, P.; Hruška, J.; Büntgen, U.; Rybníček, M. Pollution control enhanced spruce growth in the "Black Triangle" near the Czech–Polish border. *Sci. Total Environ.* **2015**, *538*, 703–711. [CrossRef]
2. Biały, W. Influence of Mining Exploitation on Soil Chemistry on the Example of Ostrava-Karviná Hard Coal District. *Qual. Prod. Improv.* **2021**, *3*, 270–282.
3. Stalmachová, B.; Sierka, E. *Managed Succession in Reclamation of Postmining Landscape*, 1st ed.; Technická Univerzita v Košiciach: Košice, Slovakia, 2014; pp. 4–58.
4. Čablík, V.; Hlavata, M.; Janáková, I.; Tora, B. Coal industry in Czech Republic. In Proceedings of the IOP Conference Series: Materials Science and Engineering, Szczyrk, Poland, 25–27 March 2019.
5. Martinec, P.; Holub, K.; Hortvík, K.; Kaláb, Z.; Knejzlík, M.; Konečný, P.; Krůl, M.; Lacina, J.; Latová, A.; Maníček, J.; et al. *Termination of Underground Coal Mining and Its Impact on the Environment*, 1st ed.; Landrová, P., Machač, J., Eds.; ANAGRAM: Ostrava, Czech Republic, 2006; pp. 22–26, 33–36, 40–41, 77.
6. Projekt Stará Karvinná. Available online: https://www.archives.cz/web/SK/index.php (accessed on 12 July 2023).
7. Hajzlerová, I. *Zaniklý Svět—Historie Staré Karvinné*, 1st ed.; Státní Ozkresní Archiv Karviná: Karviná, Czech Republic, 2002; p. 24.
8. Czech Statistical Institute. Available online: https://www.czso.cz/csu/czso/vysledky-scitani-2021-otevrena-data (accessed on 12 July 2023).
9. Martinat, S.; Dvořák, P.; Frantal, B.; Klusáček, P.; Kunc, J.; Navrátil, J.; Osman, R.; Turečková, K.; Reed, M. Sustainable urban development in a city affected by heavy industry and mining? Case study of brownfields in Karvina, Czech Republic. *J. Clean. Prod.* **2016**, *118*, 78–87. [CrossRef]
10. Marschalko, M.; Yilmaz, I.; Lamich, D.; Heviánková, S.; Kyncl, M.; Dirner, V.; Andráš, P. Morphological variations in subsidence basin and importance for land use planning: Undermined Karvina region (Czech Republic). *Carpathian J. Earth Environ. Sci.* **2014**, *9*, 187–197.

11. Popelková, R.; Mulková, M. The mining landscape of the Ostrava-Karviná coalfield: Processes of landscape change from the 1830s to the beginning of the 21st century. *Appl. Geogr.* **2018**, *90*, 28–43. [CrossRef]
12. Kupka, J. Pojď'me naplňovat potřeby krajiny Karvinska. Nejen ona si to totiž zaslouží! *Veronica: Časopis Ochránců Přírody* **2022**, *2*, 20–22.
13. Neset, K. *Vlivy poddolování: Důlní měřictví IV*, 1st ed.; SNTL: Praha, Czech Republic, 1984; pp. 20–71.
14. Konvička, M. Postindustriální stanoviště z pohledu ekologické vědy a ochrany přírody. In *Bezobratlí Postindustriálních Stanovišť: Význam, Ochrana a Management*, 1st ed.; Tropek, R., Řehounek, J., Eds.; Calla: České Budějovice, Czech Republic, 2011; pp. 9–17.
15. Komitee des Allgemeinen Bergmannstages. *Die Mineralkohlen Österreichs*; Verlag des Zentralvereins der Bergwerksbesitzer Österreichs: Vienna, Austria, 1903; pp. 377–416.
16. Stolařík, I.; Štika, J.; Tomolová, V. *Těšínsko 1. díl*, 1st ed.; Muzeum Těšínska and Valašské Muzeum v Přírodě: Český Těšín a Rožnov pod Radhoštěm, Czech Republic, 1997; pp. 73–102.
17. Skaloš, J.; Kašparovaá, I. Landscape memory and landscape change in relation to mining. *Ecol. Eng.* **2012**, *12*, 60–69. [CrossRef]
18. Husserl, E. *Idea Fenomenologie*, 1st ed.; OIKOYMENH: Praha, Czech Republic, 2001; Volume 44, pp. 21–329.
19. Kostelíčky z Našich Cest. Available online: https://www.kostelikyzcest.cz/index.php/kostel-sv-hedviky-doubrava/kostel-sv-hedviky-doubrava (accessed on 12 July 2023).
20. Kabourková, P. *Kostel sv. Hedviky v Doubravě*, 1st ed.; Obec Doubrava: Doubrava, Czech Republic, 2009; p. 56.

Disclaimer/Publisher's Note: The statements, opinions and data contained in all publications are solely those of the individual author(s) and contributor(s) and not of MDPI and/or the editor(s). MDPI and/or the editor(s) disclaim responsibility for any injury to people or property resulting from any ideas, methods, instructions or products referred to in the content.

Proceeding Paper

Determining the Presence of Micro-Particles in Drinking Water in the Czech Republic—An Exploratory Study Focusing on Microplastics and Additives [†]

Jan Halfar *, Kateřina Brožová, Karla Placová and Miroslav Kyncl

Department of Environmental Engineering, Faculty of Mining and Geology, VŠB—Technical University of Ostrava, 17. Listopadu 2172/15, 708 00 Ostrava, Czech Republic
* Correspondence: jan.halfar@vsb.cz
[†] Presented at the 4th International Conference on Advances in Environmental Engineering, Ostrava, Czech Republic, 20–22 November 2023.

Abstract: With the increasing prevalence of monitored micro-pollutants in the environment, new challenges arise in their determination. Currently, the issue of microplastics in the environment is a highly debated topic. This article focuses on the assessment of micro-particles in drinking water samples collected in the Czech Republic. The samples for analysis were collected at two locations, twice a day. Particle separation was achieved through vacuum filtration, and particle identification was performed using the μ-FTIR method. A total of 40 particles were analyzed, and their shape, color, and material composition were determined. The study results demonstrated the presence of microplastics and additives used in plastic production in the drinking water samples. The findings of this article contribute to understanding the issue of micro-particles in drinking water and raise questions for further research and measures to protect public health and the environment.

Keywords: microplastic; drinking water; FTIR; particles; additives

Citation: Halfar, J.; Brožová, K.; Placová, K.; Kyncl, M. Determining the Presence of Micro-Particles in Drinking Water in the Czech Republic—An Exploratory Study Focusing on Microplastics and Additives. *Eng. Proc.* **2023**, *57*, 16. https://doi.org/10.3390/engproc2023057016

Academic Editors: Adriana Estokova, Natalia Junakova, Tomas Dvorsky, Vojtech Vaclavik and Magdalena Balintova

Published: 1 December 2023

Copyright: © 2023 by the authors. Licensee MDPI, Basel, Switzerland. This article is an open access article distributed under the terms and conditions of the Creative Commons Attribution (CC BY) license (https://creativecommons.org/licenses/by/4.0/).

1. Introduction

Plastic materials are utilized by humans on a daily basis in nearly all their activities, and without them, we would not know the life we have on this planet. The quantity of produced plastics exhibits a consistently growing trend, with the exception of a stagnation due to the pandemic in 2020, thereby increasing the significance of addressing the issue of plastic waste. In 2021, over 390 million metric tons (Mt) of plastics were produced globally (57.2 Mt in Europe), of which 90% originated from fossil sources and only 8.3% originated from recycled materials [1]. However, the escalating production of plastics also entails serious risks as unprocessed plastic materials often end up in the environment, including water bodies [2]. In 2016, up to 10% of produced plastics found their way into aquatic environments, and by 2030, it is projected to reach as high as 53 Mt annually [3]. Such a quantity currently approaches the amount of plastics produced across Europe. Some plastic materials themselves and the fragments present in the aquatic environment are persistent. However, they are subject to chemical, biological, and mechanical processes and fragment into smaller pieces, such as microplastics [4]. This leads to the formation of microplastics and, further down, nanoplastics. Microplastics are defined in the literature as particles smaller than 5mm [5], although alternative size definitions have also been observed [6], which complicates the interpretation of study results. Microplastics were found in air [7–9], water [10–13], and soil [14–18]. The current climate situation also places significant demands on ensuring an adequate water supply for the population, agriculture, and industry. New sources of drinking water are being sought in various places, including old closed coal mines with excess mine water [19]. However, the presence of micropollutants often causes water quality problems [20]. Moreover, their removal is often technologically challenging, allowing them to permeate into drinking water during conventional treatment processes. Secondary microplastics, generated from the degradation of larger plastic pieces, sometimes subsequently coexist in water with primary microplastics intentionally manufactured for their specific effects. Nevertheless, legislative measures in some countries already restrict the addition of primary microplastics to products, which should gradually lead to a reduction in their occurrence [21]. However, due to the enormous amount of unprocessed plastic waste, an increase in the amount of mismanaged plastic pieces in aquatic environments is expected [22]. The removal of these particles should be preceded by a standardized methodology for their determination, which is currently lacking and needs to be addressed at the legislative level. Among the various methods currently in use, it is necessary to select one that can be employed in a conventionally equipped water quality laboratory. Based on the obtained data, the actual effectiveness of individual water treatment and purification technologies can then be monitored.

This article focuses on the analysis of four samples of drinking water, which underwent optical analysis to determine their shape and color. Subsequently, the material composition of the particles was identified using the micro-FTIR method. This method is frequently utilized for detecting the presence of microplastics in aquatic environments. As there are still many unanswered questions in this field, further intensive research on microplastics is necessary.

2. Materials and Methods

As mentioned above, the methodology for determining microplastics in aquatic environments suffers from significant shortcomings. There are different procedures for sample collection and handling, diverse instrumental methods for analysis, and inconsistent interpretations of the obtained data.

2.1. Sample Collection

The samples consist of drinking water collected at a sampling point within households (Figure 1). Sampling was conducted twice a day, using a volume of 50 mL. Prior to sampling, the water from the faucet was allowed to run for 10 s. Two samples were collected from the same location, once in the morning and once in the evening, providing insights into potential variations in the presence of microplastics over time. The sampling containers are made of glass with a metal lid to minimize the risk of contamination from the sampling apparatus.

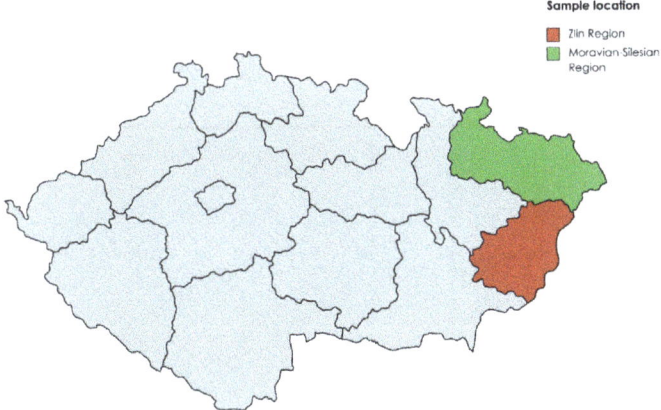

Figure 1. Sampling locations in Czech Republic.

2.2. Optical Analysis

All collected samples of drinking water underwent microscopic analysis, where each filter was examined. The filter was placed in a Petri dish, transferred from the desiccator to the microscope. Preparation and focusing of the microscope were carried out before opening the Petri dish with the filter to minimize potential contamination. The entire surface of the filter was observed, starting from the upper left corner and progressing across the entire surface until reaching the lower right corner.

This step is also crucial for assessing possible atmospheric contamination. Each blank filter was observed under the microscope, and the particles present on it were counted and subsequently analyzed using micro-FTIR. The number of particles on the blank filters was then subtracted from the total number of microplastics found in the samples for evaluation.

2.3. Infrared Spectroscopy

In order to ascertain the material composition of the particles captured on the filters, an essential analysis is conducted. Micro-FTIR analysis is employed in this particular case, utilizing Fourier-transform infrared spectroscopy in conjunction with a microscope-equipped instrument. This enables the measurement of particles down to a critical size of 10 µm, with the ultimate limit contingent upon their morphological characteristics. The Nicolet iN10 instrument, manufactured by Thermo Fisher Scientific (Waltham, MA, USA), is employed for these analytical investigations. Considering the particle size and the filtration medium employed, reflectance was selected as the optimal measurement mode, employing a liquid nitrogen-cooled mercury cadmium telluride (MCT) detector. The measurement parameters are standardized across all samples, encompassing the acquisition of spectra within the wavenumber range of 650–4000 cm^{-1}.

Prior to the commencement of the measurements, ten particles are meticulously selected through manual manipulation utilizing an optical stereomicroscope, and subsequently, they are transferred onto glass slides employing fine tweezers. The morphology and color of each particle are diligently documented throughout this transfer process. Subsequently, the individual particles undergo thorough analysis, and their visual documentation is duly recorded.

The acquired spectra are processed using Omnic Picta software. These spectra are subsequently subjected to comparison with available spectral databases. If the measured spectrum demonstrates an agreement exceeding 60% with a library entry, the resultant material composition of the particle, the percentage match, and the corresponding library spectra are meticulously documented. In cases where the agreement falls within the range of 50% to 60%, manual identification procedures are employed, entailing the scrutiny of characteristic spectral peaks. However, a careful determination is made regarding the adequacy of the match for accurate material identification, based on a comprehensive comparison with the spectral library.

3. Results and Discussion

A total of 40 particles from four samples of drinking water collected in the Moravian-Silesian and Zlín regions in the Czech Republic were analyzed. Since the presence of microplastics in drinking water has already been confirmed in multiple studies [23,24], they were also detected in the analyzed samples. A total of 17 different types of materials were identified and divided into two categories. The first category includes polymers and materials used in the plastic industry as additives or other substances that influence the properties of plastic products. The second group comprises substances detected in drinking water that are unrelated to plastics. In the first group, there were a total of nine materials with a combined count of 26 particles (Figure 2). The most frequent materials were wood with melamine–formaldehyde resin (eight) and paper with coating (eight), followed by polyethylene terephthalate (PET), polyvinyl alcohol (PVA) and viscose (two). Polyvinyl alcohol, viscose, and polyethylene terephthalate were also identified in the study conducted by Shruti et al. [25]. The second group includes eight materials with 14 particles, with cellulose (three) and cotton + flax (three) being the most prominent ones (Figure 3).

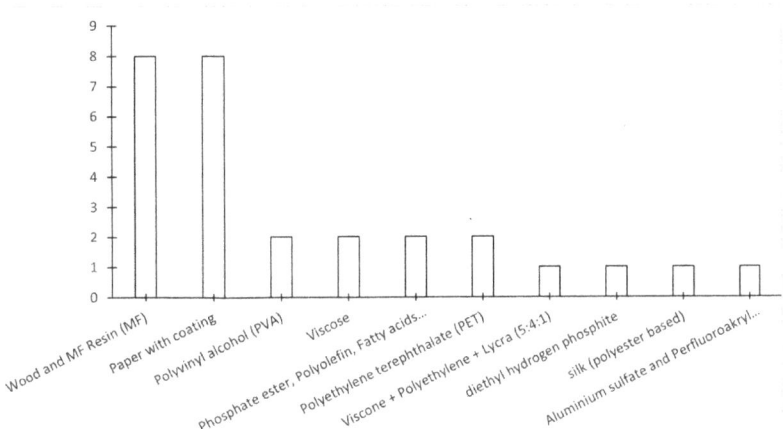

Figure 2. Group 1; number of particles for each material.

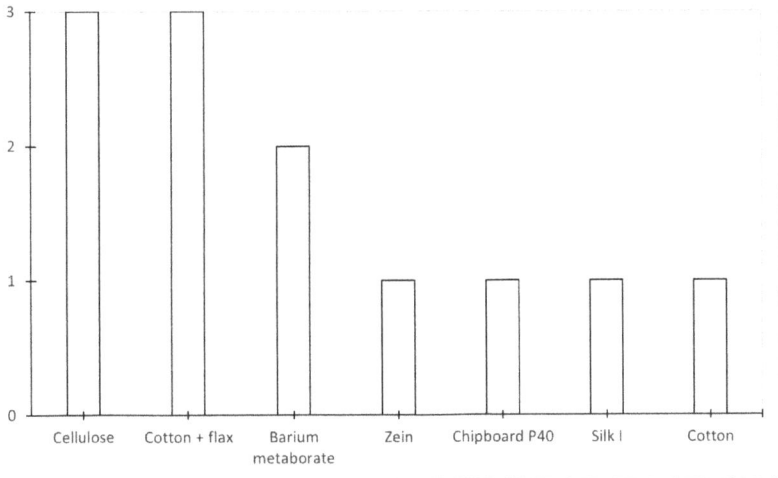

Figure 3. Group 2; number of particles for each material.

In addition to material composition, the shape of individual particles was also determined. They were sorted into four groups using optical analysis, as shown in Figure 4: Fibre (47%; 19 pcs.), Particle (30%; 12 pcs.), Fragment (18%; 7 pcs.), and Film (5%; 2 pcs.).

Similar to this study, Pérez-Guevara et al. also confirmed the most frequent occurrence of fibers in samples of drinking water [26]. The Fibre category included particles with a length at least 3 times greater than their width. The Particle category consisted of particles with regular shapes such as flakes or spheres. Deformed particles were classified as Fragments, while thin particles were categorized as Films. In the analyzed samples of drinking water, fibers were the most prevalent, followed by particles, fragments, and films.

All analyzed particles were also recorded for their color. A total of 11 different colors were observed, with the highest frequency being brown (9), followed by red (5), blue (5), white (5), and transparent (5). All colors, along with the respective particle counts, are presented in Figure 5.

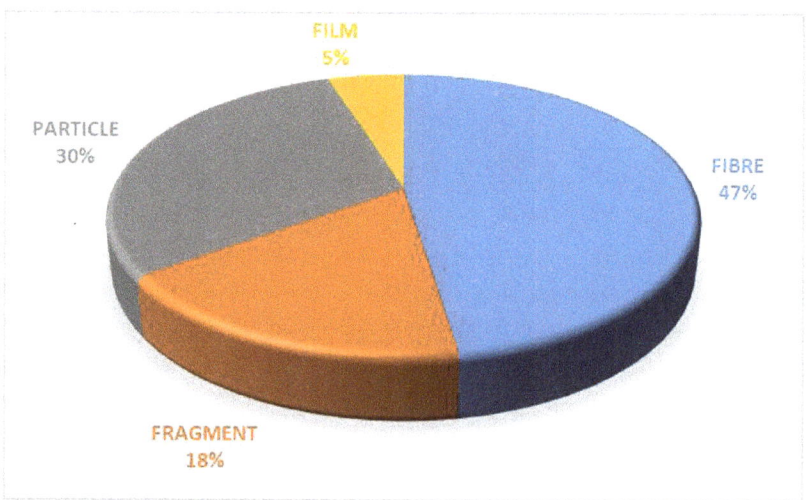

Figure 4. Particles sorted by shape.

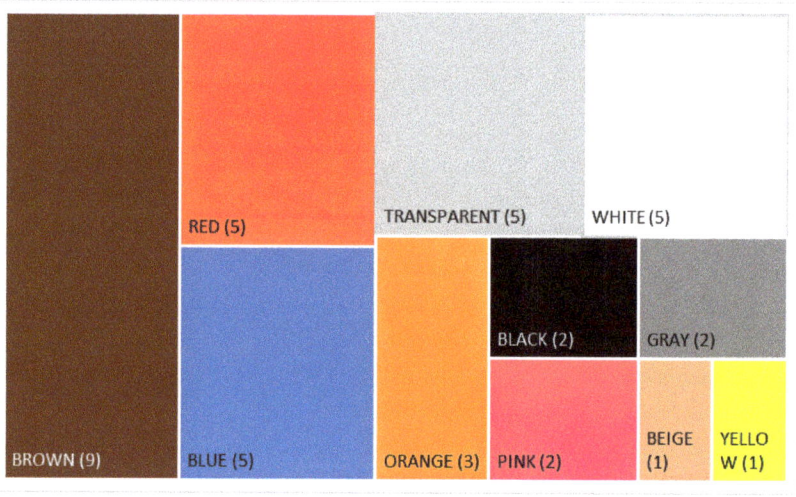

Figure 5. Particles sorted by color.

The escalating production of plastics and their accumulation in the environment will inevitably lead to a rising prevalence of microplastics in the future. Addressing this issue requires exploring strategies to mitigate the potential proliferation of microplastics. One effective approach involves the reutilization of macroplastics, wherein products can be meticulously recycled and upcycled. As an example, tires, which contribute significantly to secondary microplastic sources, could potentially find application in the production of cement composites [26,27]. This innovative reimagining of tire usage presents a promising avenue for curtailing the microplastic menace.

4. Conclusions

This study presents an initial analysis of particles in four samples of drinking water collected from the territory of the Czech Republic. The selected analytical methodology demonstrates its suitability and allows for the examination of a larger number of samples.

A pivotal step involved a systematic reduction in contamination influence, enabling the verification of proper timing for particle introduction during analysis. However, in light of impending legislative changes concerning microplastics, there emerges a pressing need to refine and standardize the methodology for their determination. The objective is to achieve the feasibility of conducting such analyses with optimal time and financial investment, while adhering to stringent result quality requirements and minimizing potential contamination. Nevertheless, the issue of microplastics' presence in drinking water remains a complex challenge, necessitating further comprehensive research for its resolution.

Author Contributions: Conceptualization, J.H.; methodology, J.H.; formal analysis, K.B.; investigation, J.H. and K.B.; writing—original draft preparation, J.H.; writing—review and editing, K.P.; visualization, J.H.; supervision, M.K.; funding acquisition, K.P. All authors have read and agreed to the published version of the manuscript.

Funding: This research was funded by the Student Grant Competition financed by VSB—Technical University of Ostrava within project "Research and detection of micropollutants in drinking and waste water from sewage treatment plants." (no. SP2023/045). This article has been produced with the financial support of the European Union under the REFRESH—Research Excellence For REgion Sustainability and High-tech Industries project number CZ.10.03.01/00/22_003/0000048 via the Operational Programme Just Transition.

Institutional Review Board Statement: Not applicable.

Informed Consent Statement: Not applicable.

Data Availability Statement: Data sets are contained within the article. Further data and materials requests should be addressed to jan.halfar@vsb.cz (J.H.).

Conflicts of Interest: The authors declare no conflict of interest.

References

1. Plastics Europe. *Plastics—The Facts 2022*; Plastics Europe: Brussels, Belgium, 2022.
2. García Rellán, A.; Vázquez Ares, D.; Vázquez Brea, C.; Francisco López, A.; Bello Bugallo, P.M. Sources, Sinks and Transformations of Plastics in Our Oceans: Review, Management Strategies and Modelling. *Sci. Total Environ.* **2023**, *854*, 158745. [CrossRef] [PubMed]
3. Borrelle, S.B.; Ringma, J.; Law, K.L.; Monnahan, C.C.; Lebreton, L.; McGivern, A.; Murphy, E.; Jambeck, J.; Leonard, G.H.; Hilleary, M.A.; et al. Predicted Growth in Plastic Waste Exceeds Efforts to Mitigate Plastic Pollution. *Science* **2020**, *369*, 1515–1518. [CrossRef] [PubMed]
4. Zhang, K.; Hamidian, A.H.; Tubić, A.; Zhang, Y.; Fang, J.K.H.; Wu, C.; Lam, P.K.S. Understanding Plastic Degradation and Microplastic Formation in the Environment: A Review. *Environ. Pollut.* **2021**, *274*, 116554. [CrossRef] [PubMed]
5. Arthur, C.; Baker, J.; Bamford, H. (Eds.) Proceedings of the International Research Workshop on the Occurrence, Effects, and Fate of Microplastic Marine Debris, 9–11 September 2008; NOAA Technical Memorandum NOS-OR&R-30. Available online: https://marinedebris.noaa.gov/proceedings-international-research-workshop-microplastic-marine-debris (accessed on 30 November 2023).
6. Hartmann, N.B.; Hüffer, T.; Thompson, R.C.; Hassellöv, M.; Verschoor, A.; Daugaard, A.E.; Rist, S.; Karlsson, T.; Brennholt, N.; Cole, M.; et al. Are We Speaking the Same Language? Recommendations for a Definition and Categorization Framework for Plastic Debris. *Environ. Sci. Technol.* **2019**, *53*, 1039–1047. [CrossRef] [PubMed]
7. Gaston, E.; Woo, M.; Steele, C.; Sukumaran, S.; Anderson, S. Microplastics Differ Between Indoor and Outdoor Air Masses: Insights from Multiple Microscopy Methodologies. *Appl. Spectrosc.* **2020**, *74*, 1079–1098. [CrossRef] [PubMed]
8. O'Brien, S.; Rauert, C.; Ribeiro, F.; Okoffo, E.D.; Burrows, S.D.; O'Brien, J.W.; Wang, X.; Wright, S.L.; Thomas, K.V. There's Something in the Air: A Review of Sources, Prevalence and Behaviour of Microplastics in the Atmosphere. *Sci. Total Environ.* **2023**, *874*, 162193. [CrossRef]
9. Xie, Y.; Li, Y.; Feng, Y.; Cheng, W.; Wang, Y. Inhalable Microplastics Prevails in Air: Exploring the Size Detection Limit. *Environ. Int.* **2022**, *162*, 107151. [CrossRef]
10. Cincinelli, A.; Scopetani, C.; Chelazzi, D.; Lombardini, E.; Martellini, T.; Katsoyiannis, A.; Fossi, M.C.; Corsolini, S. Microplastic in the Surface Waters of the Ross Sea (Antarctica): Occurrence, Distribution and Characterization by FTIR. *Chemosphere* **2017**, *175*, 391–400. [CrossRef]
11. Diez-Pérez, D.B.; Arenas, I.; Maidana, E.; López-Rosales, A.; Andrade, J.M.; Muniategui-Lorenzo, S. Microplastics in Surface Water of the Bay of Asunción, Paraguay. *Mar. Pollut. Bull.* **2023**, *192*, 115075. [CrossRef]

12. He, D.; Chen, X.; Zhao, W.; Zhu, Z.; Qi, X.; Zhou, L.; Chen, W.; Wan, C.; Li, D.; Zou, X.; et al. Microplastics Contamination in the Surface Water of the Yangtze River from Upstream to Estuary Based on Different Sampling Methods. *Environ. Res.* **2021**, *196*, 110908. [CrossRef]
13. Kirstein, I.V.; Hensel, F.; Gomiero, A.; Iordachescu, L.; Vianello, A.; Wittgren, H.B.; Vollertsen, J. Drinking Plastics?—Quantification and Qualification of Microplastics in Drinking Water Distribution Systems by MFTIR and Py-GCMS. *Water Res.* **2021**, *188*, 116519. [CrossRef] [PubMed]
14. Corradini, F.; Casado, F.; Leiva, V.; Huerta-Lwanga, E.; Geissen, V. Microplastics Occurrence and Frequency in Soils under Different Land Uses on a Regional Scale. *Sci. Total Environ.* **2021**, *752*, 141917. [CrossRef] [PubMed]
15. Corradini, F.; Meza, P.; Eguiluz, R.; Casado, F.; Huerta-Lwanga, E.; Geissen, V. Evidence of Microplastic Accumulation in Agricultural Soils from Sewage Sludge Disposal. *Sci. Total Environ.* **2019**, *671*, 411–420. [CrossRef] [PubMed]
16. Hernández-Arenas, R.; Beltrán-Sanahuja, A.; Navarro-Quirant, P.; Sanz-Lazaro, C. The Effect of Sewage Sludge Containing Microplastics on Growth and Fruit Development of Tomato Plants. *Environ. Pollut.* **2021**, *268*, 115779. [CrossRef]
17. Nizzetto, L.; Futter, M.; Langaas, S. Are Agricultural Soils Dumps for Microplastics of Urban Origin? *Environ. Sci. Technol.* **2016**, *50*, 10777–10779. [CrossRef]
18. Yang, L.; Zhang, Y.; Kang, S.; Wang, Z.; Wu, C. Microplastics in Soil: A Review on Methods, Occurrence, Sources, and Potential Risk. *Sci. Total Environ.* **2021**, *780*, 146546. [CrossRef]
19. Dvořacek, J.; Malikova, P.; Sousedikova, R.; Heviankova, S.; Rys, P.; Osickova, I. Water Production as an Option for Utilizing Closed Underground Mines. *J. South. Afr. Inst. Min. Metall.* **2022**, *122*, 571–578. [CrossRef]
20. Khan, F.S.A.; Mubarak, N.M.; Khalid, M.; Tan, Y.H.; Abdullah, E.C.; Rahman, M.E.; Karri, R.R. A Comprehensive Review on Micropollutants Removal Using Carbon Nanotubes-Based Adsorbents and Membranes. *J. Environ. Chem. Eng.* **2021**, *9*, 106647. [CrossRef]
21. Halfar, J.; Brožová, K.; Čabanová, K.; Heviánková, S.; Kašpárková, A.; Olšovská, E. Disparities in Methods Used to Determine Microplastics in the Aquatic Environment: A Review of Legislation, Sampling Process and Instrumental Analysis. *Int. J. Environ. Res. Public. Health* **2021**, *18*, 7608. [CrossRef]
22. Jambeck, J.R.; Geyer, R.; Wilcox, C.; Siegler, T.R.; Perryman, M.; Andrady, A.; Narayan, R.; Law, K.L. Plastic Waste Inputs from Land into the Ocean. *Science* **2015**, *347*, 768–771. [CrossRef]
23. Kosuth, M.; Mason, S.A.; Wattenberg, E.V. Anthropogenic Contamination of Tap Water, Beer, and Sea Salt. *PLoS ONE* **2018**, *13*, e0194970. [CrossRef] [PubMed]
24. Zhou, G.; Wu, Q.; Wei, X.-F.; Chen, C.; Ma, J.; Crittenden, J.C.; Liu, B. Tracing Microplastics in Rural Drinking Water in Chongqing, China: Their Presence and Pathways from Source to Tap. *J. Hazard. Mater.* **2023**, *459*, 132206. [CrossRef] [PubMed]
25. Shruti, V.C.; Kutralam-Muniasamy, G.; Pérez-Guevara, F.; Roy, P.D.; Elizalde-Martínez, I. Free, but Not Microplastic-Free, Drinking Water from Outdoor Refill Kiosks: A Challenge and a Wake-up Call for Urban Management. *Environ. Pollut.* **2022**, *309*, 119800. [CrossRef] [PubMed]
26. Svoboda, J.; Václavík, V.; Dvorský, T.; Klus, L.; Botula, J. The Utilization of a Combination of Recycled Rubber from Waste Tires and Waste Waters from a Concrete Plant in the Production of Cement Composites. *Key Eng. Mater.* **2020**, *838*, 59–66. [CrossRef]
27. Svoboda, J.; Dvorský, T.; Václavík, V.; Charvát, J.; Máčalová, K.; Heviánková, S.; Janurová, E. Sound-Absorbing and Thermal-Insulating Properties of Cement Composite Based on Recycled Rubber from Waste Tires. *Appl. Sci. Switz.* **2021**, *11*, 2725. [CrossRef]

Disclaimer/Publisher's Note: The statements, opinions and data contained in all publications are solely those of the individual author(s) and contributor(s) and not of MDPI and/or the editor(s). MDPI and/or the editor(s) disclaim responsibility for any injury to people or property resulting from any ideas, methods, instructions or products referred to in the content.

Proceeding Paper

Watercourses and Their Geodetic Mapping for Water Management [†]

Petr Jadviscok * and **Tereza Gottvaldova**

Department of Geodesy and Mine Surveying, Faculty of Mining and Geology, VSB—Technical University of Ostrava, 17. Listopadu 2172/15, 708 00 Ostrava-Poruba, Czech Republic; tereza.gottvaldova.st@vsb.cz

* Correspondence: petr.jadviscok@vsb.cz; Tel.: +420-596-993-326

[†] Presented at the 4th International Conference on Advances in Environmental Engineering, Ostrava, Czech Republic, 20–22 November 2023.

Abstract: This contribution deals with surveyor activities that are associated with the measurement of watercourses and their closest terrain. It clarifies the specifics and complications of the solutions of the complex tasks of water management in geodetic practice. The article is devoted to activities that are directly connected with the realization of measuring, used technologies, and interpretation results' outputs in digital form.

Keywords: maps of watercourse; GNSS; geodesy; digital terrain model

1. Introduction

The task of preparing mapping documentation is generally a common activity of surveyors in the field of engineering geodesy. This equally applies to mapping documentation for project purposes, for land modifications, or only for preparing an elaborate map for record-keeping purposes. However, in practice, surveyors do not commonly encounter preparations when mapping grounds for watercourse bed areas. Unlike other cases more commonly seen in practice, where the mapping locality may be represented, for example, by company premises, areas planned for the construction of transport or other infrastructure (roads, highways, railways, pavements, gas pipelines, water mains, drainage systems, etc.), civic or residential building objects, old premises with planned modernization and reconstruction, and many others.

The results of detailed surveying are also applied in localities where watercourse lowering occurs. Due to mining activities, changes particularly occur in the natural height profile of watercourse beds. Based on repeated measurements of the segments of watercourses affected by lowering, the condition of the bed can be observed, and the measurement results can be analyzed. Watercourse mapping and other activities essentially focus essentially on describing the terrain of the mapped area in a suitable manner, so the resulting mapping work includes all substantial aspects. At the same time, it should be clear and topographically adequate, and it should provide sufficient geometric accuracy. Achieving the optimal balance of the resulting mapping work to make sure it approaches the ideal parameters of the above mapping documentation is not quite simple, and, therefore, sufficient attention must be particularly paid to the requirements for mapping documentation both during terrain measurements and subsequent processing.

2. Activities Related to Preparatory Task and Terrain Recognition

When orders of this type are implemented, sufficient time must be devoted, more than in all other cases, to the activities of preparing and planning a suitable process for the measurement tasks. In particular, the purpose of this stage is to obtain and gather all the ground materials needed for the surveying works themselves and, furthermore, for subsequent computational and graphic processing. The expected choice of the method

---

Citation: Jadviscok, P.; Gottvaldova, T. Watercourses and Their Geodetic Mapping for Water Management. *Eng. Proc.* **2023**, *57*, 17. https://doi.org/10.3390/engproc2023057017

Academic Editors: Adriana Estokova, Natalia Junakova, Tomas Dvorsky, Vojtech Vaclavik and Magdalena Balintova

Published: 1 December 2023

Copyright: © 2023 by the authors. Licensee MDPI, Basel, Switzerland. This article is an open access article distributed under the terms and conditions of the Creative Commons Attribution (CC BY) license (https://creativecommons.org/licenses/by/4.0/).

and technology of the connection measurement should be determined in the scope of the preparatory works. Based on the performed preparatory tasks, the surveyor should gain knowledge of the availability of the geometric base points of the measurement and of the possibility to apply the envisaged positional or altitudinal measurement method.

As mentioned above, a particular emphasis when addressing such cases is applied due to terrain recognition. Activities related to terrain recognition particularly include exploration and becoming familiar with the terrain. Recognition also includes the search for a geometric base point of the measurement and the verification of whether the locality can be connected to binding reference systems in accordance with the original plan, regardless of whether it is possible to apply GNSS technology to connect to the coordinate system.

Figure 1 shows an illustration of scans obtained during terrain recognition in the process of Stonavka River bed surveying in the cadastral territory of Tranovice. It is also important to consider a suitable layout of the survey network points as part of terrain recognition. When choosing the layout of the survey network points, the spatial nature of the model locality must be respected, built-up areas and vegetation must be considered, and operation on roads or other obstacles that could restrict, burden, or even render impossible any survey works should be taken into account. Minimizing the risk of destruction is no less important, given that mapping requires more time in rather large localities. Last but not least, the very position of the geometric base points must be taken into account, unless GNSS technology methods are used for connecting the locality [1].

Figure 1. The watercourse of geodetic measured.

Terrain recognition should be a natural part of all surveying activities, representing a very important geodesic activity in the field. Duly carried out recognition enables the surveyor to acquire a better overview of the model locality and, perhaps, to conceive a more effective solution for the task. In the end, the surveyor saves a lot of work with respect to surveying activities and subsequent processing. Sufficient room should be devoted to terrain recognition, particularly in areas where surveying work is hindered by omnipresent vegetation, built-up zones, or other specific factors. The importance of recognition is even higher in cases like these. Here, the conditions are made more difficult not only by the very subject of mapping and by the watercourse bed but also by the dense vegetation along the watercourse. The effectiveness of surveying works should, therefore, be maximized.

3. Connecting the Locality to Binding Reference Systems of the Czech Republic and Their Usability in Addressing Task

Connecting a locality to binding positional or altitudinal systems should be understood as determining the position and altitude of a set of auxiliary points, i.e., the geometric base of the detailed positional and altitudinal surveying of the situation in the field. In surveying practice, the position of a point is expressed using the coordinate system of the uniform trigonometric cadastral network (S-JTSK), and its altitude is provided in the altitudinal Baltic system after adjustment (Bpv). Both these systems belong to binding reference geodesic systems of the Czech Republic, pursuant to Government Regulation No. 430/2006 Coll [1].

Both methods, based on Global Navigation Satellite Systems (GNSS) technology and classical terrestrial methods, can be used to determine the coordinates of the survey network points. At present, an effort can be seen to apply GNSS technology as much as possible, due to its low time and labor demands. However, methods using GNSS technology cannot always be used, and, in some cases, this possibility of connection is virtually excluded. Precisely mapping watercourses is one of the cases where the use of this technology is limited. In particular, for the common and omnipresent bushy vegetation along watercourses and in their near surroundings, where GNSS technology is applied to connect a locality, the fast static method is especially used and sometimes also the kinematic method is used in real time, the so-called real-time kinematic (RTK) method. The static method should be preferred to RTK when determining the position of the survey network points to be used as the geometric base for detailed surveying. The main reasons include the accuracy, quality, and relevancy of the connection [1].

Relative methods based on GNSS technology are conditioned by simultaneous measurement using at least two apparatuses. One apparatus observes a known point (called the reference), and the other is at the point to be determined (called the rover). The principle of the relative determination of the spatial position of the points assumes the same effect of measurement errors both at the reference point and at the point to be determined. Corrections can be determined by comparing the measurement results and the known spatial coordinates of the reference point; subsequently, the corrections can be introduced in the calculation of the positions of the points to be determined. Corrections can be introduced in real time, through radio or other communication technologies, or when the surveying activity has been completed and processed after the survey in an adequate software—the so-called "post-processing" [1]. The relative method can provide vector determination accuracy in the order of millimeters. Figure 2 presents the surveying set for simultaneous measurement. The reference station for observation at a point of known position is shown on the left, and the rover (the survey network point) apparatus is shown on the right, in Figure 2.

Figure 2. Fast static method; reference station (**left**); rover (**right**).

Before the survey itself, the position of the survey network points, whose position is to be determined using GNSS technology, should be suitably chosen. The position of a point should be duly considered, given that the use of GNSS technology entails a lot of pitfalls during implementation. In particular, the so-called multipath (multiple signal spreading) should be minimized through a sufficient distance from potentially reflective areas (water level, objects, cars in the immediate vicinity of the antenna, etc.) [1].

GNSS technology has its limitations: the spatial position of the points cannot be determined, irrespective of the spatial nature of the given locality. The use of GNSS technology results for geodesic purposes is conditioned by performing two independent measurements. Some measurements may seem independent but may not actually be sufficiently independent. Provided that the two measurements are carried out on two days, always at the same time of the day, the configuration of the satellites is the same for both measurements. In such a case, the layout of the satellites used for the two implemented measurements is the same at the same time. It follows from the above that measurements should be completed on various days and also at various times of the day.

The determined positions of points using GNSS technology result in spatial coordinates in the coordinate system ETRS-89. The position of the survey network points expressed in the coordinate system S-JTSK is, thus, given by the subsequent spatial transformation using a transformation key. The use of a local transformation key is advisable. Methods based on GNSS technology provide limited use. In many cases, classical terrestrial methods or their combinations must be used to connect a locality. The most commonly used methods for connection measurements include a polygon series complemented with regions (auxiliary points). The altitudinal connection of a locality can be completed using the geometric leveling method, which provides technical-grade or accurate leveling (based on the requirements for connection accuracy). The accuracy of determining the survey network point's altitudes is often also completed using trigonometry for these purposes [1,2].

The combined fast statistical method, with its own reference station and a polygon series with bilateral connection and bilateral orientation, was used for the survey works whose results are presented herein, to connect the locality into the coordinate system S-JTSK. The purpose was to determine the position and altitude of two points in the northern and southern parts of the locality of concern. These points were used as the initial points for the polygon series. The polygon series ensured coverage with survey network points within the whole range of the mapped locality. The altitudes of the survey network points were then determined using trigonometry methods, as part of building the survey network points using the polygon series [2].

The summary below compares the methods used to implement the survey works. The measurement method based on GNSS technology undoubtedly provides the advantage of a relatively short measurement time with no need for visibility among the points to be determined. GNSS technology has its limits, and it cannot be applied everywhere. Permanent signal reception is needed, which cannot be achieved in some localities (forests and dense vegetation). In addition, another limitation is related to adherence to the principles to eliminate those factors that have a negative impact on the measurement. The polygon series provides the particular advantage of adjustment to the local situation in the field. When the polygon series is combined with regions (auxiliary points), the position and altitude of a survey network point can be determined virtually anywhere. Visibility between the station and the point to be determined is the only condition. The time demands of the measurement are the disadvantage. GNSS technology seems to be suitable in a clear, uncomplicated field, with no immediate presence of elements that could restrict the use of the technology. GNSS technology cannot compete with the polygon series within the whole scope. However, an effective solution of connection was achieved by combining GNSS technology and the classical geodesic method [2].

4. Detailed Survey of River Bed and Surrounding Areas

A detailed topographic and altitudinal situation survey of the actual situation of the locality is the most demanding stage of surveying activities, in terms of time. The subject of mapping the topography or altitudinal situation is represented by the bank lines of the

Stonavka River bed, including all left and right tributaries (the upper and lower edges of the bed, including axial points).

Furthermore, there are objects situated within the watercourse (weirs, horizontal steps, bridges, supporting bridge foundations, and fords), roads including forest ways, reinforced or non-reinforced ways, buildings (residential houses, agricultural buildings, and huts), land culture borderlines, fences, service and utility lines over the ground (power line poles, etc.), elements of underground utility networks found over the ground (drainage trapdoors, shafts, drains, hydrants, etc.), field edges, gutters, and grooves. Figure 3 shows photos of the measured river and used total station, a Leica MS 60. In geodetic practice, we have recently often used robotic total stations. The advantage is that it is a so-called one-man system. A second person is not required to measure, as the total station is remotely controlled via the controller, as shown in Figure 3. The Leica Nova MS60, a multistation, enables a person to perform all measurement tasks with one instrument. Combining the total station's capabilities and 3D laser scanning enables scans, for example, of the details of objects on a watercourse [3,4].

Figure 3. Detailed surveying of the river using a total station.

The selection of detailed points whose position and altitude should be surveyed depends on many factors. In particular, it depends on the resulting scale of the mapping output and on the spatial characteristics of the model locality. Together with the selection of detailed points, the number of surveyed detailed points should also be balanced. Although certain general principles or rules for selecting detailed points do exist, any individual mapped locality must be individually approached during the process of selection. The selection of detailed points is affected by the subjective view and ability of the surveyor to capture the actual terrain. No mapping can do without a certain higher or lower degree of generalization (simplification) of the terrain. Detailed points should be selected at the determining lines of the terrain skeleton, where the slope of the terrain shows a marked change, while the more regular intervals of points can be considered in not very indented areas. The detailed positional and altitudinal surveying of the given locality was completed in the classical terrestrial mode (i.e., using the total station), using the tachymetry method. At the same time, an RTK method was tested, based on GNSS technology. The purpose was to find out the usability and accuracy of this method in addressing the given problem. Testing was completed at the selected detailed points. [5]

The final evaluation of the detailed surveying of the topography and altitudinal situation consists of a comparison of both methods and of determining their advantages and disadvantages. The purpose was to test the RTK method in a part of the mapped area, in its southern part. Here the terrain was covered by not very dense vegetation, and the local conditions were assumed to allow for surveying an integral area using the RTK method. These assumptions were not confirmed, and dead zones occurred. The RTK method is very suitable for mapping uncomplicated terrains where no elements are present that would prevent signal reception. In such a case, the measurement period can be considerably shortened, making the work easier. For localities of this type, the RTK method cannot compete with classical tachymetry. Tachymetry can be adapted to the terrain virtually anywhere.

5. Mapping and Other Documentation

The detailed surveying of the watercourse results, for obtaining the mapping documentation of the actual condition of the Stonavka River bed, includes supporting documentation. A scale of 1:500 was chosen based on the requirements, following the purpose of the map and considering the spatial characteristics of the locality. The additional documentation includes the longitudinal profile of the Stonavka River together with cross-sections of the bed. A scale of 1:2000 was chosen for the lengths, and a scale 10 times higher was chosen for the altitudes, i.e., 1:200. The mapping documentation is documented in Figure 4.

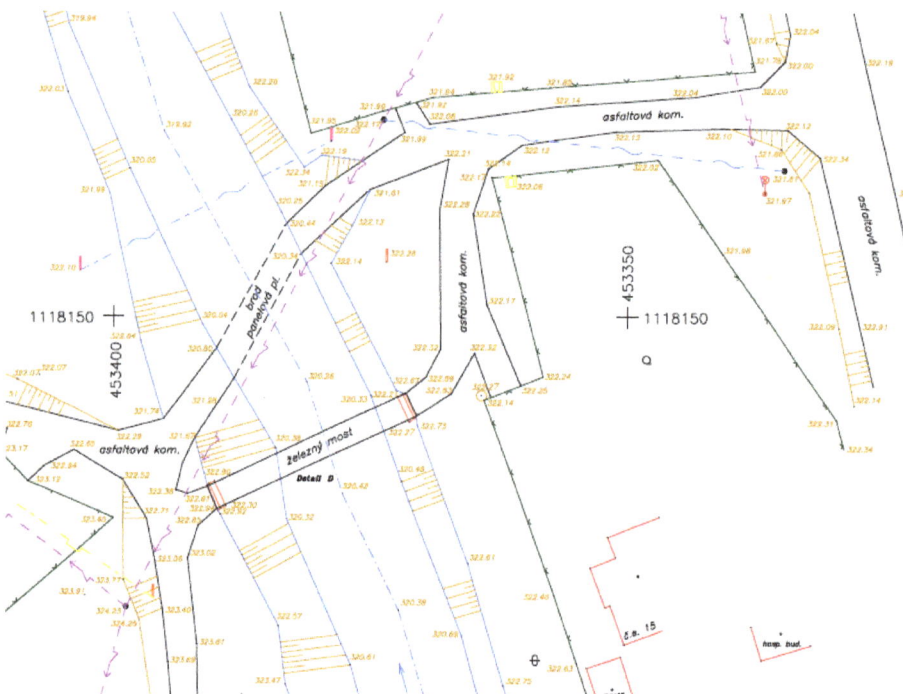

Figure 4. Mapping documentation section of the actual survey (1:500).

6. Determination of the Bed Capacity Potential–Cubic Capacity and Digital Model of Terrain

The mapping results also included an approximate determination of the capacity potential of the watercourse bed in the mapped segment. The cubic capacity of the Stonavka River bed was calculated using an approximate method. This method is based on dividing the entire watercourse segment into partial cross-sections that form bodies (prisms) defined by any pair of adjacent cross-sections and the distances between them (the height of the body).

$$V = \sum_{i=1}^{n} \left(\frac{P_i + P_{i+1}}{2} \right) \times s_{i,i+1} \tag{1}$$

Figure 5 illustrates the body used to replace the irregular shape of the watercourse bed. In the relationship for calculating the partial cubic capacity, V is the cubic capacity, P is the section area, and s is the distance of the sections [5].

Figure 5. Diagram to determine the cubic capacity of the watercourse bed (1).

The use of the approximate method of calculating the volume of the watercourse bed using cross-sections was related to the customer's request. The calculation was used to compare the volume of individual sections calculated on the basis of cross-sections and with the use of a digital terrain model. The customer only requested to verify the calculation of the cubic volume determined by the classic method of cross-sections using a digital terrain model and the created cross-sections.

The cross-sections of the Stonavka River bed and of the right tributary Mlynka are presented in Figure 6. The cross-sections are completed with the sizes of the flow profile area.

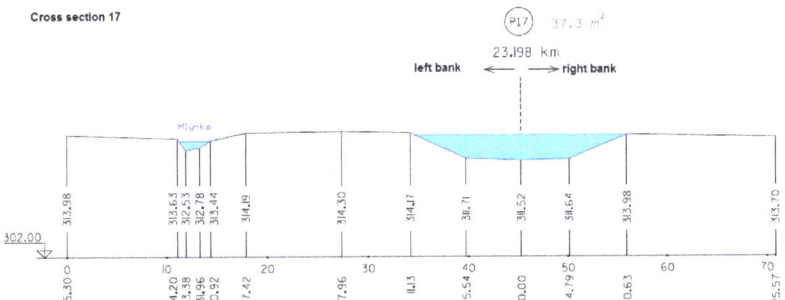

Figure 6. Diagram to determine the cubic capacity of the watercourse bed (2).

The digital terrain model (Figure 7) is used for the spatial interpretation of the result of the detailed survey of the locality of concern. A digital terrain model is a digital representation of the Earth's surface, composed of data and an interpolation algorithm. The digital terrain model allows to determine the heights of intermediate points. Using the digital terrain model, we were able to describe the subject of focus in detail. ATLAS DMT software was used to create a digital terrain model. The ATLAS DMT program with commonly available hardware can process large-scale models based on an input list of the coordinates of detailed points obtained by terrestrial surveying. This enables the creation, modification, and display of a digital terrain model in a special graphic editor, including contours, hypsometry, slopes, descriptions, general and special graphic objects, rasters, 2D/3D DXF export, 3D terrain view options, relief correction, DMT generalization, and densification DMT. A sample of the digital terrain model is shown in Figure 7. The accuracy of the digital terrain model can be increased by defining the mandatory edges.

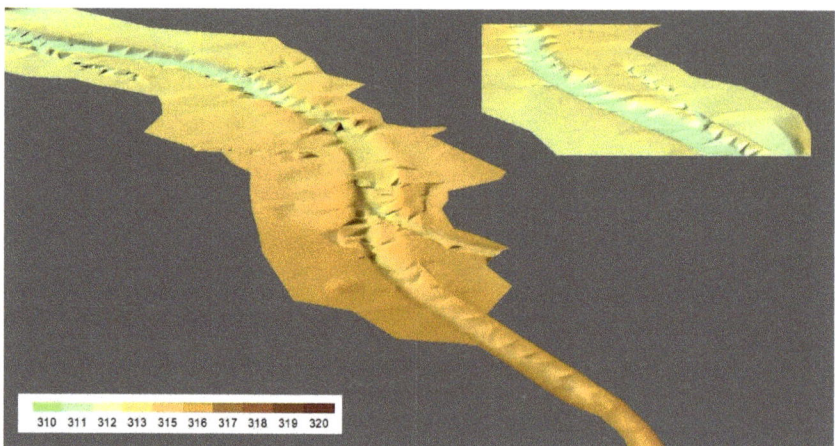

Figure 7. Digital terrain model.

In today's practice, the digital terrain model is also very often connected with the use of unmanned aerial vehicles, so-called drones. However, watercourses are often covered with vegetation. Therefore, in the case of using a drone, mapping is necessary in vegetation rest.

7. Conclusions

The aim was to implement a detailed situation survey of the actual status of the Stonavka River bed in the cadastral territory of Tranovice and to prepare the appropriate graphic and computational documentation. This paper provides insight and explains and describes the activities associated with the preparatory tasks, terrain recognition, and the measurement. This paper also discusses the connection measurement and the survey of the model locality. The documentation can be used, for example, for project purposes (designing modifications of the watercourse bed, anti-flood features, or other technical systems in the watercourse such as footbridges, dams, etc.).

The mapping documentation is completed with longitudinal sections and cross-sections and also with the digital terrain model. Potential users, thus, have the possibility to obtain a more realistic idea of the mapped locality. Project activity in water management, in some cases, requires, in addition to map documentation, the 3D results of terrestrial surveying. An example of 3D map documentation is shown in Figure 8. Drawing the documentation is also important when creating a digital terrain model, which defines the unique terrain edges in 3D.

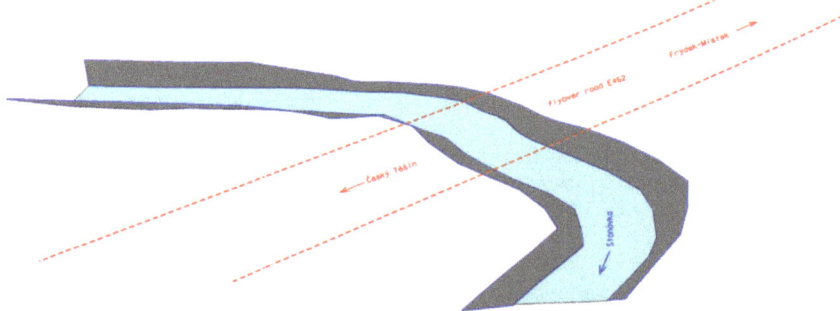

Figure 8. Example of a 3D drawing of a watercourse bed.

Author Contributions: Conceptualization, P.J. and T.G.; methodology, P.J.; software, P.J.; validation, P.J. and T.G.; formal analysis, P.J.; investigation, T.G.; resources, T.G.; data curation, P.J.; writing—original draft preparation, P.J.; writing—review and editing, T.G.; visualization, P.J.; supervision, P.J.; project administration, P.J.; funding acquisition, P.J. All authors have read and agreed to the published version of the manuscript.

Funding: This research was funded by SGS, grant number SP2023/041.

Institutional Review Board Statement: Not applicable.

Informed Consent Statement: Not applicable.

Data Availability Statement: Data is contained within the article.

Conflicts of Interest: The authors declare no conflict of interest.

References

1. Mikoláš, M.; Dandoš, R.; Subiková, M. Measuring shifts base to calibrate test equipment gnss. *Geod. Cartogr.* **2013**, *39*, 1–6. [CrossRef]
2. Staňková, H.; Černota, P. A principle of forming and developing geodetic bases in the Czech Republic. *Geod. Cartogr.* **2010**, *36*, 103–112. [CrossRef]
3. Shamim, S.; Jafri, S.R.u.N. A Comparative Study of Multiple 2D Laser Scanners for Outdoor Measurements. *Eng. Proc.* **2023**, *32*, 16. [CrossRef]
4. Sokol, Š.; Bajtala, M.; Jezko, J.; Černota, P. Testing the accuracy of determining 3D cartesian coordinates using the measuring station S8 Trimble DR Plus ROBOTIC. *Inz. Miner.* **2014**, *15*, 85–90.
5. Vrublová, D.; Kapica, R.; Gibesová, B.; Mudruňka, J.; Struś, A. Application of GNSS technology in surface mining. *Geod. Cartogr.* **2016**, *42*, 122–128. [CrossRef]

Disclaimer/Publisher's Note: The statements, opinions and data contained in all publications are solely those of the individual author(s) and contributor(s) and not of MDPI and/or the editor(s). MDPI and/or the editor(s) disclaim responsibility for any injury to people or property resulting from any ideas, methods, instructions or products referred to in the content.

Proceeding Paper

Problematic Perspectives of Units of Military Fortification in Landscape Management (Teschen Silesia, Czech Republic) †

Jiří Kupka [1,*], Adéla Brázdová [2] and Jana Vodová [1]

1. Department of Environmental Engineering, Faculty of Mining and Geology, VŠB—Technical University of Ostrava, 708 00 Ostrava-Poruba, Czech Republic; jana.prymusova@gmail.com
2. Department of Building Materials and Diagnostics of Structures, Faculty of Civil Engineering, VŠB—Technical University of Ostrava, 708 00 Ostrava-Poruba, Czech Republic; adela.brazdova@vsb.cz
* Correspondence: jiri.kupka@vsb.cz
† Presented at the 4th International Conference on Advances in Environmental Engineering, Ostrava, Czech Republic, 20–22 November 2023.

Abstract: This paper is focused on problematic perspectives of former units of casemates with enhanced fortification of the Czech borderlands landscape. These units represent military brownfields but also a functional system that interacts with surrounding nature, landscape character, and human society and has value in themselves. This paper presents various perspectives of a multidimensional approach to these objects (brownfields, ecological functions, etc.).

Keywords: brownfields; socio-economic sphere; faunistic survey; functional potential; genius loci

1. Introduction

This paper focuses on the issue of brownfields, specifically military brownfields, which can be approached from different perspectives. Different points of view on these military objects allow us to realize that some of them, although "abandoned and unused" cannot be considered insignificant but they can also have other functions. Military objects (especially units of military fortification complex—UMF) can then be viewed not only from the historical perspective (socio-economic sphere as cultural heritage) but also from the perspective of natural conditions (living and non-living nature, natural processes, and effects), functional relationship (between man and landscape, socio-cultural sphere) or genius loci (spirit of place, or landscape, interaction and associated emotions between man and landscape) [1]. In fact, however, they represent a functional system that interacts with the surrounding nature, landscape, and human society and are "valuable", even though they may not serve the purpose for which they were created or which they are intended to commemorate. To illustrate this issue (the function and potential of UMFs), one specific example of casemates with enhanced fortifications (UMF-C) was selected from Teschen Silesia (in the Moravian-Silesian region) based on four landscape approaches.

The same object can be seen as "an object without an obvious use, or use is being sought for it" (definition of brownfields) but from another point of view, this may not be the case. In the framework of this study, a large dataset has been collected. Within this dataset, different parameters of UMF-Cs were analyzed: building conditions, ownership, natural conditions, presence of animals or plants, history, or stories (investigation of users' experiences). This paper presents possible perspectives of a multidimensional approach to UMF-Cs, not only as potential brownfields but also as landscape elements that fulfill different functions. The UMF-C was chosen as a model object to present a comprehensive approach to post-military objects as brownfields and to search for their functional significance.

Brownfields are generally considered a threat due to the potential ecological burden and associated socio-economic impacts [2,3]. Their definition implies that they are "…a property, the expansion, redevelopment, or reuse of which may be complicated by the

Citation: Kupka, J.; Brázdová, A.; Vodová, J. Problematic Perspectives of Units of Military Fortification in Landscape Management (Teschen Silesia, Czech Republic). *Eng. Proc.* **2023**, *57*, 18. https://doi.org/10.3390/engproc2023057018

Academic Editors: Adriana Estokova, Natalia Junakova, Tomas Dvorsky, Vojtech Vaclavik and Magdalena Balintova

Published: 1 December 2023

Copyright: © 2023 by the authors. Licensee MDPI, Basel, Switzerland. This article is an open access article distributed under the terms and conditions of the Creative Commons Attribution (CC BY) license (https://creativecommons.org/licenses/by/4.0/).

presence or potential presence of a hazardous substance, pollutant, or contaminant" [4]. In general, the view of brownfields can be summarized as always presenting some challenges. A comprehensive approach to brownfields, illustrated by the example of UMF-C, highlighting the possible positive functions they can fulfill in the landscape can therefore be considered rather unique.

2. Background

From the point of view of the situation in Central Europe before WWII (1935–1938), there was a fear of a war conflict in the former Czechoslovakia [5]. For this reason, the units of the military fortification complex (UMF) were built, following the model of the French fortifications (so-called Maginot Line) [5]. The whole UMF consisted of objects divided according to their function into casemates with enhanced fortification (UMF-C), heavy fortification, and artillery forts. Systematic development of the UMF was started in the territory of today's Moravian-Silesian region (due to geomorphological and geographical conditions). But due to the deterioration of the political situation in former Czechoslovakia (cession of border areas to Germany and Poland in 1938 and the beginning of WWII in 1939), the UMF was never fully completed (this also applies to the territory of present Moravian-Silesian region) [5–7]. The part of the UMF that is located on the Czech side of the Teschen Silesia is specific in that it ran that it was run along an artificial border that divided the sovereign historical entity of the Duchy of Teschen (1281–1918) into two parts in 1920 [8].

During WWII were UMF objects used by the German army. At the end of the war, some of them even served as shelters and key defensive points for retreating German troops against the Soviet army [5–7,9]. After WWII, the vast majority of UMFs lost their strategic and military importance, and over half were removed [10,11]. A total of 9089 UMF-Cs were built in the border area of the former Czechoslovakia, from which 4683 objects have remained [12]. Some found temporary use as storage facilities for military material, or as storage for various materials (e.g., fruits and vegetables, fertilizers, or sprays) [13–15]. Most of the remaining UMFs are abandoned and unused [13–15]. However, it was not until the 1980s and 1990s that the general public became more interested in the phenomenon of the Czechoslovak Units of Military Fortification [5].

3. Materials and Methods

Part of the Teschen Silesia is also located in today's Poland (Województwo Śląskie) [8]. The selected UMF-C is situated in the municipality of Chotěbuz, cadastral area Zpupná Lhota (WGS84 N 49°46.09255′, E 18°35.73995′), at an altitude of 284 m above sea level and is situated approximately 3.5 km as the crow flies from the city center of Český Těšín [16].

The object is signed as XIX/432/A-140Z and was concreted in July 1938 under the auspices of the IV Army Corps, in the construction section Louky (designation XIX) and was part of the so-called "hook on the Olza river" [9,16]. This UMF-C was built in type A-140 with reinforced resistance (designation A-140Z) [16]. This type refers to construction type A, where the proper axes of both gunports are at 140° and the letter "Z" indicates a reinforced plate/front wall [5,16]. In the expected direction of attack, it was provided with a protective backfill (1.5 m thick rock cover, covered with vegetation) [5]. The following approaches were considered in our analysis: description of the current situation (socio-economic sphere), faunistical survey (natural conditions) and search for stories (genius loci). In most cases, the different perspectives are interconnected.

The description of the current situation is based on a field study carried out on 14 February 2023. The survey focused on the current state of construction or the degree of potential damage to the building, ownership relations, and the nature of the interior and character of the surrounding environment (the general description of UMF-C as brownfield). Several of the datasets obtained from the field survey on February 2023 were compared to the results from the field survey on 14 February 2014 (Appendix A).

During the process of examining UMF-C in the Moravian-Silesian Region as a specific type of brownfields, guided public interviews have been conducted continuously since 2020, and a total of 74 respondents were queried (results of the search of stories). For our purposes, respondents answered questions from a variety of fields (e.g., "Can you describe in what context you first encountered the term 'military bunker' or when you first visited it? Which Czechoslovak fortification object did you personally visit and when? What was the main reason for your visit? What motivated you to visit?"). The results of this survey also help us to get an idea of whether a given place has a specific genius loci. For illustrative purposes, the results include the responses of one respondent that relate to a named UMF-C.

One of the other parameters of a comprehensive approach to UMF-Cs as a specific type of brownfield is the study of natural conditions. For this conference paper, we present only part of the natural condition's characteristics through a faunistical survey. The faunistical survey was carried out in the interior of UMF-C No. XIX/432, focusing on the monitoring of overwintering butterfly species (Results of Lepidoptera survey in UMF-C). This survey was carried out on 14th February 2014 and 14th February 2023. Overwintering Lepidoptera in the indoor space of the UMF-C was studied by using conventional flashlights. Recorded species were examined on the walls and ceiling or on the floor. The abundance of the butterfly species found was recorded. In this case, butterflies serve as a model group because there are only a small number of butterflies overwintering in the adult stage, they are easy to determine, and the various butterfly species actively choose suitable places to overwinter.

The obtained results were consequently discussed in terms of the functional potential of the object under study and thus as a specific type of brownfield.

4. Results

This chapter presents results that provide insight into the issues surrounding our approach to brownfields.

4.1. The General Description of UMF-C No. XIX/432 as a Brownfield

In terms of the categorization of brownfields, UMF-Cs (or UMF objects in general) can be considered military brownfields, which in general can represent some of the environmental problems [17,18]. According to publicly available information, UMF-C no. XIX/432 is located on parcel no. 611/30 in the cadastral territory of Zpupná Lhota (municipality Chotěbuz) [19]. This parcel is divided between a total of four private owners but UMF-C no. XIX/432 itself belongs to only one owner (other than the owners of the parcel under UMF-C) [19]. The surrounding parcel 611/4 is divided between 2 private owners; however, none of them is even a shareholder of the parcel under the UMF-C no. XIX/432 or the UMF-C no. XIX/432 itself [19].

UMF-C no. XIX/432 is situated in a pasture. The southwest side is visibly but only slightly damaged; the entrance bar and the armored door are not preserved. Steel rods for attaching camouflage nets or barbed wire (to secure the building) are preserved on the ceiling. The loopholes are intact but without covers. Nothing remained of the internal equipment except the grenade drop muzzle with cover, brackets for mounting equipment, and periscope tubes. Water is leaking into the UMF-C through the periscope tubes (alternatively, water is leaking into the building through a structural defect). The building contains waste of various kinds (stones, wood, remnants of agricultural equipment). A more detailed description including selected parameters from the two visits to the site is provided in the sample UMF-C passport survey, including a comparison of the results from the various visits to UMF-C No XIX/432; see Appendix B.

4.2. Results of the Search of Stories of UMF-C No. XIX/432

As a part of the search for stories, the following statement by one respondent was recorded in the UMF-C no. XIX/432 (quote): "I can see it like today. It was in the autumn of the 1980s and I was on an expedition with my father as a young boy in the vicinity of Těšín. It was getting dark when a strange structure with a black hole as an entrance appeared in the field in front of me. Seeing the desire in my eyes, my dad didn't hesitate for a moment: he took some papers out of his wallet (he always had a wallet full of them), lit them on fire, and with the help of this light we slowly walked inside... This experience is so imprinted in my memory that it influenced my lifelong interest in bunkers and various underground spaces. I wanted to return to this bunker myself. I managed to "find" it again when I was 14 years old. A group of local boys had a clubhouse in the bunker. We fought fierce battles with these boys and these battles were often quite dangerous (because we used homemade slingshots). During the battles, which were incredibly real for us, we mostly "played" on partisans and Germans..."

At the same time, during the survey of UMF-C no. XIX/432, signs were also found of non-military use of the building, i.e., not as part of the history of the site but activities related to the use of the genius loci. These included the inscription "Klubovna" ("Clubhouse") next to the warning sign "Vstup zakázán, nebezpečí ohrožení života" ("Entry forbidden, danger to life"), "TGM" (an abbreviation of the name of the first Czechoslovak president Tomáš Garrigue Masaryk) or the remains of simple furniture with which the "Clubhouse" was probably equipped (some of the inscriptions are shown on Appendix C. There is also rubbish inside the building.

4.3. Results of Lepidoptera Survey in UMF-C No. XIX/432

A total of 6 species of overwintering butterflies have been documented in UMF-C no. XIX/432 in a total of 35 live individuals during the 2023 survey (see Appendix D). The European peacock (*Inachis io*) was the most abundant butterfly, with a total of 19 individuals (compared to 2014 when 14 individuals were recorded). *Scoliopteryx libatrix* (7 and 9 individuals, respectively) and *Aglais urticae* (7 and 1 individual, respectively) were also more abundantly represented. A general overview of all overwintering butterfly species found each year is given in the graphs in Appendix E.

5. Discussion

UMF-C no. XIX/432 from a general point of view represents a military brownfield that is not associated with a specific ecological burden. Considering the total number of remaining UMF-Cs on the territory of the Czech Republic and the fact that no significant wartime operations occurred on this site during WWII, this object is not considered historically or architecturally valuable. Although this object may seem to be completely abandoned and unused (corresponding to the usual definition of a brownfield), and although it has its owners, closer investigation has shown that this does not correspond to reality—it interacts not only with human beings but also with nature—its hidden functional potential has been identified. Based on the field surveys carried out in 2014 and 2023, it can be concluded that there have been no significant changes, particularly in terms of the potential devastation of the building (see Appendix B). The immediate surroundings of the object have been cleared of landfill. If the surrounding environment continues to be used as an extensive pasture or field, then the continued overgrowth of the backfill and the formation of an "island in the landscape" can be assumed (the object itself as a "cliff with a cave"; the sharp contrast of the vegetation on the ceiling of the object and on the backfill in relation to the surrounding uniform vegetation, etc., this is the subject of further study) (see Appendix A). The table presented in Appendix B, which is an example of the passporting of UMF-C objects, may represent a suitable elementary tool for the passporting of these objects in the context of landscape management.

Based on the on-site investigations realized on the UMF-C No. XIX/432 and from the guided interviews, there were signs found of human use of the property (see Appendix B), and by securing the statements of the respondents, it can be concluded that the property is or has been used as part of a hidden curriculum of land use (clubhouse) (see Appendix C) [1]. From this point of view, although this property can be considered as a brownfield site, it still has its 'users' (although without ownership or tenancy) and therefore fulfills a positive social function. In spite of, or perhaps because of, the fact that at the entrance it says, 'Military facility, no entry, danger to life', and war is clearly associated with danger, as it evokes just the right emotion in the imagination of these 'users'—the use of genius loci in play activities. The use of this building as a clubhouse cannot be considered anything other than positive based on the analysis carried out (it is not a squat, but an experience of 'military romance', there is no damage to the building). In this context, it is important to mention that this particular hidden curriculum of human use of a landscape element is associated with the history of the place. The history of the place, in connection with the genius loci, evokes the corresponding ideas of the users. In this way, children, for example, may develop a future interest not only in these objects but also in history in general. For example, the nearby UMF-Cs XIX/425, XIX/428, and XIX/429 are currently managed by the Chotěbuz Museum of History. "Playing and having fun" has thus become a local museum for the wider public. In view of the above facts, it is possible to say about the studied object UMF-C no. XIX/432 that although it does not yet fulfill this added value for the public, it fulfills another social function. Of course, the UMF-C buildings are currently used for a variety of other purposes (for example, as cottages, hotels or 'works of art' [20–22].

UMF-C objects can also fulfill an ecological function. For example, the results of the survey of overwintering butterflies as a model animal group of the interior environment of the UMF-C XIX/432 for the purpose of this study show that the presence of common species was recorded. Among the more interesting records is that of *Nymphalis polychloros*, which, although a relatively widespread species, is not abundant anywhere. Moreover, its abundance fluctuates considerably over the years, and so it may become quite rare in some years. During the vegetation period, it stays in the tree canopy, which to some extent affects the information on its actual distribution. It is therefore understandable that the number of all species of overwintering adult butterflies may vary considerably from year to year. However, the fact that these objects are important for overwintering is aptly illustrated by this group of animals. The butterfly overwintering site, in this case, a bunker, is discovered by individuals at random and used for these purposes. Overwintering butterflies are an appropriate group to illustrate the interaction between living nature and brownfields. Other suitable model groups may include spiders, overwintering bipeds, mollusks, but also lichens, and bryophytes.

6. Conclusions

Our research shows that UMF-Cs provide different types of interactions between humans and the landscape. The classical view of brownfields, which is usually based on the underuse of buildings or their disturbance to the landscape, is partly disputed [17]. It is questionable whether it is not appropriate to consider changes in the categorization of brownfields and to take this comprehensive approach to brownfields into account in landscape management.

Our research focuses on the functional potential of the UMF-C no. XIX/432 and is intended to contribute to changes in the classification of brownfields. The question is if various objects should still be considered brownfields when their significance is proven. In the case of the UMF-Cs, which play a role in the hidden curriculum of human use of the landscape and are used for overwintering butterflies. Considering the total number of UMF-Cs in the Czech Republic, it is difficult to imagine that all of them will be used in the future within the framework of the usual or modified categorization of brownfields. On the other hand, their damage or removal could be considered as the destruction of a historical legacy and perceived negatively by the public (see, for example, the large wave of negative reactions from the general public to the cutting of UMF-Cs with a diamond rope; however, this act is considered by the author himself as a "work of art" [20]).

It remains an open question whether brownfields (or at least some types of brownfields) should be viewed in different ways since in the context of this paper there is no further need to find new uses for them.

Author Contributions: Conceptualization, J.K. and A.B.; methodology, J.K. and A.B.; investigation, J.K., A.B., and J.V.; resources, J.K. and A.B.; data curation, J.K., A.B. and J.V.; writing—original draft preparation, J.K. and A.B.; writing—review and editing, J.K. and A.B.; visualization, A.B.; supervision, J.K. All authors have read and agreed to the published version of the manuscript.

Funding: This research received no external funding, and the APC was funded by the Faculty of Mining and Geology, Department of Environmental Engineering.

Institutional Review Board Statement: Not applicable.

Informed Consent Statement: Not applicable.

Data Availability Statement: Data are contained within the article.

Conflicts of Interest: The authors declare no conflict of interest.

Appendix A

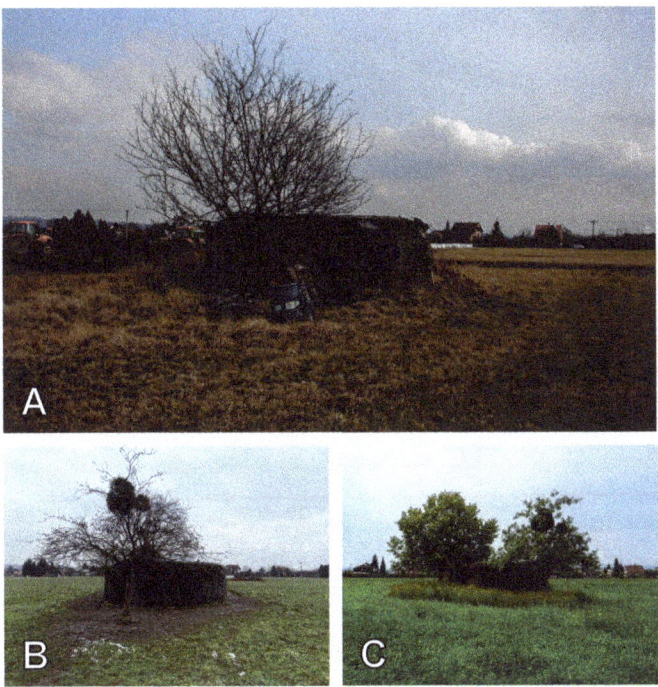

Figure A1. The bunker can also be seen as an "island" in a pasture. UMF-C no. XIX/432 in different years and at different seasons: (**A**) winter 2014; (**B**) winter 2023; (**C**) spring 2023. Authors: Jiří Kupka (**A**,**B**); Adéla Brázdová (**C**).

Appendix B

Table A1. Example of passporting UMF-Cs on the example of UMF-C no. XIX/432, including comparison of results from different dates of visits.

Date		Description of UMF-C No. XIX/432
14 February 2014		**Interior**
	floor	dry, none or only isolated organic and inorganic material on the floor
	walls	without woodwork and unplastered, with white painting; both main loopholes are open and with red preserve painting
	ceiling	without woodwork and unplastered, with white painting, sign of wood paneling on the ceiling; both holes for periscope fitting accessible
	usage	high intensity of use, the building serves as a clubhouse
	accessibility	the building is easily accessible; entrance and loopholes are accessible (respectively, not inaccessible due to bushes or burial)
		Surrounding environment
	object surroundings	there is rubbish in front of the entrance; apart from the building itself, two trees are as dominant (*Tilia cordata* and *Malvus domestica*)
	type of habitat	extensively used pasture
14 February 2023		**Interior**
	floor	flooded, none or only isolated organic material and significant amount of waste of inorganic material on the floor
	walls	without woodwork and unplastered, with white painting; both main loopholes are open and with red preserve painting
	ceiling	without woodwork and unplastered, with white painting; both holes for periscope fitting accessible
	usage	the object bears only small traces of human use; findings of airsoft pellets
	accessibility	the building is easily accessible; entrance and loopholes are accessible (respectively not inaccessible due to bushes or burial)
		Surrounding environment
	object surroundings	apart from the building itself, two trees are as dominant (*Tilia cordata* and *Malvus domestica*); stands of blackberry (*Rubus* sp.) on the backfill
	type of habitat	extensively used pasture

Appendix C

Figure A2. Inscriptions "Klubovna" ("Clubhouse") next to the warning sign "Vstup zakázán, nebezpečí ohrožení života" ("Entry forbidden, danger to life") can be seen like UMF-C's "hidden curriculum". Author: Adéla Brázdová.

Appendix D

Figure A3. An example of selected common species of overwintering butterflies (Lepidoptera) in UMF-C no. XIX/432 found in 2023: (**A**) *Aglais urticae*, (**B**) *Hypena rostralis*. Author: Jiří Kupka (**A**,**B**).

Appendix E

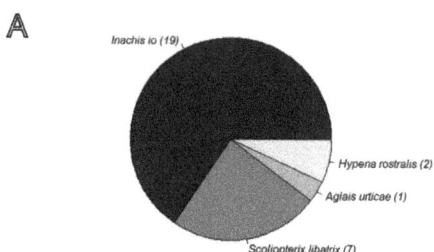

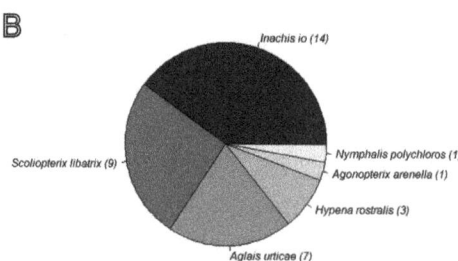

Figure A4. Schemes show results of overwintering butterflies in the UMF-C no. XIX/432 and their distribution into individual species in years (**A**) 2014, (**B**) 2023.

References

1. Kupka, J.; Brázdová, A.; Vodová, J. Units of Military Fortification Complex as Phenomenon Elements of the Czech Borderlands Landscape. *Land* **2022**, *11*, 79. [CrossRef]
2. Davis, T.S. *Brownfields—A Comprehensive Quide to Redeveloping Contaminated Property*, 2nd ed.; American Bar Association: Chicago, IL, USA, 2003; 1136p.
3. Tang, Y.-T.; Nathanail, C.P. Sticks and Stones: The Impact of the Definitions of Brownfield in Policies on Socio-Economic Sustainability. *Sustainability* **2012**, *4*, 840–862. [CrossRef]
4. Government Printing Office. Small Business Liability Relief and Brownfields Revitalization Act. U.S. Available online: https://www.congress.gov/107/plaws/publ118/PLAW-107publ118.pdf (accessed on 2 October 2023).
5. Kupka, V.; Čtverák, V.; Durdík, T.; Lutovský, M.; Stehlík, E. Československý pevnostní systém z let 1935–1938. In *Pevnosti a opevnění v Čechách, na Moravě a ve Slezsku*, 2nd ed.; Libri: Praha, Czech Republic, 2002; pp. 268–414.
6. Kuchař, P. *Guide to the Hlučín-Darkovičky Czechoslovak Fortification Complex*, 1st ed.; Retis Group: Ostrava, Czech Republic, 2012; p. 28.
7. Aron, L. *Československé Opevnění 1935–1938*, 1st ed.; Okresní Muzeum Náchod: Náchod, Czech Republic, 1990; p. 199.
8. Stolařík, I.; Štika, J.; Tomolová, V. *Těšínsko 1. díl*, 1st ed.; Muzeum Těšínska and Valašské Muzeum v Přírodě: Český Těšín, Czech Republic; Rožnov pod Radhoštěm, Czech Republic, 1997; pp. 73–102.
9. Rucki, M.; Durčák, J.; Hejda, M. *Opevnění IV. sboru: Část 1—Záchytný hák na řece Olši*, 1st ed.; AVE Centrum: Opava, Czech Republic, 2007; p. 152.
10. Natura Opava—Unplanned Use of Border Fortification Bunkers in the Hlučín and Opava Regions. Available online: http://www.natura-opava.org/opavsko/zpravy/neplanovane-vyuziti-bunkru-hranicniho-opevneni-na-hlucinsku-a-opavsku-2.html (accessed on 10 June 2023).
11. Kolář, O. *Armáda 4—Ostravská Operace 1945*, 1st ed.; Jakab: Bučovice, Czech Republic, 2019; p. 52.
12. Svoboda, T.; Lakosil, J.; Čermák, L. *Velká Kniha o Malých Bunkrech—Československé Opevnění 1936–1938*, 1st ed.; Mladá Fronta: Praha, Czech Republic, 2011; p. 321.
13. Břeclavský Deník—Klub Vojenské Historie Sídlí na Pohansku. Available online: https://breclavsky.denik.cz/zpravy_region/pohansko_bunkry20070525.html (accessed on 10 June 2023).
14. Česká Televize—Během několika let Vystavělo Československo Systém Pevností. Měly Zbrzdit Němce, než Přijdou Na Pomoc Spojenci. Available online: https://ct24.ceskatelevize.cz/domaci/2588257-bunkry-na-hranicich-pred-80-lety-marne-cekaly-na-vyzyvatele-dnes-lakaji-turisty (accessed on 10 June 2023).

15. Oživlý Svět Technických Památek: Betonové Pevnosti. Available online: http://www.technicke-pamatky.cz/sekce/38/betonov-pevnosti/ (accessed on 10 June 2023).
16. Řopíky.net—Informace o Lehkém Opevnění z Let 1936–38. Available online: http://www.ropiky.net/dbase_objekt.php?id=1075727562 (accessed on 10 June 2023).
17. Oliver, L.; Ferber, U.; Grimski, D.; Millar, K.; Nathanail, P. The scale and nature of european brownfields. In Proceedings of the CABERNET 2005—International Conference on Managing Urban Land LQM Ltd., Belfast, Northern Ireland, UK, 13–15 April 2005.
18. Vojvodíková, B.; Fojtík, R.; Tichá, I. Design and Verification of a Simple Approach to Brownfields Categorization. *Sustainability* **2021**, *13*, 11206. [CrossRef]
19. Český úřad Katastrálně Zeměměřičský—Nahlížení do Katastru Nemovitostí. Available online: https://nahlizenidokn.cuzk.cz/VyberParcelu/Parcela/InformaceO (accessed on 13 June 2023).
20. iDNES.cz—Umělec Odkrojil kus Betonového Bunkru. Nevkusné, Zlobí se Fanoušci Řopíků. Available online: https://www.idnes.cz/brno/zpravy/umelec-odkrojil-kus-bunkru-vratenin-kritika-ondrej-belica.A170504_2323167_brno-zpravy_krut (accessed on 10 June 2023).
21. E-magazín E15—Z Řopíku se za Padesát Tisíc Korun Stala Atypická Chatka. Available online: https://www.e15.cz/magazin/z-ropiku-se-za-padesat-tisic-korun-stala-atypicka-chatka-1374175 (accessed on 10 June 2023).
22. Bunkering.cz—Nevšední Zážitek pro Každého. Available online: https://bunkering.cz/ (accessed on 10 June 2023).

Disclaimer/Publisher's Note: The statements, opinions and data contained in all publications are solely those of the individual author(s) and contributor(s) and not of MDPI and/or the editor(s). MDPI and/or the editor(s) disclaim responsibility for any injury to people or property resulting from any ideas, methods, instructions or products referred to in the content.

Proceeding Paper

Revitalization of Small Watercourse [†]

Jitka Chromíková *, Veronika Rogozná and Tomáš Dvorský

Faculty of Mining and Geology, Department of Environmental Engineering, VSB—Technical University of Ostrava, 17. Listopadu 2172/15, 70800 Ostrava, Czech Republic; veronika.rogozna@seznam.cz (V.R.); tomas.dvorsky@vsb.cz (T.D.)
* Correspondence: jitka.chromikova@vsb.cz
[†] Presented at the 4th International Conference on Advances in Environmental Engineering, Ostrava, Czech Republic, 20–22 November 2023.

Abstract: The paper deals with a proposal for a small-watercourse revitalization measure, which is currently regulated by a post-local treatment. The sections of the watercourse are relatively straight and direct. The current shape of the stream does not retain water in the landscape. Any surface water in the watercourse is quickly transferred to the receiving watercourse, which is the important Hvozdnice watercourse. The practical part includes the design of revitalization measures, a new longitudinal profile, and sections. The watercourse is loosened by curves and the total length of the watercourse has been extended by 161.56 m. The route includes a wetland, four pools, and two pond wedges. An economic estimate of the proposed measure was made.

Keywords: revitalization; small watercourse; longitudinal profile; watercourses channel route

1. Introduction

Revitalization in Europe, as well as in other parts of the world, is increasingly including considering physical processes, such as bank erosion, sediment transport, channel incision, and water flow patterns, as necessary conditions for improving river conditions and promoting channel recovery. Today, it is primarily a process-based approach, aimed at restoring natural geomorphic processes and promoting spontaneous physical diversity rather than solving problems using local interventions. Revitalization is mainly concerned with retaining water in catchments, the slowing down of water flows during minimal floods and flooding, and increasing groundwater near water streams [1,2]. This involves the re-meandering of the straight channel of streams [3].

Watercourse revitalization has many benefits. It can improve water quality by eliminating excess nutrients, heavy metals, and organic compounds from river water [4,5]. Revitalization is also one of the strategies for drought adaptation, which is one of the most important challenges for society nowadays. In the Czech Republic, it is mainly hydrological drought which has caused precipitation decline as well as reduced flow rates in watercourses and groundwater levels. When there is no water supply in the soil and other subsurface layers, evapotranspiration and the related cooling effect will be reduced. Changes in the hydrological cycle and water quality, particularly in agricultural irrigation may disrupt the functioning of water infrastructure and increase demand for water abstraction [6,7].

The Czech Republic has adopted a climate-change adaptation strategy in line with the EU adaptation strategy [6,7]. The implementation document is the National Climate Change Adaptation Plan (Action Plan) which defines the adaptation measures in many areas such as forestry, agriculture, urbanization, and landscape water management. In the field of water management, this includes, among other things, the planned support for watercourses and flood plains, such as revitalization of watercourses and floodplains, restoration of natural spillways, and slowing down water flow [8].

It is well known that the water balance of large rivers is based on the flow of small rivers. The hydrology of small rivers, hydrochemistry, and water quality are closely related to the geological–geomorphological, soil-plant conditions and thropogenic processes that occur in the catchment area. The formation of small rivers and their basins is determined by surface flow from regions and landscape complexes, so the impact of natural and human factors on these areas manifests itself more rapidly and clearly [9].

The article deals with the proposal of revitalization measures for an unnamed stream located in the cadastral area of Slavkov near Opava, which is currently modified into straight and linear sections. The character of this stream does not retain water in the landscape; all surface waters are rapidly drained away.

2. Materials and Methods

2.1. Location Description

The watercourse 10214099 (Figures 1 and 2) is an unnamed stream that can also be defined as a left-bank tributary of the Hvozdnice watercourse, 4.185 km. This small watercourse, ID 10214099, has a total length of 2.057 km and flows through the municipality of Slavkov near Opava in the Moravian–Silisian region. The watercourse is located in HOD_0390-Hvozdnice unit from the source to the mouth of the Moravice River. The watercourse is managed by the Odra River Basin, a state-owned enterprise. The watercourse is located in HP 2-02-02-0940-0-00 with a catchment area of 1.58 km^2.

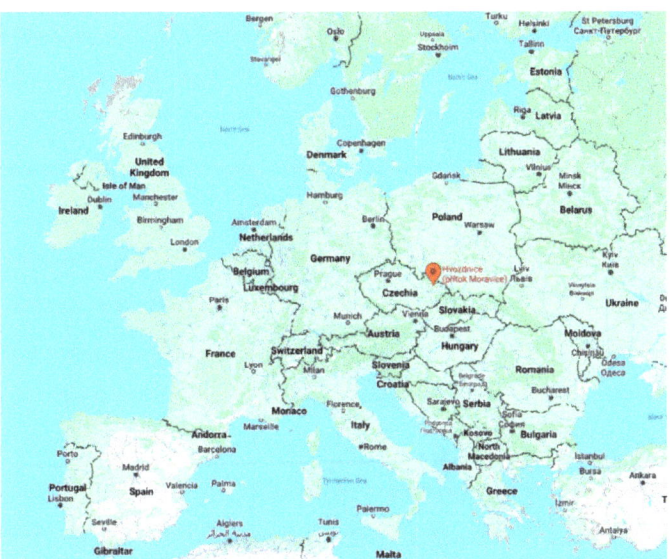

Figure 1. The simplified sketch map—localization in Czech Republic (modified by [10]).

The watercourse flows through two small water reservoirs (ponds), which serve for the rest of the local inhabitants and cross two field culverts and the Olomoucka Road II/46. From the lower pond, Pod Kovalem, the stream was dammed, and after 115 m beyond the railway crossing, it flows into the important Hvozdnice River at 4.185 km.

Figure 2. Watercourses ID 10214099, left—bank tributary of the Hvozdnice watercourse (modified by [11]).

Based on the evaluation of data from the drainage constriction and irrigation information system (ISMS), the watercourse can be concluded to serve as major drainage recipient [12]. According to the assessment of the drainage structures, it is a modified watercourse with open and closed sections [11].

2.2. Watercourses Section

The whole watercourse is divided into five sections (Figure 3). The sections are divided by specific objects—dams, reservoirs, culverts, reservoir, and spring. Points and river miles are marked from the mouth to the source, according to CSN 75 2120 [12].

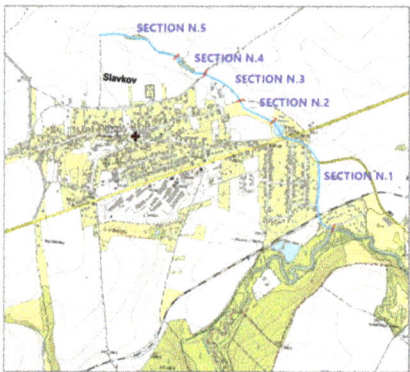

Figure 3. Watercourse sections (modified by [11]).

Section 1: About 810 m long from the mouth of the Hvozdnice River to the Pod Kovalem pond (Figure 4). The watercourse crosses the railway line for about 115 m from the mouth, then runs in a straight open channel to the local road III/461 called Otická. After that, the course of the route is in parallel with this road in the arched part. Then, it follows the dammed stream under the junction of roads II/46 and III/461 (Olomoucká x Otická). In the arched part, the flow is led through the outlet of the pond through an open two-lane fireplace. In the above-described section of the watercourse, no revitalization is planned.

Figure 4. Section 1—Pod Kovalem pond (photo by Rogozná).

Section 2: The revitalized part of the watercourse begins with section 2. It is located between the outlet of the Pod Kovalem pond (km 0.810) and the Na Lůčky culvert (km 1.060). The stream runs above the pond between the fields. In this case, the pollution and degradation of the stream is most evident in agricultural activity, where the stream is cultivated to the banks of both sides (Figure 5). Not far from the Na Lůčky is a junction with a road ditch which drains surface water. The Na Lůčky culvert (Figure 5) consists of a DN800, concrete pipe fitted with a concrete face on both sides. The flow is almost straight in section 2. The channel bed width is 0.3 to 0.5 m. The channel is overgrown with herbaceous vegetation, and, in places, the channel consists only of stones and soil.

Figure 5. Section 2 (photo by Rogozná).

Section 3: Between the Na Lůčky culvert (1.060 km) and Bocianova Street culvert (km 1.335). The Bociana Srteet culvert is a concrete pipe DN800, equipped with a concrete head. The stream flows through forest on the Na Lůčky culvert. On the banks and directly in the stream bed there are mature trees. The left bank of this section of watercourse is cultivated from the adjacent meadow to the shore (Figure 6).

Figure 6. Section 3 (photo by Rogozná).

Section 4: Begins with a field culvert near Bocianovy Street (1.335 km) and ends with a small water reservoir/pond (1.540 km). In the field culvert, the watercourse flows in a concentrated channel into the watercourse, and then flows upstream through dense forest cover. At the top of the forest there is a confluence of surface waters from the adjacent alluvial fields (Figure 7). Above this confluence is the outlet of the pond. The water is discharged from the pond through the DN 600 downstream, which flows into the stream at km 1.660. The water is discharged to this outlet using an open single-slot spillway. In the southern corner of the pond is a safety spillway paved with quarry stone. The small water reservoir (Figure 7) was revitalized between 2017 and 2018. It is a flow-through reservoir with a volume of 1690 m^3 and a length of 70 m (Revitalization of pond parcel No. 1426—Slavkov: Approval for the use of the construction, 2019, Opava). Today, the reservoir is used for capturing surface water, for recreational and landscape purposes.

Figure 7. Section 4 (photo by Rogozná).

Section 5. The last section begins from the outlet to the above-mentioned small water reservoir (1.540 km) and ends with the spring area (2.057 km). The entrance to the pond is 1.737 km. In this part of the stream, the area's agricultural activity is, again, very evident. Between 1.831 and 1.900 km, the stream flows through a field and takes a very sharp trapezoidal channel. The countercurrent from this part of the stream is the last area of vegetation with a very wide channel. In the last 120 m up to the source, the stream channel has a slope of approximately 1:1. (Figure 8).

Figure 8. Section 5 (photo by Rogozná).

2.3. Revitalization Measures

Revitalization methods may include some natural techniques generally used in flood-control demand [2]. As part of the technical revitalization solution, the flow course must be modified by the rotation of the route, creating of the longitudinal profile of the existing channel, and the design of a new longitudinal profile [13]. Small-scale river ecosystems are more vulnerable both to the direct impact of pollutants on them and to the indirect impact. The revitalization of small channel streams can involve re-meandering, during which a sinuous channel is artificially created, and then, to some extent, adjusted, or through a cessation of activities such as dredging and bank stabilization, which simplify or limit the channels [14,15]. In addition to the proposal to change the channel, the design of wetlands, pond systems, and pond wedges is also related, which contributes to biodiversity development and provides more space for shelter for small animals living in the area.

3. Results

3.1. Design of a New Watercourse Channel Route

The design of the new watercourses channel routs (Figure S1) was carried out in accordance with the requirements for revitalization of watercourse channels. The existing route is mainly straight, and the channel passes through inappropriate terrain conditions. The channel is being degraded by landowners adjacent to the watercourse channel and farmers from adjacent parcels. The watercourse was thus regulated, loosened according to the possible spatial arrangement, and thus extended by 161.56 m. The positive impact of expanding the channel's route is to reduce the flow of water flows from the watercourse and to hold more water.

According to the new design, the area above the pond, Pod Kovalem, can be effectively used on the left bank, where the flow is more relaxed and a pool can be placed. On the right bank, the rest of the material from the excavation of the new channel will be flooded. This leads to the formation of occasional pools in the old channel; these pools are filled during the rainy season and dry during the dry season. Na Lůčky Street culvert was moved. In the meadow above this relocated culvert, the most extensive loosening and shifting of the channel away from the houses occurs. The flow through the upper culvert has been maintained in the reservoir and the field culvert near Stork Street. From the outlet to the pool, above the small reservoir, and near the spring, the new channel is mostly guided in the existing valley. This proposal will reduce the amount of landwork in the most important part of the stream and thus reduce the cost of revitalisation.

The design of the new route is directionally similar to the existing stream-channel route, with the addition of pools and wedges added. The design of the new channel consists of alternating counter curves.

3.2. New Longitudinal Profile Design

The new longitudinal profile (Figure S2) is based on the current situation of maintaining three existing structures on the stream. These objects are the end of the Pod Kovalem pond, the preservation of the Bocianovy Street culvert, and a small water reservoir near the Bocianovy Street culvert. The new longitudinal profile incorporates a flowing pool to mitigate the longitudinal slope of the stream bed.

The original average longitudinal profile slope was 19.36‰. The flow axe in the spring area was 301.17 m above sea level, and the outlet of the Pod Kovalem pond was 277.74 m above sea level. The average slope is calculated as simply (301.17−277.74)/1210 (the length of the original channel), and then multiplied by 1000 to obtain the result. In a similar calculation, the length of the stream was changed to 1371.56 m, and the average slope of the new longitudinal profile was 17.08‰.

3.3. Cross-Section Design

The original stream channel has a slope gradient of approximately 1:1, except for 1:2, and the stream channel is straight in many sections and strongly regulated. There are sinkholes or plant deposits in places that form a barrier to flow.

The revitalization design includes 32 cross sections of 50 m each (Figure S3A,B). Two examples of cross-sectional profiles are shown in Figure 9. The proposal has a trapezoidal shape and a slope gradient of 1:3. Due to the reduction in permanent land occupied by private entities and the simplified settlement of property ownership, these slopes are easily linked to the existing terrain. The width of the channel at the bottom is 0.8 and 0.5 m. The bottom of the channel with a width of 0.5 m can be found in the area of the source to the field culvert on the Bocianovy street, and then the bottom of the channel with a width of 0.8 can be found from the outlet to the Pod Kovalem pond.

The flow capacity is designed to be Q10 or more depending on the terrain morphology. The water level of the stream is 0.5 m at Q10 flow.

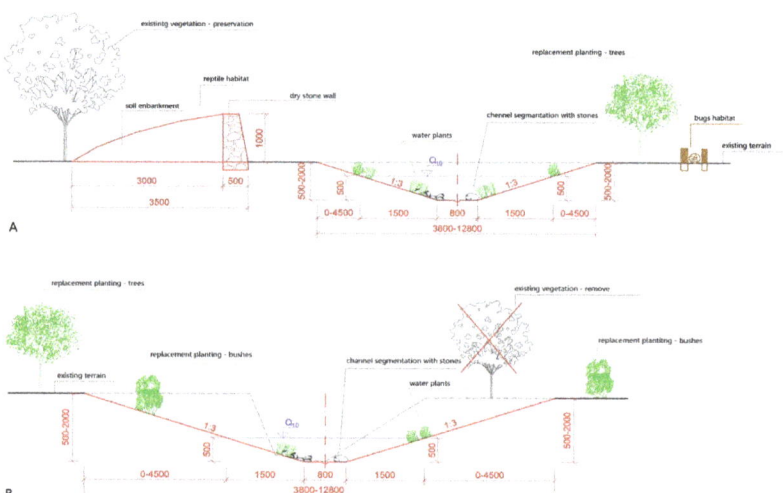

Figure 9. Sample cross profiles (**A**) and Sample cross profile (**B**).

3.4. Pools, Wetland, and Vegeation

The proposed stream length of 1.372 km, including a design for four pools, located at 0.140–0.500 km, 0.350–0.400 km, 0.450–0.500 km, and 0.871–0.900 km. The pool is deep, up to a maximum depth of 0.5 m from the flow level. During normal flows, the normal water level is 0.5 m or higher. Such a permanently high water level in the pool can be achieved both by landscaping the water hole and by subsidizing the water from a small reservoir. At the Q10 flow in the stream, the water in the water pools has a water level of 1 m. The pool is designed with a slope gradient of 1:2–1:8. The cross sections of the pools are shown in the Figures S4–S7. The pools are heavily planted with vegetation ranging from aquatic plants to coastal plants on the pool banks, and the root system stabilizes the pool banks. The choice of another fortification may not be visually and functionally suitable for the landscape. Pools are an appropriate location to catch sediments and protect small reservoirs from excessive sedimentation. Over time, sediments in pools will have to be removed due to siltation.

In the source area, a wetland of 2574 m^2 was proposed, and the terrain conditions were not very favorable. The area will be planted with wetland trees and plants, for example, Sedge, Iris, Chrastice, etc., are suitable plants. Wetland plants grow along the stream banks, shallow pools, and in water-draining areas. Wetland will remain for natural development.

Replacement planting should be established not only near pools, but also along the watercourse according to the geobiogens that are well developed for the Czech Republic. According to maps, the interest area belongs to the Polonska sub-province and, according to geobiocenes, is classified as the Oak vegetation phase where this phase dominates completely [16]. The map documentation indicates that the area of interest can currently be considered to be Linden Oak Forest as it has potential natural vegetation. In the field of interest, we can see more specifically L2.2 and K2.1 habitat types. The biotope L2.2 indicates that there were originally valley ash and alder bushes. The biotope K2.1 indicates that the area's native grass is of clay and sandy willow grass. This is mainly a grass of various species of willow.

3.5. Economic Cost

Actual prices may vary depending on time and location. Other aspects that determine the final cost of changes are, for example, the implementation company, the work time required, material costs, and the removal and processing of materials depending on their composition. We can include the Documentation for the Zoning Decision (DZD), Building Permit documentation (DBP), or joint proceedings.

Other costs may include extensive mapping of the revitalized watercourse in the area, research on pedological and hydrological conditions, new surveys, and geodesic plans for land acquisition. Furthermore, construction works carried out by a specific company specializing in watercourse revitalization and channel improvement. These works include all landworks, as well as the removal and processing of materials (based on laboratory assessment), the relocation of the culvert, the replacement of the plant, and the excavation and the seeding of a new stream path adjacent to the plant.

Other items are the preparation and securing of the development, the organization's facility which plans construction, and engineering activities. All the above indicative costs are presented in Table 1. The cost table does not include the cost of land acquisition, as this is determined on the basis of an expert's report, which varies for each individual plot.

Table 1. Cost estimation.

Cost Estimation	Price without VET (Thousands of EUR)
Exploration and project surveys	
Exploration works	7.50
Project documentation	13.10
Geodetic measurement	4.20
DZD+DBP	17.40
Engineering activity	22.80
DPB	5.0
Exploration and project surveys total	70.00
Construction work	
Site preparation, site equipment, site layout	3.30
Landworks	663.00
Disposal of excavated material	82.90
Replacement planting	10.40
Incidental organisational costs ~10%	76.00
Construction work total	835.60
Finishing activities	
Geodetic survey of the actual construction	8.30
DBP	4.10
Finishing activities total	12.40
Total coast	918.00

4. Conclusions

The length of the stream was increased from 1210 m to 1371.56 m. The transverse profile was changed to achieve a gentler bank slope of 1:3. Along the route are four pools, as well as large wetland. In this work, the treatment of vegetation and replacement planting are proposed in accordance with the applicable geobiocene. However, this work also included an economic estimate, but did not include the purchase price of the land. Revitalization is designed so that after implementation it does not further burden the administrative stream, the revitalization investor, or the water-company owner. Revitalization will initiate the process of naturalization, leaving channel and vegetation to develop in a natural processes.

Supplementary Materials: The following supporting information can be downloaded at: https://www.mdpi.com/article/10.3390/engproc2023057019/s1, Figure S1: Design of a new watercourse channel route; Figure S2: Design of new longitudinal profile of the watercourses 10214099; Figure S3: Cross profiles; Figure S4: Pool N.1; Figure S5: Pool N.2; Figure S6: Pool N.3; Figure S7: Pool N.4.

Author Contributions: J.C., T.D. and V.R. have been involved in all aspects of the manuscript such as conceptualization, methodology, software, validation, formal analysis, investigation, resources, data curation, writing—original preparation, writing—review and editing, visualization, supervision, and project administration. All authors have read and agreed to the published version of the manuscript.

Funding: This article has been produced with the financial support of the European Union under the REFRESH—Research Excellence for Region Sustainability and High-tech Industries project number CZ.10.03.01/00/22_003/0000048 via the Operational Programme Just Transition.

Institutional Review Board Statement: Not applicable.

Informed Consent Statement: Not applicable.

Data Availability Statement: Data sets are contained within the article. Further data and materials requests should be addressed to jitka.chromikova@vsb.cz (J.C.).

Conflicts of Interest: The authors declare no conflict of interest. The funders had no role in the design of the study; in the collection, analyses, or interpretation of data; in the writing of the manuscript; or in the decision to publish the results.

References

1. Bennett, S.J.; Simon, A.; Castro, J.M.; Atkinson, J.F.; Bronner, C.E.; Blersch, S.S.; Rabideau, A.J. The evolving science of stream restoration. In *Stream Restoration in Dynamic Fluvial Systems: Scientific Approaches, Analyses, and Tools*; Simon, A., Bennet, S.J., Castro, J.M., Eds.; AGU: Washington, DC, USA, 2011; Volume 194, pp. 1–8.
2. Choudhury, M.; Pervez, A.; Sharma, A.; Mehta, J. Chapter 15. Human-Induced Stresses on the Rivers beyond Their Assimilation and Regeneration Capacity. In *Ecological Significance of River Ecosystems*; Madhav, S., Kanhaiya, S., Srivasta, v.A., Singh, V., Singh, P., Eds.; Elsevier: Amsterdam, The Netherlands, 2022; pp. 281–298.
3. Feld, C.L.; Birk, S.; Bradley, D.C.; Hering, D.; Kail, J.; Marzin, A.; Melcher, A.; Nemitz, D.; Pedersen, M.L.; Pletterbauer, F.; et al. Chapter Three—From Natural to Degraded Rivers and Back Again: A Test of Restoration Ecology Theory and Practice. In *Advances in Ecological Research*; Woodward, G., Ed.; Academic Press: Cambridge, MA, USA, 2011; Volume 44, pp. 119–209.
4. Darwiche-Criado, N.; Comín, F.A.; Masip, A.; García, M.; Eismann, S.G.; Sorando, R. Effects of wetland restoration on nitrate removal in an irrigated agricultural area: The role of in-stream and off-stream wetlands. *Ecol. Eng.* 2017, *103*, 426–435. [CrossRef]
5. Hester, E.T.; Lin, A.Y.C.; Tsai, C.W. Effect of floodplain restoration on photolytic removal of pharmaceuticals. *Environ. Sci. Technol.* 2020, *54*, 3278–3287. [CrossRef] [PubMed]
6. European Commission. *Communication from the Commission to the European Parliament, the Council, the European Economic and Social Committee and the Committee of the Regions*; An EU Strategy on Adaptation to Climate Change; European Commission: Brussels, Belgium, 2013.
7. Strategy on Adaptation to Climate Change in the Czech Republic. Ministry of the Environment of the Czech Republic: Prague, Czech Republic, 2015. Available online: https://www.mzp.cz/cz/zmena_klimatu_adaptacni_strategie (accessed on 18 August 2023).
8. National Action Plan on Adaptation to Climate Change. Ministry of the Environment of the Czech Republic: Prague, Czech Republic, 2017. Available online: https://www.mzp.cz/cz/narodni_akcni_plan_zmena_klimatu (accessed on 18 August 2023).
9. Alokhina, T. Rivers revitalisation: Approaches to decision. *E3S Web Conf.* 2020, *166*, 01010. [CrossRef]
10. Google. (n.d.). Available online: https://www.google.cz/maps/place/Hvozdnice+(p%25C5%2599%25C3%25ADtok+Moravice)/@46.6102219,6.6950384,5z/data=!4m6!3m5!1s0x4713d9b4483adfc5:0x9be255a0b5ab2b9c!8m2!3d49.9091563!4d17.7005926!16s%252Fg%252F120wkwnj?hl=en&entry=ttu (accessed on 18 August 2023).
11. HEIS VÚV—TGM WRI. Hydroecological Information System; T. G. Masaryk Water Research Institute, Public Research Institution. Available online: https://heis.vuv.cz/data/webmap/isapi.dll?map=mp_heis_voda&TMPL=MAPWND_MAIN (accessed on 18 August 2023).
12. *ČSN 75 2120*; Mileage of Water Stream and Storage. Czech Standard Institute: Prague, Czech Republic, 2019.
13. RRC. *Manual of River Restoration Techniques*; The River Restoration Centre: Cranfield, UK, 2002.
14. Wohl, E.; Lane, S.N.; Wilcox, A.C. The science and practice of river restoration. *Water Resour. Res.* 2015, *51*, 5974–5997. [CrossRef]
15. Rinaldi, M.; Wyżga, B.; Dufour, S.; Bertoldi, W.; Gurnell, A. River processes and implications for fluvial ecogeomorphology: A European perspective. In *Treatise on Geomorphology*; Shroder, J., Butler, D.R., Hupp, C.R., Eds.; Academic Press: San Diego, CA, USA, 2013; Volume 12, pp. 37–52.
16. NCA. Available online: https://aopkcr.maps.arcgis.com/apps/webappviewer/index.html?id=ee190990a1be4ac685d5f7c69c637ae4&fbclid=IwAR2l07QEIgvF5sO6bSXahI9uKsiejbNotygku7VHJxIinQHc2vxWDlNjWew (accessed on 18 August 2023).

Disclaimer/Publisher's Note: The statements, opinions and data contained in all publications are solely those of the individual author(s) and contributor(s) and not of MDPI and/or the editor(s). MDPI and/or the editor(s) disclaim responsibility for any injury to people or property resulting from any ideas, methods, instructions or products referred to in the content.

Proceeding Paper

Risk Management in the Water Industry [†]

Zuzana Zemanova *, Sarka Krocova and Patrik Sirotiak

Faculty of Safety Engineering, VSB—Technical University of Ostrava, 700 30 Ostrava, Czech Republic; sarka.krocova@vsb.cz (S.K.); patrik.sirotiak@vsb.cz (P.S.)
* Correspondence: zuzana.zemanova.st@vsb.cz
[†] Presented at the 4th International Conference on Advances in Environmental Engineering, Ostrava, Czech Republic, 20–22 November 2023.

Abstract: Extraordinary events have always threatened us. The continuity of drinking water supplies cannot be exposed to such risks. When dimensioning water supply systems, it is necessary to consider the use of preventive measures, which include risk management. The following article is the basic scope of this issue and presents risk management procedures in the water industry.

Keywords: risk management; water supply systems; water industry

1. Introduction

Water industry is a prerequisite for the sustainable development of life on Earth [1]. However, the supply of water to villages and towns cannot be taken for granted, as the infrastructure is vulnerable to several risks, both natural and anthropogenic, that can affect water supply. Events threaten the continuity of drinking water supplies are numerous every year around the world [2]. An example is the situation in 1998 in Australia, when the treated water for Sydney had high cryptostoadia values. This event let to the introduction of regulations that require risk analysis to identify potential hazards in the drinking water supply system [3]. Some areas are susceptible to drought, others to flooding, and each may be at risk of a shortage of drinking water needed for life. For these reasons, the issue of water has become a hot topic in the Czech and global environment [4].

Scientific knowledge and technical possibilities must be used to cope with the situations that arise. In the Czech Republic, Kročová and Pokorný [5] and Tuhovčák and Ručka [6] have addressed the issue of water supply infrastructure. The comparison of drinking water supply systems and water supply systems was discussed in the article by Czech and international authors [7], in which they confirm the importance of searching for best practices leading to better results. It is important in the future to consider that water supply may be compromised, which would have a negative impact on the population. Therefore, it is necessary to prepare for such events by appropriate planning or construction and rehabilitation of infrastructure [8].

To protect citizens, facilities, or production systems, we need to know the local risks. These can be identified, and subsequent recommendations and measures can be formulated to prevent or reduce damage, based on risk management. The aim of risk management is, according to Kročová [9], "to keep the system under consideration functional and within the given scope for as long as possible". This will ensure the protection of protected interests such as the life and health of a person but also reduce economic losses. The result can then be the faster recovery of infrastructures important for the functioning of society [9].

2. Operational Safety Risks to Drinking Water Supplies—Description of the Problem

Safety is the state of being able to withstand threats that can be expected with a certain probability. However, its protection is not absolute; there will always be some risk of events that threaten the system [10]. Our world and the threats are evolving, which requires us, citizens, and those who protect us, namely the security system and the integrated rescue system, to constantly adapt [11].

We can see risk at every point in the distribution system. The population depends on the supply of drinking water, and if it cannot be delivered, an emergency or crisis situation arises [9]. The functionality of the system and the supply of drinking water can be compromised for a variety of reasons. Natural factors have affected the technical infrastructure since its inception. In terms of natural hazards, these are mainly water movements, run-off, or landslides. In the context of climate change, it is clear that they will increase in intensity and frequency, as pointed out, for example, by the authors of [12].

Partial Conclusions

The response to risk is a key part of the protection of critical infrastructure, especially for natural events [13]. In contrast, anthropogenic hazards cannot be forecast in this way, so more consideration needs to be given to alternatives for maintaining operations in the extraordinary events. The hazard level of an emergency, as well as the type, is specific to each region of a country.

3. Risk Analysis in the Water Industry

The water supply infrastructure system is susceptible to a number of natural and anthropogenic risks at every stage that cannot be predicted. The continuity of supply cannot be exposed to such risks when we do not know what may happen. A more reliable approach is needed. Mathematical models are often used in the design and operation of water supply systems, and risk analysis methods are often applied [14].

Currently, several methods for risk analysis have been developed for different technological systems. Examples include the checklist method, ETA or FTA, the FMEA method, HACCP, HAZOP, the risk matrix method, or SWOT.

It is important that the methods meet the needs of the analysis for the government department, municipalities, and water system operators. The used method must provide results in a form that will present the character of the risk and how to deal with it and must ensure the repetition of its process, the monitoring of the analysis process, and the verifiability of the results.

Risk analysis has a brief history, particularly in the water industry, dating back to the turn of the millennium. As a result of the extreme floods in 1997, 2002, or 2013 in Europe [14,15], especially in the Czech Republic [16], the issue of drinking water supply in crisis situations started to receive more attention. The risk assessment issues were addressed and regular problems and failures of the individual elements of the water supply infrastructure were also considered.

Partial Conclusions

An important prerequisite for the selection of a risk analysis method concerns what results are expected from the method. This problem can be seen particularly in the neglect of the water sector at the present time, when crisis planning experts are focusing on other elements of critical infrastructure. The authors Řehák and Hromada [17] deal with the resilience of critical infrastructure and focus on strengthening the resilience of the energy sector. The water sector also needs to continue to be addressed, investigated, and protected.

4. Supporting Methodologies

Supporting methodologies have been developed for the needs of the water supply industry, which serve to perform a risk analysis of the water supply system.

4.1. Water Safety Plans

The WHO is an organization with the goal of spreading health care around the world to give everyone a chance at life and wellbeing. According to their priorities, the organization has been promoting Water Safety Plans (WSP) since 2004 [18], where the authors describe this procedure. The WSP methodology is similar to the HACCP method, which has been used mainly in the food industry since the 1960s. However, the question remains as to why food safety has been prioritized over the safety of the drinking water supply.

Currently, WSP is implemented in a number of European and non-European countries such as Iceland, Bangladesh, New Zealand, and England [19,20]. With these examples, there is a demonstration of use in both developed and developing countries.

The implementation of WSP in the Czech Republic is the focus of a study [21], which specifies the factors that facilitate the implementation of WSP. As a result, it was found that the implementation is influenced by the quality of the process of developing these procedures. The expected benefits of risk assessment include improved water quality, reduced accidents, the improved protection of water resources, reduced disease in the drinking water user population, improved monitoring, the improved understanding of the entire supply system, the reduced costs of corrective action, and the better positioning of supervisory authorities to control the operator. The disadvantages are mainly for the operator, who incurs higher economic and administrative costs.

4.2. WaterRisk

The issue of risks in drinking water supply has been discussed for many years. However, the food sector has always been given more emphasis. When the issues of water quality and water control started to emerge, it was certain which area would lead the way. However, the HACCP method was not simply applied to the drinking water supply. For this reason, water experts began to address the issue of developing a new risk analysis method and created the WaterRisk project. The aim of the project was to create a methodology for the development of WSP in the conditions of the Czech Republic. Risk analysis in the field of water supply is a time-consuming and professionally demanding process; that is why the Methodological Proposal for the preparation of risk assessment of drinking water supply systems according to the Act on the Protection of Public Health [22] was created and together with it a software application with the same name, in which the created methodology was digitalized. The aim was to simplify the risk analysis process for water supply operators, as it was clear that one day it would be necessary to prepare it [23].

The advantage of the method is that specific information about the water system is not needed. However, some input information is required, which is known to the operator. If some input information is not available or is incorrect, this is not a problem for the method. Then, it is necessary to determine which objects will be analyzed. Their description must be provided. The method contains predefined undesirable states that are linked to a specific element of the system under consideration. It is a demonstration of a system fault. A diagram of the risk analysis process can be seen in Figure 1. The method defines fifty-eight undesirable conditions, which are divided according to their nature into natural, social, and technical hazards. Hazard identification is conducted for each object separately and then risk analysis is carried out, where hazard values K1 are determined as negligible and K5 as very high. Based on the resulting K values, corrective measures are determined.

The advantage of using WaterRisk is, in particular, the possibility of using an application whose outputs is exported to a PDF document, so it can be easily attached to the operating rules. The application process itself is shown in Figure 2.

The proposed risk mitigation measures can be exported from the application and can be further worked on, especially in connection with the development of the recovery financing plan. The advantage can also be seen in the improvement of problem areas and undesirable conditions. Another advantage is the creation of a risk matrix when using the application for risk analysis, which shows the initial state of risks in the system before and after the implementation of measures. The selected procedures try to free the worker from their subjective actions and use a multicriteria approach [24].

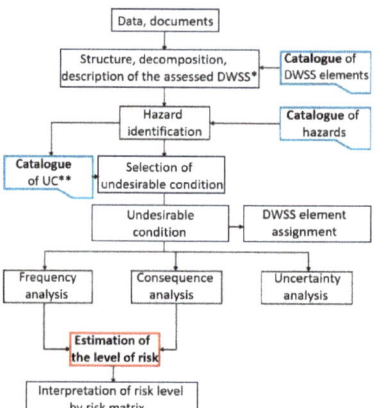

Figure 1. Diagram of the WaterRisk analysis ([24], edited). * Drinking water supply system; ** undesirable condition.

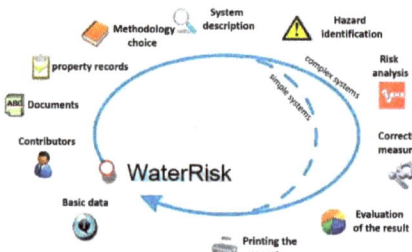

Figure 2. WaterRisk user interface ([24], edited).

5. Discussion

The purpose of risk analysis is to create assumptions to prepare for events that have a negative impact on the system. Methodologies that are suitable for water systems have been described above. The authors report the positive impacts of performing risk analysis in the drinking water supply sector in [9,16]. By conducting a risk analysis and making it mandatory in national legislation, there has been a reduction in cases where established water supply practices have not been followed and thus a reduction in cases of public illness. Conducting risk analysis in the water sector is an important tool for ensuring continuous improvement of water quality. Since risk analysis is performed with minor modifications in an equivalent manner worldwide, it is easier to understand local problems.

However, it is also necessary to consider the potential pitfalls in the actual development of the risk analysis. These are mainly economic pressure, the desire to simplify problems, subjective aspects, and others, as mentioned, for example, in [25]. The authors focus on the pitfalls of implementing risk analysis in the water industry in this article. Ensuring the functionality of the water system is necessary even though it is not economically attractive. According to their research, the solution to a given risk analysis should be sufficiently familiar with the problem since risks are identified and evaluated by a specific person or group of people. For this reason, it is necessary to ensure that the subjective bias of the problem is eliminated. Failure to do so could result in the production of a non-descript document that will not lead to further action by the risk management system.

According to [20], the key prerequisites for managing a successful risk analysis are the participation of the governing bodies of the public administration and the simultaneous participation of the work teams. Some authors believe that safety in the water sector can be ensured via the consistent enforcement of legislation with the cooperation of all stakeholders [21].

In the water sector, the methods commonly used for risk analysis have been modified, as it is a very specific field. The development of the risk analysis itself does not need to be an obstacle if carried out correctly. It is important to reflect on the fact that the water industry is not just a normal business, but a service to us, the population, who depend on water, and therefore, it is not appropriate to take an economic approach.

If, based on a risk analysis, the preconditions for emergencies are already in place, it is important to prepare for them from a crisis management perspective. If staff ignore risks, conceal them, or fail to assess them properly, this could have catastrophic consequences.

6. Conclusions

The risks cannot be completely avoided. They can only be reduced to an acceptable level by applying preventive measures. The acceptable level is variable for different systems. However, it is always important to protect life and health. In the field of water infrastructure, the primary objective is therefore to transport water in the quantity and quality required to the public. And while sometimes precautionary measures may seem meaningless, by applying the right combination of measures, we can achieve the desired protection. This is because the system must involve both the objects and the line structures of the water supply infrastructure and those who work with and operate the infrastructure.

Although the area of natural and anthropogenic risks has been studied for many years, there is a lack of sufficient solutions in the water infrastructure. There are not many authors that deal with this issue. In the future, the issue should be adequately incorporated into emergency planning of state and local governments, as a reduction or interruption of drinking water supply is a major threat to both the population and critical infrastructure elements.

Author Contributions: Conceptualization, Z.Z.; methodology, Z.Z.; formal analysis, P.S.; investigation, Z.Z.; resources, Z.Z. and S.K.; data curation, Z.Z.; writing—original draft preparation, Z.Z. and P.S.; writing—review and editing, Z.Z. and S.K.; visualization, P.S.; supervision, S.K.; project administration, Z.Z. and P.S.; funding acquisition, P.S. All authors have read and agreed to the published version of the manuscript.

Funding: This research received no external funding.

Institutional Review Board Statement: Not applicable.

Informed Consent Statement: Not applicable.

Data Availability Statement: No new data were created or analyzed in this study. Data sharing is not applicable to this article.

Conflicts of Interest: The authors declare no conflict of interest.

References

1. Kročová, Š.; Kavan, Š. Cooperation in the Czech Republic border area on sustainability of water management. *Land Use Policy* **2019**, *86*, 351–356. [CrossRef]
2. Blake, S.; Walker, R.; Walker, R. Potable Water Issues during Disaster Response and Recovery: Lessons Learned from Recent Coastal Disasters. *Solut. Coast. Disasters* **2011**, *2012*, 779–794. [CrossRef]
3. Amjad, U.Q.; Luh, J.; Baum, R.; Bartram, J. Water safety plans: Bridges and barriers to implementation in North Carolina. *J. Water Health* **2016**, *14*, 816–826. [CrossRef] [PubMed]
4. Kubečková, D.; Kročová, Š. Long-term sustainability of the landscape in new climatic conditions. *IOP Conf. Ser. Earth Environ. Sci.* **2017**, *92*, 012032. [CrossRef]
5. Kročová, Š.; Pokorný, J. Water management development trends in the 21st century. *IOP Conf. Ser. Earth Environ. Sci.* **2020**, *444*, 012031. [CrossRef]
6. Agudelo-Vera, C.; Avvedimento, S.; Boxall, J.; Creaco, E.; De Kater, H.; Di Nardo, A.; Djukic, A.; Douterelo, I.; Fish, K.E.; Rey, P.L.; et al. Drinking Water Temperature around the Globe: Understanding, Policies, Challenges and Opportunities. *Water* **2020**, *12*, 1049. [CrossRef]

7. Sucháček, T.; Aldea, A.; Bylka, J.; Marko, I.; Tuhovčák, L. Comparative Analysis and Benchmarking of Water Supply Systems and Services in Central and Eastern Europe. *Proceedings* **2018**, *2*, 597. [CrossRef]
8. Řehák, D.; Šenovský, P.; Slivková, S. Resilience of Critical Infrastructure Elements and Its Main Factors. *Systems* **2018**, *6*, 21. [CrossRef]
9. Kročová, Š.; Václavík, V. Water-use in the context of the approaching climate change. *IOP Conf. Ser. Earth Environ. Sci.* **2017**, *92*, 012030. [CrossRef]
10. Sun, F.; Chen, J.; Tong, Q.; Zeng, S. Integrated risk assessment and screening analysis of drinking water safety of a conventional water supply system. *Water Sci. Technol.* **2007**, *56*, 47–56. [CrossRef]
11. Kavan, Š.; Kročová, Š.; Pokorný, J. Assessment of the Readiness and Resilience of Czech Society against Water-Related Crises. *Hydrology* **2021**, *8*, 14. [CrossRef]
12. Nguyen, V.T.V. Urban Water Infrastructure Design in Climate Change Context: Advances and Challenges in Developing Engineering Practice Guidelines. In Proceedings of the World Environmental and Water Resources Congress 2023, Henderson, NV, USA, 21–25 May 2023; pp. 399–404. [CrossRef]
13. Yu, F.; Li, X.Y.; Han, X.S. Risk response for urban water supply network using case-based reasoning during a natural disaster. *Saf. Sci.* **2018**, *106*, 121–139. [CrossRef]
14. Brázdil, R.; Dobrovolný, P.; Luterbacher, J.; Moberg, A.; Pfister, C.; Wheeler, D.; Zorita, E. European climate of the past 500 years: New challenges for historical climatology. *Clim. Chang.* **2010**, *101*, 7–40. [CrossRef]
15. Petrucci, O.; Aceto, L.; Bianchi, C.; Bigot, V.; Brázdil, R.; Pereira, S.; Kahraman, A.; Kılıç, Ö.; Kotroni, V.; Llasat, M.C.; et al. Flood Fatalities in Europe, 1980–2018: Variability, Features, and Lessons to Learn. *Water* **2019**, *11*, 1682. [CrossRef]
16. Brázdil, R.; Chromá, K.; Řehoř, J.; Zahradníček, P.; Dolák, L.; Řezníčková, L.; Dobrovolný, P. Potential of Documentary Evidence to Study Fatalities of Hydrological and Meteorological Events in the Czech Republic. *Water* **2019**, *11*, 2014. [CrossRef]
17. Řehák, D.; Slivková, S.; Janečková, H.; Štuberová, D.; Hromada, M. Strengthening Resilience in the Energy Critical Infrastructure: Methodological Overview. *Energies* **2022**, *15*, 5276. [CrossRef]
18. Davison, A.; Howard, G.; Stevens, M.; Callan, P.; Fewtrell, L.; Deere, D.; Bartram, J.; Water, S.; WHO. *Water Safety Plans. Managing Drinking-Water Quality from Catchment to Costumer*; World Health Organization: Geneva, Switzerland, 2005; 244p.
19. Gunnarsdottir, M.J.; Gardarsson, S.M.; Elliott, M.; Sigmundsdottir, G.; Bartram, J. Benefits of Water Safety Plans: Microbiology, Compliance, and Public Health. *Environ. Sci. Technol.* **2012**, *46*, 7782–7789. [CrossRef]
20. Nijhawan, A.; Jain, P.; Sargaonkar, A.; Labhasetwar, P.K. Implementation of water safety plan for a large-piped water supply system. *Environ. Monit. Assess* **2014**, *186*, 5547–5560. [CrossRef]
21. Baracho, R.O.; Najberg, E.; Scalize, P.S. Factors That Impact the Implementation of Water Safety Plans—A Case Study of Brazil. *Water* **2023**, *15*, 678. [CrossRef]
22. Kožíšek, F. *Metodický Návod ke Zpracování Posouzení Rizik Systémů Zásobování Pitnou Vodou Podle Zákona o Ochraně Veřejného Zdraví*; State Institute of Health: Praha, Czech Republic, 2018; Available online: https://www.szu.cz (accessed on 15 July 2023).
23. Gunnarsdóttir, M.J.; Gissurarson, L.R. HACCP and water safety plans in Icelandic water supply: Preliminary evaluation of experience. *J. Water Health* **2008**, *6*, 377–382. [CrossRef]
24. WaterRisk. 2022. Available online: www.waterrisk.cz (accessed on 15 July 2023).
25. Mudaliar, M.M. Success or failure: Demonstrating the effectiveness of a Water Safety Plan. *Water Supply* **2012**, *12*, 109–116. [CrossRef]

Disclaimer/Publisher's Note: The statements, opinions and data contained in all publications are solely those of the individual author(s) and contributor(s) and not of MDPI and/or the editor(s). MDPI and/or the editor(s) disclaim responsibility for any injury to people or property resulting from any ideas, methods, instructions or products referred to in the content.

Proceeding Paper

Post-Mining Landscape of the Karviná Region and Its Importance for Nature and Landscape Conservation [†]

Lukáš Kupka * and Barbara Stalmachová

Department of Environmental Engineering, Faculty of Mining and Geology, VŠB—Technical University of Ostrava, 708 00 Ostrava-Poruba, Czech Republic; barbara.stalmachova@vsb.cz
* Correspondence: lukas.kupka.st2@vsb.cz
† Presented at the 4th International Conference on Advances in Environmental Engineering, Ostrava, Czech Republic, 20–22 November 2023.

Abstract: The article deals with the post-mining landscape of the Karviná region from the point of view of nature and landscape protection. From different perspectives, the subsidence of Lake Kozinec is presented, which can play an important role in the protection of nature and landscape of the post-mining landscape. A faunistic and floristic survey of the model area, including a habitat survey, was carried out.

Keywords: biodiversity of subsidence basins; Kozinec; post-mining landscape; subsidence lake

1. Introduction

The aim of the research was to evaluate the natural and ecological parameters for the possibility of declaring the aquiferous subsidence of basins in the Karviná region in the category of a small-scale special protection area under Act No. 114/1992 Coll. on nature and landscape protection. The selected model area of Lake Kozinec in Doubrava is very interesting. Already at the beginning of the creation of this lake, technical reclamation measures were taken to model the terrain in such a way as to create a high diversity of micro-habitats (islands and islets for nesting birds, shallows; see Figure 1). At present, the Kozinec model area is a local biocenter within the territorial ecological stability systems.

Figure 1. Artificially created islands for nesting gulls.

Research conducted in Poland, e.g., [1–3], demonstrated that reservoirs formed in subsidence basins form an extremely important element of the natural environment, which often restores in a natural way thanks to the "forces of nature" and spontaneously creates valuable and productive ecosystems [4].

2. Materials and Methods

The research work was divided into three parts:
- Faunistic survey;
- Floristic survey;
- Habitat survey.

It is necessary to mention that these surveys were focused on the evaluation of biological qualities, not on a detailed inventory survey. For a detailed description of the flora and especially the fauna, not only longer-term research would be needed, but it would also involve a whole range of experts on individual groups and taxa. The basic sources of information for the faunistic and floristic surveys, which were carried out during the years 2021 and 2022, were field mapping and data published in the literature.

2.1. Faunistic Survey

The fauna survey was mainly focused on aquatic animals and their immediate surrounding. There is a number of methods for catching animals, collecting them, and identifying them, but the most reliable methods usually mean a significant intervention in the life of the animal, possibly even causing their injury or death (mainly applying to invertebrates). This research was based on the belief that it should not take place at the cost of injuring the animal and should interfere with its life as little as possible. The survey was specifically focused on the inventory of endangered species listed in Annex III of the Decree of the Ministry of the Environment of the Czech Republic No. 395/1992 Coll., as amended, which implements certain provisions of Act No. 114/1992 Coll., as amended. At the same time, attention was paid to the species listed in the Red Lists [5]. Basic non-invasive methods included direct observation, photography, and using camera traps and a bat detector.

The measured data were recorded in a chronological field notebook and then transcribed into the database. The determination of individual species of animals took place directly in the field; photo documentation was taken, and, if necessary, specialist literature was used for their determination [5–10].

2.2. Floristic Survey

The floristic survey was carried out regularly every 14 days during the growing season to capture both spring and summer aspects. The measured data were noted in a chronological field notebook and subsequently transcribed into a database. The determination of individual plant species was carried out directly in the field using the publication key to the flora of the Czech Republic [11], while in some cases, the mobile application, "Plant net Plant identification" (Copyright © 2014–2023 Pl@ntNet™), was used for orientation of the found plants.

2.3. Habitat Survey

The creation of the list of natural habitat types occurring in the model area was based mainly on data from the Natura 2000 habitat mapping from 2007 to 2022 and our own field survey, which followed the floristic survey. Characteristics of individual natural habitat types were then based on the Catalogue of Biotopes of the Czech Republic [12]. Although this catalog was created primarily for the purpose of mapping Natura 2000 biotopes, it was prepared in such a way that it represents essentially a complete inventory of all biotopes occurring in the territory of the Czech Republic. These are divided into nine main categorizations, the last of which includes habitats heavily influenced or created by human activity.

3. Results

3.1. Faunistic Survey

According to Decree No. 395/1992 Coll., a total of 7 species of animals are in the category of critically endangered, another 16 species of animals are in the category of highly endangered, and then 12 species of animals are in the category of endangered. According to the Red Lists of Invertebrates and Vertebrates [5,13], four species are in the critically endangered (CR) category, five species in the endangered (EN) category, eighteen species in the vulnerable (VU) category, and eleven species in the near-threatened (NT) category.

Table 1 shows the characteristics of the relationship of animals to the aquatic environment (categories A, B, and C). Table 2 shows the selected species of animals found, characterized in terms of their endangerment according to the Decree of the Ministry of Environment of the Czech Republic No. 395/1992 Coll. as amended and according to the Red Lists of Vertebrates and Invertebrates of the Czech Republic [5,13]. This list of animals was also supplemented with species whose occurrence could not be demonstrated during the orientation survey but are documented in literature sources related to the model area.

Table 1. The relationship of the species to the aquatic environment.

A	Close relationship. The habitat is permanently occupied by the species, or the habitat is by nature a prerequisite for the existence of the species at a critical stage of its biological cycle (e.g., during the breeding season, during the development of offspring, etc.) or plays an important role in it.
B	Animals that require or prefer a given habitat (e.g., foraging, nesting, shelter, etc.) but are not directly tied to that habitat type.
C	Loose-to-indifferent relationship to the biotope. They usually occur randomly in a given area, usually finding suitable conditions in surrounding biotopes or inhabiting ecotonal habitats, including species completely unrelated to any community type, occurring everywhere.

Table 2. List of proven species of animals in the model area. Abbreviations expressing the level of endangerment of IUCN Red Lists: CR—critically endangered; EN—endangered; VU—vulnerable; NT—near-threatened; LC—least concern; DD—data deficient, NE—not evaluated. Abbreviations expressing threat according to Decree 395/92 Coll.: KO—critically endangered, SO—severely endangered, O—endangered). The column " Eco. relations " was drawn up according to Table 1.

No.	Scientific Name	Eco. Relations	Decree	IUCN	Source
	MAMMALIA				
1.	Lepus europaeus	C	-	NT	own survey
2.	Lutra lutra	A	SO	NT	own survey
	AVES				
3.	Acrocephalus arundinaceus	B	SO	VU	own survey
4.	Alcedo atthis	B	SO	VU	[14]
5.	Ardea cinerea	B	-	NT	own survey
6.	Botaurus stellaris	B	KO	CR	[14]
7.	Circus aeruginosus	B	O	VU	own survey
8.	Columba oenas	C	-	VU	[14]
9.	Crex crex	C	SO	VU	[14]
10.	Dendrocopos medius	C	O	VU	[14]
11.	Ergus merganser	A	KO	CR	own survey
12.	Gallinula chloropus	B	-	NT	own survey
13.	Haliaeetus albicilla	C	KO	EN	own survey
14.	Chroicocephalus ridibundus	B	-	VU	own survey
15.	Chroicocephalus ridibundus	B	-	NE	own survey

Table 2. Cont.

No.	Scientific Name	Eco. Relations	Decree	IUCN	Source
16.	*Ixobrychus minutus*	B	KO	CR	own survey
17.	*Larus argentatus s. l.*	B	-	NA	own survey
18.	*Locustella luscinioides*	B	O	EN	[14]
19.	*Motacilla flava*	C	SO	VU	[14]
20.	*Picus canus*	C	-	VU	own survey
21.	*Podiceps cristatus*	A	O	VU	own survey
22.	*Riparia riparia*	B	O	NT	[14]
23.	*Sterna hirunda*	B	SO	EN	[14]
24.	*Tachybaptus ruficollis*	A	O	VU	own survey
25.	*Tringa totanus*	B	KO	CR	[14]
	REPTILIA				
26.	*Anguis fragilis*	C	SO	NT	own survey
27.	*Lacerta agilis*	C	SO	VU	own survey
28.	*Natrix natrix*	B	O	NT	own survey
29.	*Zootoca vivipara*	C	SO	NT	[14]
	AMPHIBIA				
30.	*Bombina bombina*	A	SO	EN	[14]
31.	*Bufo Bufo*	B	O	VU	own survey
32.	*Lissotriton vulgaris*	B	SO	VU	own survey
33.	*Pelophylax esculentus*	A	O	NT	own survey
34.	*Pelophylax ridibunda*	A	KO	NT	[14]
35.	*Rana arvalis*	B	KO	EN	[14]
36.	*Triturus alpestris*	B	SO	VU	own survey
	PISCES				
37.	*Chondrostoma nasus*	A	-	VU	[14]
38.	*Phoxinus phoxinus*	A	O	VU	[14]
	COLEOPTERA				
39.	*Cucujus cinnaberinus*	C	SO	VU	[14]
	MOLLUSCA				
40.	*Monachoides incarnatus*	C	-	NT	own survey

3.2. Floristic Survey

A total of 168 plant species were found during the flora survey, of which two species are listed as critically endangered according to Decree No. 395/1992 Coll. According to the Red List of Vascular Plants [15], two species are listed in the endangered (EN) category, and another four species are listed in the near-threatened (NT) category (see Table 3).

Table 3. List of proven plants in the model site (abbreviations expressing the level of threat of IUCN Red Lists: CR—critically endangered; EN—endangered; VU—vulnerable; NT—near-threatened; LC—least concern; DD—data deficient, NE—not evaluated; and threat according to Decree 395/92 Coll.: KO—critically endangered, SO—severely endangered, O—endangered).

No.	Scientific Name	Decree	IUCN	Source
1.	*Acer campestre*	-	LC	own survey
2.	*Dysphania botrys*	-	NT	[14]
3.	*Equisetum telmateia*	-	NT	[14]
4.	*Orobanche flava*	-	NT	own survey
5.	*Potamogeton nodosus*	-	NT	own survey
6.	*Salvinia natans*	KO	EN	own survey
7.	*Trapa natans*	KO	EN	[14]

3.3. Habitat Survey

Of the nine main habitat types [12], 30 habitat units from seven types were recorded in the model area (see Table 4). The most valuable series of habitats were found in the southern and southeastern part of the drop-off of Lake Kozinec (wetland alder, wetland willow, sedge and rush stand and in the water body itself, macrophyte vegetation of naturally eutrophic and mesotrophic standing waters, and macrophyte vegetation of shallow standing waters). Also of value was the eastern and northeastern part of the lake subsidence, which is naturally connected to the remnants of floodplain forest (hardwood meadow) and which is part of the local biocenter. None of the identified habitats is included in the Red List of habitats of the Czech Republic.

Table 4. List of natural categories and biotope units of the model area except for biotopes strongly influenced by humans [12].

Summary of recorded stream and reservoir habitat units (V) in the model area:
Macrophyte vegetation of naturally eutrophic and mesotrophic still waters (V1)
Macrophyte vegetation of water streams (V4)
Summary of recorded wetland and riparian vegetation (M) habitat units in the model area
Reed beds of eutrophic still waters (M1.1)
Riverine reed vegetation (M1.4)
Unvegetated river gravel banks (M4.1)
Muddy riverbanks (M6)
Herbaceous fringes of lowland rivers (M7)
Summary of recorded units of vernal pool and peatland habitats (R) in the model area
Forest springs without tufa formation (R1.4)
Summary of recorded secondary grassland and heathland habitat units (T) in the model area
Mesic Arrhenatherum meadows (T1.1)
Cynosurus pastures (T1.3)
Wet Filipendula grasslands (T 1.6)
Summary of recorded scrub habitat units (K) in the model area
Willow carrs (K1)
Tall mesic and xeric scrub (K3)
Summary of recorded forest habitat units (L) in the model area
Ash–alder alluvial forests (L2.2)
Hardwood forests of lowland rivers (L2.3)
Willow–poplar forests of lowland rivers (L2.4)
Polonian oak–hornbeam forests (L3.2)
Acidophilous beech forests (L5.4)

4. Discussion

The research, which included faunistic, floristic, and biotope surveys, was aimed at evaluating the model area of Lake Kozinec in the Karviná post-mining landscape for designation as a small-scale special protection area under Act No. 114/1992 Coll. on Nature and Landscape Protection.

In the Czech Republic, there are a number of sites of importance for nature and landscape protection that are located in landscapes that have been significantly affected by mining and mineral raw material processing in the past, and in some cases, their creation was directly initiated by mining. Some of these important places are part of the state nature and landscape protection as small-scale special protection areas or as special areas of conservation (SACs). Understandably, such places as part of cultural heritage are listed as important cultural monuments.

In the Czech Republic, the formation of subsidence lakes results from quite specific remnants of mining, which, for example, are associated with other valuable animal and plant communities compared to those mentioned above and in which dragonflies (Odonata) or amphibians (Amphibia) find their refugia, and for migratory birds (Aves), they become important stops during their migration.

The multi-aspect and multi-dimensionality concept of biological diversity results in a variety of ways of measuring it and constructing its assessment indicators. The species richness of the recorded species—329 species—was reported in [3,4] in the area of subsidence reservoirs in Upper Silesia (Poland).

The topic of the small-scale special protection area in the Karviná post-mining landscape is made even more interesting by the fact that it has been more than thirty years since the interesting and scientifically significant nature reserve Loucké rybníky (Loučky Ponds) was closed in Louky nad Olší (today Karviná—Louky) as a result of mining activities. Today, the area has very little in common with its former state, although it still retains some importance (e.g., a stopover for many rare bird species). It is, therefore, be an interesting and certainly a worthy legacy after more than two centuries of coal mining in the Karviná region if the formerly important Loucké rybníky Nature Reserve in the Olza River floodplain (after all, these ponds were also created by human activity) found its "counterpart" in, for example, the newly declared small-scale special protection area, which in turn was created as a result of mining activity (and coincidentally also in the Olza River floodplain).

Subsequently, surveys of fauna and flora, including habitat surveys, were carried out in the area of this model area. The results of these surveys are presented in Section 3. Already in the early days of the creation of the lake subsidence, technical reclamation measures were taken to model the terrain in such a way as to create a high diversity of microhabitats (islands and islets for nesting birds, shallows).

An indicative survey of the fauna showed a total of 113 species of animals, the most important of which are according to Decree No. 395/1992 Coll. from the category of critically endangered species: Botaurus stellaris, Ergus merganser, Haliaeetus albicilla, Ixobrychus minutus, Pelo-phylax ridibunda, Rana arvalis, Tringa tota-nus; according to the Red Lists of Invertebrates and Vertebrates [5,13], these are four species in the critically endangered (CR) category: Botaurus stellaris, Ergus merganser, Ixobrychus minutus, and Tringa totanus.

A total of 168 plant species were found during the orientation survey of the flora, of which the most important according to Decree No. 395/1992 Coll. are the critically endangered species: Salvinia natans and Trapa natans; according to the Red List of Vascular Plants [15], two species are in the endangered (EN) category: Salvinia natans and Trapa natans; four other species are in the near-threatened (NT) category: Dysphania botrys, Equisetum telmateia, Orobanche flava, and Potamogeton nodosus.

Of the nine major habitat categories [12], 31 habitat units from seven categories were recorded in the model area. However, none of the documented biotopes belongs to the Red List of biotopes of the Czech Republic. The macrophyte vegetation of naturally eutrophic and mesotrophic standing waters (V4) and macrophyte vegetation of shallow standing waters (V2) that have been documented in the model area, although man-made, are so close to nature in their condition that they were not categorized as heavily impacted by man (X) in the research.

On the basis of indicative surveys, it can be stated that the model area of Lake Kozinec is very interesting in terms of the occurrence of animals and plants. Further surveys of the flora and fauna carried out by experts on individual taxa can yield a number of other interesting findings.

The model area of Lake Kozinec and its surroundings can become a small-scale special protection area in the category of natural monuments (PP Poklesové jezero Kozinec). Given its anthropogenic origin, this category seems much more appropriate than the nature reserve category. Alternatively, the site may be designated as a Natura 2000 bird area (54 bird species have been recorded in the model area). The possibility of designating the model area as a site of a special area of conservation (SAC Poklesové jezero Kozinec) can also be considered. Changing the status of the model area (e.g., declaring it as a small-scale special protection area) is not only important from the point of view of nature and landscape protection interests but also, as already outlined, is important for the mining landscape of Karviná as such, for its prestige in the eyes of the population.

One of the current intentions in the model area (in its southwestern part) is that part of the Lake Kozinec subsidence should be used by the public (fishermen, recreational users). In the public part, a beach with a gently sloping bottom should be created, as the bottom in this part is steep. This arrangement is not the most suitable for a potential change in the status of the model area, not only in terms of disturbance to birds, for example, but also in terms of hygiene for the visitors themselves. The so-called Karviná Sea is suitable for recreational purposes, and there is a public swimming pool in the village of Doubrava itself, not far from Kozinec.

Mining subsidence reservoirs are important for the development of the post-industrial areas of the coal basins, which will be able to be sensibly used for the development of post-mining areas under the condition of a general change in the way of thinking about the environment. The consequence of this change (and not the source!) should be the transformation of the existing economic models of the world and their reasonable application [16].

The mountain landscape in the Erzgebirge has already "lived to see" its recognition and is now a UNESCO World Heritage Site as the "Mining Cultural Landscape of the Erzgebirge". It is perhaps possible to look to the future with some hope that the mining landscape of the Karviná region (and Ostrava region) may also be heading toward a similar goal. It has the potential to do so!

5. Conclusions

The aim of the research was to evaluate the natural and ecological parameters of the model area of Lake Kozinec in Doubrava for the possibility of its declaration as a small-scale special protection area according to Act No. 114/1992 Coll. on Nature and Landscape Protection.

The second section, "Material and Methodology", presents the methods of work (data collection of the fauna and flora orientation survey, habitat survey) and the material used.

In the third section, the results of the surveys of fauna and flora and the evaluation and elaboration of the list of habitat types of the model area are presented.

From the results carried out, it follows that 168 species of plants, 113 species of animals, and 31 habitat units from seven categories of habitats from the Catalogue of Habitats of the Czech Republic could be found or documented in the literature.

Section 4 summarizes the findings and discusses the future of the model area. The content of this discussion and one of the objectives of the research include, among other things, a presentation of the model area of Lake Kozinec (lake subsidence) as a place that can become a small-scale special protection area (nature monument) or a Natura 2000 protected area (bird area or a site of European importance) in the future. However, this is a very long run. The technical and biological reclamation phase, which is essentially entirely in the hands of humans, will be replaced by a third phase, the naturalization phase, which will be entirely "in the hands" of nature, but it is up to people to decide what the final approach to this model area will be. The question is whether the change in its status will ultimately succeed. This area does indeed have this potential, and so, it is more than symbolic that it was mining activity that contributed to the creation of the "blue-green" pearl of Karviná!

Author Contributions: Conceptualization, L.K. and B.S.; methodology, L.K. and B.S.; investigation, L.K.; resources, L.K. and B.S.; data curation, L.K. and B.S.; writing—original draft preparation, L.K. and B.S.; writing—review and editing, L.K. and B.S.; visualization, L.K.; supervision, B.S.; funding acquisition, B.S. All authors have read and agreed to the published version of the manuscript.

Funding: This research was funded by the EU Programme LIFE, project No. LIFE20 IPC/CZ/000004 "IP LIFE for Coal Mining Landscape Adaptation".

Institutional Review Board Statement: Not applicable.

Informed Consent Statement: Not applicable.

Data Availability Statement: Data are contained within article.

Conflicts of Interest: The authors declare no conflict of interest.

References

1. Tokarska-Guzik, B.; Rostański, A. Rola zatopisk (zalewisk) pogórniczych w renaturalizacji przemysłowego krajobrazu Górnego Śląska. *Przegląd Przyr.* **1996**, *7*, 267–272.
2. Sierka, E.; Molenda, T.; Chmura, D. Environmental repercussion of subsidence reservoirs reclamation. *J. Water Land Dev.* **2009**, *13a*, 41–52. [CrossRef]
3. Sierka, E.; Stalmachová, B.; Molenda, T.; Chmura, D.; Pierzchała, Ł. *Environmental and Socio-Economic Importance of Mining Subsidance Reservoirs*; BEN-Technicka Literatura: Praha, Czech Republic, 2012; 112p.
4. Woźniak, G. *Diversity of Vegetation on Coalmine Heaps of the Upper Silesia (Poland)*; Szafer Institute of Botany, Polish Academy of Science: Kraków, Poland, 2010; 319p.
5. Hejda, R.; Farkač, J.; Chobot, K. (Eds.) *Červený seznam ohrožených druhů České republiky. Bezobratlí.*; Agentura ochrany přírody a krajiny ČR: Praha, Czech Republic, 2017; p. 760.
6. Balát, F. *Klíč k určování našich ptáků v přírodě*; Ilustroval Jan DUNGEL. Academia: Praha, Czech Republic, 1986.
7. Horsák, M.; Juřičková, L.; Picka, J. *Měkkýši České a Slovenské republiky*; Kabourek: Zlín, Czech Republic, 2013; ISBN 978-80-86447-15-5.
8. Zwach, I. *Obojživelníci a plazi České republiky: Encyklopedie všech druhů, určovací klíč*; Grada: Praha, Czech Republic, 2009; ISBN 978-80-247-2509-3.
9. Waldhauser, M.; Černý, M. *Vážky České republiky: Příručka pro určování našich druhů a jejich larev*; Český svaz ochránců přírody Vlašim: Vlašim, Czech Republic, 2014; ISBN 978-80-87964-00-2.
10. Macek, J.; Laštůvka, Z.; Beneš, J.; Traxler, L. *Motýli a housenky střední Evropy*; Academia: Praha, Czech Republic, 2015; ISBN 978-80-200-2429-9.
11. Kubát, K. (Ed.) *Klíč ke květeně České republiky*; Academia: Praha, Czech Republic, 2002; ISBN 80-200-0836-5.
12. Chytrý, M. *Katalog biotopů České republiky: Habitat catalogue of the Czech Republic*, 2nd ed.; Agentura ochrany přírody a krajiny ČR: Praha, Czech Republic, 2010; ISBN 978-80-87457-02-3.
13. Chobot, K.; Němec, M. (Eds.) *Červený seznam ohrožených druhů České republiky. Obratlovci.*; Agentura ochrany přírody a krajiny: Praha, Czech Republic, 2017; p. 760.
14. Moravskoslezské Investice a Development, a.s. Koncepce rozvoje pohornické krajiny Karvinska do roku. 2030. Available online: https://poho2030.cz/wp-content/uploads/2023/10/koncepce-pohornicka-final.pdf (accessed on 29 November 2023).
15. Chobot, K. et Grulich. In *Červený seznam cévnatých rostlin České republiky*; AOPK ČR: Praha, Czech Republic, 2017; ISBN 978-88-80076-47-6, ISSN 1211-3603.
16. Sierka, E.; Radosz, Ł.; Ryś, K.; Woźniak, G. Ecosystem Services and Post-industrial Areas. In *Green Scenarios: Mining Industry Responses to Environmental Challenges of the Anthropocene Epoch*; Dyczko, A., Jagodziński, A.M., Woźniak, G., Eds.; Taylor & Francis Group: London, UK, 2022; pp. 265–274.

Disclaimer/Publisher's Note: The statements, opinions and data contained in all publications are solely those of the individual author(s) and contributor(s) and not of MDPI and/or the editor(s). MDPI and/or the editor(s) disclaim responsibility for any injury to people or property resulting from any ideas, methods, instructions or products referred to in the content.

Proceeding Paper

Removal of Humic Substances from Water with Granular Activated Carbons [†]

Danka Barloková [1,*], Ján Ilavský [1], Jana Sedláková [2] and Alena Matis [3]

1. Department of Sanitary and Environmental Engineering, Faculty of Civil Engineering, The Slovak University of Technology, Radlinského 11, 810 05 Bratislava, Slovakia; jan.ilavsky@stuba.sk
2. Podtatranská Vodárenská Prevádzková Spoločnosť, Public Limited Company, Hraničná 662/17, 058 89 Poprad, Slovakia; jana.sedlakova@pvpsas.sk
3. Water Research Institute, Nábr. arm. gen. L. Svobodu 5, 812 48 Bratislava, Slovakia; alena.matis@gmail.com
* Correspondence: danka.barlokova@stuba.sk; Tel.: +421-907796684
† Presented at the 4th International Conference on Advances in Environmental Engineering, Ostrava, Czech Republic, 20–22 November 2023.

Abstract: The article deals with the removal of humic substances from surface water in the High Tatras locality and with the reduction in the intensity of water coloration. Four different types of granular activated carbon (Norit 1240, WG12, Filtrasorb TL830, and Filtrasorb 300) were compared in the experiments. The quality of the water from the water source supplied to the filter columns, and the water at the outlet, the filtration speed, and the efficiency of sorption media were monitored. The results showed more than 70% efficiency in the removal of humic substances and COD_{Mn} from water, but only about 50% efficiency from the point of view of water color, as determined with the technology used.

Keywords: drinking water treatment; filtration; granulated activated carbon; humic substances; color of water; adsorption efficiency

1. Introduction

Humic substances (HS) often constitute the main part of natural organic pollution (NOM) in natural waters. The structure of humic substances is not yet known exactly. These are complex high-molecular organic compounds of an aromatic–aliphatic nature that contain carbon, oxygen, hydrogen, and nitrogen. Humic substances have a relative molecular weight ranging from a few hundred to tens of thousands. The elementary composition of HS is listed in Table 1 [1,2].

Table 1. Elementary composition of humic substances presented in natural waters.

Humic Substances	Elementary Composition (%)			
	C	O	H	N
humic acids	52–62	30–39	2.5–5.8	2.6–5.1
fulvic acids	43–52	42–51	3.3–6.0	1.0–6.0

The highest representation expressed in % has organic carbon in humic substances, representing approximately 50%, and fulvic acids have almost the same content of oxygen as carbon does, while for humic acids, this value is lower by approximately 20%. Humic substances are characterized by the presence of carboxyl and hydroxyl functional groups (phenolic; alcoholic), as well as methoxyl and carbonyl groups. These groups are attached both to the aromatic cores and to the side aliphatic chains [3].

The properties of humic substances have a very close connection with their composition, which is influenced by humification processes, with the size of molecules (molecular

weight) and particles, their degree of dispersion (true or colloidal solutions), their polarity, which is determined by the character of the skeleton (aromatic; aliphatic), and mainly, the type, number, and dissociation ability of functional groups. The function of the composition of humic substances is their solubility in water (fulvic acids have higher solubility than humic acids do, and therefore natural waters contain, on average, 87% of fulvic acids), ability to aggregate, ability to dissociate (mainly carboxyl functional groups), and related charge ratios (zeta potential of particles). Some properties of humic substances are also influenced by the composition of the water and its pH. Figure 1 shows basic properties of humic substances [3].

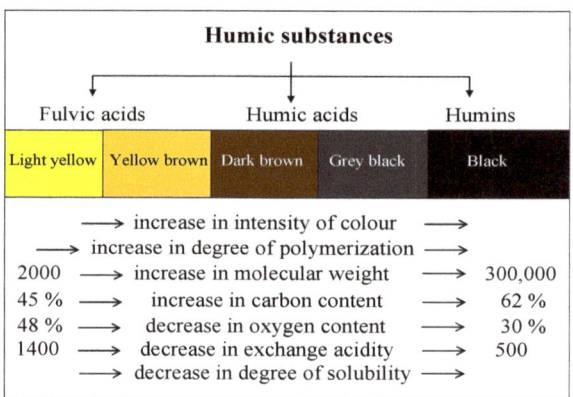

Figure 1. Basic properties of humic substances.

The undesirable effect of humic substances in terms of their influence on water quality and its treatment can be summarized as follows:

- An increase in the intensity of the color of the water;
- An increase the acidity of the water;
- An effect on the smell and taste of water;
- An effect on the formation of metal complexes (e.g., with Fe, Mn, Al, and Cu);
- An effect on the adsorption of organic compounds (e.g., pesticides, PCBs, and phthalates);
- An effect on the formation of trihalomethanes (THMs) during water chlorination [4–6].

The presence of humic substances in waters is manifested by their coloring, e.g., fulvic acids have a light yellow to yellow-brown color depending on the concentration, while humic acids are dark brown. Due to their origins, these are the natural colors of the waters. Usually, a concentration of humic substances of 2.5 mg·L^{-1} corresponds approximately to the water color of 20 mg·L^{-1} Pt. However, the color changes considerably with the pH value of the water (solutions with a higher pH are more colored). In addition, the color of water depends on the composition of humic substances, the size of dispersed particles, etc., which can be different in different surface waters. Therefore, a generally valid linear relationship between the concentration of humic substances and the color of water cannot be obtained [6].

Humic substances are characterized by complex-forming properties [7,8]. The formation of complexes is one of the reasons for the leachability of heavy metals from the soil by humic substances (their content in water increases) and the reason for the increased concentration of Fe and Mn in peat waters.

The content of humic substances in raw water causes a problem in the individual technological stages of water treatment [9,10]; higher doses of coagulant and disinfectants are applied, but mainly, they act as precursors for the formation of halogen compounds.

Important precursors of organohalogene compounds in water are primarily fulvic acids and humic acids. It has been experimentally confirmed that approximately 60% more organochlorine compounds are formed from fulvic acids compared to humic acids [11,12]. At the same time, fulvic acids have higher solubility in water than humic acids do, which is why natural waters contain an average of 87% fulvic acids. To limit the formation of chlorinated hydrocarbons, the content of humic substances in the water must be reduced as much as possible or the method of disinfection must be changed.

In Slovakia, the determination of humic substances in drinking water is not regulated by legislation [13]. It is based on the COD_{Mn} value, absorbance (A^{254}), or color of the water. Exceeding the limit value is the reason for the decision of whether or not to determine humic substances. The limit value for COD_{Mn} is 3.0 mg·L^{-1}.

In the older standard [14], humic substances were among the physico-chemical indicators with a limit value of 2.5 mg·L^{-1}. Exceeding the value of 2.5 mg·L^{-1} indicates the possible presence of THMs in water supplied with chlorine.

Requirements for the quality of raw water and limit values of water quality indicators for individual categories of standard methods for the treatment of raw water to drinking water are established by decree [15]. This decree contains recommended water treatment procedures for individual categories, A1, A2, and A3. The categories take into account the technological complexity and effectiveness of the treatment.

The requirements for the quality of surface water and the quality objectives of surface water intended for the abstraction of drinking water are determined by the regulation [16]. The assessment of surface water quality and its classification into individual categories is carried out on the basis of selected indicators, which include COD_{Mn} and TOC, turbidity, and water color and absorbance (Table 2).

Table 2. Classification of surface water into individual categories based on selected parameters.

Parameter	Unit	Category A1		Category A2		Category A3	
		OH	MH	OH	MH	OH	MH
COD_{Mn}	mg·L^{-1}	2	3	5	7	8	10
TOC	mg·L^{-1}	<5.0	5.0	5.0	7.0	8.0	10.0
Turbidity	FNU	2	5	-	30	-	50
Color	mg·L^{-1} Pt	10	20	50	100		
Absorbance (254 nm)		-	0.08	-	0.15	-	0.30

The lowest concentration of humic substances is found in groundwater (up to 0.1 mg·L^{-1}). In surface waters, the concentration of them varies from 0.1 to 20 mg·L^{-1} and it is more than that in seawater since the freshwater environment possesses a greater fraction of organic material. In waters from peatlands, the concentration of humic substances fluctuates in a wide range, mostly in tens of mg·L^{-1}. In some stagnant waters, it is possible to determine the concentration of humic substances in amounts up to 500 mg·L^{-1}. Humic substances can be found in municipal wastewater with a concentration varying from 118 to 228 mg/g, and approximately 42% of these humic substances represent HS [6,17].

1.1. Removal of Humic Substances

Humic substances (HS) can be removed from drinking water via several treatment processes, but different treatment processes can have different efficiencies in terms of their removal and drinking water safety. Coagulation, flocculation, adsorption, oxidation, ion exchange, membrane filtration, and biological and electrochemical processes or advanced oxidation processes (AOPs) can be used to remove HS from water [10,18–21].

Coagulation is one of the most widely used processes for treating water from surface sources. The application of coagulation to remove humic substances from drinking water sources has received much attention from researchers worldwide, because it has been effective in helping to prevent the formation of disinfection byproducts (DBPs). Nevertheless, with the increased fluctuation of humic substances in water (in terms of concentration and composition), the efficiency of conventional coagulation has significantly decreased, and therefore it is necessary to develop improved coagulation processes by optimizing operating conditions, and to develop more effective inorganic or organic coagulants, as well as to combine coagulation with other water treatment technologies [21].

Only high-molecular humic acids are removed from water via coagulation, but in the case of low-molecular fulvic acids (with a relative molecular weight of up to 500), the efficiency is significantly lower, and some fractions cannot be removed at all. The best efficiency is achieved in the treatment of humic waters via clarification in the acidic region (at pH from 4 to 6), when large and easily separable aggregates are formed [22].

Different methods are used to remove humic substances from water; among them is adsorption, which is effective and also, compared to classic methods, simpler in terms of operation and the equipment used—filters with an adsorption bed. This technology is especially suitable for small water treatment plants, or where it is not possible to use conventional water treatment associated with coagulation.

The adsorbent affects the efficiency of the adsorption process. The porous structure and surface area of the adsorbent influences adsorption and the kinetics of adsorption itself. The larger the surface area of the adsorbent, the more efficient the adsorption process. For this reason, we are constantly looking for materials—adsorbents with a large surface area and thus faster kinetics of pollutant removal.

The adsorption of pollutants is also influenced by physico-chemical properties of treated water, such as pH value, the initial concentration of pollutants, the type and size distribution of molecules, and water temperature.

Activated carbon is a versatile adsorbent that can remove diverse types of pollutants such as metal ions, dyes, phenols, and a number of other organic and inorganic compounds and bio-organisms. However, its use is sometimes restricted due to its higher cost. Due to the higher cost of activated carbon, attempts are being made to regenerate spent activated carbon. Chemical as well as thermal regeneration methods are used for this purpose. However, these procedures are not very cheap and also produce additional effluents and result in a considerable loss of the adsorbent. Therefore, in situations where cost factors play a major role, scientists are looking for low-cost adsorbents for the removal of water pollution. A wide variety of materials have been investigated for this purpose and they can be classified into three categories: (i) natural materials, (ii) agricultural wastes, and (iii) industrial wastes. These materials are generally available free of cost or at a low cost as compared to that of activated carbons [23].

The removal of pollutants from waters utilizing biological materials is a relatively recent advancement. It was only in the 1990s that a new technology, biosorption, developed that could also help the removal of heavy metals and other pollutants from waters. Various biosorbents [23–25] have been tested for the removal of pollutants, especially metal ions, with very encouraging results.

1.2. WTP Nový Smokovec

The water treatment plant (WTP) was put into operation in 1972. The capacity of the WTP is 20 $L \cdot s^{-1}$. The water treatment plant treats water from the Štiavnik stream in Nový Smokovec. The stream is also called Červený potok due to the fact that, in certain periods (snow melting, spring periods, long-term rains, and torrential rains), the water is colored red due to the humic substances from the peat soils, and shows an above-limit COD_{Mn} value, low pH (around 6.0), increased water color (up to 30 $mg \cdot L^{-1}$), and also turbidity (max. 5.8 FNU).

Figure 2 shows the COD_{Mn} values in the period from 2012 to 2022.

The water is slightly mineralized water; the total amount of dissolved substances at 105 °C is between 37 and 88 $mg \cdot L^{-1}$, the conductivity is 5.0 $mS \cdot m^{-1}$, the average concentration of calcium is 4.3 $mg \cdot L^{-1}$, the average concentration of magnesium is 1.2 $mg \cdot L^{-1}$, the sum of Ca + Mg is 0.14 $mmol \cdot L^{-1}$, the absorbance is 0.057, $ANC_{4.5}$ is in a concentration of

0.38 mmol·L^{-1}, aggressive CO_2 according to Heyer is in a concentration of 9.5 mg·L^{-1}, the average manganese concentration is 0.055 mg·L^{-1}, the iron concentration is 0.04 mg·L^{-1}, the turbidity is 1.2 FNU, the water color is 10 mg·L^{-1} Pt, the water temperature is between 1.8 and 13.1 °C, and the pH is from 5.9 to 8.7. The indicated values are at the entrance to the water treatment plant.

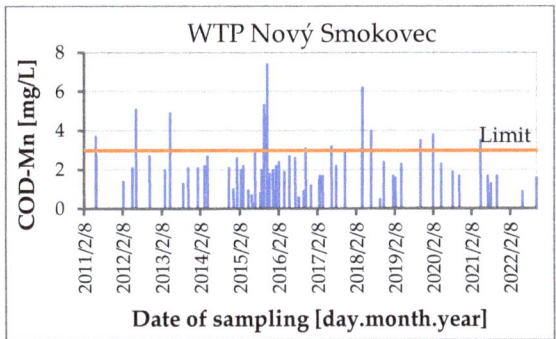

Figure 2. Values of COD_{Mn} in raw water from Štiavnik stream.

The treatment technology consists of a sampling object, sedimentation via flow through a stilling and sedimentation tank, filtration on three open rapid sand filters and deacidification on filters with a PVD filling. Treated water is accumulated in a reservoir. Currently, the water treatment plant is used as an additional water source.

2. Materials and Methods

The experiments were carried out at the Nový Smokovec plant. Two filter columns (with an inner diameter of 5.0 cm, height of 150 cm, and height of filter filling of 90–100 cm) filled with granular activated carbon Norit 1240, WG12, Filtrasorb 300, and Filtrasorb TL830 were used to verify the efficiency of humic substance elimination from water. Their basic characteristics are listed in Table 3.

Table 3. Properties of utilized granulated activated carbon [26–29].

Parameter	Norit 1240	WG12	F300	F830
Iodine number (mg·g^{-1})	min. 1020	min. 1000	min. 1000	min. 900
Methylene blue (mg·g^{-1})	min. 230	min. 30	min. 245	min. 260
Specific surface (BET) (m^2·g^{-1})	1150	min 1000	1050	1100
Particle size (mm)	0.85–2.0	1.0–1.5	0.85–2.0	0.6–2.36
Median diameter of particles (mm)	1.4	1.2	1.4	1.6
Operational density (g·gcm^{-3})	0.480	0.450 ± 30	0.450	0.460
Abrasion (-)	75	85	96	95
Hardness (-)	97	95	75	78
Coefficient of uniformity	1.6	1.3	1.4	2.1
Humidity (wt.%)	max. 2	max. 2	max. 3	max. 2

During the experiments, the quality of raw and treated water at the outlet from the filter columns and the water flow were monitored. The filtration conditions are summarized in Table 4.

For the determination of humic acids (HA), a spectrophotometric method in the visible region (at a wavelength of 420 nm) was employed, which used the extraction of humic substances at a low pH value into pentanol and subsequent re-extraction from pentanol with a NaOH solution [30,31]. To convert absorbance into concentration, we used an empirical coefficient (valid for peat waters in Slovakia).

$$\text{HL}\left(\text{mg} \cdot \text{L}^{-1}\right) = \frac{A \times 68 \times 250}{250 \times 2.8} \tag{1}$$

where HL is the determined concentration of humic substances, A is absorbance at 420 nm, 250 mL is the volume of water used for extraction, 68 is the empirical coefficient and 2.8 is the width of the cuvette used for analysis (in cm).

Table 4. Filtration conditions.

Parameter	Norit 1240	WG12	F300	F830
Height of filtration bed (cm)	92	92	100	100
Weight of the bed (g)	1127.4	907.5	1418.3	1350.8
Avg. flow through the column (mL·min^{-1})	203.14	208.86	205.94	207.94
Avg. filtration velocity (m·h^{-1})	6.207	6.382	6.293	6.354
Bed contact time (min)	8.89	8.65	9.53	9.44

A Hach DR2800 spectrophotometer (Hach Lange GmbH, Düsseldorf, Germany) was used for the analysis of humic substances, water color, and other water quality parameters, while turbidity was determined using the Hach 2100Q instrument (Hach Company, Loveland, CO, USA), and the pH values were determined using Hach SensION+ ph3 Benchtop Meter (Hach Lange GmbH). Chemical oxygen demand was determined using the method of Kübel–Tieman (COD-Mn) in accordance with STN EN ISO 8467:2001 [32].

3. Results and Discussion

Based on the obtained data, we compared the effectiveness of granulated activated carbon (Norit 1240, WG12, Filtrasorb TL830, and Filtrasorb 300) in removing humic substances and COD$_{Mn}$ from water (Figures 3–5), and at the same time, we also monitored other physico-chemical parameters defined in Decree Ministry of Health SR no. 91/2023 Coll. for drinking water.

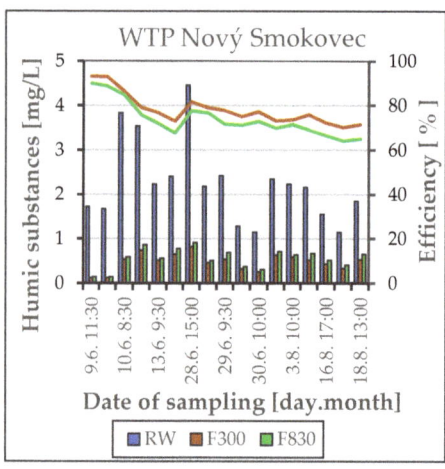
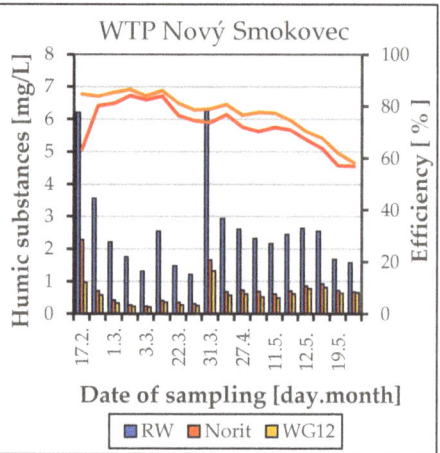

Figure 3. Concentration of humic substances in raw and treated water and the efficiency of sorption materials.

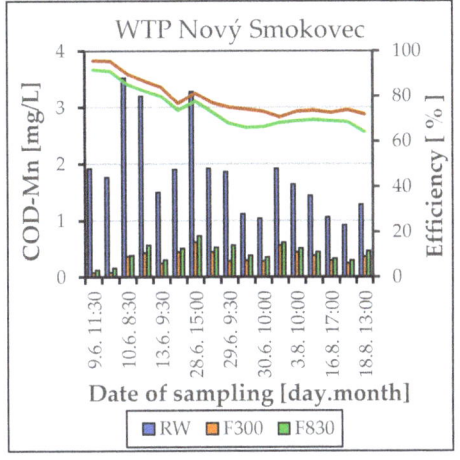 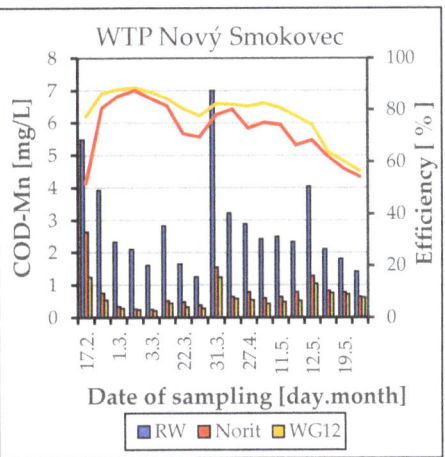

Figure 4. Concentration of COD$_{Mn}$ in raw and treated water, and the efficiency of sorption materials.

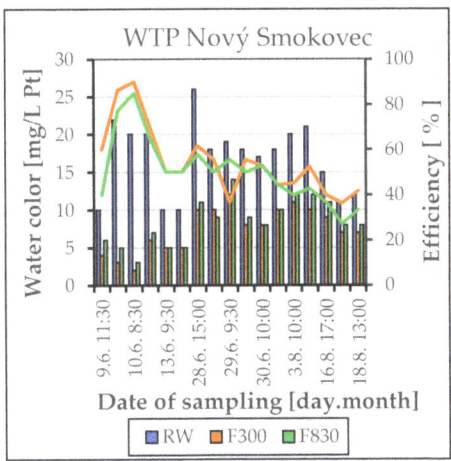 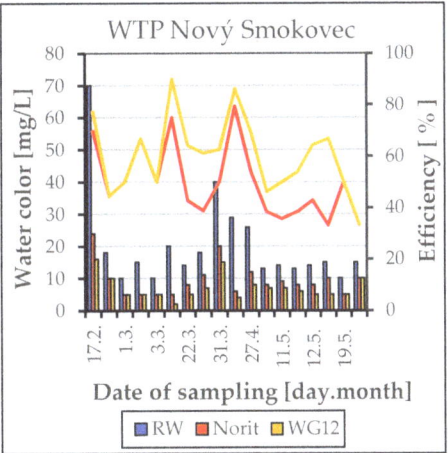

Figure 5. Color of water in raw and treated water, and the efficiency of sorption materials.

4. Conclusions

Humic substances (humic acids; fulvic acids) are created via the decomposition of organic matter and are a common part of natural waters. If they are not removed in the water treatment process and come into contact with disinfectants, the reaction with them leads to the formation of carcinogens and toxic substances. It is necessary to remove HS from the water to prevent the formation of these harmful substances.

Adsorption is a very effective method of removing humic acids from water. Activated carbon, zeolites, modified clays, iron-based sorption materials, materials obtained from various types of waste, and biological adsorbents are used as adsorption materials. Adsorption efficiency is affected by several factors, such as pH, initial HS concentration, type of adsorbent, and contact time.

This article presents results from pilot plant experiments of HS removal via adsorption. Different types of activated carbon (Norit 1240, WG 12, F300, and F830) were used as sorption materials, and the efficiency of the materials used was monitored and compared. The experiments were carried out at the Nový Smokovec plant. The average efficiency of removing humic substances from water was 78.10% in the case of Filtrasorb 300, 74.13% in the case of Filtrasorb TL830, 72.15% in the case of Norit 1240, and 77.37% in the case of WG12. The average water color reduction efficiency using granular activated carbon was 60.32% for WG12, 49.61% for Norit 1240, 54.62% for Filtrasorb 300, and 50.47% for Filtrasorb 830TL. The adsorption capacity of activated carbon was not exhausted (Figure 3).

The experiments were carried out with breaks from February 17 to August 18. The effort was to capture the deteriorated water quality with a pronounced red color and a high content of humic substances, which did not occur during the given period. Therefore, the experiments will continue in 2023.

On the basis of the pilot tests, granular activated carbon will be proposed for the overall modernization of the WTP in Nový Smokovec.

Author Contributions: D.B. and J.S. worked out a concept and plan of experiments. J.I. and D.B. ensured the installation of all equipment, and the assembly and verification of used technologies. D.B., J.I. and J.S. performed all experiments and water sampling. J.I. analyzed water samples (part of the water samples were analyzed in a laboratory of the water company in Poprad). D.B. and A.M. evaluated the obtained results from the experiments. A.M. processed the literature research on the given issue. D.B. and J.I. wrote the paper with the contribution of all co-authors. All authors have read and agreed to the published version of the manuscript.

Funding: The experiments were financially supported by Slovak Research and Development Agency of the Slovak Republic (Projects APVV-18-0205 and APVV-22-0610) and by the Ministry of Education, Science, Research and Sports of the Slovak Republic (Project VEGA 1/0825/21).

Institutional Review Board Statement: Not applicable.

Informed Consent Statement: Not applicable.

Data Availability Statement: Data will be available on request.

Acknowledgments: Thanks go to the workers of the water company in Poprad and at the WTP in Nový Smokovec.

Conflicts of Interest: The authors declare that the research was conducted in the absence of any commercial or financial relationships that could be construed as a potential conflict of interest.

References

1. Schnitzer, M.; Khan, S.U. *Humic Substances in the Environment*; M. Dekker: New York, NY, USA, 1972.
2. Aiken, G.R.; McKnight, D.M.; Wershaw, R.L.; MacCarthy, P. (Eds.) *Humic Substances in Soil, Sediment and Water: Geochemistry, Isolation and Characterization*; Wiley-Interscience: New York, NY, USA, 1985; 692p.
3. Stevenson, F.J. *Humus Chemistry: Genesis, Composition, Reactions*; John Wiley & Sons: New York, NY, USA, 1982; 443p.
4. MacCarthy, P. The principles of Humic Substances. *Soil Sci.* **2001**, *166*, 738–751. [CrossRef]
5. Suffet, I.H.; MacCarthy, P. (Eds.) *Aquatic Humic Substances: Influence on Fate and Treatment of Pollutants*; American Chemical Society: Washington, DC, USA, 1989.
6. Pitter, P. *Hydrochemie*, 5th ed.; Vydavatelství VŠCHT Praha: Praha, Czech Republic, 2015; 792p. (In Czech)
7. Singer, P.C. *Trace Metals and Metal-Organic Interactions in Natural Water*; Ann Arbor Science Publishers: Ann Arbor, MI, USA, 1973.
8. Nriagu, J.O.; Coker, R.D. Trace Metals in Humic and Fulvic Acids from Lake Ontario Sediments. *Environ. Sci. Technol.* **1990**, *14*, 443. [CrossRef] [PubMed]
9. Snoeying, V.L.; Jenkins, D. *Water Chemistry*; Wiley: New York, NY, USA, 1991; 480p.
10. Sillanpää, M. *Natural Organic Matter in Water, Characterization and Treatment Methods*, 1st ed.; Elsevier: Amsterdam, The Netherlands, 2015; 382p.
11. Grünwald, A.; Janda, V.; Fišar, P.; Bížová, J.; Štastný, B. Evaluation of the potential of THM formation in humic waters. In Proceedings of the Conference Voda Zlín 2002, Zlín, Czech Republic, 26–27 March 2002; pp. 67–71. (In Czech)
12. Clark, R.M.; Thurnau, R.C.; Sivaganesan, M.; Ringhand, P. Prediction the formation of Chlorinated and Brominated By-Products. *J. Environ. Eng.* **2001**, *6*, 493–501. [CrossRef]
13. *Decree of the Ministry of Health of the Slovak Republic no. 91/2023 Coll. for Indicators and Limit Values of Drinking Water Quality and Hot Water Quality, Drinking Water Monitoring Procedure, Drinking Water Supply Risk Management, and Risk Management of Domestic Distribution Systems*; The Office of the Government of the Slovak Republic: Bratislava, Slovakia, 2023. (In Slovak)
14. *STN 75 7111*; Water Quality, Drinking Water. Slovak Institute for Normalization, Metrology and Testing: Bratislava, Slovakia, 1998. (In Slovak)

15. *Decree of the Ministry of the Environment of the Slovak Republic no. 636/2004 Coll. Annex 1 about the Requirements for the Quality of Raw Water and Monitoring Water Quality in Water Supply Systems*; The Office of the Government of the Slovak Republic: Bratislava, Slovakia, 2023.
16. *Regulation of the Government of the Slovak Republic No. 269/2010 Coll. Annex 1 and 2, Laying Down the Requirements to Achieve a Good Condition of Water*; The Office of the Government of the Slovak Republic: Bratislava, Slovakia, 2022.
17. Basumallick, S.; Santra, S. Monitoring of ppm level humic acid in surface water using ZnO–chitosan nanocomposite as fluorescence probe. *Appl. Water Sci.* **2017**, *7*, 1025–1031. [CrossRef]
18. Crittenden, J.; Trussell, R.R.; Hand, D.; Howe, K.; Tchobanoglous, G. *MWH's Water Treatment: Principles and Design*, 3rd ed.; John Wiley & Sons: Hoboken, NJ, USA, 2012; 1907p.
19. Binnie, C.; Kimber, M. *Basic Water Treatment*, 5th ed.; ICE Publishing: London, UK, 2013; 316p.
20. Yue, Y.; An, G.; Liu, L.; Lin, L.; Jiao, R.; Wang, D. Pre-aggregation of Al13 in optimizing coagulation for removal of humic acid. *Chemosphere* **2021**, *277*, 130268. [CrossRef] [PubMed]
21. Bhatnagar, A.; Sillanpää, M. Removal of natural organic matter (NOM) and its constituents from water by adsorption—A review. *Chemosphere* **2017**, *166*, 497–510. [CrossRef] [PubMed]
22. Alomar, T.; Qiblawey, H.; Almomani, F.; Al-Raoush, R.I.; Han, D.S.; Ahmad, N.M. Recent advances on humic acid removal from wastewater using adsorption process. *J. Water Process Eng.* **2023**, *53*, 103679. [CrossRef]
23. Bhatnagar, A.; Minocha, A.K. Conventional and non-conventional adsorbents for removal of pollutants from water—A review. *Indian J. Chem. Technol.* **2006**, *13*, 203–217.
24. Pertile, E.; Dvorský, T.; Václavík, V.; Heviánková, S. Use of Different Types of Biosorbents to Remove Cr (VI) from Aqueous Solution. *Life* **2021**, *11*, 240. [CrossRef] [PubMed]
25. Pertile, E.; Václavík, V.; Dvorský, T.; Heviánková, S. The Removal of Residual Concentration of Hazardous Metals in Wastewater from a Neutralization Station Using Biosorbent—A Case Study Company Gutra, Czech Republic. *Int. J. Environ. Res. Public Health* **2020**, *17*, 7225. [CrossRef] [PubMed]
26. Filtrasorb 300: Granular Activated Carbon, CalgonCarbon [online]. 2023. Available online: https://www.calgoncarbon.com/app/uploads/DS-FILTRA30019-EIN-E1.pdf (accessed on 14 August 2023).
27. Filtrasorb TL830: Granular Activated Carbon, Chemwiron Carbon [online]. 2023. Available online: https://www.treitelonline.com/wp-content/uploads/2021/03/PDS_FILTRASORB-TL830.pdf (accessed on 2 July 2023).
28. Norit GAC 1240: Cabot Corporation, Norit Granulated Activated Carbon. Available online: https://www.ulprospector.com/en/na/Food/Detail/17187/420404/Norit-GAC-1240 (accessed on 15 September 2018).
29. Envi-Pur. Granular Activated Carbon [Online]. 2023. Available online: Doc_108_granulovane_aktivni_uhli_wg12_cz.pdf (accessed on 14 August 2023). (In Czech)
30. *STN 75 7567*; Water Quality, Determination of Humic Substances, Photometric Method. Slovak Institute of Technical Normalization: Bratislava, Slovakia, 2007. (In Slovak)
31. Standard Methods for the Examination of Water and Wastewater. *5510 Aquatic Humic Substances*, 19th ed.; American Public Health Association: Washington, DC, USA, 1995.
32. *STN EN ISO 8467:2001*; Water Quality, Determination of Chemical Oxygen Consumption Permanganate. Slovak Institute for Normalization, Metrology and Testing: Bratislava, Slovakia, 2020.

Disclaimer/Publisher's Note: The statements, opinions and data contained in all publications are solely those of the individual author(s) and contributor(s) and not of MDPI and/or the editor(s). MDPI and/or the editor(s) disclaim responsibility for any injury to people or property resulting from any ideas, methods, instructions or products referred to in the content.

Proceeding Paper

The Traditions and Technologies of Ecological Construction in Portugal [†]

Svitlana Delehan [1,*], Hanna Melehanych [2] and Andrii Khorolskyi [3]

1. Centre for Interdisciplinary Research at UzhNU, Uzzhorod National University, 88000 Uzhhorod, Ukraine
2. Resource Centre for Sustainable Development, Uzzhorod National University, 88000 Uzhhorod, Ukraine; hanna.melehanych@uzhnu.edu.ua
3. Branch for Physics of Mining Processes of the National Academy of Sciences of Ukraine, 49005 Dnipro, Ukraine; andreykh918@gmail.com

* Correspondence: svitlana.delehan-kokaiko@uzhnu.edu.ua; Tel.: +380-666947664

† Presented at the 4th International Conference on Advances in Environmental Engineering, Ostrava, Czech Republic, 20–22 November 2023.

Abstract: This paper identifies the main factors that shape energy poverty. Based on the analysis of statistical, regulatory documentation, as well as the state of the housing stock in Portugal, a number of new factors have been identified that are inherent in developed countries. The factors that shape energy poverty include the low energy efficiency of buildings, lack of access to energy-efficient sources, imperfect regulatory framework in the construction sector, and low level of public awareness of loans to cover energy-related needs. In view of this, it can be argued that energy poverty is not a problem of Third World countries only but is a global problem that will become even more acute as we move to clean and energy-efficient resources. To overcome the problems of energy poverty, the experience of Portugal was analysed. It was found that the practice of granting loans to reimburse the cost of energy purchases is not effective. Based on the analysis of government programmes and plans in Portugal, it was found that one of the most effective ways to overcome energy poverty is to ensure the energy certification of buildings. The incentive for the transition to certification is a set of state programmes on construction loans, allocation of funds for modernisation, etc. The identified factors and tools for stimulating energy efficiency improvement allow us to formulate a strategy for overcoming energy poverty, which includes energy certification of buildings, transition to more efficient and energy-saving heating means, lending to the population, and allocation of funds for modernisation.

Keywords: energy poverty; certification; energy carrier; modernisation

1. Introduction

In Europe, energy poverty is a serious problem that leads to serious negative consequences for health, well-being, and social inclusion. The COVID-19 pandemic has further highlighted the importance of energy services in everyday life, as many people have spent much more time at home than before. Energy poverty should be understood as the lack of access to modern energy services [1]. However, the problem of energy poverty should not be considered in relation to the country's raw material potential, the level of energy resources, etc. Energy poverty is primarily explained by the level of modern energy services. In other words, a developed, modern society should strive to use safe and efficient energy supply systems.

Energy services are becoming increasingly important. It covers various aspects of life, such as work, education, entertainment, and a comfortable stay at home. We believe that the level of consumption and demand for energy services will continue to grow, which can be explained as follows:

- Many people have been forced to work from home, which has led to an increase in the consumption of electricity and other energy resources;
- Remote learning and online entertainment have become the norm, which has also increased the need for reliable and affordable energy services.

It is worth noting that energy poverty can affect people in different social categories, especially low-income groups, who may face difficulties in paying energy bills and maintaining comfortable conditions at home. This can lead to a deterioration in living conditions and negative impacts on health [2,3].

Before proceeding to the main material, the justification of the research objectives, it is worth analysing the view of the definition of "energy poverty". The term energy poverty first appeared in the early 1990s of the 20th century [3]. And at the initial stage, it was seen exclusively as the lack of sufficient capacity for heating and cooling homes. However, over time, views on this problem have changed. Today, energy poverty is a system of complex systemic inequalities. This inequality is explained by obstacles to the provision of modern energy at an affordable price. At the same time, this problem is difficult to measure because it is dynamic and changes depending on time and space. Today, it is a problem not only of a social dimension but also of a cultural dimension, as it determines the quality of life and must meet basic needs.

Even now, energy poverty is fundamental to improving quality of life and is the basis for economic development. Ensuring access to affordable, reliable, sustainable, and modern energy for all is one of the Sustainable Development Goals [4]. This issue is addressed not only by the governmental but also by supranational institutions and organisations [5–7].

Therefore, to combat energy poverty, it is important to develop policies aimed at ensuring access to efficient and affordable energy services for all segments of the population. Social support programmes and energy efficiency can help to reduce the burden on household budgets and provide more stable living conditions for people. It is a mistake to assume that environmental poverty is unique to the Third World. Given the problems of urbanisation, industrialisation, and the inadequacy of public policy to meet the needs of society, this problem exists in the European Union as well.

Given the global dimension of the problem, as well as the developed regulations [8,9] and practices [10,11], we understand how important it is to study the traditions and technologies of green building in the European Union [12–14]. This requires us to identify the key factors that contribute to the formation of energy poverty. The study of traditions and construction technologies (in the example of Portugal) will allow us to analyse existing approaches to overcoming energy poverty. This study will allow us to recommend the most reliable practices for the construction of houses, which will facilitate access to quality energy resources, which in turn will improve the quality of life. In addition, recommendations for construction will be developed. All of this together allows us to develop a strategy to overcome energy poverty.

2. Methodology

The research methods used were the analysis of reporting documentation on the use of energy types for heating buildings in the European Union [15–19]. We analysed Eurostat data on transport [20–23] and electricity supply to buildings in Portugal [24,25]. The analysis of this documentation allowed us to identify key indicators of energy poverty. Considerable attention was paid to the study of the living conditions of Portuguese residents, which allowed us to identify the types of energy sources used for domestic needs. In addition, we analysed the regulatory framework for construction certification. The results obtained allow us to assess the level of energy poverty, study the balance of energy resources, and formulate recommendations for improving living conditions in Portugal, as well as formulate recommendations on construction technologies that can be used not only in Portugal but also in the European Union.

We analysed the "Casa Eficiente 2020" [26], "The IFRRU 2020 Programme" [27], The National Energy and Climate Plan for the Period of 2021–2030 (PNEC 2030) [28], and The Plan for Recovery and Resilience (PRR 2021–2026) [29].

The final recommendations for overcoming energy poverty through the use of the latest building technologies are formed using the method of iteration and comparison. It is important to note that we aimed to propose strategies that can be implemented in the European Union.

3. Results

3.1. Identification of Key Indicators of Energy Poverty Based on the Analysis of the Use of Residential Energy Sources

In Portugal, the processes of industrialisation and urbanisation started with a certain delay, and public housing policy has not always been properly targeted. The shortage of housing in the second half of the 20th century led to an increase in illegal construction and self-builds, and the number of abandoned buildings increased, especially in Lisbon and Porto. Around the 1970s, around 40% of residential buildings in the country were unlicensed [30,31].

In order to make housing more affordable, a rent freeze was introduced during the Estado Novo dictatorship, but its extension had negative impacts, including a lack of investment by landlords, which led to a deterioration in housing quality. Compared to other European countries, where welfare policies have actively promoted the availability, affordability, and quality of housing, in Portugal, the requirements for thermal insulation of residential buildings began to be regulated by law only in the 1990s, and before that, the level of thermal insulation of the housing stock was much lower [32,33].

Due to the poor quality of construction of residential buildings, many households are in a difficult financial situation, which prevents them from carrying out repairs. Problems exist not only with keeping homes warm. According to Eurostat data for 2020, a quarter of Portugal's population (25.2%) lived in homes with leaks, dampness, or rot on windows or floors. The map below (Figure 1) shows that Portugal is one of the leading European countries in this area.

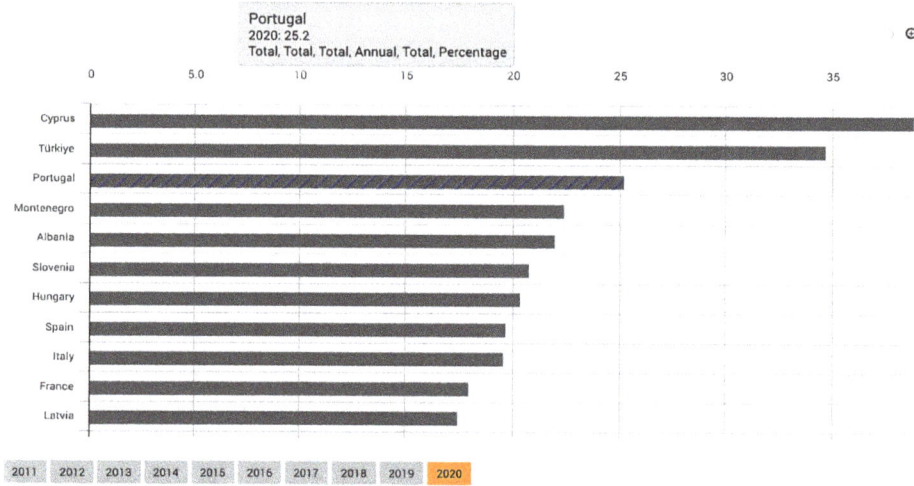

Figure 1. Information on unsatisfactory living conditions among the countries of the European Union.

In Portugal, it is common for houses to be built without heating systems or with only a fireplace, which is extremely inefficient as more than 80% of the heat produced is lost, and such a system requires considerable physical effort to maintain. This is an approach to construction that is rarely seen in other European Union countries (with the exception of Malta and Spain), where almost all houses have central heating systems or other fixed systems.

In the EU, natural gas is the most commonly used fuel for home heating, but in Portugal, the majority of the population does not have this option. The gas distribution network covers only 34% of households in the country, and it is mostly located in large urban areas (Figure 2). The economically less developed segments of the population, in particular, still use bottled gas, which is more expensive than piped gas and is not sold at preferential tariffs.

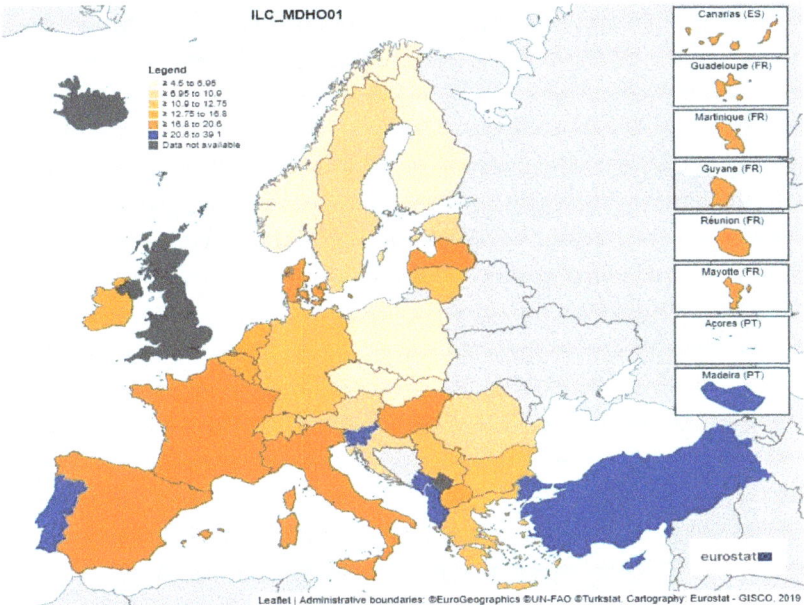

Figure 2. Access of the population to centralised gas distribution networks.

In the absence of stationary heating systems, a significant proportion of the population uses inexpensive portable electric heaters. Obviously, the low cost of such devices means that they are of limited efficiency and have high operating costs. Some people are concerned about high electricity bills and refuse to use heaters at all. In turn, low-income Portuguese, especially, use their electrical appliances until they stop working, even if their efficiency is already reduced, which is often the case with refrigerators, freezers, and televisions. When purchasing appliances, the poorest always prefer the cheapest models without considering future energy costs.

The use of electrical devices to combat the heat is handled with great caution. Despite the fact that since the mid-2000s, Portugal has been actively installing air conditioners in homes (pre-installation of this equipment in new buildings is required), they are still not a common solution among citizens, especially among the most economically vulnerable.

According to Eurostat data for 2021 [34], compared to other EU countries, Portugal was ranked second from the bottom in terms of the share of energy used by households to heat their homes (30.8%). For comparison, the highest rate was observed at that time in Luxembourg (80.3%).

In recent years, electricity prices for households in Portugal have been among the highest in the European Union. According to Eurostat, in the second half of 2018, Portugal ranked first in terms of electricity prices for households and was also among the leaders in gas prices (when comparing prices at purchasing power parity). At that time, taxes and fees included in electricity bills in Portugal accounted for 55% of the total cost.

The liberalisation of the energy market and the ongoing reforms in the sector have led to a more complex situation in the supply of energy, as the number of energy suppliers has increased and the contractual terms and conditions and differences between distribution networks and suppliers have become more diverse. As a result, this situation has created uncertainty and vulnerability, especially among the most vulnerable segments of the population.

These factors contribute to the identification of key indicators of energy poverty.

1. A significant proportion of the Portuguese population (18.9% or 1.9 million people) has limited ability to ensure an adequate level of heat in their homes;
2. Approximately 752,956 households (1.9 million people) benefited from preferential electricity tariffs;
3. Approximately 34,709 households (87,000 people) benefited from preferential gas tariffs;
4. About 1,202,567 households (3 million people) face energy costs that account for 10% or more of their income;
5. The energy efficiency of residential buildings is relatively low, with 69.6% of all dwellings in Portugal classified as C or below;
6. Approximately 4.3 per cent of the population (440,000 people) is in arrears for utilities.

3.2. Analysis of the Practice of Applying State Policy to Cover Energy Costs

Based on the definition of energy poverty, analysis of selected key indicators, and comparison with social assistance and income information, it is estimated that between 1.9 and 3 million people in Portugal are in energy poverty, based on their housing conditions and the ratio of income to energy consumption.

In 2010, as part of the National Energy Strategy, a special preferential tariff for electricity was introduced (under Decree Law 128-A/2010), and the following year, in 2011, a preferential tariff for natural gas was also introduced (under Decree Law 101/2011). These preferential tariffs are aimed at helping the most economically vulnerable segments of the population. The preferential electricity tariff is available to citizens who receive various social benefits and whose total annual income does not exceed EUR 5808. However, access to the preferential tariff for natural gas is more limited, and there are still no preferential tariffs for bottled gas. Amid the pandemic crisis, in 2020, social tariffs for electricity and natural gas became available to all unemployed people.

According to data for 2020, the percentage of households that benefited from the preferential electricity tariff was approximately 14% of the total number of households, while for the social tariff for natural gas, this figure was 2% of the total number of households that consumed natural gas. The dynamics of the number of people receiving preferential tariffs are shown in the above graph (Figure 3).

While measures taken to support a large number of citizens in covering their energy costs are important, they do not address the root causes of energy poverty.

3.3. Study the Effectiveness of Introducing Energy Certification of Buildings to Overcome Energy Poverty

In 2020, Portugal adopted the National Energy and Climate Plan for the period 2021–2030 (PNEC 2030) [28]. The plan aims to develop a long-term strategy to combat energy poverty, including programmes to promote energy efficiency and the integration of renewable energy sources.

In March 2021, a set of measures aimed at combating energy poverty was included in the Recovery and Resilience Plan (PRR 2021–2026). In particular, it was decided to allocate 100,000 cheques as direct assistance to the most vulnerable families to improve their energy supply in residential buildings. These and other measures and policies were later developed and expanded as part of the National Long-Term Strategy for Combating

Energy Poverty for the period 2021–2050. This strategy foresees, among other things, the allocation of at least EUR 300 million for measures to improve the energy efficiency of residential buildings.

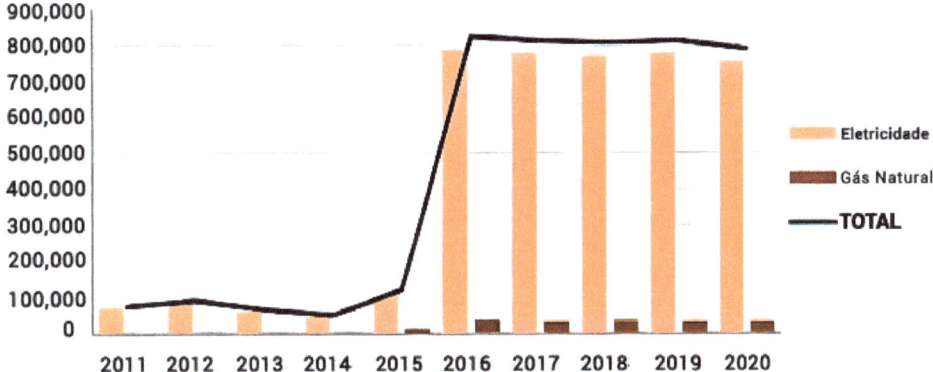

Figure 3. Dynamics of population distribution by types of resources that benefited from preferential tariffs.

In 1990, Portugal first established requirements for assessing the thermal insulation of residential buildings and preventing overheating. In 2006, the Energy Certification System for Buildings (SCE) was approved, which allows for the assessment of the energy efficiency of buildings based on an eight-step scale (from A+, very efficient, to F, very inefficient) and provides owners with up-to-date information on the impact of this classification on comfort, health, and energy consumption. Each building's energy class and energy efficiency, including the contribution of renewable energy sources, corresponds to a specific colour in the figure. They are presented in comparison to a reference value and are calculated under standard conditions. The energy performance rating is an indicator of the overall efficiency of a building. The higher the rating, the higher the energy efficiency and the lower the operating costs. The same scale is used to determine the environmental impact of a building.

Between 2014 and 2020, approximately 1.3 million energy certificates were issued, but only 12.3% of residential buildings were classified as very efficient (A and A+). Approximately 70% of certified residential buildings have an efficiency class of C or lower, as shown in the chart (Figure 4).

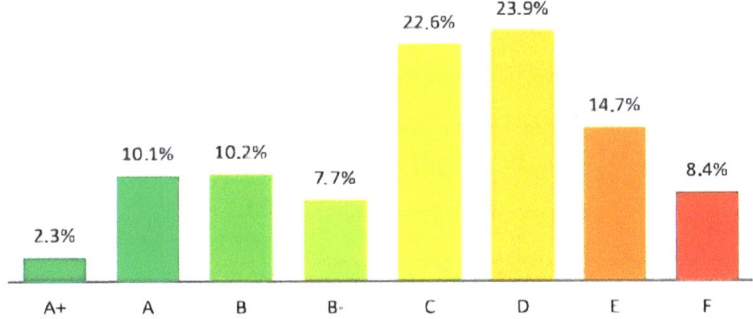

Figure 4. Distribution of residential dwellings in Portugal by energy efficiency level (based on SCE assessment) where A+ corresponds to the difference between the calculated specific heat losses and the maximum permissible value of A+ 0–25%; A 26–50%; B 51–75%; B⁻ 76–100%; C 101–150%; D 151–200%; E 201–250%; F more than 251%.

Based on the analysis of energy certificates, it can be concluded that the national housing stock does not provide all its residents with adequate living conditions, such as proper heat, sound insulation, and indoor air quality.

Much of the housing stock is outdated and in need of repair. The issue of improving the energy efficiency of buildings during such repairs remains open and requires careful consideration.

It is mandatory to obtain an energy certificate in these cases:

1. New buildings under construction;
2. Secondary real estate subject to major repairs, i.e., repairs where the total cost of works is more than 25% of the total cost of the building based on the average construction cost published annually;
3. Real estate for sale or lease;
4. Buildings that are subject to financing programmes if energy certification is required for the implementation of the programmes;
5. Owners of buildings who are eligible for tax benefits if energy certification is required to obtain these benefits.

To obtain an energy certificate, it is enough to contact one of the many organisations that provide such services, such as SCE. You need to prepare a package of documents and ensure that experts have access to the premises. Once the experts have assessed the building, you should wait for their decision and obtain an energy certificate. It is worth noting that energy efficiency certificates have validity periods that depend on the purpose of the building.

4. Discussion

Based on an analysis of the Portuguese housing stock and government policies in the building and energy sectors, it was found that a significant proportion of Portuguese households (18.9%) have limited capacity to meet their energy needs. This is due not so much to low income but mainly to the low level of energy efficiency of buildings (70%). This problem has arisen due to a number of circumstances that are factors in the formation of energy poverty:

1. The state is lagging behind in the development of the regulatory framework in the form of construction requirements. This provoked unsatisfactory quality of housing sanitary and hygienic conditions in part of the Portuguese population (25.2%). By this indicator, Portugal is one of the leaders in the European Union. Based on this, it can be assumed that the delay of the country's leadership and lack of attention to construction requirements is one of the factors behind the formation of energy poverty;
2. The low energy efficiency of buildings and imperfect heating systems. The example of Portugal shows that more than 80% of dwellings use fireplaces as a source of heating. These are rather imperfect heating systems. In addition, portable electric heaters, which have low energy efficiency, are used as an alternative to centralised heating. This can be explained not only by the lack of resources on the part of the population but also by the lack of access to more economically advantageous resources, such as natural gas;
3. Ineffective state policy in the area of combating energy poverty. This is reflected in the fact that a part of the population does not use the right to loans to cover energy costs.

Thus, we have identified the factors that contribute to energy poverty, which is expressed not only in terms of access to resources but also in terms of access to more efficient heating, electricity, etc. One of the most important triggers is government policy. It is the state that sets the requirements for construction, which subsequently affects the condition of buildings and energy efficiency. This allowed us to further analyse the existing regulatory framework in Portugal and establish that the key to overcoming energy poverty is the introduction of energy certification in the form of energy certificates.

An energy certificate (Figure 5) provides a number of benefits to the property owner. First, a building with a higher energy class becomes a competitive advantage in the real estate market. In addition, a certificate can help to attract financing at more favourable rates or to benefit from IMI or IMT property tax exemptions.

Figure 5. Sample energy certificate.

A variety of tools are available to property owners to support the implementation of the measures specified in the energy certificate. Using these tools, financing or incentives can be obtained to improve and renovate the property.

The Casa Eficiente 2020 programme provides concessional loans for activities aimed at improving the environmental performance of private residential buildings, in particular, in the areas of energy efficiency, water conservation, and municipal waste management. Renovations can cover both interior and exterior works of the building. The programme is open to owners of residential buildings, their sections, and related condominiums, and buildings can be located in any region of the country.

The IFRRU 2020 programme is designed to invest in urban renewal and energy efficiency across the country. This instrument promotes the comprehensive rehabilitation of buildings, including multi-apartment and private houses over 30 years old or younger, but with a low level of conservation (2 or less according to the criteria set out in Decree-Law no. 266-B/2012 of 31 December).

Real estate tax incentives (IMI) allow municipalities to set a reduction of up to 25% of the municipal real estate tax rate for energy-efficient urban buildings for certain years. These incentives apply if the buildings have an energy class A or show improvements two levels higher than before.

Real estate transfer tax (IMT) exemptions provide tax relief for the purchase of property intended for renovation if the relevant renovation work is started within three years of the purchase date and other conditions are met.

The use of these tools allows for sustainable improvements in energy efficiency and is an important step in the fight against energy poverty and the promotion of environmental sustainability in real estate.

5. Conclusions

This paper is a comprehensive study that addresses one of the most important issues—overcoming energy poverty. Based on the data of leading organisations (Eurostat, reports on energy certification in Portugal, etc.), we have identified a number of factors that shape energy poverty. These factors include an imperfect regulatory framework for construction, which causes buildings to not meet sanitary, hygienic, and ergonomic requirements; public awareness of the possibilities of reimbursement of energy costs; and the cost of energy and the efficiency of their use for domestic needs. The low level of the energy efficiency of buildings, as well as the population's access to quality energy, are also worth noting. All of this shows that energy poverty is not a problem for the Third World alone but is a global issue that will become even more acute as we move towards clean and energy-efficient resources.

To overcome the problems of energy poverty, the experience of Portugal was analysed. It was found that the practice of granting loans to reimburse energy costs is not effective. This is due to the lack of awareness of the population about the possibilities, as well as the unsatisfactory condition of the housing stock.

An analysis of Portugal's regulatory documents and government programmes has shown that energy certification of buildings is a prerequisite for overcoming energy poverty. The transition to energy efficiency is stimulated by a set of government programmes on construction loans, allocation of funds for modernisation, etc.

A study of the factors that shape energy poverty, as well as an analysis of state policy in the field of green building technologies, can offer a deeper understanding of the problem.

Author Contributions: Conceptualization, S.D., H.M., and A.K.; methodology, H.M.; validation, S.D.; formal analysis, S.D. and A.K.; investigation, A.K.; resources, S.D.; data curation, S.D.; writing—original draft, S.D. and A.K.; writing—review and editing, H.M., S.D., and A.K.; visualization, S.D.; supervision, H.M.; project administration, A.K. and S.D.; funding acquisition, H.M. All authors have read and agreed to the published version of the manuscript.

Funding: This research was funded by Erasmus+, grant number No. 2021-1-SK01-KA220-HED-000023274.

Institutional Review Board Statement: Not applicable.

Informed Consent Statement: Not applicable.

Data Availability Statement: Data are contained within the article.

Acknowledgments: Acknowledgement of the scientific team of the Erasmus+ project 2021-1-SK01-KA220-HED-000023274 "Support of higher education system in a context of climate change mitigation through regional-level of carbon footprint caused by a product, building and organization".

Conflicts of Interest: The authors declare no conflict of interest.

References

1. Burgess, M.G.; Langendorf, R.E.; Moyer, J.D.; Dancer, A.; Hughes, B.B.; Tilman, D. Multidecadal Dynamics Project Slow 21st-Century Economic Growth and Income Convergence. *Commun. Earth Environ.* **2023**, *4*, 220. [CrossRef]
2. Zhang, L.; Yin, J.; Li, J.; Sun, H.; Liu, Y.; Yang, J. Association between Dietary Caffeine Intake and Severe Headache or Migraine in US Adults. *Sci. Rep.* **2023**, *13*, 10220. [CrossRef] [PubMed]
3. Lozano, F.J.; Lozano, R.; Lozano-García, D.F.; Flores-Tlacuahuac, A. Reducing Energy Poverty in Small Rural Communities through in Situ Electricity Generation. *Discov. Sustain.* **2023**, *4*, 13. [CrossRef]
4. Di Falco, S.; Lynam, G. New Evidence on the Rural Poverty and Energy Choice Relationship. *Sci. Rep.* **2023**, *13*, 3320. [CrossRef] [PubMed]
5. Mukhtarov, S.; Mikayilov, J.I. Could Financial Development Eliminate Energy Poverty through Renewable Energy in Poland? *Energy Policy* **2023**, *182*, 113747. [CrossRef]
6. Energy Consumption in Households—Statistics Explained. Available online: https://ec.europa.eu/eurostat/statistics-explained/index.php?title=Energy_consumption_in_households#Energy_consumption_in_households_by_type_of_end-use (accessed on 16 August 2023).

7. Jiglau, G.; Bouzarovski, S.; Dubois, U.; Feenstra, M.; Gouveia, J.P.; Grossmann, K.; Guyet, R.; Herrero, S.T.; Hesselman, M.; Robic, S.; et al. Looking Back to Look Forward: Reflections from Networked Research on Energy Poverty. *iScience* **2023**, *26*, 106083. [CrossRef] [PubMed]
8. Marí-Dell'Olmo, M.; Oliveras, L.; Barón-Miras, L.E.; Borrell, C.; Montalvo, T.; Ariza, C.; Ventayol, I.; Mercuriali, L.; Sheehan, M.; Gómez-Gutiérrez, A.; et al. Climate Change and Health in Urban Areas with a Mediterranean Climate: A Conceptual Framework with a Social and Climate Justice Approach. *Int. J. Environ. Res. Public Health* **2022**, *19*, 12764. [CrossRef]
9. Koengkan, M.; Fuinhas, J.A.; Auza, A.; Ursavaş, U. The Impact of Energy Efficiency Regulations on Energy Poverty in Residential Dwellings in the Lisbon Metropolitan Area: An Empirical Investigation. *Sustainability* **2023**, *15*, 4214. [CrossRef]
10. Stojilovska, A.; Dokupilová, D.; Gouveia, J.P.; Bajomi, A.Z.; Tirado-Herrero, S.; Feldmár, N.; Kyprianou, I.; Feenstra, M. As Essential as Bread: Fuelwood Use as a Cultural Practice to Cope with Energy Poverty in Europe. *Energy Res. Soc. Sci.* **2023**, *97*, 102987. [CrossRef]
11. Gabriel, M.F.; Cardoso, J.P.; Felgueiras, F.; Azeredo, J.; Filipe, D.; Conradie, P.; Van Hove, S.; Mourão, Z.; Anagnostopoulos, F.; Azevedo, I. Opportunities for Promoting Healthy Homes and Long-Lasting Energy-Efficient Behaviour among Families with Children in Portugal. *Energies* **2023**, *16*, 1872. [CrossRef]
12. Ogut, O.; Bartolucci, B.; Parracha, J.L.; Bertolin, C.; Tzortzi, J.N.; Frasca, F.; Siani, A.M.; Mendes, M.P.; Flores-Colen, I. Energy Poverty in Portugal, Italy, and Norway: Awareness, Short-Term Driving Forces, and Barriers in the Built Environment. *IOP Conf. Ser. Earth Environ. Sci.* **2023**, *1176*, 012023. [CrossRef]
13. Bulkeley, H.; Marvin, S.; Palgan, Y.V.; McCormick, K.; Breitfuss-Loidl, M.; Mai, L.; von Wirth, T.; Frantzeskaki, N. Urban Living Laboratories: Conducting the Experimental City? *Eur. Urban Reg. Stud.* **2019**, *26*, 317–335. [CrossRef]
14. Barbosa, R.; Vicente, R.; Santos, R. Climate Change and Thermal Comfort in Southern Europe Housing: A Case Study from Lisbon. *Build. Environ.* **2015**, *92*, 440–451. [CrossRef]
15. Bouzarovski, S.; Petrova, S. A Global Perspective on Domestic Energy Deprivation: Overcoming the Energy Poverty–Fuel Poverty Binary. *Energy Res. Soc. Sci.* **2015**, *10*, 31–40. [CrossRef]
16. EUR-Lex—52019DC0640—EN—EUR-Lex. Available online: https://eur-lex.europa.eu/legal-content/PT/TXT/?uri=celex%25 3A52019DC0640 (accessed on 16 August 2023).
17. Grossmann, K.; Jiglau, G.; Dubois, U.; Sinea, A.; Martín-Consuegra, F.; Dereniowska, M.; Franke, R.; Guyet, R.; Horta, A.; Katman, F.; et al. The Critical Role of Trust in Experiencing and Coping with Energy Poverty: Evidence from across Europe. *Energy Res. Soc. Sci.* **2021**, *76*, 102064. [CrossRef]
18. Desigualdade do Rendimento e Pobreza em Portugal: As Consequências Sociais do Programa de Ajustamento. Available online: https://aeaveiro.pt/biblioteca/index.php?page=13&id=4659&db= (accessed on 16 August 2023).
19. Khorolskyi, A.; Hrinov, V.; Kaliushenko, O. Network Models for Searching for Optimal Economic and Environmental Strategies for Field Development. *Procedia Environ. Sci. Eng. Manag.* **2019**, *6*, 463–471.
20. Duarte, G.; Silva, A.; Baptista, P. Assessment of Wireless Charging Impacts Based on Real-World Driving Patterns: Case Study in Lisbon, Portugal. *Sustain. Cities Soc.* **2021**, *71*, 102952. [CrossRef]
21. Onat, N.C.; Abdella, G.M.; Kucukvar, M.; Kutty, A.A.; Al-Nuaimi, M.; Kumbaroğlu, G.; Bulu, M. How Eco-Efficient Are Electric Vehicles across Europe? A Regionalized Life Cycle Assessment-Based Eco-Efficiency Analysis. *Sustain. Dev.* **2021**, *29*, 941–956. [CrossRef]
22. Martins, F.; Moura, P.; de Almeida, A.T. The Role of Electrification in the Decarbonization of the Energy Sector in Portugal. *Energies* **2022**, *15*, 1759. [CrossRef]
23. Domagała, J.; Kadłubek, M. Economic, Energy and Environmental Efficiency of Road Freight Transportation Sector in the EU. *Energies* **2023**, *16*, 461. [CrossRef]
24. Pedro, A.; Krutnik, M.; Yadack, V.M.; Pereira, L.; Morais, H. Opportunities and Challenges for Small-Scale Flexibility in European Electricity Markets. *Util. Policy* **2023**, *80*, 101477. [CrossRef]
25. Silvestre, I.; Pastor, R.; Neto, R.C. Power Losses in Natural Gas and Hydrogen Transmission in the Portuguese High-Pressure Network. *Energy* **2023**, *272*, 127136. [CrossRef]
26. Casa Eficiente 2020. Available online: https://casaeficiente2020.pt/ (accessed on 16 August 2023).
27. HOME—IFRRU. Available online: https://ifrru.ihru.pt/home-en1 (accessed on 16 August 2023).
28. Mendes, P.-J.M.; Brandão, J.; Pinto, R.V.; Aparício, H. National Energy and Climate Plan 2030 | Towards a Carbon Neutral Future. Lexology. Available online: https://www.lexology.com/library/detail.aspx?g=8fd21e2d-50a2-4fed-a99e-7c7e59fd6995 (accessed on 16 August 2023).
29. Recovery and Resilience Facility. Available online: https://commission.europa.eu/business-economy-euro/economic-recovery/recovery-and-resilience-facility_en (accessed on 16 August 2023).
30. Ramos, C.; Reis, E. Floods in Southern Portugal: Their Physical and Human Causes, Impacts and Human Response. *Mitig. Adapt. Strateg. Glob. Change* **2002**, *7*, 267–284. [CrossRef]
31. Pinheiro, M.D. Urban Sustainability Assessment System—The Portuguese Scheme, Lidera Approach and Two Urban Application Examples. In *Urban Planning: Practices, Challenges and Benefits*; Nova Science Publishers: New York, NY, USA, 2014; pp. 207–272.
32. Silva, P. *Designing Urban Rules from Emergent Patterns: Co-Evolving Paths of Informal and Formal Urban Systems—The Case of Portugal*; IOP Publishing: Bristol, UK, 2018; Volume 158. [CrossRef]

33. Nogueira, M.; Soares, P.M.M. A Surface Modelling Approach for Attribution and Disentanglement of the Effects of Global Warming from Urbanization in Temperature Extremes: Application to Lisbon. *Environ. Res. Lett.* **2019**, *14*, 114023. [CrossRef]
34. Energy Consumption in Households. Available online: https://ec.europa.eu/eurostat/statistics-explained/index.php?title=Energy_consumption_in_households (accessed on 16 August 2023).

Disclaimer/Publisher's Note: The statements, opinions and data contained in all publications are solely those of the individual author(s) and contributor(s) and not of MDPI and/or the editor(s). MDPI and/or the editor(s) disclaim responsibility for any injury to people or property resulting from any ideas, methods, instructions or products referred to in the content.

Proceeding Paper

Effect of Ash from Biomass Combustion on Tailings pH [†]

Lukas Balcarik, Bohdana Simackova, Samaneh Shaghaghi * and Lucie Syrova

Department of Environmental Engineering, Faculty of Mining and Geology, VŠB–Technical University of Ostrava, 17. listopadu 2172/15, Poruba, 708 00 Ostrava, Czech Republic; lukas.balcarik.st@vsb.cz (L.B.); bohdana.simackova.st@vsb.cz (B.S.); lucie.syrova.st@vsb.cz (L.S.)
* Correspondence: shaghaghisamaneh@gmail.com
[†] Presented at the 4th International Conference on Advances in Environmental Engineering, Ostrava, Czech Republic, 20–22 November 2023.

Abstract: This article deals with the use of ash from biomass burning for the remediation of thermally active dump in Heřmanice, Czech Republic. Nowadays, various chemical, physical, or biological methods of remediation are used for the remediation of dumps. The authors discuss the complex use of ash from biomass as a possibility for biological remediation of the Heřmanice dump. The main advantage of obtaining ash when burning biomass is primarily the fact that it is a renewable energy source, which produces electricity and large amounts of ash, which can be used, for example, for the remediation of the Heřmanice dump. Tailings are characterized by their acidity, while fly ash is characterized by high alkalinity. This study deals with which ratio (tailings:ash) would achieve the necessary neutral values in order to prevent the release of heavy metals into the surroundings of the Heřmanice dump. The value of the active soil reaction (pH/H_2O), the value of the exchange soil reaction (pH/$CaCl_2$), the value of hydrolytic acidity (H_a), together with the value of soluble salts in the tailings, i.e., electrical conductivity, were also studied. Based on the obtained results, it can be concluded that the addition of biomass combustion ash had a positive effect on the pH value of the tailings. Based on this fact, Al^{3+} is excreted more slowly into the environment. A higher content of aluminum is toxic to plants, while in a smaller amount, its content is necessary and at the same time an important factor in the process of plant growth. The mixing of alkaline ash with unburned surface tailings from the thermally active Heřmanice dump significantly influenced its acidity, which had a positive effect on increasing the active and exchange acidity.

Keywords: thermally active dumps; tailing; biomass combustion ash

1. Introduction

During the deep mining of black coal and its subsequent processing, waste is generated, which used to be deposited in dumps. This waste consisted of carboniferous rocks originating from mining, preparatory works, shaft sinking, roof rocks (sandstones, shales, conglomerates) coated with coal substance, waste material produced during coal washing, and coal slurry [1–3]. Carboniferous rocks are characterized by the presence of finely dispersed coal substance, which is very difficult to separate using conventional beneficiation methods. It is precisely this characteristic of carboniferous rocks that is the reason why the material deposited in the Ostrava-Karviná region (OKR), Czech Republic, contains a large amount of coal substance. The spoil heaps in the OKR were not covered or protected, and therefore they were exposed to mechanical and chemical weathering. Thermal processes, triggered by the oxidation or combustion of coal substance or other combustible materials (wood residues, plastics, etc.) deposited on the spoil heaps for a relatively long period, were a common occurrence. Endogenous combustion is still taking place in some spoil heaps. Currently, in the Czech Republic, within the OKR territory, there are three such thermally active spoil heaps of black coal waste: Heřmanice, Hedvika, and Ema, the latter of which can already be classified as a burning landfill [4,5].

The thermal activity of the dump causes the release of harmful gases into the surroundings and also disrupts the stability of the spoil heap structure (posing a risk of sinkholes forming in areas of hot spots) [6]. Residues of pyrite, which is present in the spoil material, can be oxidized by bacteria, particularly the *Acidithiobacillus ferrooxidans* genus [7–9]. Sulfuric acid, formed during this process of pyrite oxidation, is the reason for the acidification of the spoil. Due to the very low content of basic rocks (particularly carbonates) in the spoil of the Ostrava-Karviná region (OKR), the neutralization of the resulting acids occurs very slowly. As a result of the acidity of the spoil, there can be leaching of hazardous metals from the spoil heap into the immediate surroundings [2,4]. However, the main problem affecting the immediate surroundings of the spoil heap is the ongoing thermal activity, which is closely associated with gas emissions, potential suspension of dust particles, and the presence of hazardous elements that may be present in the deposited material. Thermally active spoil heaps can thus represent a significant ecological burden [10,11].

The remediation of such a dump and its surroundings is necessary to minimize the negative impacts of the thermal process, not only on the environment but also on the health of people living in the immediate vicinity. Currently, physical, chemical, and biological methods are used for the remediation of dumps, each with its advantages and disadvantages (economic costs, inefficiency, or negative impacts on the surrounding environment). During the 1990s, mining activities gradually declined, resulting in a significant reduction in the amount of mining waste (tailing). Nowadays, tailing is considered a useful material for further utilization. It finds application in construction (road construction) or as a material for terrain modifications (remediation, reclamation). However, the use of mining tailing as a construction material in road infrastructure comes with many challenges. Tailing can only be used in road embankments if it does not contain trace amounts of coal. A higher coal content in spoil can cause endogenous combustion and subsequent self-ignition of the road embankment. Tailing is also utilized in the reclamation of mining landscapes. For these purposes, it can be used regardless of its petrographic composition, grain size, or water content [1,2,4].

Within legislation and societal thinking, there is increasing pressure for the reutilization and reduction of disposed waste, especially through landfilling. The focus of discussion lies primarily in finding ways to transform waste, whether newly generated or stored in dumps, into resources that can be utilized in further development such as construction and reclamation. As mentioned earlier, the main issue with spoil material is its acidity, which poses a challenge for biological reclamation. One potential remediation method that could address the acidity issue of the spoil is the utilization of the alkaline properties of biomass ash resulting from biomass combustion. Our objective was to contribute to a more efficient and sustainable remediation of thermally active dump heaps through the use of biomass ash as an alternative material, as demonstrated by this pilot study.

2. Material and Methods

2.1. Characterization of Sampeling Site

The Heřmanice dump was formed during deep coal mining in the 19th and 20th centuries. It received tailing from the Ida deep mine and industrial waste from former mining companies and coking plants. The Heřmanice dump covers an area of approximately 103 hectares and is located within the cadastral territory of Slezská Ostrava in the municipal districts of Hrušov and Heřmanice, Czech Republic. Currently, the area consists of a reclaimed section of the dump in the west and remediated areas in the south. In the central part of the Heřmanice dump, there are three sludge ponds, the thermally active portion, and a secured landfill for chemical waste owned by Ostrava-Karviná Coke Plants, a joint-stock company. The amount of tailing imported to the storage area for mining waste is approximately 20 million cubic meters. The unaffected part of the tailing, still affected by thermal activity, contains approximately 12 million tons of spoil [12].

Over the past few years, a series of remediation interventions have been carried out in the area. The first remediation works took place in the late 1990s when a more extensive thermal process was observed. A reduction of thermal activity was attempted by layering ash combined with clay soil on the surface of the thermally active area. However, this attempt proved to be ineffective. Subsequent efforts were made but also proved unsuccessful, resulting in the removal of all vegetation from the ridge of the thermally active dump. The most effective method to prevent thermal activity has been determined to be the removal of the mining tailing. However, thermal processes are unpredictable, and it cannot be guaranteed that complete cessation of activity will occur [13].

2.2. Methodology of Sample Collection and Preparation

Samples of the tailing were collected from the thermally active part of the Heřmanice dump. The sampling locations were chosen to be as representative as possible of the spoil in that area. A total of 20 sampling points were selected, encompassing the sloping terrain and flat surfaces (see Figure 1). Initially, the top layer of the dump, approximately 10–20 cm thick, which was not part of the collected sample, was removed. The sampling was conducted from a depth of 20–50 cm. After cooling down, the spoil samples from all sampling points were homogenized to create a composite sample, which was stored in a sealable plastic container.

Figure 1. Sampling locations at the Heřmanice dump.

The tailing samples were mechanically processed in the laboratory using a jaw crusher, Retsch type BB200 WC (Haan, Germany). After homogenization, the tailing was further size-reduced using a Retsch jaw crusher, type BB200 WC (Haan, Germany), to approximately 2 mm. Subsequently, the samples were dried to a constant weight in a vacuum dryer, model VO29 MEMMERT (Schwabach, Germany). The dried tailing samples were stored in a desiccator.

The sample of grate ash from biomass combustion of plant origin was obtained from an industrial boiler in a thermal power plant. At the time of sampling, wood chips were being burned. The ash was stored in a sealable plastic container with a lid to prevent air and moisture infiltration. The grate ash was size-reduced to a particle size of 2 mm using a Retsch jaw crusher, type BB200 WC (Haan, Germany). The dried ash samples were stored in a desiccator.

The samples of ash and tailing were blended in ratios of 1:6, 1:7, and 1:8, ensuring an excess of the tailing sample.

2.3. Determination of Monitored Parameters

The active acidity (pH/H$_2$O) and passive acidity (pH/CaCl$_2$) were determined following the ČSN EN ISO 10390 (836221) [14]. The measurements were performed using an inoLab® pH 7110 laboratory pH meter from Xylem Analytics Germany Sales GmbH (Weilheim, Germany). To assess the active and passive acidity, criteria specified in the Table 1 were utilized.

Table 1. Criteria for assessing active and passive acidity [15].

pH/H$_2$O	Interpretation	pH/CaCl$_2$	Interpretation
<5.4	Strongly acidic	<4.8	Strongly acidic
5.5–6.9	Moderately acidic	4.8–5.2	Moderately high acidic
6.5–6.9	Slightly acidic	5.2–5.5	Moderately acidic
7.0	Neutral	5.5–7.5	Moderately acidic to sightly alkaline
7.1–7.5	Slightly alkaline	>7.5	Moderately to strongly alkaline
7.6–8.3	Moderately alkaline	-	-
>8.4	Strongly alkaline	-	-

The determination of soluble salts based on electrical conductivity (EC) was carried out according to ČSN ISO 11265 (836210) [16] using the WTW Multi 3320 instrument (WTW Germany, Weilheim, Germany). Water-soluble electrolytes were extracted from the soil in a ratio of soil to water of 1:5, and their concentration was determined based on the increase in specific electrical conductivity of the filtered extract. The criteria for assessing electrical conductivity were applied as presented in Table 2.

Table 2. Criteria for assessing active and passive acidity [17].

Electrical Conductivity mS cm^{-1}	Soils
<0.70	Non-saline
0.71–1.20	Moderately saline
>1.20	Saline

The chemical composition of the samples was determined semi-quantitatively by X-ray fluorescence on the XEPOS (Spectro, Kleve, Germany) energy dispersion spectrometer. After trituration, the samples were placed in a plastic cuvette with a Mylar protective foil and then analyzed in a protective atmosphere (He). The phase composition and microstructural properties were determined using the X-ray powder diffraction (XRD) technique. XRD patterns were obtained using a Rigaku SmartLab diffractometer (Rigaku, Tokyo, Japan) with a D/teX Ultra 250 detector. The measured XRD patterns were evaluated using PDXL 2 software (version 2.4.2.0) and compared with the PDF-2 database, 2015 release (ICDD, Newton Square, Worcester, MA, USA).

3. Results and Discussion

From a legislative perspective, tailing is classified as mining waste and must be handled accordingly. In the Czech Republic, this matter is addressed by Act No. 157/2009 Coll., which establishes relevant regulations for the management of mining waste. The law also sets rules for mitigating the impacts on water, air, soil, plants, animals, and the landscape that may result from the handling of mining waste [18]. However, the law does not apply to mining waste generated during mineral deposit exploration, extraction, and processing. This exception also includes water containing substances resulting from the extraction of oil and minerals, as well as waste generated during mineral deposit exploration and mining activities not directly related to these operations. The mentioned law also does not apply to materials obtained during the extraction and processing of non-metallic minerals, which are used for reclamation and remediation, or for the stabilization and disposal of mining works. The aforementioned law is supplemented by Decree No. 428/2009 Coll. and No. 429/2009 Coll., which regulate the requirements for a mining waste management plan, as specified in Decree No. 281/2018 Coll. These decrees fall within the framework of environmental protection regulations and aim to minimize the risk of soil and water pollution and improve the quality of the environment [19–21]. Additionally, within the Waste Catalog, specifically in the Decree No. 8/2021 Coll. of the Ministry of the Environment of the Czech Republic, waste from mining activities is classified in Group 01, which includes all waste from geological exploration, mining, processing, and further treatment of minerals and stones [22]. Under Czech legislation, waste generated from mining activities and activities carried out by mining methods, which are deposited in spoil heaps and tailings ponds, are not part of waste management under Act No. 541/2020 Coll., but are subject to Act No. 44/1988 Coll. (Mining Act) [23,24].

In 2001, the question of whether mining waste (tailing) should be considered as waste was discussed. After discussions among state authorities and mining organizations, mining waste was eventually recognized as a filling material for reclamation projects, such as spoil heaps, embankments, or road construction, according to Act No. 22/1997 Coll. As a result, mining waste was removed from the category of waste and is now only utilized with the permission of a certificate and conformity confirmation [25]. These certificates are regularly renewed through tests conducted in accredited laboratories, and individual mines perform regular sampling and analysis of the mining waste to compare its parameters with those stated in the certificate. Currently, mining waste can be used for road construction, but without endangering the environment. From an environmental protection perspective, the material can be considered suitable and non-threatening if the concentrations of leaching substances do not exceed the values specified in Appendix C of the ČSN 73 6133 standard [26]. When using mining waste for reclamation and remediation, it is necessary to adhere to the rules outlined in Decree No. 428/2009 Coll., which also regulates the requirements that must be included in the project documentation for remediation [19].

The tailing from thermally active dumps can also contain various hazardous organic substances, such as polychlorinated biphenyls (PCBs), polycyclic aromatic hydrocarbons (PAHs), and others. These substances can be a source of contamination in the immediate surroundings, and therefore their remediation is necessary [10]. The procedures and methods for removing these hazardous substances from soil and tailing are specified in Decree No. 429/2009 Coll., which is regulated in terms of the requirements for the waste management plan by Decree No. 281/2018 Coll.

Knowledge of the mineralogical–petrographic characteristics of tailing is crucial for assessing its potential impact on the environment. The petrographic composition of tailing in the Ostrava-Karviná region (Czech Republic) is practically identical among individual dumps. The most prevalent rock types are various types of aleuropelites, accompanied by fine-grained and medium to coarse-grained sandstones in significant and variable quantities, which are more resistant to weathering. Calcite is rare in the rocks of the Ostrava-Karviná region. Small amounts of clay minerals (such as mixed illite–smectite structures) may exhibit swelling when in contact with water. Based on the elemental analysis (see Table 3), it can be observed that elements such as Si (27%), Fe (25%), and Al (11%) were predominant in the tailing. Pertile et al. state that considering the percentage of silicon, it can be assumed that acidification will not occur due to its loss, but rather due to the deficiency of basic cations during the decomposition of aluminosilicates [2]. The sulfur content in the dry coal samples from the Ostrava-Karviná region is generally low. Pešek et al. report that the total sulfur values vary between 0.4% and 4.8% with an average of 2.1% across different layers and areas [3]. This was confirmed by the sample of tailing from the Heřmanice dump, which had a total sulfur content of around 3.5%. Generally, the content of heavy metals in the carbonaceous rocks of the Ostrava-Karviná region is low and does not exceed the background levels of other industrial emissions from an ecological perspective. The composition of ash from biomass combustion depends on various factors, with the main ones being the origin of the biomass and the method of combustion.

Table 3. Elemental analysis of tailings and cyclone ash samples using XRF method in % [2].

%	Tailing	Ash	%	Tailing	Ash	%	Tailing	Ash	%	Tailing	Ash
Na	<0.01	<0.01	Mo	0.001	0.001	Pr	<0.0002	0.005	Te	<0.0003	<0.0003
Ag	<0.0002	<0.0002	Nb	0.005	0.0001	Cs	0.014	<0.0004	Sr	0.053	0.09
Al	**11**	0.82	Nd	0.029	0.01	Cu	0.03	0.007	Ta	0.01	0.008
As	0.02	0.0001	Ni	0.03	0.01	Fe	**25**	1.2	Sn	0.0008	0.0001
Ba	0.68	0.15	P	0.3	2.3	Ga	0.009	0.001	Th	0.008	0.001
Bi	<0.0001	<0.0001	Pb	0.03	0.001	Ge	0.0002	0.0002	Ti	1.3	0.13
Br	0.008	0.004	Cr	0.04	0.005	Hf	0.002	0.0001	Tl	0.0004	0.0001
Ca	0.68	**20**	Rb	0.054	0.02	Hg	0.006	<0.0001	U	0.002	<0.0001
Cd	0.0005	0.0001	S	3.5	2.3	I	0.06	<0.0003	V	0.05	0.002
Ce	0.01	0.02	Sb	0.0004	<0.0003	K	**5.5**	**25**	W	0.002	<0.0001
Cl	0.49	1.3	Se	0.0006	0.0001	La	0.01	0.01	Y	0.01	0.001
Co	0.005	0.0003	Si	**27**	**28**	Mg	0.2	1.4	Zn	0.05	0.03

Note: Values of elements exceeding 5% are highlighted in bold.

During the biomass combustion process, organic matter decomposes into various components, including carbon oxides, water vapor, nitrogen oxides, sulfur oxides, and other gases, which are then emitted from the combustion process. Ash is formed, which contains nutrients and micronutrients. For example, ash from wood combustion is rich in calcium, while ash from straw and grains is rich in potassium. Magnesium, calcium, and potassium are present in the form of carbonates, which are formed during the mineralization of organic compounds by the conversion of cations to oxides, which then hydrate and convert to carbonates under atmospheric conditions [27]. The nitrogen content in ash is very low and practically negligible because at higher temperatures during the combustion process, it transforms into oxides and escapes into the atmosphere. Risky metals can also be present in ash, but their occurrence is closely related to the origin of the burned biomass and the temperature during combustion. Generally, ash from biomass combustion exhibits alkaline properties.

The high content of alkaline metals and their form are the main factors contributing to these properties. The pH value of biomass ash increases primarily due to the loss of organic acids during the combustion process and the formation of soluble oxides, hydroxides, carbonates, and bicarbonates of calcium, magnesium, potassium, and sodium. The pH value is further influenced by the combustion temperature and storage time, with alkalinity decreasing as both factors increase. During prolonged storage, the hydroxides in the ash can convert to carbonates. For example, ash from woody biomass generally has a higher pH value due to a higher calcium content and lower sulfur content compared to ash from straw and grains [28]. Based on elemental analysis, the bottom ash from biomass combustion also exhibits low levels of heavy metals. The silicon content in bottom ash (28%) is practically the same as in the original material (27%).

The samples of ash and overburden were mixed in ratios of 1:6, 1:7, and 1:8 to achieve a pH value within the neutral range of 6.1–7.0, considering the research objectives. One of the parameters monitored in the samples was their active acidity, which is determined by hydrogen ions present in the solution. These hydrogen ions originate from dissociated mineral and organic acids. The values of active acidity (pH/H_2O) provide information about the current state of acidity and alkalinity of the material [29]. Passive acidity (pH/$CaCl_2$) is formed by adsorbed H^+ and Al^{3+} (Fe^{3+}) ions, which can be released into the solution through the exchange with basic cations from neutral salt solutions in the soil. Passive acidity is considered a more important and commonly used indicator for evaluating acidity, as it undergoes less significant changes compared to active acidity over time. The values of passive acidity are generally lower than active acidity, ranging from a pH difference of 0.2–1.0 [30]. If the mixture of overburden and ash is to be used in the biological remediation of the dump, particularly in combination with the importation of soil substrate, the active acidity becomes significant as it fundamentally affects the biochemical processes occurring in the soil and the nutrient uptake processes by autotrophic organisms [15,30,31]. The values of active and passive acidity, along with the evaluation, are presented in Table 4.

Table 4. Active and passive acidity in the studied samples.

Sample	Active Acidity		Passive Acidity	
	pH/H$_2$O	Interpretation	pH/CaCl$_2$	Interpretation
Mixture of Ash/Tailing 1:6	8.2	Moderately alkaline	6.7	Moderately acidic to slightly alkaline
Mixture of Ash/Tailing 1:7	7.1	Slightly alkaline	5.8	Moderately acidic to slightly alkaline
Mixture of Ash/Tailing 1:8	6.6	Slightly acidic	5.7	Moderately acidic to slightly alkaline
Tailing	4.5	Strongly acidic	4.6	Strongly acidic
Ash	11.2	Strongly alkaline	10.2	Moderately to strongly alkaline

Acidic tailings can limit microbial activity, reduce the availability of essential nutrients, and cause aluminum formation in the subsurface, slowing root growth and limiting access to water and nutrients [15]. By combining strongly alkaline grate ash with strongly acidic tailings, the expected significant modification of tailing acidity occurred. Based on the determined values of both active and passive acidity, the mixture of grate ash and tailings in a ratio of 1:6 can be considered the most suitable. It is a realistic assumption that the use of this mixture for remediation will be much more effective than if the grate ash was applied to the tailings in layers.

The evaluation of soluble salts in the soil based on electrical conductivity (EC) is used to determine the amount of salts in the samples. Excessive concentration of ions from any group of salts can lead to the weakening of plant growth or even their complete death. The values of electrical conductivity, along with the evaluation, are presented in Table 5.

Table 5. Values of specific conductivity of the studied material.

Sample	Specific Conductivity	
	mS cm^{-1}	Interpretation
Mixture of Ash/Tailing 1:6	3.5	Salty
Mixture of Ash/Tailing 1:7	3.0	Salty
Mixture of Ash/Tailing 1:8	3.0	Salty
Tailing	0.2	Unsalted
Ash	4.6	Salty

Based on the values of specific conductivity in all samples of the mixture as well as in the ash sample alone, it can be concluded that these are saline samples. Considering the composition of the ash, it primarily consists of calcium, silicon, and iron salts, which are not environmentally hazardous. Conversely, the content of calcium salts will contribute to increasing the pH value of the strongly acidic tailings.

4. Conclusions

The mixing of alkaline ash with unburned surface tailings from the thermally active Heřmanice dump significantly influenced its acidity, which had a positive effect on increasing the active and exchange acidity. The ash is also a rich source of various basic cations that can enrich the tailings. This can ultimately accelerate and positively impact the reclamation of landscapes affected by deep mining. The use of a mixture of tailings and ash will certainly be more effective in the remediation of the thermally active dump and its surroundings than the mere layering of these materials.

Author Contributions: Conceptualization, L.B.; methodology, L.B., B.S. and L.S.; software, S.S.; validation, L.B. and S.S.; formal analysis, L.S.; investigation, L.B.; resources, B.S.; data curation, L.B. and L.S.; writing—original draft preparation, L.B., L.S. and B.S.; writing—review and editing, S.S., L.B. and L.S.; visualization, S.S.; supervision, L.B.; project administration, L.S.; funding acquisition, L.S. All authors have read and agreed to the published version of the manuscript.

Funding: This research was funded by: VSB-TUO, Faculty of Mining and Geology—grant number SP2022/57; VSB-TUO, Faculty of Mining and Geology—grant number SP2023/017. Project CZ.11.4.120/0.0/0.0/15_006/0000074 TERDUMP Cooperation VŠB-TUO/GIG Katowice on the survey of burning dumps on both sides of the common border.

Institutional Review Board Statement: Not applicable.

Informed Consent Statement: Not applicable.

Data Availability Statement: Data are contained within the article.

Conflicts of Interest: The authors declare no conflict of interest.

References

1. Abramowicz, A.; Rahmonov, O.; Chybiorz, R. Environmental Management and Landscape Transformation on Self-Heating Coal-Waste Dumps in the Upper Silesian Coal Basin. *Land* **2020**, *10*, 23. [CrossRef]
2. Pertile, E.; Dvorský, T.; Václavík, V.; Syrová, L.; Charvát, J.; Máčalová, K.; Balcařík, L. The Use of Construction Waste to Remediate a Thermally Active Spoil Heap. *Appl. Sci.* **2023**, *13*, 7123. [CrossRef]
3. Pešek, J.; Sýkorová, I.; Michna, O.; Forstová, J.; Martínek, K.; Vašíček, M.; Havelcová, M. Major and Minor Elements in the Hard Coal in the Czech Upper Paleozoi. Available online: https://asep.lib.cas.cz/arl-cav/cs/detail/?&idx=cav_un_epca-1*0357 902 (accessed on 3 July 2023).
4. Surovka, D.; Pertile, E.; Dombek, V.; Vastyl, M.; Leher, V. Monitoring of Thermal and Gas Activities in Mining Dump Hedvika, Czech Republic. *IOP Conf. Ser. Earth Environ. Sci.* **2017**, *92*, 012060. [CrossRef]
5. Pertile, E.; Surovka, D.; Sarčáková, E.; Božoň, A. Monitoring of Pollutants in an Active Mining Dump Ema, Czech Republic. *Inz. Miner.* **2017**, *2017*, 45–50.
6. Smoliński, A.; Dombek, V.; Pertile, E.; Drobek, L.; Gogola, K.; Żechowska, S.W.; Magdziarczyk, M. An Analysis of Self-Ignition of Mine Waste Dumps in Terms of Environmental Protection in Industrial Areas in Poland. *Sci. Rep.* **2021**, *11*, 8851. [CrossRef] [PubMed]
7. Jablonka, R.; Vojtkova, H.; Kasakova, H.; Qian, L. The Adaptation of Acidithiobacillus Ferrooxidans for the Treatment of Hazardous Waste. *Int. Multidiscip. Sci. GeoConf. SGEM* **2012**, *5*, 853.
8. Šimonovičová, A.; Ferianc, P.; Vojtková, H.; Pangallo, D.; Hanajík, P.; Kraková, L.; Feketeová, Z.; Čerňanský, S.; Okenicová, L.; Žemberyová, M.; et al. Alkaline Technosol Contaminated by Former Mining Activity and Its Culturable Autochthonous Microbiota. *Chemosphere* **2017**, *171*, 89–96. [CrossRef] [PubMed]
9. Vojtková, H.; Janulková, R.; Švanová, P. Phenotypic Characterization of Pseudomonas Bacteria Isolated from Polluted Sites of Ostrava, Czech Republic. *Geosci. Eng.* **2012**, *58*, 52–57. [CrossRef]
10. Mi, J.; Yang, Y.; Zhang, S.; An, S.; Hou, H.; Hua, Y.; Chen, F. Tracking the Land Use/Land Cover Change in an Area with Underground Mining and Reforestation via Continuous Landsat Classification. *Remote Sens.* **2019**, *11*, 1719. [CrossRef]
11. Pertile, E.; Surovka, D.; Božoň, A. The Study of Occurrences of Selected PAHs Adsorbed on PM10 Particles in Coal Mine Waste Dumps Heřmanice and Hrabůvka (Czech Republic). *Int. Multidiscip. Sci. GeoConf. SGEM* **2016**, *3*, 161–168.
12. DIAMO s.p. About Heap—Characteristic | DIAMO, s.p. Available online: https://www.diamo.cz/hermanickahalda/o-halde/popis (accessed on 1 August 2023).
13. DIAMO s.p. About Heap—History | DIAMO, s.p. Available online: https://www.diamo.cz/hermanickahalda/o-halde/historie (accessed on 1 August 2023).
14. *ČSN 10390 (836221)*; Soils, Treated Biowaste and Sludge—Determination of pH. Czech Standardization Institute: Prague, Czech Republic, 2022.

15. Apal, A.L. *Soil_Test_Interpretation_Guide 08062023*; North Country Organics: Bradford, ON, Canada, 1999.
16. ČSN ISO 11265 (836210); Soil Quality—Determination of Electrical Conductivity. Czech Standardization Institute: Prague, Czech Republic, 1996.
17. FAO. *Mapping of Salt-Affected Soils—Technical Manual*; FAO: Rome, Italy, 2020.
18. *Czech Republic Act No. 157/2009 Coll.*; Act on the Management of Mining Waste and on the Amendment of Certain Acts. Parliament of the Czech Republic: Prague, Czech Republic, 2006.
19. *Czech Republic 428/2009 Coll.*; Decree on the Implementation of Certain Provisions of the Act on the Management of Mining Waste. Parliament of the Czech Republic: Prague, Czech Republic, 2009.
20. *Czech Republic 429/2009 Coll.*; Decree on the Determination of Requirements for the Plan for the Management of Mining Waste, Including the Evaluation of Its Properties and Some Other Details for the Implementation of the Act on the Management of Mining Waste. Parliament of the Czech Republic: Prague, Czech Republic, 2009.
21. *Czech Republic 281/2018 Coll.*; Amending Decree No. 429/2009 Coll., on the Specification of Requirements for the Plan for Handling Mining Waste, Including the Evaluation of Its Properties and Certain Other Details for the Implementation of the Act on Handling Mining Waste. Ministry of the Environment of the Czech Republic: Prague, Czech Republic, 2018.
22. *Czech Republic 8/2021 Coll.*; Decree on the Waste Catalogue and Assessment of Waste Properties (Waste Catalogue). Ministry of the Environment of the Czech Republic: Prague, Czech Republic, 2021.
23. *Czech Republic 541/2020 Coll.*; Waste Act. Parliament of the Czech Republic: Prague, Czech Republic, 2021.
24. *Czech Republic 44/1988 Coll.*; Act on the Protection and Use of Mineral Resources (Mining Act). Federal Assembly of the Czech Republic: Prague, Czech Republic, 1988.
25. *Czech Republic 22/1997 Coll.*; Act on Technical Requirements for Products and on the Amendment of Certain Acts. Parliament of the Czech Republic: Prague, Czech Republic, 1997.
26. ČSN 73 6133 (736133); Design and Construction of Earthworks for Ground Transportation. Czech Standardization Institute: Prague, Czech Republic, 2010.
27. Demeyer, A.; Voundi Nkana, J.C.; Verloo, M.G. Characteristics of Wood Ash and Influence on Soil Properties and Nutrient Uptake: An Overview. *Bioresour. Technol.* **2001**, *77*, 287–295. [CrossRef] [PubMed]
28. Pana, H.; Eberhard, T.L. Characterization of Fly Ash from the Gasification of Wood and Assessment for Its Application as a Soil Amendment. *BioResources* **2011**, *6*, 3987–4004. [CrossRef]
29. Wherry, E.T. Soil Acidity and a Field Method for Its Measurement. *Ecology* **1920**, *1*, 160–173. [CrossRef]
30. Sparks, D.L. Environmental Soil Chemistry—2nd Edition. Available online: https://shop.elsevier.com/books/environmental-soil-chemistry/sparks/978-0-12-656446-4 (accessed on 3 August 2023).
31. Hernández, T. Acidity. In *Environmental Geology*; Encyclopedia of Earth Science; Kluwer Academic Publishers: Dordrecht, The Netherlands, 1999; p. 6.

Disclaimer/Publisher's Note: The statements, opinions and data contained in all publications are solely those of the individual author(s) and contributor(s) and not of MDPI and/or the editor(s). MDPI and/or the editor(s) disclaim responsibility for any injury to people or property resulting from any ideas, methods, instructions or products referred to in the content.

Proceeding Paper

The Resistance of an Enamelled Material to Biochemical Leaching †

Vladislav Blažek * and Jaroslav Závada

Private Laboratory, 739 34 Šenov, Czech Republic; centurian@seznam.cz
* Correspondence: vla.blazek@seznam.cz; Tel.: +420-775882148
† Presented at the 4th International Conference on Advances in Environmental Engineering, Ostrava, Czech Republic, 20–22 November 2023.

Abstract: This article describes the resistance of chemically durable enamelled surfaces against the effects of chemical leachate and enzymes of bacteria *Acidithiobacillus ferrooxidans*, which are commonly used in biohydrometallurgy today during the biochemical leaching of ore. For many years now, the Vítkovice ENVI company has successfully used steel parts protected by the tested enamel in the construction of wastewater treatment plants and biogas stations. This article summarizes the results of a study that dealt with verifying the possibility of usage of widely applied enamel in the field of biochemical leaching during the treatment of raw materials. From the performed experiments, it follows that enamel can endure the effects of leaching solutions, but on poorly treated areas such as the edges and corners of parts, a quick onset of corrosion occurs.

Keywords: leaching; biochemical leaching; enamelled material; *Acidithiobacillus ferrooxidans*

1. Introduction

In an effort to expand the product range of Vítkovice ENVI beyond storage tanks, WWTPs and biogas stations, there was a need to add tanks designed for processing lean ores and metal-bearing waste using chemical and biochemical leaching. It was therefore necessary to verify the resistance of the enamel commonly used in the company's current production to leaching media and enzymes produced by microorganisms. The initial verification experiment took place in laboratory conditions, simulating leaching in both the presence and absence of microorganisms. In today's practice, tanks made of steel or concrete scalded with a plastic liner, which protects these materials against corrosion and biocorrosion, are used to leach poor ores. Enamel could be widely used in this industry (in terms of its durability and price).

2. Theory

Testing enamel's resistance to leachates used in biohydrometallurgy is not mentioned in the literature. There is only some limited information about the leaching of enamel in order to determine which metals may be released during its use, especially during food preparation [1].

The issue of the resistance of the enamelled surface is addressed by the author Rolf Lorentz [2], who already in 1983 exposed enamelled parts of various qualities to several polar solvents (distilled water, steam, water mixed with Na_2CO_3 and water mixed with $CaCl_2$ and HCl) at different temperatures and pressures. Subsequently, he evaluated the resistance of the enamelled surface on the basis of its weight loss. Based on the performed experiments, he came to the conclusion that distilled water does not cause the corrosion of the enamel at normal temperatures; however, tap water does cause a slight corrosion on the enamelled surface. When the pH deviates, whether towards an acidic or basic area, a corrosion of the enamelled surface occurs. The author states that the pH value and the temperature of the solution are decisive for the corrosion of the enamel.

The effect of the leachates and enzymes produced on the enamelled surface itself has not yet been addressed.

3. Materials and Methods

Experimentally, samples of enamelled sheet metal were exposed to four potential leaching solutions with and without microorganisms (Table 1).

Table 1. Distribution of samples and composition of leaching solutions.

Sample 1	9K nutrient medium with *Acidithiobacillus ferrooxidans* bacteria
Sample 2	
Sample 3	Sterile nutrient medium 9K
Sample 4	Demineralized water acidified with H_2SO_4
Sample 5	$Fe_2(SO_4)_3$ solution acidified with H_2SO_4

Sheets with the trade name KOSMALT E 300 were manufactured by VSŽ Košice and are currently conventionally used for the production of enamelled parts in the Vítkovice ENVI company. The dimensions of the sheet samples used in the experiment were 120 × 120 mm. These samples were subsequently enamelled by Vítkovice ENVI. The enamelled surface produced is referred to by the manufacturer as a chemically highly resistant enamelled surface. The enamel layer in this case consists of a cover and a base layer.

The base enamel was made from a slurry prepared from a 22EC frit manufactured by Ferro. This combination has been used successfully for many years at Vítkovice ENVI. The covering enamel surface was fired from an enamel slurry, which was ground up from a PP40821 frit manufactured by Mefrit together with clay, water and pigment (RAL 6009). According to its colour, this enamel is referred to internally as "green enamel". The samples produced in this way are commonly used by this company to simulate various situations to which enamelled parts will be exposed in practice.

The solutions used were prepared from chemicals of analytical purity. The pH of all solutions was adjusted with sulfuric acid. *Acidithiobacillus ferrooxidans* bacteria used in the experiment were obtained from the database of microorganisms of the Institute of Environmental Engineering VŠB-TU Ostrava. The distribution of the samples and the composition of the individual leaching solutions are given in Table 1.

4. Experiment

Sample 1 experiments were performed in laboratories at 20 °C. The samples of enamelled sheets were leached for 64 weeks in plastic beakers with the prepared media. The leach solution was aerated during the experiment with a small air pump to bring the experiment as close as possible to the practical conditions of aerobic leaching and also to provide sufficient oxygen for the bacteria *Acidithiobacillus ferrooxidans*.

The pH was maintained at 1.5 during the experiment. Sulfuric acid was used in the experiment to correct the pH. The temperature of the leaching media was maintained between 22 and 26 °C throughout the experiment.

During and after the leaching process, the condition of the enamelled surface was visually evaluated. All samples were weighed before the start of the experiment. Samples 1 and 2 were exposed to 9K leaching medium in the presence of said bacterial inoculum, while sample 3 was exposed to sterile 9K medium. Samples 4 and 5 were exposed to a solution of ferric sulphate acidified with sulfuric acid.

Two samples were exposed to the effects of the 9K medium, in the presence of the microorganism *Acidithiobacillus ferrooxidans*, due to the fact that in practice it is one of the most commonly used leaching media for leaching lean ores. This reduced the uncertainty of the results obtained for the possible practical use of enamelled surfaces in the construction of leaching tanks. The remaining three samples had the character of rather comparative samples without the presence of the mentioned microorganisms (Figure 1).

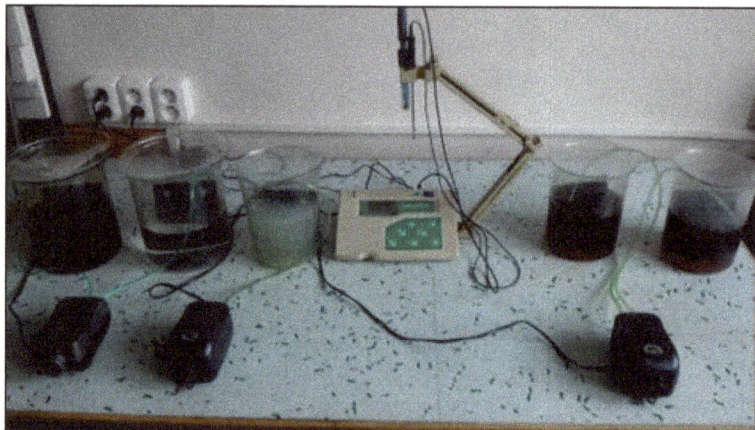

Figure 1. Resistance testing of enamelled samples in leachates.

5. Results

The surface of the enamelled parts showed no visual wear during the experiment. The enamel sufficiently protected the metal from the corrosive effects of the leaching solutions. However, this did not apply to the enamelled surface on the edges of the panels of samples 1, 2 and 5, which were exposed to solutions containing Fe^{3+} ions (Figures 2–4), where the initial visible deterioration of the enamel already occurred after 15 days of leaching. In the case of samples 3 and 4, there was an initial peeling of the enamel (also on the edges), which did not appear until after 30 days of leaching; these samples were exposed to solutions containing only Fe^{2+} ions (medium 9K) and water acidified with H_2SO_4 (Figure 5).

Figure 2. Damage to the edges of samples 1 to 3 (from left to right) after 15 days of leaching.

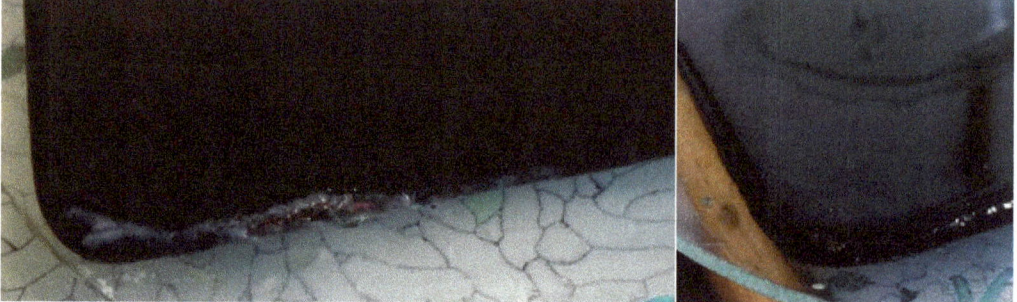

Figure 3. Detail of edge damage for samples 1 and 2 (from left to right) after 15 days of leaching.

Figure 4. Detail of sample 3's damaged edges after 15 days of leaching.

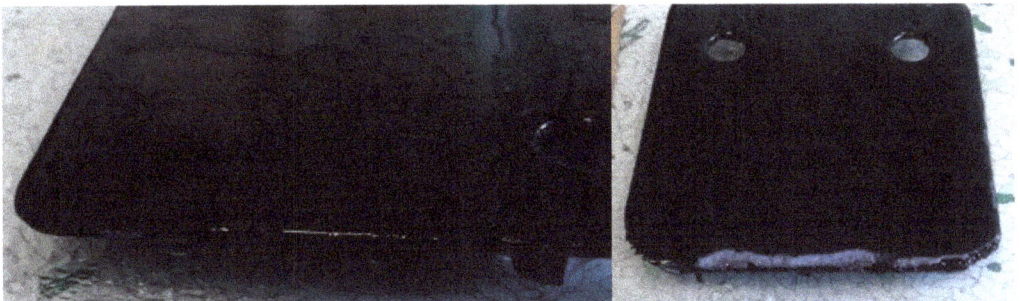

Figure 5. Sample 4 and 5's damaged edges after 30 days of leaching.

In contrast to the enamel on the surface of the part, the enamel on the edges of the samples is always less resistant and may show small pores (it is not inspected by the poroscope, compared to the surface of the sample) and cracks caused by sample handling. Despite the fact that during the production process, the edges of the parts to be enamelled are rounded, the enamel on the edges of the parts and on the circumference of the holes necessary for the assembly of the tanks is always less durable.

A rapid process of metal oxidation followed from the damaged places on the edges of the parts. It is clear that as the corroded area became larger, the rate of metal oxidation increased. It is also worth noting that during the progressive corrosion of the metal, it was possible to observe on the individual samples the formation of small overhangs (Figure 6) formed by the enamelled surface, which remained intact. The resulting overhangs were very fragile and easily crumbled.

Figure 6. Portals formed by the enamel during the degradation of the base plate.

During the next 60 weeks of leaching, a considerable corrosion of the enamelled parts occurred (Figures 7 and 8), but the initial points of corrosion were always the edges of the enamelled parts, from where the corrosion spread below the amorphous enamel phase. The enamelled surface withstood all leachates used in the experiment. The metal protected by the layer of enamel was free of signs of corrosion in all samples (after cutting) even after 15 months of exposure to leachates (Figure 9).

Figure 7. Portals formed by the enamel during the degradation of the base plate.

Figure 8. Extensive corrosion of samples 4 and 5 after almost 15 months of leaching.

Figure 9. View of the cut sample.

The obtained results show that the proportion of Fe^{3+} ions in the solution (whether formed purely chemically or biochemically) has the greatest influence on the destruction of metal in the areas where the enamel was damaged. The influence of direct enzymatic processes of *Acidthiobacillus ferrooxidans* on metallic material, in this experiment, cannot be unambiguously demonstrated and evaluated.

During the construction of tanks, the Vítkovice ENVI company uses only sheets without enamelled edges, because these do not come into contact with the solutions present in the tank. All joints are covered with sealant, and the degree of resistance of the sealant to the liquids used in the experiment has not yet been determined. It is obvious that in the case of minor damage to the tank, the corrosion proceeds briskly through the enamelled part. This fact must be taken into account when considering the possibility of using enamelled parts for the construction of tanks for the leaching of metal-bearing raw materials, where high abrasion can be expected, especially under kinetic leaching conditions.

6. Conclusions and Recommendations

Experiments have shown that sheets with an enamelled surface of chemically resistant enamel, manufactured by Vítkovice ENVI, resist the effects of the 9K nutrient medium, both with and without the presence of the microorganism *Acidithiobacillus ferrooxidans*, provided that the edges of these parts (although enamelled) do not come into contact with this solution. Enamelled edges and especially corners are subject to the effects of leaching due to the natural occurrence of pores and defects on the edges. However, in practice, the edges of the sheets do not come into contact with the solutions inside the tanks, because they are separated from the solutions by high-quality sealants.

The part would also be destroyed in the event of damage to the enamelled surface, such as abrasion. The extent of this risk needs to be further investigated, and it is questionable how long it would take for such damage to the enamel to occur during the leaching process, especially in kinetic leaching.

Before planning the production of these tanks, it would also be necessary to determine whether the sealants used will be resistant to the leachates used.

Author Contributions: The authors of the article, V.B. and J.Z., participated equally in all research areas (methodology, validation, formal analysis, investigation, sources, data processing, and writing). All authors have read and agreed to the published version of the manuscript.

Funding: This research received no external funding.

Institutional Review Board Statement: For our study, it is not relevant.

Informed Consent Statement: Informed consent was obtained from Jaroslav Závada, who was involved in the study.

Data Availability Statement: No new statistic data were created or analyzed in this study. Data sharing is not applicable to this article.

Conflicts of Interest: The authors declare no conflict of interest.

References

1. Goul, J.H.; Butler, S.W.; Boyer, K.W.; Steele, E.A. Hot leaching of ceramic and enameled cookware: Collaborative study. *J. Assoc. Off. Anal. Chem.* **1983**, *66*, 610–619.
2. Lorentz, R. Corrosion of Enamel for Chemical Industry by Neutral Water Medium. *Mitteilungen Des. Ver. Dtsch. Emailfachleute E.V.* **1986**, *34*, 65–76.

Disclaimer/Publisher's Note: The statements, opinions and data contained in all publications are solely those of the individual author(s) and contributor(s) and not of MDPI and/or the editor(s). MDPI and/or the editor(s) disclaim responsibility for any injury to people or property resulting from any ideas, methods, instructions or products referred to in the content.

Proceeding Paper

Evaluation of the Effect of De-icing Materials on Soil Quality in Selected Areas of the Moravian-Silesian Region [†]

Lucie Syrová *, Bohdana Šimáčková, Lukáš Balcařík and Samaneh Shaghaghi

Department of Environmental Engineering, Faculty of Mining and Geology, VŠB–Technical University of Ostrava, 17. Listopadu 2172/15, Poruba, 708 00 Ostrava, Czech Republic; bohdana.simackova.st@vsb.cz (B.Š.); lukas.balcarik.st@vsb.cz (L.B.); shaghaghisamaneh@gmail.com (S.S.)
* Correspondence: lucie.syrova.st@vsb.cz
[†] Presented at the 4th International Conference on Advances in Environmental Engineering, Ostrava, Czech Republic, 20–22 November 2023.

Abstract: This study focuses on the evaluation of the influence of sodium chloride on soil quality in selected locations in the Moravian-Silesian Region. A pilot study was performed in which soil samples were collected at distances of 5 m, 10 m and 20 m perpendicular to the road. The soil samples were analyzed in laboratories to demonstrate the occurrence and possible influence of de-icing material on soil and environmental quality in the vicinity of roads.

Keywords: soils; de-icing; NaCl; sodium chloride

1. Introduction

Inert gritting material (gravel) or chemically active material is used to make roads more passable. By far the most commonly used chemically active material for winter road maintenance is sodium chloride (NaCl). In addition to its positive impacts, the use of de-icing salts can have unintended environmental consequences. In addition to retaining Na^+ and Cl^-, road maintenance with NaCl can also affect the chemical and physical properties of the soil [1]. Sodium chloride causes an increase in pH or release of ammoniacal N [2]. Ion exchange processes can lead to replacement and subsequent leaching of base cations and plant nutrients such as Ca^{2+}, Mg^{2+} and K^+ if excessive amounts of Na^+ enter the system [1]. The accumulation of NaCl in soil can also lead to impaired water permeability, poor aeration, surface crusting and increased alkalinity [3]. Bäckström et al. [4] stated in their study that pH in soil solutions decreased by one unit due to ion exchange. In their study, Baraza and Hasenmueller [1] found that soils temporarily store salts, either in porewater or adsorbed on soil particles. Chloride ions in soils increase osmotic pressure, which affects the ability of the soil to retain water. They also affect soil organisms, reducing overall soil productivity [5]. Among microorganisms, salinization has been shown to weaken the competitiveness of surviving populations. In their study, Ke et al. [6] observed that the higher the salt content applied, the greater the similarity in fungal and bacterial populations. The similarity of fungi increased from 32% to 75% and the similarity of bacteria from 35% to 83%. Chloride and sodium ions also cause the release and transport of metals such as Pb and Pd into the soil [7]. Bäckström et al. [4] in a study of mobilization of heavy metals by de-icing salts in a roadside environment found that the heavy metal content in soil increased during winter season. The concentration of Cd in the aqueous phase increased in response to the ion exchange process. Zn concentration increased due to ion exchange with calcium. Amrhein et al. [8], in their study on the effect of de-icing salts on metal and organic matter mobilization in roadside soils, showed that NaCl released metals (Cr, Pb, Ni, Fe and Cu) bound to organic matter and colloids. They also found that the release is most significant at high adsorbed sodium content and low ionic strength.

If sodium ions are present in the soil in optimal concentrations, they have a positive effect on the soil and plants. Sodium ions help to maintain moisture in the soil and help plants to survive in low moisture areas [9]. However, most plants cannot tolerate higher levels of soluble salts in water [10]. Salt ions cause osmotic and ionic stress to plants. Soil salinization interferes with water and nutrient uptake by roots [11]. It also affects plant morphology and anatomy, such as reduction in root growth and leaf development. High concentrations of Na^+ in soils can reduce the availability of other nutrients to plants and can lead to reduced plant growth [10]. In woody plants, sodium uptake is usually more effectively regulated than chloride uptake. The concentrations of these ions are also higher in older leaves than in young leaves [12]. Plants detect Na^+ through specific mechanisms (root halotropism). The mechanisms are activated when they detect salt exposure, which triggers the Salt Overly Sensitive (SOS) pathway [13].

However, some plant species, especially halophytes, benefit from higher concentrations of sodium and chloride ions. Halophytes live their life cycle in extremely saline conditions, which has led them to develop adaptive responses to stress conditions in order to survive. Adaptive conditions in salt-tolerant plants have evolved as the ability to tolerate stress by establishing different mechanisms. Successful adaptation to salt tolerance in halophytes occurs through the synthesis of a wide range of antioxidants (secondary metabolites) [9]. Plants also use tissue tolerance, sodium exclusion, tolerance to osmotic stress, and tissue-specific sodium sequestration, with many transport genes involved in Na^+ partitioning processes in plants [13].

The aim of the pilot study was to evaluate the effect of chemically active de-icing material (NaCl) on soils in Bruntál, Nebory and Ropice with a comparison to soil that does not undergo winter maintenance in Bystřice.

2. Methods and Materials

2.1. Sample Collection and Preparation

Soil samples were collected within the Moravian-Silesian Region, Czech Republic, specifically in Bruntál—III/0451 (49°58′34″ N 17°27′08″ E), Nebory—II/474 (49°40′34.4″ N 18°37′05.5″ E), Ropice—I/11 (49°42′52.4″ N 18°37′05.8″ E) and Bystřice—III/01143 (49°37′56.9″ N 18°43′32.9″ E) at distances of 5 m, 10 m and 20 m perpendicular to the road. Samples were collected monthly from December 2021 at three sites that were maintained with NaCl (Bruntál, Nebory and Ropice) and one site without NaCl maintenance (Bystřice). Given the objective to demonstrate the presence of chlorides in soils, rather than its distribution at depth, disturbed soil samples were collected. Approximately 1 kg of material was collected from each sampling site, depending on soil fraction and moisture content. The samples were then dried, homogenized and sieved through a 2 mm diameter analytical stainless steel sieve Preciselekt, s.r.o. (Malhostovice, Czech Republic).

2.2. Methods of Determined Parameters

Exchangeable acidity (pH/$CaCl_2$) was determined according to ČSN EN ISO 10390 (836221) Soil, treated biowaste and sludge—Determination of pH [14]. The principle of the method is the determination of pH in a calcium chloride solution with a concentration of 0.01 mol L^{-1} (pH/$CaCl_2$). The solid-to-liquid ratio is 1:5 (volume fraction). The standard allows a range of extraction duration from 2 h to 24 h. The values were measured using a laboratory pH meter inoLab® pH 7110 by Xylem Analytics Germany Sales GmbH (Weilheim, Germany). The results were interpreted according to Table 1.

The dissolved salt content was determined on the basis of electrical conductivity according to standard ČSN EN 27888 (757344) Water quality—determination of electrical conductivity (ISO 7888:1985) [16]. The principle of the method is to measure the conductivity in a filtrate prepared from a 1:5 soil/distilled water leachate, which has been shaken for 5 min in the incubator shaker KS 4000i Control, IKA® (Staufen, Germany). The values were measured with the laboratory conductivity meter inoLab® Cond 7110 by Xylem Analytics

Germany Sales GmbH (Weilheim, Germany). The results were interpreted according to Table 2.

Table 1. Soil classification based on exchangeable acidity—pH/CaCl$_2$ [15].

pH/CaCl$_2$	Soil Evaluation	pH/CaCl$_2$	Soil Evaluation	pH/CaCl$_2$	Soil Evaluation
<3	Extremely acidic	6.0–6.9	Slightly acidic	9.1–10.0	Strongly alkaline
3.0–3.9	Very strongly acidic	7.0	Neutral	10.1–11.0	Very strongly alkaline
4.0–4.9	Strongly acidic	7.1–8.0	Slightly alkaline	>11.0	Extremely alkaline
5.0–5.9	Moderately acidic	8.1–9.0	Moderately alkaline		

Table 2. FAO (USDA) classification for soil salinity assessment [17].

Conductivity (mS·cm^{-1})	Salinity	Conductivity (mS·cm^{-1})	Salinity
<0.75	None	4.0–8.0	Strong
0.75–2.0	Slight	8.0–15.0	Very strong
2.0–4.0	Moderate	>15	Extreme

Soil leachate was prepared with distilled water in the ratio of 1:10 for chloride determination. The soil leachate was shaken for 24 h on the incubator shaker KS 4000i Control, IKA® (Germany) and then filtered through a PRAGOPOR 6 membrane filter with a pore size of 0.4 µm (PRAGOCHEMA, Prague, Czech Republic). Chlorides were determined in the filtrate by titration with silver nitrate volumetric solution in neutral medium to potassium chromate indicator according to norm ISO 9297 (757420) Water quality. Determination of chloride. Silver nitrate titration with chromate indicator (Mohr's method) [18]. The principle of the method is titration of the sample with a volumetric solution of silver nitrate (AgNO$_3$) to form a precipitate of silver chloride (Ag$_2$CrO$_4$). Potassium chromate (K$_2$CrO$_4$) is used as an indicator to visually indicate the equivalence point.

3. Results and Discussion

An important physico-chemical property of soils is the soil reaction (pH)—the soil acidity. It consists of two parts: active acidity and exchangeable (passive) acidity. This is the type of acidity in which the concentration of free H$^+$ in the soil solution varies only slightly compared to exchangeable acidity. Exchangeable acidity is a measurement of H$^+$ and Al^{3+} ions retained or fixed on the soil colloid [19]. The observed values including their evaluation are presented in Table 3.

The soils located near the roads in Nebory and Ropice, which are maintained chemically with NaCl in winter, can be classified as slightly acidic based on the average pH/CaCl$_2$ value and moderately acidic in Bruntál. The soil near the road in Bystřice, which does not undergo chemically active winter maintenance, can be classified as slightly acidic. Based on the measured pH values (Table 3), it can be concluded that the pH values of the soils in Nebory, Ropice, Bruntál and Bystřice do not show a seasonal dependence. Distance from the road also has no effect on soil pH. Equiza et al. [3] in their study reported that soil pH values at roadside sites were significantly higher than at control sites, ranging from pH 7.6 to 8.5 due to the use of de-icing salts. Bäckström et al. [4] in their study reported that the pH of the soil solutions decreased by one unit due to ion exchange as a result of the application of de-icing material (NaCl) during the winter season. However, neither of these trends were observed in this pilot study. Soil pH did not change significantly as a result of chemically active road treatment. The soil conductivity value is used to determine soil salinity, which is proportional to the amount of dissolved salts present in the soil. It is usually higher in soils with a higher content of mineral salts such as calcium, magnesium and sodium. It can be affected by various factors such as soil texture, organic matter content, pH value and water content. The values observed, including their evaluation, are presented in Table 4.

Table 3. pH/CaCl$_2$ of conducted soil samples from December 2021 to March 2023.

DATE	Nebory			Ropice			Bruntál			Bystřice		
	5 m	10 m	20 m	5 m	10 m	20 m	5 m	10 m	20 m	5 m	10 m	20 m
XII/21	6.90	6.58	6.64	7.05	7.01	6.97	6.40	6.20	6.10	6.25	6.31	6.27
I/22	6.81	6.56	6.61	7.08	7.03	6.99	5.70	5.50	5.60	6.20	6.16	6.19
II/22	6.82	6.54	6.59	6.94	6.87	6.85	5.70	5.60	5.50	6.38	6.44	6.34
III/22	6.73	6.49	6.53	6.89	6.83	6.79	5.80	5.60	5.60	6.30	6.39	6.30
IV/22	6.69	6.48	6.52	6.81	6.77	6.72	5.80	5.70	5.60	6.21	6.12	6.19
V/22	6.64	6.51	6.44	6.79	6.73	6.71	5.70	5.50	5.50	6.27	6.20	6.23
VI/22	6.63	6.42	6.49	6.83	6.86	6.74	5.70	5.60	5.50	6.33	6.24	6.17
VII/22	6.68	6.49	6.43	6.80	6.78	6.77	5.70	5.60	5.50	6.22	6.25	6.18
VIII/22	6.64	6.57	6.51	6.90	6.85	6.84	5.80	5.60	5.60	6.37	6.32	6.29
IX/22	6.72	6.54	6.59	6.87	6.83	6.82	5.80	5.70	5.70	6.13	6.20	6.16
X/22	6.69	6.57	6.58	6.81	6.89	6.86	5.80	5.70	5.60	6.21	6.36	6.27
XI/22	6.81	6.62	6.48	7.00	6.93	6.91	6.00	5.60	5.80	6.28	6.19	6.23
XII/22	6.80	6.50	6.64	7.00	7.02	6.95	6.30	5.90	6.10	6.15	6.26	6.20
I/23	6.74	6.48	6.47	7.09	6.95	7.00	6.20	6.00	5.90	6.24	6.20	6.30
II/23	6.85	6.73	6.56	6.95	6.84	6.85	5.80	5.60	5.50	6.40	6.50	6.30
III/23	6.93	6.50	6.67	6.84	6.85	6.66	5.70	5.50	5.60	6.34	6.38	6.35
IV/23	6.60	6.42	6.40	6.76	6.70	6.68	5.80	5.70	5.50	6.20	6.25	6.18
AVERAGE	6.75	6.53	6.54	6.91	6.87	6.83	5.86	5.68	5.66	6.26	6.28	6.24
MAX	6.93	6.73	6.67	7.09	7.03	7.00	6.40	6.20	6.10	6.40	6.50	6.35
MIN	6.60	6.42	6.40	6.91	6.87	6.83	5.70	5.50	5.50	6.13	6.12	6.16

Table 4. Conductivity of conducted soil samples from December 2021 to March 2023 in mS·cm^{-1}.

DATE	Nebory			Ropice			Bruntál			Bystřice		
	5 m	10 m	20 m	5 m	10 m	20 m	5 m	10 m	20 m	5 m	10 m	20 m
XII/21	0.23	0.17	0.21	0.21	0.18	0.15	0.11	0.09	0.07	0.11	0.11	0.13
I/22	0.20	0.15	0.19	0.18	0.17	0.13	0.10	0.07	0.06	0.17	0.16	0.14
II/22	0.19	0.18	0.13	0.13	0.11	0.10	0.06	0.05	0.05	0.17	0.12	0.13
III/22	0.16	0.18	0.14	0.14	0.16	0.12	0.06	0.04	0.04	0.18	0.15	0.14
IV/22	0.20	0.14	0.17	0.15	0.18	0.15	0.06	0.04	0.04	0.18	0.14	0.10
V/22	0.18	0.13	0.15	0.13	0.17	0.15	0.06	0.05	0.04	0.12	0.11	0.13
VI/22	0.16	0.16	0.13	0.16	0.14	0.10	0.05	0.05	0.04	0.12	0.12	0.15
VII/22	0.14	0.15	0.13	0.18	0.15	0.11	0.04	0.04	0.04	0.14	0.13	0.14
VIII/22	0.19	0.17	0.13	0.12	0.12	0.15	0.04	0.04	0.04	0.12	0.12	0.13
IX/22	0.21	0.23	0.18	0.16	0.17	0.13	0.04	0.05	0.04	0.16	0.14	0.15
X/22	0.19	0.13	0.15	0.19	0.14	0.11	0.06	0.04	0.04	0.11	0.13	0.12
XI/22	0.21	0.14	0.17	0.21	0.16	0.14	0.05	0.05	0.04	0.14	0.13	0.12
XII/22	0.22	0.16	0.20	0.20	0.17	0.12	0.13	0.10	0.10	0.10	0.10	0.11
I/23	0.19	0.13	0.17	0.18	0.11	0.11	0.10	0.08	0.07	0.18	0.18	0.12
II/23	0.18	0.14	0.13	0.09	0.18	0.11	0.07	0.06	0.05	0.17	0.14	0.19
III/23	0.18	0.12	0.13	0.12	0.15	0.1	0.06	0.04	0.04	0.12	0.12	0.12
IV/23	0.25	0.15	0.2	0.2	0.14	0.11	0.06	0.04	0.04	0.14	0.13	0.17
AVERAGE	0.19	0.15	0.16	0.16	0.15	0.12	0.07	0.06	0.05	0.14	0.13	0.13
MAX	0.25	0.23	0.21	0.21	0.18	0.15	0.13	0.10	0.10	0.18	0.18	0.19
MIN	0.14	0.12	0.13	0.09	0.11	0.10	0.04	0.04	0.04	0.10	0.10	0.10

The soils located near the road in Nebory, Ropice and Bruntál, which are maintained by chemically active material NaCl in winter, are classified as non-saline based on the average conductivity value. The soil near the road in Bystřice, which is not maintained by chemically active material, can be classified as non-saline. Based on the measured values, it can be concluded that the conductivity of the soils in Nebory, Ropice, Bruntál and Bystřice does not show a seasonal dependence. In their study, Bäckström et al. [4] reported that the conductivity values increased significantly due to chemically active road maintenance, and with increasing distance from the road, the conductivity values decreased according to the measurements, with an impact distance of approximately 10 m. However, this pilot study failed to demonstrate the effect of chemically active winter maintenance on the measured conductivity values (Table 4) in soils in the selected locations. The impact of distance was also not demonstrated, i.e., distance from the road also has no effect on soil conductivity.

In addition to pH and conductivity, chloride ions were also determined in soil samples. Chloride ions were determined below the detection limit at all sites.

The effect of de-icing materials on soil quality depends on many factors. Among the most important factors are the amount and type of de-icing material used, the duration of application of the chemically active material, the frequency of application, the type of soil in the vicinity and, particularly, the current weather. Sodium chloride helps to keep roads passable, ensuring traffic safety during the winter months. The results so far indicate that the advantages of the chemically active material, sodium chloride, outweigh the disadvantages in selected locations in the Moravian-Silesian Region. In all three locations (Nebory, Ropice and Bruntál), where there is regular chemical winter road maintenance, sodium chloride is used as a de-icing material. The pilot study should be complemented by other investigation at sites that could help to complete the overall overview of the potential impact of sodium chloride on roads in the Moravian-Silesian Region.

4. Conclusions

This pilot study evaluated the effect of de-icing materials on soil quality in selected locations in the Moravian-Silesian Region.

Author Contributions: Conceptualization, L.S.; methodology, B.Š., L.S., S.S. and L.B.; software, S.S.; validation, L.S. and L.B.; formal analysis, B.Š.; investigation, L.S.; resources, B.Š.; data curation, L.S. and B.Š.; writing—original draft preparation, L.S., B.Š., S.S. and L.B.; writing—review and editing, L.S., S.S., B.Š. and L.B.; visualization, S.S.; supervision, L.S.; project administration, S.S.; funding acquisition, L.B. All authors have read and agreed to the published version of the manuscript.

Funding: This research received no external funding.

Institutional Review Board Statement: Not applicable.

Informed Consent Statement: Not applicable.

Data Availability Statement: Data are contained within the article.

Conflicts of Interest: The authors declare no conflict of interest.

References

1. Baraza, T.; Hasenmueller, E.A. Road salt retention and transport through vadose zone soils to shallow groundwater. *Sci. Total Environ.* **2021**, *755*, 142240. [CrossRef] [PubMed]
2. Lancaster, N.A.; Bushey, J.T.; Tobias, C.R.; Song, B.; Vadas, T.M. Impact of chloride on denitrification potential in roadside wetlands. *Environ. Pollut.* **2016**, *212*, 216–223. [CrossRef] [PubMed]
3. Equiza, M.A.; Calvo-Polanco, M.; Cirelli, D.; Señorans, J.; Wartenbe, M.; Saunders, C.; Zwiazek, J.J. Long-term impact of road salt (NaCl) on soil and urban trees in Edmonton, Canada. *Urban For. Urban Green.* **2017**, *21*, 16–28. [CrossRef]
4. Bäckström, M.; Karlsson, S.; Bäckman, L.; Folkeson, L.; Lind, B. Mobilisation of heavy metals by deicing salts in a roadside environment. *Water Res.* **2004**, *38*, 720–732. [CrossRef] [PubMed]
5. Rietz, D.N.; Haynes, R.J. Effects of irrigation-induced salinity and sodicity on soil microbial activity. *Soil Biol. Biochem.* **2003**, *35*, 845–854. [CrossRef]
6. Ke, C.; Li, Z.; Liang, Y.; Tao, W.; Du, M. Impacts of chloride de-icing salt on bulk soils, fungi, and bacterial populations surrounding the plant rhizosphere. *Appl. Soil Ecol.* **2013**, *72*, 69–78. [CrossRef]
7. Pahlavan, F.; Ghasemi, H.; Yazdani, H.; Fini, E.H. Soil amended with Algal Biochar Reduces Mobility of deicing salt contaminants in the environment: An atomistic insight. *Chemosphere* **2023**, *323*, 138172. [CrossRef] [PubMed]

8. Amrhein, C.; Strong, J.E.; Mosher, P.A. Effect of deicing salts on metal and organic matter mobilization in roadside soils. *Environ. Sci. Technol.* **1992**, *26*, 703–709. [CrossRef]
9. Khandare, S.D.; Singh, A.; Chaudhary, D.R. Halophytes: A potential source of antioxidants. In *Marine Antioxidants*; Elsevier: Amsterdam, The Netherlands, 2023; pp. 185–196. [CrossRef]
10. Parida, A.K.; Das, A.B. Salt tolerance and salinity effects on plants: A review. *Ecotoxicol. Environ. Saf.* **2005**, *60*, 324–349. [CrossRef] [PubMed]
11. Munck, I.A.; Bennett, C.M.; Camilli, K.S.; Nowak, R.S. Long-term impact of de-icing salts on tree health in the Lake Tahoe Basin: Environmental influences and interactions with insects and diseases. *For. Ecol. Manag.* **2010**, *260*, 1218–1229. [CrossRef]
12. Hanslin, H.M. Short-term effects of alternative de-icing chemicals on tree sapling performance. *Urban For. Urban Green.* **2011**, *10*, 53–59. [CrossRef]
13. Venkataraman, G.; Shabala, S.; Véry, A.A.; Hariharan, G.N.; Somasundaram, S.; Pulipati, S.; Sellamuthu, G.; Harikrishnan, M.; Kumari, K.; Shabala, L.; et al. To exclude or to accumulate? Revealing the role of the sodium HKT1;5 transporter in plant adaptive responses to varying soil salinity. *Plant Physiol. Biochem.* **2021**, *169*, 333–342. [CrossRef] [PubMed]
14. *ČSN EN ISO 10390 (836221)*; Soil, Treated Biowaste and Sludge—Determination of pH. Czech Standardization Agency: Prague, Czech Republic, 2022.
15. Schachtschabel, P.; Scheffer, F. *Lehrbuch der Bodenkunde: Unter Mitarb. von W. R. Fischer*; Ferdinand Enke Verlag: Stuttgart, Germany, 1992; ISBN 3432847734.
16. *ČSN EN 27888 (757344)*; Water Quality. Determination of Electrical Conductivity (ISO 7888:1985). Czech Standards Institute: Prague, Czech Republic, 1996.
17. Food and Agriculture Organization of the United Nations. *Mapping of Salt-Affected Soils: Technical Manual*; FAO: Rome, Italy, 2020; ISBN 978-92-5-132687-9. Available online: https://www.fao.org/3/ca9215en/ca9215en.pdf (accessed on 22 June 2023).
18. *ČSN ISO 9297 (757420)*; Water Quality. Determination of Chloride. Silver Nitrate Titration with Chromate Indicator (Mohr's Method). 02/1996. Czech Standards Institute: Prague, Czech Republic, 1996.
19. Onwuka, M.I.; Ozurumba, U.V.; Nkwocha, O.S. Changes in Soil pH and Exchangeable Acidity of Selected Parent Materials as Influenced by Amendments in South East of Nigeria. *J. Geosci. Environ. Prot.* **2016**, *4*, 80–88. [CrossRef]

Disclaimer/Publisher's Note: The statements, opinions and data contained in all publications are solely those of the individual author(s) and contributor(s) and not of MDPI and/or the editor(s). MDPI and/or the editor(s) disclaim responsibility for any injury to people or property resulting from any ideas, methods, instructions or products referred to in the content.

Proceeding Paper

Assessment of the Risks to the Drinking Water Supply System of the Nový Malín Communal Waterworks [†]

Silvie Drabinová *, Miroslav Kyncl and Martin Minář

Faculty of Mining and Geology, VŠB-Technical University of Ostrava, 17. listopadu 2172/15, 708 00 Ostrava-Poruba, Czech Republic; miroslav.kyncl@vsb.cz (M.K.); martin.minar.st@vsb.cz (M.M.)
* Correspondence: silvie.drabinova@vsb.cz
[†] Presented at the 4th International Conference on Advances in Environmental Engineering, Ostrava, Czech Republic, 20–22 November 2023.

Abstract: This article addresses an important task for the operators of public water supply systems. It is to ensure the long-term security, reliability, and provision of the required quantity and prescribed quality of drinking water systems. To fulfill this task, a program called "Risk Assessment" has been developed. The risk assessment program for the drinking water supply system is incorporated into current legislation, which mandates that all operators of public water supply systems assess the waterworks that they operate. The objective of this article is to assess the risks of the drinking water supply system of the Nový Malín communal waterworks in accordance with the applicable legislation and recommended methodology. These waterworks constitute communal waterworks and encompass 7 underground water sources serving 3 municipalities and 7 settlements totaling 3500 residents. The results of this analysis should serve as a basis for updating the operational regulations of the waterworks and for assessing whether the methodology used is suitable for this type of water supply system.

Keywords: risk assessment; supply system; safe operation; drinking water

Citation: Drabinová, S.; Kyncl, M.; Minář, M. Assessment of the Risks to the Drinking Water Supply System of the Nový Malín Communal Waterworks. *Eng. Proc.* **2023**, *57*, 27. https://doi.org/10.3390/engproc2023057027

Academic Editors: Adriana Estokova, Natalia Junakova, Tomas Dvorsky, Vojtech Vaclavik and Magdalena Balintova

Published: 6 December 2023

Copyright: © 2023 by the authors. Licensee MDPI, Basel, Switzerland. This article is an open access article distributed under the terms and conditions of the Creative Commons Attribution (CC BY) license (https://creativecommons.org/licenses/by/4.0/).

1. Introduction

Since 2004, the World Health Organization (WHO) has proposed a concept known as "Water Safety Plans" or "Safe Drinking Water Supply Plans". This concept was introduced in the third edition of the "Guidelines for Drinking-water Quality". This methodology is based on over 50 years of experience and serves as the foundation for national regulations and standards for drinking water safety. In accordance with Council Directive 98/83/EC on the quality of water intended for human consumption, as specified in Commission Directive (EU) 2015/1787, this approach is referred to as "risk assessment." The International Water Association (IWA), comprising water utilities and experts from around the world, supports this strategy. This approach also forms the basis of current Czech legislation, where the term "risk assessment" has been used since 2017 [1,2].

The risk assessment of water supply systems is already established and conducted in more than ten European and several non-European countries. The main anticipated benefits of this process include the improvement of drinking water quality, reduction of the number and severity of incidents, enhanced protection of water resources, a decrease in acute illnesses among the population, better monitoring of the water supply system's operation, gaining a comprehensive understanding of the entire supply system's functioning, corrective measure cost reduction, and improved oversight and control by regulatory authorities [3].

However, it is important to note that for water utility operators, performing a risk analysis can entail increased administrative complexity and associated costs.

An amendment to Public Health Protection Act No. 258/2000 Coll., effective from 1 November 2017, imposed a new requirement on producers and suppliers of drinking water for public distribution to prepare risk assessments, also known as "water safety plans." These assessments will be part of the operational regulations of public water supply systems and the monitoring program, thus complying with Council Directive 98/83/EC on the quality of water intended for human consumption [4].

2. Legislation of the Risk Assessment Process

The entire drinking water supply system in the Czech Republic is based on several key legal regulations. The foundation includes the Water Act (Act No. 254/2001 Coll.), the Act on Water Supply and Sewerage Systems for Public Needs (Act No. 274/2001 Coll.), and the implementing decree (Decree No. 428/2001 Coll.). Furthermore, for risk assessment, Act No. 258/2000 Coll. on Public Health Protection and its implementing decree No. 252/2004 Coll. are essential, as they define the indicators of drinking water quality.

2.1. Act No. 258/2000 Coll. on Public Health Protection

The risk assessment of drinking water is an obligation for entities such as water utility operators or commercial wells that supply drinking water to public water supply systems or produce it from an individual source as part of their business activities, for example, farms, hotels, and guesthouses where drinking water is used for operational purposes.

The risk assessment is methodically processed as a document that comprehensively describes the risk analysis process within the drinking water supply system. This document also includes proposals for corrective and monitoring measures designed to address risk situations.

The risk assessment must be integrated into the operational regulations, as stated in Section 3c, Paragraph 1 of Act No. 258/2000 Coll. [4], as amended. The operator of the drinking water supply system is obliged to keep these operational regulations up to date and submit them for approval to the relevant public health protection authority [5].

The operator of the drinking water supply system is required to prepare (or update) the operational regulations, including the risk assessment of the drinking water supply system, and submit them for approval to the relevant public health protection authority no later than 31 October 2025 (see Article IV of Act No. 202/2017 Coll., as amended) [5]. These regulations must be approved by the appropriate hygiene station. The State Health Institute, the Czech Water Association, and the Association of Water Supply and Sewerage Field play significant roles in the professional oversight of this process [6,7].

2.2. Decree No. 252/2004 Coll. Laying down Hygiene Requirements for Drinking and Hot Water, and the Frequency and Scope of Drinking Water Monitoring

The detailed procedure for risk assessment is contained in this decree, which specifies the hygiene requirements for drinking and hot water and the regularity of drinking water inspections. This decree was amended in 2018 by Decree No. 70/2018 Coll., which stipulates that the risk assessment will be conducted in accordance with Annex No. 7 to this decree. It also mandates the updating of existing operational regulations [8].

Important requirements for risk assessment include:

- Risk assessment is moderate and high in the case of non-compliance. For water utility systems with many identified risks, only those with significant consequences are considered unacceptable.
- Risk assessments are conducted independently by individual water supply systems. If they are part of a communal waterworks and have a single operator, that operator is responsible for risk assessment. If there are multiple operators for a communal waterworks, the risk assessment is carried out in such a way that individual components complement each other [8].

The individual procedures are divided into 8 steps, which were applied in a specific location, the Nový Malín communal waterworks (Section 4).

3. Methodological Guide for Conducting Risk Assessments of Drinking Water Supply Systems in Accordance with the Public Health Protection Act

A methodology has also been developed by the SZÚ for risk assessment, which corresponds to Annex No. 7 of Decree No. 252/2004 Coll. This methodology is based on international standards and previous studies with the aim of ensuring the delivery of safe and high-quality drinking water that meets the requirements of all users. It is essential for this water not only to comply with hygiene standards but also with organoleptic parameters such as taste, odor, and color which influence the initial impression when drinking tap water [6]. This methodology can be continuously supplemented and updated by processors based on new insights and experiences [6].

The methodological guide for the risk assessment of drinking water supply systems is based on the Public Health Protection Act. It encompasses systematic procedures to ensure the delivery of safe water from its source to consumers. This integrated approach is key to securing high-quality and safe drinking water for all [9].

As part of the risk assessment, the operator of the drinking water supply system forms a team with the goal of verifying whether the operated drinking water supply system is documented (from the water source to the consumer). Another task of the team is to identify hazardous events that have previously jeopardized or could jeopardize the quality or quantity of the supplied drinking water [10].

The objective is to assess the level of risk for a given hazardous event (incident) and subsequently manage the risks (determine corrective and control measures, verification, etc.) in a way that minimizes the danger to the quality and delivery of drinking water [5].

4. Application of the Risk Assessment Methodology at a Specific Location

The selected water supply systems to be described are located in the Šumperk district in the cadastral areas of the municipalities of Nový Malín, Hrabišín, and Dolní Studénky. The land area of the Nový Malín municipality is approximately 30.1 km^2 and consists of three parts: the Mladoňov u Oskavy settlement, the Plechy part, and the Nový Malín part. The land area of the Hrabišín municipality is 13.9 km^2, with individual parts being Dolní Olešná, Horní Olešná, and Loučky. The land area of the Třemešek satellite is less than 1 km^2 [11]. The total amount of water produced in the group water supply system Nový Malín in individual sources in the years 2018–2022 can be seen in Table 1.

Table 1. Water produced in the Nový Malín communal waterworks [11].

Source Name	2018 *	2019 *	2020 *	2021 *	2022 *
Luže	57.7	61.9	41.5	55.4	72.3
Malínský les	150.3	141.8	218.1	197.9	142.1
Pod lesem	48.0	24.9	10.9	14.9	39.7
Vrt Mladoňov	9.4	9.6	9.2	9.3	9.9

* in thousands of cubic meters. Sources of drinking water.

As of 31 December 2022, the Mladoňov settlement had 209 residents and is supplied with water from a single public water supply. All water connections are equipped with FLOW 2200 ultrasonic water meters (Kamstrup, Roskilde, Denmark) with remote reading capability and water leak detection. The water supply infrastructure was built between 2005 and 2007, along with sewage and wastewater treatment facilities. In the Mladoňov settlement, there is a predominance of family houses, recreational cottages, hospitality establishments, and an environmental education center. The highest point of the water supply system is located at an altitude of 597 m above sea level, and the lowest point is at 460 m above sea level.

The main water supply network is situated in the Nový Malín municipality, where there were 3459 residents as of 31 December 2022. This network comprises three catchment areas: Malínský les, Lokalita pod lesem, and Luže. The history of the water supply system in the municipality dates back to 1929. In the Malínský les area, a small hydroelectric power plant was built on the feeder. The municipality of Nový Malín owns three water reservoirs that serve for water distribution. The water supply is divided into three pressure zones, and pressure ratios are controlled through the use of water reservoirs. The highest point in the municipality is Kamenný vrch, with an elevation of 947 m above sea level, while the populated part of the municipality lies between 316 m above sea level and 464 m above sea level [11].

In the Třemešek area, there is an industrial zone and approximately 30 family houses. These houses are supplied with water from a water supply network that is operationally integrated into the Nový Malín water supply, although it falls within the cadastral territory of the Dolní Studénky municipality [11].

The Hrabišín municipality owns a single water reservoir where water is pumped from the reservoir in Nový Malín. As of 31 December 2022, the municipality had 836 residents. Water connections have been provided to most houses, but approximately half of them do not use the supplied water as they have their own wells. In total, 210 water meters are installed, of which 200 are residential water meters with remote reading and leak detection on Multical Flow 2200 connections [11].

The minimum usable quantity from all four underground water sources (Malinský les, Pod lesy—boreholes no. 1, 2, and 3, Luže—boreholes no. 17 and 20, and Mladoňov catchment areas) owned by the Nový Malín municipality is 29.78 L per second. It can be concluded that for the Nový Malín, Hrabišín, and Třemešek communal waterworks, there is a sufficient quantity of water available for supplying the population. There is also enough water to supply the surrounding municipalities, even after accounting for reserves in case of potential climate changes or other natural disasters (in Table 2). The existing underground sources ensure an adequate water supply even during dry periods [11,12].

Table 2. Clear information about the water sources of the Nový Malín municipality [11].

Source Name	Annual Permitted Abstraction (m^3)	Permit for Water Handling Into:	Water Source Origin	Maximum Permitted Abstraction (L/s)
Malínský les	157,721	31 January 2056	shallow groundwater	10.0
Vrt č.1	9999	31 December 2030	groundwater	1.5
Vrt č.2	94,600	31 December 2030	groundwater	3.5
Vrt č.3	94,608	31 December 2050	groundwater	5.0
Vrt č.17	78,840	31 January 2056	groundwater	7.0
Vrt č.20	78,840	31 January 2056	groundwater	7.0
Mladoňov	20,004	31 December 2031	groundwater	1.5

The water from all sources, in terms of quality, complies with Decree No. 252/2004 Coll., which sets out the hygiene requirements for drinking and hot water. Only disinfection for water safety during its transportation is provided. Disinfection is ensured by the dosing of sodium chlorite.

4.1. Water Reservoirs

The water reservoirs in Nový Malín, Mladoňov, and Hrabišín are always underground tanks with a water volume designed for the seamless supply of water. A list of water reservoirs located in the subject area is provided in Table 3. Their main purpose is to accumulate drinking water and create suitable pressure conditions at the point of use. Chlorination may also take place in these reservoirs. The water reservoirs are fully functional without any apparent defects [11].

Table 3. Clear information about the water reservoirs [11].

Name of the Water Reservoir	Annual Permitted Abstraction (m^3)	Permit for Water Handling Into:	Water Source Origin
VDJ * Mladoňov	593	597	50
VDJ * Hrabišín	407	410	100
VDJ * 40	476	479	40
VDJ * 200	411	415	200
VDJ * 300	375	379	300

* VDJ—water reservoir.

4.2. Distribution Water Mains

The distribution mains for the supply and delivery of water in the group waterworks of Nový Malín, Hrabišín, and the Třemešek satellite have been under construction since 1928 up to the present day (in Table 4). The renovation of water mains is being carried out according to the approved renovation plan from 2022.

Table 4. Basic information about the water supply networks of the Nový Malín group waterworks [11].

Water Supply Name	Owner	DN	Length in Kilometers	Material	Number of Connections
Water Supply Mladoňov	village Nový Malín	3.5	63 a 90	HDPE	89
Water Supply Nový Malín	village Nový Malín	29.6	from 50 up to 150	1/2 cast iron a 1/2 plastic	1022
Water Supply Hrabišín	village Hrabišín	11.4	90 a 110	PVC	275
Water Supply satelit Třemešek	village Dolní Studénky	1.0	90	HDPE	38

In 2022, a water loss of 11.5% was identified across the entire pipeline network, which is assessed as a good condition, well below the Czech Republic's average [11].

On the Hrabišín and Mladoňov water supply systems, pressure-reducing valves are installed in control shafts. In Hrabišín, an ATS (automatic transfer switch) is also installed to secure pressure conditions in the upper pressure zone of the Hrabišín municipality towards the Libina municipality [11].

The potential risks that may arise within the operation of the group water supply are:

- Natural disasters;
- Unauthorized entry into protective zones or buildings and water contamination;
- Exceeding the prescribed dosage of disinfectant;
- Inadequate hygiene protection of drinking water;
- Agricultural activities in the vicinity of raw water sources;
- Malfunction of source pumps or ATS;
- Infiltration of surface water into sources and contamination of raw water.

5. Conclusions

When evaluating the drinking water supply system of the Nový Malín group waterworks company, it was found that there were 21 risks (in Table 5) with low danger levels, for which possible corrective and monitoring measures were proposed. A total of 5 identified risks had parameters indicating a medium level of risk, for which corrective measures and a date for implementation were established (in Table S1). No risk was assessed as high, indicating a low level of threat to the operational system.

The obtained results were incorporated into the operational regulations and monitoring program. Furthermore, it was determined that the methodology used is suitable for this type of group water supply.

Table 5. Summary of identified hazards and associated risks according to the individual parts of the supply system risk matrix as per [6].

Part of the System	Risk Level			Total
	High	Medium	Low	
Sources	–	2	11	13
accumulation and treatment	–	2	5	7
distribution	–	1	5	6
Total	0	5	21	26

A detailed schedule is attached.

Supplementary Materials: The following supporting information can be downloaded at: https://www.mdpi.com/article/10.3390/engproc2023057027/s1, Table S1: Medium risk level and corrective measures.

Author Contributions: Conceptualization, S.D.; writing—review and editing, M.K.; investigation, M.M. All authors have read and agreed to the published version of the manuscript.

Funding: This research received no external funding.

Institutional Review Board Statement: Not applicable.

Informed Consent Statement: Not applicable.

Data Availability Statement: Data supporting reported results can be found at Supplementary Materials.

Conflicts of Interest: The authors declare no conflict of interest.

References

1. World Health Organization. *Guidelines for Drinking Water Quality: Fourth Edition Incorporating the First and Second Addenda*, 4th ed. + 1st add + 2nd add; World Health Organization: Geneva, Switzerland, 2022. Available online: https://apps.who.int/iris/handle/10665/352532 (accessed on 3 August 2023).
2. Tuhovčák, L.; Ručka, J.; Kučera, T.; Bouda, R.; Kolářová, L.; Turčínek, J. *Risk Assessment of Public Water Supply Systems within the Scope of SmVaK Ostrava*; SOVAK: Praha, Czech Republic, 2020; No. 5, ISSN 1210-3039.
3. Tuhovčák, L.; Kučera, T.; Ručka, J. Posouzení Rizik Jako Součást Provozních Řádů Veřejných Vodovodů; VUT FAST Brno, Institute of Municipal Water Management. 2020. Available online: https://voda.tzb-info.cz/20077-posouzeni-rizik-jako-soucast-provoznich-radu-verejnych-vodovodu (accessed on 1 August 2023). (In Czech)
4. Czech Republic. Act No. 258/200 Coll., on the Protection of Public Health and on the Amendment of Certain Related Acts. Available online: https://www.zakonyprolidi.cz/cs/2000-258 (accessed on 25 April 2023). (In Czech)
5. Risk Assessment of the Drinking Water Supply System. Available online: https://www.vodarizika.cz/posouzeni-rizik/ (accessed on 7 June 2023).
6. Kožíšek, F.; Pumann, P.; Šašek, J.; Jeligová, H. *Methodological Guide for Assessing Risks of Drinking Water Supply Systems According to the Public Health Protection Act. Version 2*; National Institute of Public Health: Prague, Czech Republic, 2018.
7. Paul, J.; Kožíšek, F.; Hloušek, T. Risk Assessment—Challenges and Potential Errors. Vodní Hospodářství, 2021. Available online: https://vodnihospodarstvi.cz/posouzeni-rizik-uskali-a%25E2%2580%25AFmozne-chyby/ (accessed on 22 April 2023).
8. Czech Republic. The Decree No. 252/2004 Coll. Laying Down the Sanitary Requirements for Drinking and Hot Water and the Frequency and Scope of Inspection of Drinking Water. Available online: https://www.zakonyprolidi.cz/cs/2004-252 (accessed on 28 April 2023). (In Czech)
9. Tuhovčák, L. *WaterRisk: Analýza Rizik Veřejných Vodovodů*; Akademické Nakladatelství, CERM: Brno, Czech Republic, 2010; ISBN 978-80-7204-676-8.
10. Tuhovčák, L.; Ručka, J. Hazard identification and risk analysis of water supply systems. In *Strategic Asset Management of Water Supply and Wastewater Infrastructures*, 1st ed.; IWA Publishing: Londýn, UK, 2009; pp. 287–298, ISBN 1-84339-186-4.
11. Operation Nový Malín s.r.o. (PNM). Internal documentation: Unpublished, 2023, Czech Republic.
12. Kyncl, M.; Heviánková, S.; Nguien, T.L.C. Study of supply of drinking water in dry seasons in the Czech Republic. *IOP Conf. Ser. Earth Environ. Sci.* **2017**, *92*, 012036. [CrossRef]

Disclaimer/Publisher's Note: The statements, opinions and data contained in all publications are solely those of the individual author(s) and contributor(s) and not of MDPI and/or the editor(s). MDPI and/or the editor(s) disclaim responsibility for any injury to people or property resulting from any ideas, methods, instructions or products referred to in the content.

Proceeding Paper

Using Bio-Monitors to Determine the Mercury Air Pollution in a Former Mining Area [†]

Lenka Demková [1,*], Lenka Bobuľská [1], Ľuboš Harangozo [2] and Július Árvay [2]

[1] Department of Ecology, Faculty of Humanities and Natural Sciences, University of Presov, 17. Novembra 1, 080 01 Presov, Slovakia; lenka.bobulska@unipo.sk
[2] Institute of Food Sciences, Faculty of Biotechnology and Food Sciences, Slovak University of Agriculture in Nitra, Tr. A. Hlinku 2, 949 76 Nitra, Slovakia; lubos.harangozo@uniag.sk (Ľ.H.); julius.arvay@uniag.sk (J.Á.)
* Correspondence: lenka.demkova@unipo.sk; Tel.: +421-51-757-207
[†] Presented at the 4th International Conference on Advances in Environmental Engineering, Ostrava, Czech Republic, 20–22 November 2023.

Abstract: Total mercury air pollution was evaluated in the former mining area of Gelnica (Slovakia) using tree bark, mosses (*Climacium* sp., *Pleurosium* sp.), and lichen (*Pseudevernia* sp.). Samples were collected (tree bark) and exposed (moss and lichen bags) on the heaps and near the mines. Additionally, the internal parts of the mines were evaluated. The mercury content in the bio-monitors was evaluated using an AMA-254 analyzer. The results showed significant differences in tree bark mercury content between the mines and heaps. The Hg content in mosses and lichens was not influenced by the type of mining work. The lichen *Pseudevernia* sp. was found to be the best Hg accumulator compared with mosses.

Keywords: tree bark; moss and lichen bag technique; environmental monitoring; open mining pits; heaps of waste material

Citation: Demková, L.; Bobuľská, L.; Harangozo, Ľ.; Árvay, J. Using Bio-Monitors to Determine the Mercury Air Pollution in a Former Mining Area. *Eng. Proc.* **2023**, *57*, 28. https://doi.org/10.3390/engproc2023057028

Academic Editors: Adriana Estokova, Natalia Junakova, Tomas Dvorsky, Vojtech Vaclavik and Magdalena Balintova

Published: 6 December 2023

Copyright: © 2023 by the authors. Licensee MDPI, Basel, Switzerland. This article is an open access article distributed under the terms and conditions of the Creative Commons Attribution (CC BY) license (https://creativecommons.org/licenses/by/4.0/).

1. Introduction

Air pollution is currently understood as an urgent regional and global problem. The large number of pollutants released from various emission sources contributes to the deterioration of air quality, soil quality, purity of resources, agricultural production, and human health [1]. The high content of hazardous elements in the air, when absorbed into the human body, can lead to direct poisoning or chronic intoxication, depending on the exposure [2]. Mercury is considered the most toxic non-essential metal in humans. Its presence in the environment is associated with several industrial activities including mining [3]. Mercury has extremely dangerous effects on the human body and causes a wide range of diseases [4]. Among all anthropogenic activities, mining and industrial activities that focus on the processing of ore materials are among the most important producers of dust and aerosol emissions [5]. Underground mining and the blasting of the upper layers retain solid particles, which are similar in composition to the substrate of the igneous, sedimentary, or mineral upper layers [6]. However, it is not only the areas where mining and processing activities are currently taking place that are problematic, since the dust particles produced during mining operations are carried by wind and rain, it is easy to pollute the wide-ranging surroundings of mining areas [7]. Methods based on bioindicators have become popular to evaluate the state and quality of air in areas with different types of environmental loads [8–10]. This method uses the ability of living organisms (both plants and animals) to respond sensitively to stress caused by changes in their natural environment. Undoubtedly, these changes include an increased content of risk elements in environmental components. A changed or disturbed environment manifests as changes (disruptions) in the physiological and biochemical reactions of the bioindicators [11]. Owing to their excellent accumulation capabilities, mosses and lichens are considered to be among the most suitable and frequently used bioindicators of air quality [12]. According to previous studies, the ability of these organisms to accumulate hazardous elements in their insoles is much higher than that of other organisms [13]. Biomonitoring methods based on mosses and lichens have several advantages over classical methods. They are cheap, available, highly sensitive, can also be used in different areas of the world, and are very suitable for monitoring [14]. Moreover, it is possible to simultaneously evaluate several hazardous substances in the air [15]. Thus far, mosses and lichens have been successfully used to monitor air quality in various types of environments, such as traffic, parking lots, urban areas, mining areas, and the interior spaces of buildings [16–20]. Bioindicator methods aimed at assessing soil and air pollution, as well as monitoring environmental changes, include tree bark. Because the outer parts of the bark are no longer physiologically active and do not have disruptive growth cycles or metabolic processes, tree bark is an ideal bioindicator [21]. Different types of trees can serve as bioindicators of different types of pollutants. In European countries, oak, pine, plane tree, ash, and elm are most often used [22]. Studies using tree bark as a bioindicator have the potential to contribute to the better understanding and monitoring of environmental problems and issues in the formulation of environmental protection measures.

The aims of this study were (i) to compare the accumulation capacity of mosses and lichens, (ii) to evaluate the suitability of tree bark for mercury air pollution, (iii) to compare ambient air pollution depending on the type of mine work, and (iv) to evaluate the state of mercury air pollution in different internal parts of the mining pits.

2. Material and Methods

2.1. Collection of Tree Bark Samples and the Methods of Their Evaluation

Tree bark was taken mainly from deciduous trees (oak, beech) at a height of approximately 1.5–2.0 m. Using a chisel, four bark samples 1 × 1 cm in size were taken from each tree, each from a different cardinal direction. From the four samples collected from each tree, one mixed sample was created and placed in a PE bag. Bark samples were collected from the vicinity of five mines (with three samples collected from the vicinity of each mine) and from the vicinity of four heaps (two samples from each heap). Tree bark was maintained in a Memmert UF 110 m forced-air laboratory furnace (Memmert GmbH & Co. KG; Schwabach, Germany) at 40 °C for 22 h. The samples were homogenized in an IKA A 10 basic rotary homogenizer (IKA Werke GmbH & Co. KG, Staufen, Germany) and stored in resealable PE bags before analysis. The total mercury content was determined using an AMA-254 instrument (AlTec Spol. s r.o., Prague, Czech Republic).

2.2. Moss and Lichen Collection, Preparation, Exposure, and Evaluation

Two mosses (*Climacium* sp. and *Pleurosium* sp.) and one lichen (*Pseudevernia* sp.) were collected from the Slanské Vrchy Mountains in places that were free of environmental burdens, at least 1 km from main roads, and at least 0.5 km from forest roads. Approximately 500 g of material was collected from each taxon, stored in a paper bag, and transported to the laboratory, where the samples were manually cleaned of plant parts, needles, and soil particles. The mosses and lichen were washed three times in deionized water for 5, 10, and 20 min (approximately 10 L of deionized water per 100 g dry weight of mosses and lichen). After washing, the samples were manually wrapped and dried at 40 °C for 24 h (Venticell 111; BMT, Czech Republic). Approximately 5 g of each taxon was wrapped and tied in nylon mesh (2 mm) and cut into pieces (10 × 10 cm). Each taxon was stored in the laboratory as a control sample (to determine its initial condition). Subsequently, each taxon was exposed to sites of interest in two replications. The samples were exposed to the internal environment of five mines (always at the beginning, middle, and end of the mine, and one series outside in front of the mine) and four heaps of mining material. The mosses and lichen bags were exposed for 6 weeks. The analysis of the samples for the presence of Hg was carried out in the same way as in the case of the tree bark samples, as described above. The relative accumulation factor (RAF) was used to evaluate the Hg content in moss and lichen samples, which was calculated as follows:

$$RAF = (C_{exp} - C_{cont})/C_{cont}$$

where C_{exp} is the mercury content measured after exposure, C_{cont} is the mercury content measured before exposure (in the control sample).

2.3. Statistical Evaluation of the Obtained Data

All statistical analyses were performed using the PAST program [23]. All data were logarithmically transformed prior to analysis. The non-parametric Mann–Whitney U test was used to compare the values of Hg (in the bark) and RAF (for moss and lichen bags) between the types of mine works (mines, heaps). The non-parametric Kruskal–Wallis test was used to compare the accumulation abilities of individual taxa and to compare the content or use in moss and lichen bags depending on the place of exposure (beginning, middle, end, and external environment of the mine).

3. Results and Discussion

3.1. Content of Mercury in Tree Bark

The mercury content (min-max (average ± standard deviation)) in tree bark samples ranged between (0.009–0.166 (0.041 ± 0.06) mg/kg). Preasetia et al. [24], who evaluated the mercury content in tree bark in Indonesia around gold mines, found that the average mercury content in *T. catappa*, *M. indica*, and *S. aromaticum*, and *L. domesticum* bark samples reached an average value of 0.0662, 0.0424, 0.0261, and 0.0154 mg/kg, respectively. Near mining areas focused on gold mining in Myanmar, the mercury content in tree bark ranged from 0.002 to 0.417 mg/kg [25]. Comparing two mining bodies it was found that the content of mercury in tree bark was significantly higher ($p < 0.05$) in the bark of trees growing next to (on the) heaps than in those growing next to the open mining pits (Figure 1). In earlier studies, it was found that the surface of mine heaps is more disturbed by weather [13] than open mine pits, which are more stable in terms of pollution.

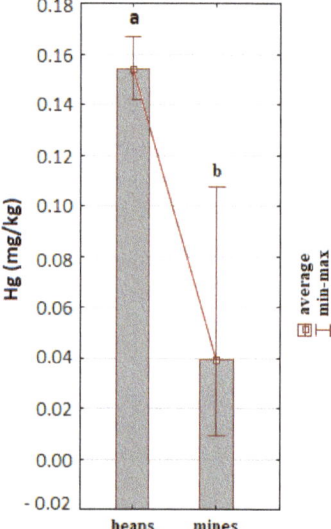

Figure 1. Comparison of tree bark mercury content between trees growing on different mining works and Mann–Whitney U test results expressing statistical differences (various letters)/similarity (same letters) between mining works in tree bark mercury content.

3.2. Content of Mercury in Moss and Lichen Bags and Comparation between Taxa

The results showed that the mercury content in moss and lichen bags (regardless of the taxa), expressed through RAF (Hg) range from (0.01–0.16 (0.09 ± 0.08)). The samples exposed near the mines reached values of (0.01–0.16 (0.09 ± 0.02)), and the samples exposed on the heaps reached values of (0.07–0.10 (0.08 ± 0.01)). The results of the non-parametric Mann–Whitney U test showed that there was no significant difference between the evaluated mining works. Because the difference between heaps and mines was confirmed for tree bark, we assume that this is because the exposure time of mosses and lichens was negligible compared to that of tree bark. Open mining pits are a source of environmental pollution, particularly during active mining. After completion, they are much more stable and are not disturbed by external factors (weather, people), as in the case of heaps. Heaps with loose surfaces are susceptible to erosion. Surface erosion can occur when natural elements, such as rainfall and wind, interact with the heaps. This can lead to the spread of contaminated material into surrounding areas and waterways.

There were some differences between the individual taxa. As shown in Table 1, the ability of lichen *Psuedevernia* sp. to accumulate mercury was significantly higher ($p < 0.05$) than that of the mosses *Climacium* sp. and *Pleurosium* sp. Several studies worldwide have compared the accumulation abilities of mosses and lichens and found that epiphytic lichens are primarily used as bio-monitors for qualitative indication as well as for the spatial and quantitative assessment of atmospheric metal contamination [26–28]. Bargagli et al. [29] compared the advantages and disadvantages of using mosses and lichens. The advantage is that, compared to mosses, epiphytic lichens are less affected by snow cover in winter, but the disadvantage is their high sensitivity to sulfur oxides in polluted areas. Lippo et al. [30] concluded that mosses more easily reflect regional differences in heavy metal deposition compared to lichens.

Table 1. The content of mercury in individual taxa expressed by RAF and the results of the non-parametric Kruskal–Wallis test expressing a statistically significant difference (various letters)/similarity (same letters) between taxa (* standard deviation).

RAF (Hg)	*Climacium* sp. [a]	*Pleurosium* sp. [a]	*Pseudevernia* sp. [b]
min	0.068	0.072	0.010
max	0.130	0.093	0.164
average	0.085	0.084	0.113
median	0.081	0.084	0.113
SD *	0.015	0.006	0.035

3.3. Comparation of Mercury Content in Moss and Lichen Taxa between Different Parts of the Mine

Moss and lichen bags exposed to different parts of the mine showed different results. The highest RAF (Hg) values were determined at the beginning of the mine and the lowest at the external part of the mine. The results of the Kruskal–Wallis test showed that RAF (Hg) values statistically differed only between the beginning (B) and the external part (EX) of the mine, as well as between the end of the mine (E) and the external part (EX) (Figure 2). The pollutant content can vary in different parts of the mining complex owing to several factors. Geological conditions, the type, and method of mining in a given part, as well as various natural and geological processes, can influence this.

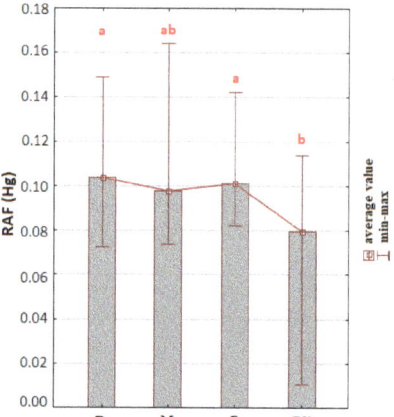

Figure 2. Mercury content in moss and lichen bags exposed in different parts of the mine (B-beginning, M-middle, E-end, EX-external part of the mine) expressed through RAF (Hg) and the results of the non-parametric Kruskal–Wallis test expressing statistical differences (various letters)/similarity (same letters) between different parts of the mine in RAF values.

4. Conclusions

Mercury is considered one of the most toxic heavy metals with a serious impact on human and ecosystem health. Its monitoring in various components of the environment as well as the implementation of measures to limit its spread mainly in risk areas should be the priority of every state. Former mining areas, which are characterized by a high number of unrehabilitated mining bodies, are very risky from the point of view of the spread of a wide spectrum of contaminants.

The use of bioindicators such as tree bark as well as moss and lichen bags has proven to be an effective alternative to traditional monitoring methods. While tree bark showed statistically significant differences between mining works in terms of mercury content, the moss and lichen bags did not differ by exposure location (different mining works). A difference was noted between different taxa. The lichen *Pseudevernia* sp. Was found to be a significantly better Hg accumulator when compared to the mosses. The difference in the mercury in the air in different parts of the mine corridors was interesting, and it was probably caused by a wide range of factors, such as geological processes, the type of mining in the given part of the mine, air flow, and many others.

Author Contributions: Conceptualization, L.D. and L.B.; methodology, L.D.; software, Ľ.H.; validation, Ľ.H. and J.Á.; formal analysis, Ľ.H. and J.Á., investigation, L.D. and L.B.; resources, L.D.; data curation, L.D.; writing—original draft preparation, L.D.; writing—review and editing, L.D., J.Á. and L.B.; visualization, L.D.; supervision, J.Á.; project administration, L.D.; funding acquisition, L.D. All authors have read and agreed to the published version of the manuscript.

Funding: This research was founded by the Scientific Grant Agency of the Ministry of Education, Science, Research and Sport of the Slovak Republic—VEGA no. 1/0213/22.

Institutional Review Board Statement: Not applicable.

Informed Consent Statement: Not applicable.

Data Availability Statement: The data presented in this study are available upon request from the corresponding author.

Conflicts of Interest: The authors declare no conflict of interest.

References

1. Ravindra, K.; Suman Mor, R.; Aggarwal, A.N. Generalized additive models: Building evidence of air pollution, climate change and human health. *Environ. Int.* **2019**, *132*, 104987. [CrossRef] [PubMed]
2. Mannucci, P.M.; Franchini, M. Health Effects of Ambient Air Pollution in Developing Countries. *Int. J. Environ. Res. Public Health* **2017**, *14*, 1048. [CrossRef] [PubMed]
3. Liu, S.; Wang, X.; Guo, G.; Yan, Z. Status and environmental management of soil mercury pollution in China: A review. *J. Environ. Manag.* **2021**, *277*, 111442. [CrossRef] [PubMed]
4. Patel, U.N.; Patel, U.D.; Khadayata, A.V.; Vaja, R.K.; Patel, H.A.; Modi, C.M. Assessment of Neurotoxicity Following Single and Co-exposure of Cadmium and Mercury in Adult Zebrafish: Behavior Alterations, Oxidative Stress, Gene Expression, and Histological Impairment in Brain. *Water Air Soil Pollut.* **2021**, *232*, 340. [CrossRef]
5. Luís, A.T.; Teixeira, P.; Almeida, S.F.P.; Matos, J.X.; Silva, E.F. Environmental impact of mining activities in the Lousal area (Portugal): Chemical and diatom characterization of metal-contaminated stream sediments and surface water of Corona stream. *Sci. Total Environ.* **2011**, *409*, 4312–4325. [CrossRef] [PubMed]
6. Rare Earth Elements: A Review of Production, Processing, Recycling, and Associated Environmental Issues; EPA: Cincinnati, OH, USA, 2012. Available online: http://www.miningwatch.ca/files/epa_ree_report_dec_2012.pdf (accessed on 20 September 2023).
7. Samara, T.; Spanos, I.; Platis, P.; Papachristou, T.G. Heavy Metal Retention by Different Forest Species Used for Restoration of Post-Mining Landscapes, N. Greece. *Sustainability* **2020**, *12*, 4453. [CrossRef]
8. Conti, M.E.; Cecchetti, G. Biological monitoring: Lichens as bioindicators of air pollution assessment—A review. *Environ. Pollut.* **2001**, *114*, 471–492. [CrossRef]
9. Molnár, V.É.; Tőzsér, D.; Szabó, S.; Tóthmérész, B.; Simon, E. Use of Leaves as Bioindicator to Assess Air Pollution Based on Composite Proxy Measure (APTI), Dust Deposition and Elemental Concentration of Metals. *Plants* **2020**, *9*, 1743. [CrossRef]
10. Azzary, M.F. Plant bioindicators of pollution in Sadat City, Western Nile Delta, Egypt. *PLoS ONE* **2020**, *15*, e0226315.
11. Holt, E.A.; Miller, S.W. Bioindicators: Using Organisms to Measure Environmental Impacts. *Nat. Educ. Knowl.* **2011**, *2*, 8.
12. Godzik, B. Use of Bioindication Methods in National, Regional and Local Monitoring in Poland—Changes in the Air Pollution Level over Several Decades. *Atmosphere* **2020**, *11*, 143. [CrossRef]
13. Ren, H.; Zhao, Y.; Xiao, W. Influence of management on vegetation restoration in coal waste dump after reclamation in semi-arid mining areas: Examining ShengLi coalfield in Inner Mongolia, China. Environ. *Sci. Pollut. Res.* **2021**, *28*, 68460–68474. [CrossRef] [PubMed]
14. Markert, B. Definitions and principles for bioindication and biomonitoring of trace metals in the environment. *J. Trace Elem. Med. Bio.* **2007**, *21*, 77–82. [CrossRef] [PubMed]
15. Barandovski, L.; Frontasyeva, M.V.; Stafilov, T. Multi-element atmospheric deposition in Macedonia studied by the moss biomonitoring technique. *Environ. Sci. Pollut. Res.* **2015**, *22*, 16077–16097. [CrossRef] [PubMed]
16. Vuković, G.; Ančić Urošević, M.; Škrivanj, S.; Vergel, K.; Tomašević, M.; Popović, A. The first survey of airborne trace elements at airport using moss bag technique. *Environ. Sci. Pollut. Res.* **2017**, *24*, 15107–15115. [CrossRef] [PubMed]
17. Vuković, G.; Urošević, M.A.; Razumenić, I.; Kuzmanoski, M.; Pergal, M.; Škrivanj, S.; Popović, A. Air quality in urban parking garages (PM10, major and trace elements, PAHs): Instrumental measurements vs. active moss biomonitoring. *Atmos. Environ.* **2014**, *85*, 31–40. [CrossRef]
18. Tretiach, M.; Adamo, P.; Bargagli, R.; Baruffo, L.; Carletti, L.; Crisafulli, P.; Giordano, S.; Modenesi, P.; Orlando, S.; Pittao, E. Lichen and moss bags as monitoring devices in urban areas, Part I: Influence of exposure on vitality. *Environ. Pollut.* **2007**, *146*, 380–390. [CrossRef]
19. Koz, B.; Cevik, U.; Akbulut, S. Heavy metal analysis around Murgul (Artvin) copper mining area of Turkey using moss and soil. *Ecol. Indic.* **2012**, *20*, 17–23. [CrossRef]
20. Demková, L.; Oboňa, J.; Árvay, J.; Michalková, J.; Lošák, T. Biomonitoring road dust pollution along streets with various traffic density. *Pol. J. Environ.* **2019**, *28*, 3687–3696. [CrossRef]
21. Walkenhorst, A.J.; Hagemeyer, J.; Breckle, W. Plants as biomonitors. In *Indicators for Heavy Metals in the Terrestrial Environment: Passive Monitoring of Air-Borne Pollutants, Particularly Trace Metals with Tree Bark*; Market, W.B., Ed.; VCH Publishers: Weinheim Germany, 1993; pp. 524–540.
22. Sawidis, T.; Breuste, J.; Mitrovic, M.; Pavlovic, P.; Tsigaridas, K. Trees as bioindicator of heavy metal pollution in three European cities. *Environ. Pollut.* **2011**, *159*, 3560–3570. [CrossRef]
23. Hammer, Ø.; Harrper, D.A.T.; Ryan, P.D. Past: Paleontological Statistics Software Package for Education and Data Analysis. *Palaeontol. Electron.* **2001**, *4*, art. 4.
24. Prasetia, H.; Sakakibara, M.; Sera, K.; Laird, J.S. Evaluation of the Total Mercury Weight Exposure Distribution Using Tree Bark Analysis in an Artisanal and Small-Scale Gold Mining Area, North Gorontalo Regency, Gorontalo Province, Indonesia. *Int. J. Environ. Res. Public Health* **2022**, *19*, 1. [CrossRef] [PubMed]
25. Soe, P.S.; Kyaw, W.T.; Arizono, K.; Ishibashi, Y.; Agusa, T. Mercury Pollution from Artisanal and Small-Scale Gold Mining in Myanmar and Other Southeast Asian Countries. *Int. J. Environ. Res. Public Health* **2022**, *19*, 6290. [CrossRef] [PubMed]
26. Aleksander-Kwaterczak, U.; Ciszewski, D. Metal Mobility in Afforested Sites of an Abandoned Zn-Pb Ore Mining Area. *Appl. Sci.* **2020**, *10*, 6041. [CrossRef]

27. Cecconi, E.; Fortuna, L.; Peplis, M. Element accumulation performance of living and dead lichens in a large-scale transplant application. Environ. Sci. Pollut. Res. **2021**, *28*, 16214–16226. [CrossRef] [PubMed]
28. Lodenius, M. Use of plants for biomonitoring of airborne mercury in contaminated areas. *Environ. Res.* **2013**, *125*, 113–123. [CrossRef] [PubMed]
29. Bargagli, R.; Monaci, F.; Borghini, F.; Bravi, F.; Agnorelli, C. Mosses and lichens as biomonitors of trace metals. A comparison study on Hypnum cupressiforme and Parmelia caperata in a former mining district in Italy. *Environ. Pollut.* **2002**, *116*, 279–287. [CrossRef] [PubMed]
30. Lippo, H.; Poikolainen, J.; Kubin, E. The use of moss, lichen, and pine bark in the nationwide monitoring of atmospheric heavy metal deposition in Finland. *Water Air Soil Pollut.* **1995**, *85*, 2241–2246. [CrossRef]

Disclaimer/Publisher's Note: The statements, opinions and data contained in all publications are solely those of the individual author(s) and contributor(s) and not of MDPI and/or the editor(s). MDPI and/or the editor(s) disclaim responsibility for any injury to people or property resulting from any ideas, methods, instructions or products referred to in the content.

Proceeding Paper

Separator Systems for Light Liquids [†]

Vojtěch Václavík *, Tomaš Dvorský and Pavlína Richtarová

Department of Environmental Engineering, Faculty of Mining and Geology, VSB—Technical University of Ostrava, 17. Listopadu 2172/15, 708 00 Ostrava-Poruba, Czech Republic; tomas.dvorsky@vsb.cz (T.D.); pavlina.richtarova.st@vsb.cz (P.R.)
* Correspondence: vojtech.vaclavik@vsb.cz
[†] Presented at the 4th International Conference on Advances in Environmental Engineering, Ostrava, Czech Republic, 20–22 November 2023.

Abstract: This article deals with separator systems for light liquids that can be used in the management of rainwater from hard and polluted surfaces. Attention is focused on the material of the separators, the smallest nominal width and the types of separators. The procedure used for determining the type and size of a light liquid separator and its operation and maintenance is presented in this article as well. An example of a light liquid separator design for an industrial area is introduced in the experimental part.

Keywords: separator for light liquids; material; nominal width; operation; maintenance; design example

1. Introduction

Separators for light liquids are an integral part of the process of managing rainwater from polluted hard surfaces contaminated with oil. The areas that are significantly contaminated and used for capturing rainwater include industrial sites, especially their parking lots and roads [1–4].

Furthermore, light liquid separators are used in the petroleum industry, in oil extraction, to separate oil and water in order to reduce extraction costs. Axial separators with multi-stage separation are used here with water flow rates ranging from 3 to 7 m^3/h and an oil input fraction below 10% [5–7]. Due to the fact that the process of the separation of light liquids from water is not simple, simulation models are developed in order to increase the separation efficiency and to select the right type of separator [8–10].

Our paper deals with the design of light liquid separators as devices to be used in the management of rainwater that is polluted with oil.

According to the CSN EN 858-1 standard [11], a light liquid separator can be defined as a device used for the treatment of industrial wastewater or rainwater containing light liquids that are not emulsified. These are liquids with a density lower than or equal to 0.95 g/cm^3 (e.g., petroleum substances). These may be present in wastewater from vehicle parking areas, garages, refuelling points, car wash boxes or industrial plants. The separator prevents such polluted water from entering the receiving stream or sewer network.

2. Structural Composition

From the design point of view, the structural composition of the separator comprises one or more tanks in which the technological equipment of the separator is installed—separating the space, sludge trap and sampling point. The tank may be of circular or rectangular design. It may be constructed on-site or prefabricated.

According to Table 1, separators can be divided into two classes based on the maximum permissible residual oil content at the outlet. In this article, the focus is on light liquid separators that belong to the first class.

Citation: Václavík, V.; Dvorský, T.; Richtarová, P. Separator Systems for Light Liquids. *Eng. Proc.* **2023**, *57*, 29. https://doi.org/10.3390/engproc2023057029

Academic Editors: Adriana Estokova, Natalia Junakova, Tomas Dvorsky, Vojtech Vaclavik and Magdalena Balintova

Published: 7 December 2023

Copyright: © 2023 by the authors. Licensee MDPI, Basel, Switzerland. This article is an open access article distributed under the terms and conditions of the Creative Commons Attribution (CC BY) license (https://creativecommons.org/licenses/by/4.0/).

Table 1. Classes of light liquid separators.

Class	Maximum Permissible Residual Oil Content (mg/L)	Typical Relieving Procedure
I	5.0	Coalescing separators
II	100.0	Gravity separators

The recommended nominal sizes for light liquid separators are 1, 3, 5, 6, 10, 15, 20, 30, 40, 50, 65, 80, 100, 125, 150, 200, 300, 400 and 500.

The smallest permissible nominal width of the inlet, outlet and eventual interconnecting pipes has to adhere to the respective nominal size value of the light liquid separators, as presented in Table 2. Provisions must be made to ensure that the inlet, outlet and interconnecting pipes can withstand any movement or settlement of the soil.

Table 2. Minimum nominal width.

Nominal Width (NS)		DN_{min} [a]
Lower or equal to NS 3		100
Higher than NS 3	up to NS 6	125
Higher than NS 6	up to NS 10	150
Higher than NS 10	up to NS 20	200
Higher than NS 20	up to NS 30	250
Higher than NS 30	up to NS 100	300
Higher than NS 100		400

[a] Nominal width can be applied to either the inner diameter or the outer diameter.

The most commonly used materials for the construction of separation facilities include the following:

- Concrete—plain concrete, concrete with dispersed reinforcement and reinforced concrete;
- Metallic materials—cast iron, steel and stainless steel;
- plastic—polyethylene, polypropylene and glass fibre-reinforced plastic.

In the case of sealing materials, the use of elastomers (rubber) and permanently elastic sealing materials is allowed. The use of cement mortar and similar sealing materials is prohibited.

The use of other materials is permitted provided that all requirements of the relevant standards are met. All materials that come into direct contact with wastewater shall be chemically resistant to its action or shall be otherwise suitably protected.

3. Separator with Sludge Trap and Separating Space

The principle is based on the different densities of light liquids and water. Wastewater contains, in addition to light liquids and suspended sediment (e.g., sand and mud). These settle at the bottom of the tank in the so-called sludge trap after the wastewater flows into the tank. The flow through the sludge trap is directed by a flow diverter. From the sludge trap, the mechanically cleaned wastewater flows into the separating space. A coalescing filter is located here, which separates light liquids from wastewater—the coalescing gravity principle. The separated light liquids are collected near the surface in an area designed to capture them. The wastewater treated in this way is discharged through a drainage channel into the discharge pipe.

There is a safety element in the form of a float cap in the space behind the coalescing filter. The cap is used to close the drainage channel when the maximum height of captured light liquids is exceeded. The closure occurs automatically due to the difference in density of the separated light liquids and water.

The sludge trap and the separating space may be in one common tank or divided into several separate tanks.

In some cases, a sludge filter is fitted between the sludge trap and the separating space.

Figure 1 shows a diagram of a light liquid separator with a sludge trap and a separating space.

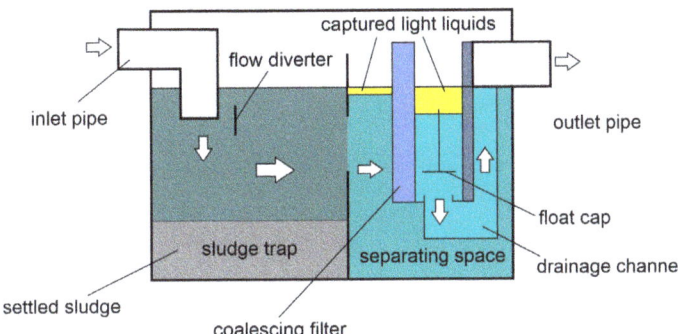

Figure 1. Light liquid separator with a sludge trap and a separating space.

4. Separator with Sludge Trap, Separating Space and Final Separating Stage

Compared to the above-mentioned type, the separator is equipped with the so-called final separating stage. A sorption filter filled with sorbent (e.g., Fibroil) is fitted here. It is used to capture the residual amount of light liquids that have not been separated in the separating space. It is used in particular when the quality requirements for the runoff water are increased—it guarantees a concentration of light liquids at an outflow of 0.2 to 0.5 mg/L. The treated water flows into the outlet pipe (see Figure 2).

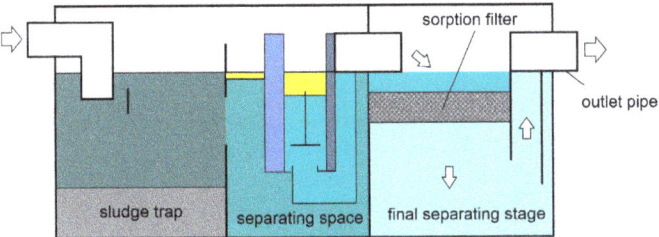

Figure 2. Light liquid separator with sludge trap, separating space and final separating stage.

The sludge trap, separating space and final separating stage can be in one common tank or divided into several separate tanks.

5. Procedure Used for the Equipment Size Designing Process

The procedure used for the light liquid separator size designing process consists of 6 points:

A. Location and catchment area

First, it is important to determine the location of the site of interest and its relevant catchment area.

B. Rainfall intensity at the given periodicity, n

This is a tabular value available in the reference sources—e.g., short-term rain intensities in the Elbe, Oder and Morava catchment areas by Josef Trupel. It is determined on the basis of the measured results of an ombrographic station. The station is selected as it is close as possible to the site of interest.

C. Runoff coefficient, φ, and calculation of the drainage area, A

The determination of the runoff coefficient is based on CSN 75 6101 [12]. It is determined according to the type of development, the type of land and the slope of the terrain.

The runoff coefficient is indirectly related to the calculation of the hard drainage area from which the rainwater will be drained.

D. Maximum rainwater runoff, Q_r

The maximum rainwater runoff is calculated using the following formula:

$$Q_r = \varphi \cdot i \cdot A \tag{1}$$

where,

Q_r—maximum rainwater runoff [l/s];
φ—runoff coefficient [–];
i—rain intensity [l/s·ha];
A—drainage area [ha].

E. Maximum wastewater runoff, Q_s

The maximum wastewater runoff is used only in the case of the treatment of water other than rainwater polluted with oil substances. Most often, this is the treatment of industrial wastewater from factories, car washes or fuel filling stations.

It is calculated using the following formula:

$$Q_s = Q_{s1} + Q_{s2} + Q_{s3} + \cdots \tag{2}$$

where,

Q_s—maximum wastewater runoff [l/s];
Q_{s1}—wastewater runoff from all runoff sites [l/s];
Q_{s2}—wastewater runoff from car wash facilities [l/s];
Q_{s3}—wastewater runoff from high pressure cleaning facilities [l/s].

F. Nominal size of NS separator

This parameter is calculated using the following formula:

$$NS = (Q_r + (f_x \cdot Q_s)) \cdot f_d \tag{3}$$

where,

NS—nominal separator width [–];
Q_r—maximum rainwater runoff [l/s];
f_x—aggravating coefficient depending on the type of runoff [–];
Q_s—maximum wastewater runoff [l/s];
f_d—density coefficient for the light liquid in question [–].

Both the aggravating coefficient and the density coefficient are tabular values available in CSN EN 858-2 [13].

6. Operation and Maintenance

According to the legal regulations of the Czech Republic, a light liquid separator is a water management facility. In order to be inspected by the water authorities and to ensure the proper functioning of the separator, it is necessary to comply with the conditions and procedures developed for its operation and maintenance. The equipment must be monitored and maintained in accordance with the manufacturers' instructions, and maintenance must be carried out in a timely and planned manner. The necessary information is given in the documents included in the technical documentation and is usually also provided by the manufacturers of the separators. These are the operating rules and the operating logbook.

Operating rules

This is basically an instruction manual for the operation of the equipment. Operation and maintenance are carried out by the equipment user and its operator. The user manages the operation of the separator, decides on major operational measures relating to the actual separation process, keeps records of laboratory monitoring and material consumption and also provides training for the operators. The operators carry out periodic tasks and keep detailed records in an operating logbook.

The activities performed by the operators can be divided according to intervals into the following:

- Once per week—visual inspection;
- Twice a month—the cleaning of coalescing and sorption filters, and the checking of the sludge trap;
- Once per six months—the removal of trapped light liquids;
- Once per year—the cleaning of the sludge trap, replacement of the sorption filter cartridge and inspection;
- No interval/as needed—sampling, and other activities listed above are performed as needed.

Operating logbook

This is used together with the operating rules. Entries are made in the operating logbook by the equipment operators on individual faults and malfunctions at the time of their occurrence and rectification, on spare parts and maintenance or maintenance needs. Entries are also made in the logbook in the event of handling. Handling in this case means, for example, the removal of sludge or the taking of control samples. In the case of sludge removal, the date of sludge removal, the amount of sludge removed in m^3, the description of the maintenance activity and who performed the operation are usually written in the record. If necessary, the logbook must be shown on request.

7. Practical Example of the Design of a Light Liquid Separator Tank

This chapter is focused on a practical example of the design of a light liquid separator system for rainwater from an industrial site in the Liberec Region. The theoretical designing procedure is presented in Section 5. The rainfall intensity for the industrial area in the Liberec Region was determined from Trupl tables. A 15 min long rainfall period with a periodicity of 0.2, or 1 × in 5 years, was chosen as the initial value for calculation. The periodicity of 0.2 is typically used for industrial areas. The f_x coefficient was chosen to be zero according to the tables. There is no need to account for adverse conditions for the separation or treatment of multiple types of wastewater. The sewerage system option proposed for the site only deals with rainwater management.

Rainwater calculation:

Runoff coefficient, φ: 0.8 (Asphalt areas)
Runoff coefficient, φ: 0.6 (Concrete pavement)
Rainfall intensity, i: 152 l/s·ha (loc. Liberec, periodicity 0.2)
Area, A: 3726 m^2 (Asphalt areas—based on the situation)
Area, A: 2141 m^2 (Concrete pavement—based on the situation)

$$Q_r = \varphi \cdot i \cdot A = (0.8 \cdot 152 \cdot 0.3726) + (0.6 \cdot 152 \cdot 0.2141) = 64.84 \text{ l/s}$$

Wastewater calculation:

$$Q_s = 0 \text{ l/s}$$

Water from parking area drainage is not polluted from car washes or high-pressure cleaning equipment.

Choice of nominal size of separator:

Coefficient f_x: 0
Coefficient f_d: 1 (do 0.85 g/cm^3)
Rainwater Q_r: 64.84 l/s
Polluted water Q_s: 0 l/s

$$NS = (Q_r + (f_x \cdot Q_s)) \cdot f_d = (64.84 + (0 \cdot 0)) \cdot 1 = 64.84$$

From the resulting value, NS = 64.84, and Table 2, it can be seen that the smallest permissible nominal width of the inlet pipe and outlet pipe is DN300.

Technical specification:

The light liquid separator from ASIO spol. s r.o. has been chosen on the basis of the calculation of the nominal tank size. The chosen type was AS-TOP 80 RC/EO PB PP. It is a class I gravity-coalescing separator of with sludge and separating space. The sludge area is designed for small quantities of sludge (100 × NS). The separator technology is designed for an influent water contamination value of C10-C40 < 4000 mg/L. The parameters of the treated water at the outlet then reach the value C10-C40 = 2–5 mg/L. This meets the discharge limit requirements for class I separators.

The separator is a plastic–concrete construction. The tank consists of a double-walled shell made of polypropylene (PP). The intermediate shell is factory-fitted with a B500B, Ø12 concrete reinforcement. At the installation site, the intermediate shell is subsequently filled with concrete of the strength class C 35/45. The plastic shell ensures the watertightness of the tank and protects the concrete filling from the negative effects of the surrounding environment. The outer shell therefore protects, for example, against the negative effects of groundwater or the ingress of ballast water. The inner shell protects against the aggressiveness of oily water. The construction of the tank does not require additional protection against corrosion.

The ceiling slab of the tank must be insulated with IPA 400H PE. The insulation layer prevents the penetration of ground moisture, and of surface and groundwater into the intermediate shell.

The basic technical parameters of the tank are presented in Table 3.

Table 3. The basic parameters of the designed light liquid separator tank.

Nominal Width NS	80
Maximum flow rate (l/s)	80
Max. quantity of captured light liquids (l)	871
Sludge trap volume (m^3)	9.88
Outer tank diameter (mm)	3430
Outer tank height (mm)	2220
Inlet height (mm)	1650
Outlet height (mm)	1550
DN inlet, outlet	300
Volume of concrete necessary for the intermediate tank shell (m^3)	4.61
Transport weight of the tank (kg)	2343

The technological baffles inside the tank are made of polypropylene. Coalescing filter inserts are made of polyurethane foam mounted in stainless steel frames. The float cap is also made of stainless steel. The technical parameters of the coalescing filter are presented in Table 4.

Table 4. Technical parameters of the coalescing polyurethane filter.

Bulk Density of Polyurethane Foam	25 kg/m^3
Tensile strength	120–135 kPa
Thermal resistance	−40 až +10 °C
Compressibility	40% of compression at 5.0 kPa
Expansibility	80–100%

The design is a single cylindrical tank that is intended to be stored below ground level. The selected tank type does not allow the tank to be placed in areas with groundwater levels below the foundation slab. The tank is self-supporting after installation without the need for concrete application.

According to the manufacturer's requirements, the maximum foundation base depth must be 5000 mm below the prepared terrain. The tank must be placed on a reinforced concrete slab with adequate bearing capacity and a flatness of ±5 mm.

From the static point of view, the tank ceiling is designed for the ground load of the road structure with vehicle traffic—a random load from the vehicle to the centre of the hatch; F = 50 kN.

8. Conclusions

This paper presents the theoretical basis for light liquid separators as one of the construction objects in the field of the management of rainwater polluted with oil substances. The theoretical part describes the design solution of light liquid separators. The focus is on the classes, the smallest nominal width according to the NS and the material. Two basic types of light liquid separators are also described in this section.

The experimental part is focused on an example of light liquid separator design for an industrial site in the Liberec Region. There is a procedure of the hydro-technical calculation of the separator design, based on which the min. DN of the piping at the inlet and outlet of the light liquid separator is designed. The AS-TOP 80 RC/EO PB PP type of light liquid separator from ASIO, an a.s. company, was selected and recommended on the basis of the determination of the min. DN. It includes the technical and design parameters of the tank as well.

Author Contributions: Conceptualisation, V.V. and P.R.; methodology, V.V.; formal analysis, T.D. and P.R.; investigation, V.V., T.D. and P.R.; resources, V.V. and T.D.; writing—original draft preparation, V.V., T.D. and P.R.; writing—review and editing, V.V. and T.D.; visualisation, V.V., T.D. and P.R.; supervision, V.V.; funding acquisition, T.D. All authors have read and agreed to the published version of the manuscript.

Funding: This article has been produced with the financial support of the European Union under the REFRESH—Research Excellence For REgion Sustainability and High-tech Industries project number CZ.10.03.01/00/22_003/0000048 via the Operational Programme Just Transition.

Institutional Review Board Statement: Not applicable.

Informed Consent Statement: Not applicable.

Data Availability Statement: The authors declare that all data supporting the results of this research are available in this article.

Conflicts of Interest: The authors declare no conflict of interest.

References

1. Krocova, S. The Water Protection Trends in the Industrial Landscape. *Inz. Miner.* **2014**, *15*, 174–177.
2. Hluštík, P.; Novotný, J. The Testing of Standard and Recyclable Filter Media to Eliminate Hydrogen Sulphide from Sewerage Systems. *Water* **2018**, *10*, 689. [CrossRef]
3. Hluštík, P.; Zeleňáková, M. Risk Analysis of Failure in Sewer Systems in Czech Municipalities. *Pol. J. Environ. Stud.* **2019**, *28*, 4183–4190. [CrossRef]
4. Hluštík, P.; Singrová, V. Efficiency of Wastewater Treatment Plants. *Int. Multidiscip. Sci. GeoConference SGEM* **2018**, *18*, 321–328. [CrossRef]
5. Zeng, X.; Zhao, L.; Fan, G.; Yan, C. Experimental Study on the Design of Light Phase Outlets for a Novel Axial Oil-Water Separator. *Chem. Eng. Res. Des.* **2021**, *165*, 308–319. [CrossRef]
6. Delfos, R.; Murphy, S.; Stanbridge, D.; Olujić, Ž.; Jansens, P.J. A Design Tool for Optimising Axial Liquid–Liquid Hydrocyclones. *Miner. Eng.* **2004**, *17*, 721–731. [CrossRef]
7. Liu, M.; Chen, J.; Cai, X.; Han, Y.; Xiong, S. Oil–Water Pre-Separation with a Novel Axial Hydrocyclone. *Chin. J. Chem. Eng.* **2018**, *26*, 60–66. [CrossRef]
8. Zeng, X.; Zhao, L.; Fan, G.; Yan, C. Numerical and Experimental Study on a New Axial Separator for Liquid-Liquid Separation. *J. Taiwan Inst. Chem. Eng.* **2021**, *123*, 104–114. [CrossRef]
9. Zeng, X.; Xu, Y.; Zhao, L.; Fan, G.; Yan, C. Numerical Investigation on Axial Liquid-Liquid Separators with Different Swirl Chambers. *Chem. Eng. Process. Process Intensif.* **2021**, *161*, 108324. [CrossRef]

10. Li, T.; Sun, Z.; Geng, K.; Sun, M.; Wang, Z. Numerical Analysis of a Novel Cascading Gas–Liquid Cyclone Separator. *Chem. Eng. Sci.* **2023**, *270*, 118518. [CrossRef]
11. *CSN EN 858-1*; Separator Systems for Light Liquids (e.g., Oil and Petrol)—Part 1: Principles of Product Design, Performance and Testing, Marking and Quality Control. ČSN: Praha, Czech Republic, 2003.
12. *CSN 756101*; Sewer Systems and House Connections. ČSN: Praha, Czech Republic, 2012.
13. *CSN EN 858-2*; Separator Systems for Light Liquids (e.g., Oil and Petrol)—Part 2: Selection of Nominal Size, Installation, Operation and Maintenance. ČSN: Praha, Czech Republic, 2003.

Disclaimer/Publisher's Note: The statements, opinions and data contained in all publications are solely those of the individual author(s) and contributor(s) and not of MDPI and/or the editor(s). MDPI and/or the editor(s) disclaim responsibility for any injury to people or property resulting from any ideas, methods, instructions or products referred to in the content.

Proceeding Paper

Phytotoxicity Assessment of Wastewater from Industrial Pulp Production [†]

Oto Novak

Department of Environmental Engineering, Faculty of Mining and Geology, VŠB–Technical University of Ostrava, 17. Listopadu 2172/15, 708 00 Ostrava, Czech Republic; oto.novak@vsb.cz

[†] Presented at the 4th International Conference on Advances in Environmental Engineering, Ostrava, Czech Republic, 20–22 November 2023.

Abstract: During the production of pulp from natural raw materials, such as wood, large amounts of waste are generated. When they are reused, e.g., in agriculture, ecological requirements towards the environment must be met, and the associated risks must be known. In the framework of this study, the phytotoxicity of liquid waste generated during the industrial processing of pulp using the sulfite process was investigated. The results of semichronic toxicity tests on *Lepidium sativum* L. and *Sinapis alba* L. showed a direct effect on the growth and development of plants and proved that the undiluted samples were phytotoxic. However, the phytotoxicity of the investigated waste samples was influenced by their dilution ratio. Thus, finding the optimal solution ratio is crucial for the reuse of liquid waste. A nonphytotoxic effect on the roots of the tested plants was proven with a solution in a ratio of 1:100. In addition, the control samples diluted with a medium manifested phytotoxicity due to the activity of internal microorganisms.

Keywords: phytotoxicity; wastewater; pulp; *Sinapis alba* L.; *Lepidium sativum* L.

1. Introduction

The paper industry is one of the largest global polluters of the environment. Large amounts of gaseous, liquid, and solid waste are released into the environment, especially during the sulfite process of pulp production. The biggest problems in terms of environmental pollution include wastewater from production and emissions into the air [1]. Wastewater from pulp production mainly contains organic compounds (resins and organic acids), chlorinated compounds, nitrates, and phosphates, while toxic metals and metalloids are found in low quantities [2].

Due to the relatively high number of organic substances in waste, it can be assumed that the use of pulp production waste may be suitable for agricultural purposes and the production of horticultural substrates. However, during the sulfite process, large amounts of chemicals are used, and some of them can be toxic to the environment. Therefore, the application of pulp waste in agriculture can represent a serious environmental risk. In particular, it can inhibit or stimulate plant growth. Ultimately, it can also affect the quality and species diversity of soil microorganisms [3,4].

A suitable alternative could be the introduction of technology that uses inorganic matter to reduce the content of contaminants without causing phytotoxicity. Such methods include, for example, biosorption, where waste plant biomass is used as a biosorbent, the basic building blocks of which are cellulose, hemicellulose, and lignin. This technology does not depend on the metabolism of microorganism cells; rather, it is based on physicochemical interactions between metal and functional groups of plant cell walls. The process is characterized by relatively high speed and the possibility of recovering adsorbed metal from the biosorbent using reverse desorption processes [5,6].

The chemical composition of waste generated during the sulfite process may vary, and it is thus advisable to regularly assess the waste from the point of view of ecotoxicity [7].

Citation: Novak, O. Phytotoxicity Assessment of Wastewater from Industrial Pulp Production. *Eng. Proc.* **2023**, *57*, 30. https://doi.org/10.3390/engproc2023057030

Academic Editors: Adriana Estokova, Natalia Junakova, Tomas Dvorsky, Vojtech Vaclavik and Magdalena Balintova

Published: 7 December 2023

Copyright: © 2023 by the author. Licensee MDPI, Basel, Switzerland. This article is an open access article distributed under the terms and conditions of the Creative Commons Attribution (CC BY) license (https://creativecommons.org/licenses/by/4.0/).

Through acute toxicity tests, it is possible to assess its direct effect on the biotic component of the environment [8]. From the point of view of practical use of the waste in agriculture, it is essential to carry out regular monitoring of its phytotoxic effect, which is assessed using tests on selected plants [9].

The aim of this research was to assess the phytotoxicity of liquid waste (wastewater) generated during the production of pulp and to search for optimal dilution of wastewater that reduces the inhibitory effect on plant growth.

2. Materials and Methods

2.1. Tested Samples and Their Composition

The tested waste samples from the sulfite pulping process were a mixture of waste and water containing mixtures of waste fibers (FB), sludge from mechanical processing and sorting (SL), and industrial wastewater produced during the pulping process (Lenzing Biocel Paskov a.s., Paskov, Czech Republic). The composition of individual samples is shown in Table 1; the control samples contained only distilled water. Regarding the composition of the samples, an increased activity of microorganisms (primarily cellulolytic microorganisms) was assumed in them. To stimulate their activity, a tryptone–soybean culture medium, TSB (M011, Soyabean Casein Digest Medium, HiMedia Laboratories, Mumbai India), was added only to sample D.

Table 1. Composition of the basic (undiluted) tested samples for evaluating the phytotoxicity of liquid waste.

Sample Identification	FB [g]	SL [g]	WW [mL]	TSB [mL]
A	50	50	300	-
B	50	100	450	-
C	100	50	450	-
D	50	50	200	100

Notes: FB: waste fibers, SL: sludge, WW: wastewater, TSB: tryptone–soybean culture medium.

Samples in Table 1 were filtered through filter paper (KA-2, Papírna Perštějn spol. s.r.o., Czech Republic) before testing. These samples were further diluted with sterile distilled water in ratios of 1:10, 1:50, and 1:100. The samples prepared in this way were then subjected to semichronic toxicity testing according to the current Czech legislation [10,11] on white mustard (*Sinapis alba* L.) and cress (*Lepidium sativum* L.).

2.2. Phytotoxicity Tests

To test the phytotoxicity of industrial wastewater, tests were carried out on mustard (*Sinapis alba* L.), and cress (*Lepidium sativum* L) plants according to standard ČSN EN ISO 18763 (836447) [11]. In order to ensure the safe use of wastewater in agriculture in accordance with Decree No. 273/2021 Coll. [11], inhibition or stimulation of root growth higher than 50% compared to the control must be confirmed within a period of 72 h. A total of 20 seeds of the test plants were spread on filter paper in Petri dishes and exposed to the tested wastewater solutions, which consisted of a mixture of waste from pulp production. Root lengths of germinating plants were measured after 72 h seed incubation, and a thermostat set at 20 °C was used, as can be seen from Figure 1. From the results of each observed concentration (undiluted sample and dilution of 1:10, 1:50, and 1:100), the arithmetic means were determined, which were later compared with the arithmetic mean from the control measurement. Based on the arithmetic means of plant root lengths, inhibition was calculated for each concentration.

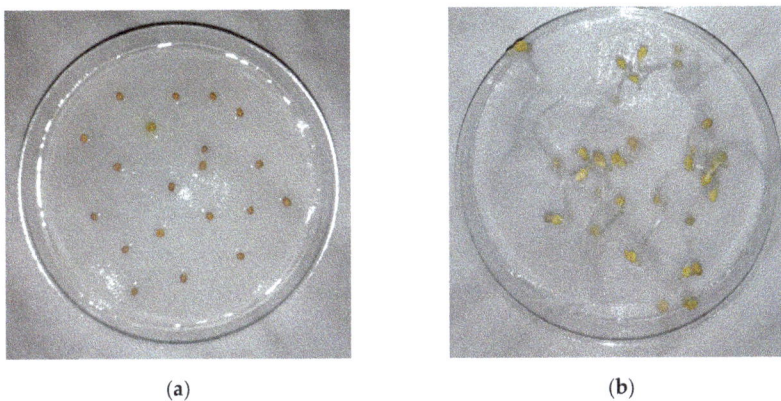

Figure 1. Example of semichronic toxicity test on mustard seeds (**a**) at the start of testing, 0 h; (**b**) at the end of testing, 72 h.

Inhibition was calculated using the following formula: (arithmetic mean length of roots in control−arithmetic mean length of roots in sample)/(arithmetic mean length of roots in control).

3. Results and Discussion

3.1. Results of the Semichronic Toxicity Test on Mustard (Sinapis alba L.)

Based on the results of the inhibition of growth of the roots of the tested plants for each sample, it can be stated that the undiluted samples of pulp waste mixtures caused inhibition of growth of the roots of the tested plants, as can be seen from Figure 2. In all examined undiluted samples (A–D), there was more than 80% inhibition of *Sinapis alba* root growth; therefore, the liquid waste was phytotoxic. Inhibition in undiluted samples ranged from 81.63% to 100%. The application of this mixture of waste in agriculture without any subsequent treatment represents an environmental risk.

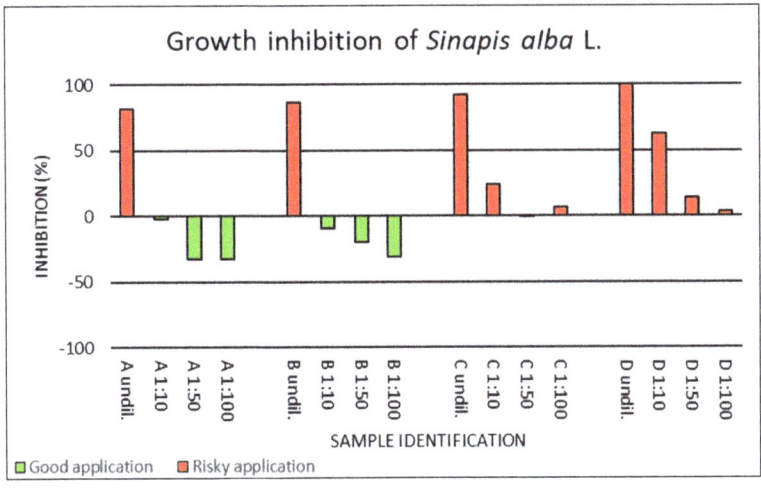

Figure 2. Graphical evaluation of the phytotoxicity of a mixture of pulp waste on mustard.

Better results were obtained with diluted samples A and B in ratios 1:50 and 1:100, where they even showed a stimulating effect on plant development. Compared to the control, stimulation reached values of 19.79–32.72%. Negative effects on growth of *Sinapis alba* were confirmed for sample D in dilution of 1:10 as the inhibition of plant development and growth was found to be 62.92%, i.e., 12.92% more than the ecotoxicity limit according to Decree No. 273/2021 Coll. [11].

3.2. Results of the Semichronic Toxicity Test on Cress (Lepidium sativum L.)

The results of measurement for *Lepidium sativum* are shown in Figure 3. The results from the undiluted samples showed inhibition of growth similar to the results for *Sinapis alba*. The measurement results also showed that in samples A, B, and C, when they were diluted 1:10, the growth of the roots of the tested plants was stimulated. However, phytotoxic results were achieved in sample A at a dilution of 1:10 and sample D at dilutions of 1:10 and 1:50. Sample A in dilution of 1:10 showed stimulation of plant growth and development by 86.63% compared to control samples, which, however, was in accordance with the legislative requirement for evaluation tests according to Decree 273/2021 Coll. [11]. The results must be evaluated as an inhibition phenomenon as the detected stimulation exceeded the standard 50% limit value by 36.63%. Samples B and C diluted in a ratio of 1:10 had stimulating effects on the growth of *Lepidium sativum* roots. More diluted samples B and C inhibited the growth of the root, with the inhibition reaching a value up to 19.19%. The most toxic was again sample D, which contained a nutrient medium for microorganisms.

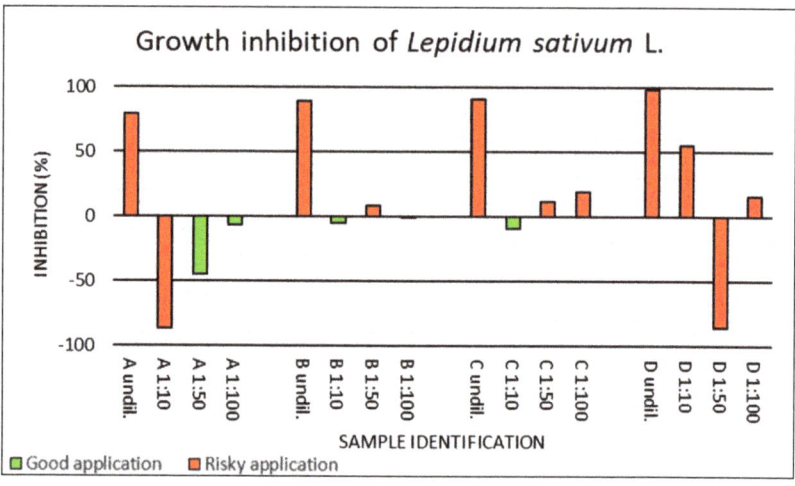

Figure 3. Graphical evaluation of the phytotoxicity of a mixture of pulp waste on cress.

3.3. Discusion of the Results

Overall, the results showed that the undiluted samples of the pulp waste mixture had a negative effect on root growth of mustard and cress, and according to the Czech legislation, Decree 273/2021 Coll. [11], we consider these samples phytotoxic due to the above-limit results (more than 50% inhibition or stimulation). The overall results are shown in Table 2. In sample A with 1:10 dilution, negative effects on the growth of cress were also detected, and we can also consider this sample of waste mixture as phytotoxic. In the other cases of sample A, no phytotoxic effects of waste mixtures on the examined plants were detected, as shown in Table 2.

Table 2. Evaluation of the phytotoxicity of a mixture of pulp waste according to Decree 273/2021 Coll.

Sample Identification	Dilution	Mustard Results	Cress Results	Complete Results
A	Undiluted	×	×	×
	1:10	✔	×	×
	1:50	✔	✔	✔
	1:100	✔	✔	✔
B	Undiluted	×	×	×
	1:10	✔	✔	✔
	1:50	✔	✔	✔
	1:100	✔	✔	✔
C	Undiluted	×	×	×
	1:10	✔	✔	✔
	1:50	✔	✔	✔
	1:100	✔	✔	✔
D	Undiluted	×	×	×
	1:10	×	×	×
	1:50	✔	×	×
	1:100	✔	✔	✔

Notes: [✔] nonphytotoxic, [×] phytotoxic.

Samples diluted with tryptone–soybean culture medium, undiluted samples, and samples diluted in ratios of 1:10 and 1:50 were found to be very risky. Because the tryptone-soybean broth supports the growth of indigenous microorganisms, their increased activity can be assumed in all samples in the D series that were diluted with the culture medium. Nutritional support for the growth of microorganisms in the tested samples will be shown by increased production of their secondary metabolites. In accordance with previously published studies [12–15], it can be assumed that these metabolites exhibit a high level of diversity in their properties, such as structures, phytotoxic activities, and toxicity modes. In particular, secondary metabolites of fungi are poisonous substances to plants produced through naturally occurring biochemical reactions of these microorganisms [16]. Therefore, the phytotoxicity of the mentioned samples of the D series was probably increased by the metabolites of the microorganisms found in this medium. In addition, some expert studies have shown that synergistic effects of the toxic biomolecules produced can occur in different microorganisms, as evidenced, for example, in [16].

4. Conclusions

This research confirms the necessity of monitoring waste from pulp production, especially from the point of view of its reuse in other sectors, e.g., agriculture. The research also showed that in some cases, the application of pulp waste as fertilizer can lead to significant deterioration in the quality and quantity of flora. Undiluted pulp waste was found to slow the growth of mustard (*Sinapis alba* L.) and cress (*Lepidium sativum* L.) plants. Choosing the optimal dilution of waste with water can be key to preventing its negative effect. This research confirmed the negative effect of undiluted pulp waste samples. In the case of sample A diluted in a ratio of 1:10, phytotoxic effects on cress were observed. Samples diluted with tryptone–soybean medium also showed phytotoxicity due to the activity of internal microorganisms.

Funding: The research was funded by the Project for Specific University Research (SGS) No. SP2023/4 by the Faculty of Mining and Geology of VŠB, Technical University of Ostrava.

Institutional Review Board Statement: Not applicable.

Informed Consent Statement: Not applicable.

Data Availability Statement: Data are contained within the article.

Conflicts of Interest: The author declares no conflict of interest.

References

1. Pertile, E.; Surovka, D.; Balcařík, L. Potential use of saline wastewater for the agglomeration of fluidized fly ash. *Inz. Miner.* **2018**, *2*, 273–280. [CrossRef]
2. Pertile, E.; Václavík, V.; Dvorský, T.; Heviánková, S. The removal of residual concentration of hazardous metals in wastewater from a neutralization station using biosorbent—A case study company Gutra, Czech Republic. *Int. J. Environ. Res. Public Health* **2020**, *17*, 7225. [CrossRef] [PubMed]
3. Šimonovičová, A.; Ferianc, P.; Vojtková, H.; Pangallo, D.; Hanajík, P.; Kraková, L.; Feketeová, Z.; Čerňanský, S.; Okenicová, L.; Žemberyová, M.; et al. Alkaline Technosol contaminated by former mining activity and its culturable autochthonous microbiota. *Chemosphere* **2017**, *171*, 89–96. [CrossRef] [PubMed]
4. Lachka, M.; Soltisova, K.; Nosalova, L.; Timkova, I.; Pevna, V.; Willner, J.; Janakova, I.; Luptakova, A.; Sedlakova-Kadukova, J. Metal-containing landfills as a source of antibiotic tolerance. *Environ. Monit. Assess.* **2023**, *195*, 262. [CrossRef]
5. Pertile, E.; Zamarsky, P. An alternative method of removing Cr (VI) from aquatic solution using chemically modified cone biomass and *Fomitopsis pinicola*. In Proceedings of the IOP Conference Series: Earth and Environmental Science, Changchun, China, 21–23 August 2020; Volume 444, p. 012043. [CrossRef]
6. Pertile, E.; Dvorský, T.; Václavík, V.; Heviánková, S. Use of different types of biosorbents to remove Cr (VI) from aqueous solution. *Life* **2021**, *11*, 240. [CrossRef] [PubMed]
7. Vojtková, H.; Janáková, I. Research of waste dump water mutagenicity of bacterial detection system SOS chromotest. *Water Sci. Technol.* **2011**, *63*, 2833–2837. [CrossRef] [PubMed]
8. Antonkiewicz, J.; Baran, A.; Pelka, R.; Wisla-Świder, A.; Nowak, E.; Konieczka, P. A mixture of cellulose production waste with municipal sewage as new material for an ecological management of wastes. *Ecotoxicol. Environ. Saf.* **2019**, *169*, 607–614. [CrossRef] [PubMed]
9. Smeriglio, A.; Trombetta, D.; Cornara, L.; Valussi, M.; De Feo, V.; Caputo, L. Characterization and phytotoxicity assessment of essential oils from plant byproducts. *Molecules* **2019**, *24*, 2941. [CrossRef] [PubMed]
10. ČSN EN ISO 18763 (836447); Kvalita Půdy-Stanovení Toxických Účinků Znečišťujících Látek Na Klíčení a Růst Raných Stadií Vyšších Rostlin. Český Normalizační Institute: Praha, Czech Republic, 2020.
11. *Vyhláška č. 273/2021 Sb.: Vyhláška o Podrobnostech Nakládání s Odpady*; Ministerstvo Životního Prostředí: Praha, Czech Republic, 2021.
12. Xu, D.; Xue, M.; Shen, Z.; Jia, X.; Hou, X.; Lai, D.; Zhou, L. Phytotoxic Secondary Metabolites from Fungi. *Toxins* **2021**, *13*, 261. [CrossRef] [PubMed]
13. Evidente, A.; Cimino, A.; Masi, M. Phytotoxins produced by pathogenic fungi of agrarian plants. *Phytochem. Rev.* **2019**, *18*, 843–870. [CrossRef]
14. Šimonovičová, A.; Kupka, D.; Nosalj, S.; Kraková, L.; Drahovská, H.; Bártová, Z.; Vojtková, H.; Boturová, K.; Pangallo, D. Differences in metabolites production using the Biolog FF Microplate™ system with an emphasis on some organic acids of *Aspergillus niger* wild type strains. *Biologia* **2020**, *75*, 1537–1546. [CrossRef]
15. Tyc, O.; Song, C.; Dickschat, J.S.; Vos, M.; Garbeva, P. The ecological role of volatile and soluble secondary metabolites produced by soil bacteria. *Trends Microbiol.* **2017**, *25*, 280–292. [CrossRef] [PubMed]
16. Taba, K.; Honsho, M.; Asami, Y.; Iwasaki, H.; Nonaka, K.; Watanabe, Y.; Iwatsuki, M.; Matsui, H.; Hanaki, H.; Teruya, T.; et al. Synergistic effect of secondary metabolites isolated from Pestalotiopsis sp. FKR0115 in overcoming β-lactam resistance in MRSA. *J. Gen. Appl. Microbiol.* **2023**. [CrossRef]

Disclaimer/Publisher's Note: The statements, opinions and data contained in all publications are solely those of the individual author(s) and contributor(s) and not of MDPI and/or the editor(s). MDPI and/or the editor(s) disclaim responsibility for any injury to people or property resulting from any ideas, methods, instructions or products referred to in the content.

Proceeding Paper

The Determination of Soil Microbial Biomass Carbon and Adenosine Triphosphate Concentrations at Different Temperatures [†]

Lenka Bobuľská [1,*], Lenka Demková [1] and Tomáš Lošák [2,3]

1 Department of Ecology, Faculty of Humanities and Natural Sciences, University of Prešov, 17. November 1, 080 01 Prešov, Slovakia; lenka.demkova@unipo.sk
2 Department of Environmentalistics and Natural Resources, Faculty of Regional Development and Interna-tional Studies, Mendel University in Brno, Zemědělská 1, 613 00 Brno, Czech Republic; tomas.losak@mendelu.cz or tlosak@fzt.jcu.cz
3 Department of Agroecosystems, Faculty of Agriculture and Technology, University of South Bohemia in Ceske Budejovice, Branišovská 1645/31A, 370 05 Ceske Budejovice, Czech Republic
* Correspondence: lenka.bobulska@unipo.sk
† Presented at the 4th International Conference on Advances in Environmental Engineering, Ostrava, Czech Republic, 20–22 November 2023.

Abstract: Freezing soil samples at subzero temperatures is a commonly employed preservation method in soil science to halt microbial activity and enzymatic processes, preserving the sample's composition and structure. The concentration of adenosine triphosphate (ATP) and the soil microbial biomass carbon content were determined from soil samples stored at +4 °C, −21 °C and −80 °C for 24 h, 7 days, and 20 days. The results showed that the ATP in the soil was not significantly affected by temperature or storage time. Larger differences were observed in the carbon content of microbial biomass, where the amount of this parameter increased by 14.2% after 20 days of the experiment.

Keywords: soil freezing; microbial activity; enzymes; soil storage

1. Introduction

To ensure the sustainability of agroecosystems, with an emphasis on soil quality and health, it is important to monitor changes in soil properties. Important factors for this assessment are physical, chemical, and biological characteristics; vegetation cover; and management practices [1]. In all microbiological studies, it is preferred to work with freshly collected soil samples; however, for practical reasons, this is not always possible. Freezing is the most frequently used method for preserving soil for microbiological analysis. Storage temperature and length can have different effects on the total number of monitored parameters in soil samples [2,3]. Soil microbial biomass is an essential component of soil and is responsible for nutrient cycling, energy flow, and the regulation of soil organic matter conversion [4]. To measure the biomass of microorganisms, the determination of soil adenosine triphosphate (ATP), which is a part of all living forms, is a universal form of energy in biological systems, and is rapidly degraded in dead cells, is required. ATP fulfills very important functions in the soil, as it is a storehouse and transports energy, participates in the synthesis of DNA and RNA, and in the biosynthesis of cells and the regulation of cell metabolism [2,5]. ATP is a critical biomolecule in soil microbiology and is indicative of soil microbial activity [6]. It serves as an energy currency for various soil microorganisms, facilitating essential biochemical processes [7]. The measurement of soil ATP can provide insights into soil health and fertility, as higher ATP levels often correlate with greater microbial activity and nutrient cycling [8]. Conversely, low ATP levels may indicate poor soil conditions or degradation [9]. The ATP content in soil can fluctuate seasonally, with higher levels typically observed during periods of increased microbial activity, such as in the spring and summer [10]. This variability underscores the dynamic nature of soil ecosystems [11]. Researchers commonly employ bioluminescence assays to quantify soil ATP levels, a method that relies on the light emission produced by ATP when it reacts with luciferase enzymes [12]. Soil management practices, such as organic matter additions and reduced tillage, can influence soil ATP levels by promoting microbial growth and activity [13]. Understanding soil ATP dynamics is essential for sustainable agriculture and soil conservation efforts, as it can guide decisions regarding nutrient management and soil health improvement [14]. However, it is crucial to note that while soil ATP is a valuable indicator, it should be used in conjunction with other soil health metrics for a comprehensive assessment [15].

The aim of this study was to monitor the amount of microbial biomass carbon and ATP concentration in soils stored at +4 °C, −21 °C and −80 °C for 24 h, 7 days, and 20 days.

2. Material and Methods

The collection of soil samples for the determination of ATP and microbial biomass carbon (C_{mic}) was carried out on research plots of Prešov University study field on a permanent grassland. The investigated location was characterized by good organic carbon content (4.1%) and a neutral soil reaction (pH 6.9). The dominant soil type within the research locality was Fluvisol, which is typical of alluvial plains, river fans, valleys, and tidal marshes on all continents and in all climate zones. The university study field focuses on the cultivation of medicinal plants used in research. The collected soil samples were manually freed of plant and animal residues, sieved (<2 mm), and adjusted to a 40% soil-holding capacity (WHC). Subsequently, these samples were incubated at three different temperature ranges (+4 °C, −21 °C and −80 °C) under aerobic conditions in the dark. Microbial biomass carbon and soil ATP concentrations were determined at all temperature ranges every time after 24 h, 7 days, and 21 days in triplicate. C_{mic} was determined by the fumigation extraction method [16], which uses chloroform vapor to completely kill microorganisms, and the released organic carbon was easily extractable with 0.5 M K_2SO_4. ATP in the soil was extracted using a TCA reagent, the main component of which was 0.5 M trichloroacetic acid, and the determination of ATP itself was carried out by scintillation using the enzyme luciferin-luciferase [17].

3. Results and Discussion

The results showed that ATP in the soil was not significantly affected by temperature or storage time. The concentration did not change significantly during incubation (Figure 1). Some studies [18,19] have found that soil ATP concentration tends to decrease with decreasing temperature, particularly in colder climates. This may be due to reduced microbial activity in cold soils. Those studies have shown that soil ATP levels can vary seasonally, with higher concentrations during the warmer months and lower concentrations during the winter. This is often linked to changes in microbial activity and plant root exudates. Some research [20,21] has demonstrated that the activity of soil microbes responsible for ATP production can be temperature dependent. Different types of microbes may respond differently to temperature changes, affecting the overall ATP levels in the soil, which was not shown in our study. Scientific investigations have examined the broader ecological consequences of temperature changes on soil ATP dynamics. Changes in temperature can influence not only ATP levels but also nutrient cycling, carbon storage, and overall ecosystem functioning in soil [22].

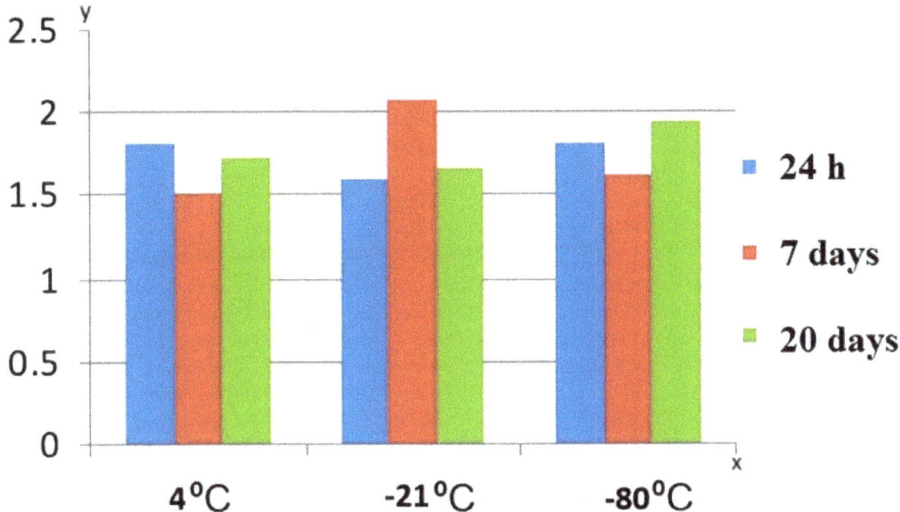

Figure 1. ATP concentration (nmol ATP g^{-1}) at the different temperature ranges.

In our study, larger differences were observed in the carbon content of microbial biomass, where the amount of this parameter increased by 14.2% after 20 days of the experiment, which could probably be due to the decomposability of organic soil matter and the subsequent synthesis of new biomass during the one-day incubation before the actual determination (Figure 2).

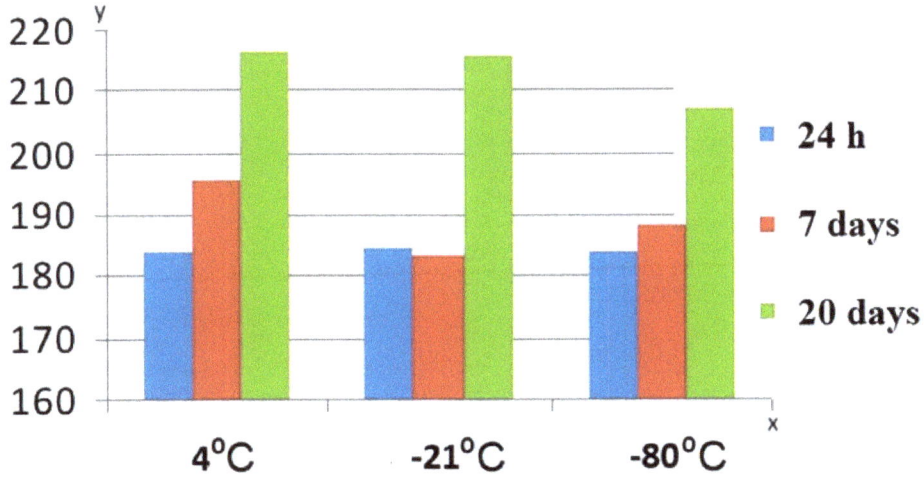

Figure 2. Microbial biomass carbon ($\mu gC.g^{-1}$) at the different temperature ranges.

Microbial biomass exhibited a pronounced response to temperature variations, with increased temperatures generally leading to higher microbial biomass production [23], while in temperate ecosystems, microbial biomass tended to peak during the spring and fall when temperatures are moderate and declined during hot summer and cold winter periods [24]. Studies focusing on arctic soils have shown that microbial biomass declines significantly as temperatures rise due to the sensitivity of cold-adapted microbial communities to warming [25]. Temperature increases in tropical ecosystems can stimulate microbial biomass growth, but this effect may be limited by nutrient availability [26]. However, at high temperatures exceeding 40 °C, microbial biomass in soils can decline sharply due to thermal stress and the denaturation of enzymes and cellular proteins [27]. Understanding the intricate relationship between temperature and microbial biomass is essential for predicting how climate change may impact ecosystem functioning and biogeochemical cycling [28].

4. Conclusions

In this study, we investigated the impact of storage conditions on soil ATP and microbial biomass, and our results consistently demonstrated that storage duration and temperature did not exert any significant effect on these parameters. Larger differences were observed in the carbon content of microbial biomass. Our findings suggest that the soil ATP content and microbial biomass remain remarkably stable over time, irrespective of storage conditions, indicating the resilience of these biological indicators to environmental variation. In summary, our research reinforces the notion that soil ATP and microbial biomass are robust and unaffected by storage conditions, confirming their utility as reliable indicators for assessing soil microbial activity and ecosystem health over time. These findings emphasize the stability and reliability of soil ATP and microbial biomass measurements under various storage conditions, which make them valuable tools for ecological and environmental research.

Author Contributions: Conceptualization, L.B.; methodology, L.B., L.D. and T.L.; formal analysis, L.D.; investigation, L.B., L.D. and T.L; resources, L.B.; data curation, L.D.; writing—original draft preparation, L.B.; writing—review and editing, L.D. and T.L. All authors have read and agreed to the published version of the manuscript.

Funding: This research was funded by Slovak Scientific Agency VEGA No. 2/0018/20 and Slovak Research and Development Agency APVV-20-0140.

Institutional Review Board Statement: Not applicable.

Informed Consent Statement: Not applicable.

Data Availability Statement: Not applicable.

Conflicts of Interest: The authors declare no conflict of interest.

References

1. Stenberg, B.; Johansson, M.; Pell, M.; Sjödahl-Svensson, K.; Stenström, J.; Torstensson, L. Microbial biomass and activities in soil as affected by frozen and cold storage. *Soil Biol. Biochem.* **1998**, *30*, 393–402. [CrossRef]
2. Contin, M.; Corcimaru, S.; De Nobili, M.; Brookes, P.C. Temperature changes and the ATP concentration of the soil microbial biomass. *Soil Biol. Biochem.* **2000**, *32*, 1219–1225. [CrossRef]
3. Castellazzi, M.S.; Brookes, P.C.; Jenkinson, D.S. Distribution of microbial biomass down soil profiles under regenerating woodland. *Soil Biol. Biochem.* **2004**, *36*, 1485–1489. [CrossRef]
4. Li, X.; Chen, Z. Soil microbial biomass C and N along a climatic transect in the Mongolian steppe. *Biol. Fertil. Soils* **2004**, *39*, 344–351. [CrossRef]
5. Alef, K.; Nannipieri, P. *Methods in Applied Soil Microbiology and Biochemistry*, 1st ed.; Academic Press: Cambridge, MA, USA, 1995; p. 608.
6. Hargreaves, S.K.; Hofmockel, K.S. Adenosine triphosphate (ATP) as a subsurface proxy for soil microbial biomass and community metabolic activity. *Soil Biol. Biochem.* **2014**, *70*, 183–194.
7. Philippot, L.; Hallin, S.; Schloter, M. Ecology of denitrifying prokaryotes in agricultural soil. *Adv. Agron.* **2007**, *96*, 249–305.
8. Contin, M.; Todd, A.D.; Brookes, P.C. the ATP concentration in the oil microbial biomass. *Soil Biol. Biochem.* **2001**, *33*, 701–704. [CrossRef]
9. Bünemann, E.K.; Schwenke, G.D.; Van Zwieten, L. Impact of agricultural inputs on soil organisms—A review. *Aust. J. Soil Res.* **2006**, *44*, 379–406. [CrossRef]
10. Marschner, P.; Yang, C.H.; Lieberei, R.; Crowley, D.E. Soil and plant specific effects on bacterial community composition in the rhizosphere. *Soil Biol. Biochem.* **2001**, *33*, 1437–1445. [CrossRef]

11. Nannipieri, P.; Ascher, J.; Ceccherini, M.T.; Landi, L.; Pietramellara, G.; Renella, G. Microbial diversity and soil functions. *Eur. J. Soil Sci.* **2003**, *54*, 655–670. [CrossRef]
12. Karl, D.M.; Bailiff, M.D. The measurement and distribution of dissolved adenosine triphosphate in oligotrophic marine environments. *Mar. Biol.* **1989**, *101*, 611–617.
13. Bossio, D.A.; Scow, K.M.; Gunapala, N.; Graham, K.J. Determinants of soil microbial communities: Effects of agricultural management, season, and soil type on phospholipid fatty acid profiles. *Microb. Ecol.* **1998**, *36*, 1–12. [CrossRef] [PubMed]
14. Lehmann, J.; Solomon, D.; Kinyangi, J.; Dathe, L.; Wirick, S.; Jacobsen, C. Spatial complexity of soil organic matter forms at nanometre scales. *Nat. Geosci.* **2007**, *1*, 238–242. [CrossRef]
15. Creamer, R.E.; Hannula, S.E.; Leeuwen, J.P.V.; Stone, D.; Rutgers, M.; Schmelz, R.M.; de Ruiter, P.C.; Bohse Hendriksen, N.; Bolger, T.; Bouffaud, M.L.; et al. Ecological network analysis reveals the inter-connection between soil biodiversity and ecosystem function as affected by land use across Europe. *Appl. Soil Ecol.* **2016**, *97*, 112–124. [CrossRef]
16. Vance, E.D.; Brookes, P.C.; Jenkinson, D.S. An extraction method for measuring soil microbial biomass C. *Soil Biol. Biochem.* **1987**, *19*, 703–707. [CrossRef]
17. Jenkinson, D.S.; Oades, J.M. A method for measuring adenosine triphosphate in soil. *Soil Biol. Biochem.* **1979**, *11*, 193–199. [CrossRef]
18. Fioretto, A.; Papa, S.; Pellegrino, A.; Ferringo, A. Microbial activities in soils of a Mediterranean ecosystem in different successional stages. *Soil Biol. Biochem.* **2009**, *41*, 2061–2068. [CrossRef]
19. Vance, E.D.; Chapin III, F.S. Substrate limitations to microbial activity in taiga forest floors. *Soil Biol. Biochem.* **2001**, *33*, 173–188. [CrossRef]
20. Bérard, A.; Sassi, M.B.; Kaisermann, A.; Renault, P. Soil microbial community responses to heat wave components: Drought and high temperature. *Clim. Res.* **2015**, *66*, 243–264. [CrossRef]
21. Karaca, A.; Cetin, S.C.; Turgay, O.C.; Kizilkaya, R. Soil Enzymes as Indication of Soil Quality. In *Soil Enzymology. Soil Biology*; Shukla, G., Varma, A., Eds.; Springer: Berlin/Heidelberg, Germany, 2010; Volume 22, pp. 119–148.
22. Allen, A.P.; Gillooly, J.F. Towards an integration of ecological stoichiometry and the metabolic theory of ecology to better understand nutrient cycling. *Ecol. Lett.* **2009**, *12*, 369–384. [CrossRef]
23. Davidson, E.A.; Janssens, I.A. Temperature sensitivity of soil carbon decomposition and feedbacks to climate change. *Nature* **2006**, *440*, 165–173. [CrossRef]
24. Bardgett, R.D.; Lovell, R.D. Seasonal changes in soil microbial communities along a fertility gradient of temperate grasslands. *Soil Biol. Biochem.* **1999**, *31*, 1021–1030. [CrossRef]
25. Zhang, Y.; Chen, H.Y.; Reich, P.B. Forest productivity increases with evenness, species richness and trait variation: A global meta-analysis. *J. Ecol.* **2005**, *93*, 253–266. [CrossRef]
26. Cleveland, C.C.; Nemergut, D.R.; Schmidt, S.K.; Townsend, A.R. Increases in soil respiration following labile carbon additions linked to rapid shifts in soil microbial community composition. *Biogeochemistry* **2007**, *82*, 229–240. [CrossRef]
27. Alam, M.S.; Xiao, L.; Zhao, M.; Liu, G.; Zhang, Y. Changes in microbial biomass and community structure in soils of a semi-arid grassland ecosystem following a simulated extreme rainfall event. *Soil Biol. Biochem.* **2017**, *105*, 89–98.
28. Fierer, N.; Bradford, M.A.; Jackson, R.B. Toward an ecological classification of soil bacteria. *Ecology* **2007**, *88*, 1354–1364. [CrossRef] [PubMed]

Disclaimer/Publisher's Note: The statements, opinions and data contained in all publications are solely those of the individual author(s) and contributor(s) and not of MDPI and/or the editor(s). MDPI and/or the editor(s) disclaim responsibility for any injury to people or property resulting from any ideas, methods, instructions or products referred to in the content.

Proceeding Paper

The Design for the Reconstruction of Settling Tanks [†]

Pavel Buchta, Vojtěch Václavík * and Tomaš Dvorský

Department of Environmental Engineering, Faculty of Mining and Geology, VSB—Technical University of Ostrava, 17. listopadu 2172/15, 708 00 Ostrava-Poruba, Czech Republic; pavel.buchta.st@vsb.cz (P.B.); tomas.dvorsky@vsb.cz (T.D.)
* Correspondence: vojtech.vaclavik@vsb.cz
[†] Presented at the 4th International Conference on Advances in Environmental Engineering, Ostrava, Czech Republic, 20–22 November 2023.

Abstract: This article describes the design of the reconstruction of the mechanical and technological equipment of the settling tanks at the wastewater treatment plant in the municipality of Boskovice. The reconstruction was focused on the flocculation cylinder, the travelling bridge, the bridge drive, and the collection of the settling tank floating sludge.

Keywords: reconstruction; settling tank; flocculation cylinder; travelling bridge; bridge drive; floating sludge

1. Introduction

Settling tanks are the final construction objects of the biological wastewater treatment process, where active sludge is separated from wastewater and purified water flows into the receiving body, assuming that the facility in question is a mechanical–biological wastewater treatment plant. In some cases, the functionality of the settling tanks is demonstrated by numerical models [1–3]. After stabilisation [4] and dewatering, the treated sludge from the settling tanks can be used for reclamation works but also for the pyrolysis process [5].

This article deals with the design of the reconstruction of the machinery and technological equipment of the circular settling tanks of the Boskovice WWTP. It is a part of the intensification of this treatment plant, with a required capacity of 20,000 equivalent inhabitants.

The design was carried out with an emphasis on high reliability, long service life, and minimal maintenance. This innovation also included a proposal to replace the surface skimming with floating sludge spraying; the outlet of treated water into a reinforced concrete trough with a screw collector with a pump; the outlet of this into the floating sludge trough; and the outlet of the treated water using a submerged pipe with suction holes at the top.

2. Description of the Current State of the Settling Tanks before Reconstruction

The WWTP and its facilities are located in the cadastral territory of the town of Boskovice. It belongs to the Association of Water Supply and Sewerage of Towns and Municipalities. It serves for the treatment of wastewater from the town of Boskovice and some of its local parts, processing of waste from the sewage treatment, and disposal of waste from cesspools and septic tanks. Due to its size, it is also used to process sludge from the nearby small WWTPs. A separate sewerage system is used in the town. This means that sewage and rainwater are discharged separately through the sewerage network.

The existing WWTP was at the limit of its capacity and was also hydraulically overloaded by more than 250 m³/d. In accordance with the amendment to Water Act No. 254/2001 [6], it is probably necessary to increase the collection of wastewater, which already constitutes a significant part of the load. The WWTP was intensified to ensure proper treatment of wastewater from the town of Boskovice. At the same time, the collection of

wastewater from cesspits and septic tanks, as well as the treatment of liquid sludge from smaller wastewater treatment plants, had to be enabled in order to meet the requirements of Government Regulation No. 401/2015 [7]. The investment project also included the reconstruction of the two existing settling tanks, which are used to capture biological sludge from the activation tanks by sedimentation. They are of circular shape with an internal diameter of 15 m.

The technological equipment consisted of a steel central column, on the top of which a ring collector was installed; a roofed flocculation cylinder with a deflector; a travelling bridge with bottom and surface skimming; a bridge drive via a rack bar; and a rail travel. In addition, it also included the collection of floating sludge, the automatic cleaning of the overflow edges and the tank crest located on the travelling bridge, and the outflow of the purified water through a collection trough with an overflow edge and a baffle.

The investor's requirement was to address the inadequate removal of floating sludge. This sludge was generated because of the presence of certain filamentous micro-organisms, characterised by a strongly hydrophobic cell surface and the formation of biologically surface-active substances. During aeration in the activation tanks, these microorganisms caused the formation of a biological foam in which other microorganisms of the activated sludge were trapped.

The existing system consisted of fixed scraping plates, suspended from a swing bridge and arranged so that the sludge was moved by centrifugal force to the perimeter of the tank where it was carried to the floating sludge cesspool by a tilting scraper. The ST level behind the scraping plates was sprayed with partially treated water through a submersible pump and a pipe distribution system suspended from the bridge railing. The spread of floating sludge is shown in Figure 1. Despite all these measures, the drainage of the floating sludge was not satisfactory.

Figure 1. Spread of floating sludge.

3. Reconstruction Plan

In the existing settling tanks, a complete replacement of the mechanical and technological equipment with new ones was proposed. It consists of a new supporting central steel column, flocculation cylinder, rectifying deflector, travelling bridge with bottom scraping, and a system of floating sludge removal, using a surface screw and a pump with a height-adjustable overflow cesspool. The outlet flows into the floating sludge trough. The hot-dip galvanised steel structure has been replaced by stainless steel. The outflow of purified water from the ST was redesigned. Instead of the existing reinforced concrete outlet troughs, an underwater collection using submerged perforated pipes was designed.

The level in the settling tanks is found at the overflow edge in the treated water discharge manholes, where the water continues to drain to the Parshall flume.

3.1. Flocculation Cylinder

One of the key features of a flocculation ST is the regulated flow in the flocculation cylinder. The incoming activation mixture is ejected from the inflow column through the outlet openings to the wall of the flocculation cylinder. The inflowing activation mixture drops at the flocculation cylinder wall due to its density and returns upwards through the centre of the flocculation cylinder. This arrangement results in flocculation, i.e., the adherence of small flakes of activated sludge to large ones. The second characteristic feature of this arrangement is the limitation of the flow density. Below the edge of the flocculation cylinder, part of the mixture corresponding to the outflow quantity is separated and goes into the ST separation area. These features must be maintained for the proper operation and high efficiency of the ST [8].

Good flocculation is a prerequisite for effective sludge separation in the ST. By flocculation, we mean precipitation and floc formation. The diameter of the flocculation cylinder should be approximately 33% of the ST diameter. The immersion depth is recommended to be around 67% of the tank depth.

The flocculation cylinder has been designed as a lightweight structure suspended from a central column, consisting of four identical sections connected by bolted joints. The wall is made of PE foil and is reinforced around the perimeter with stainless steel profiles. The cylinder is height-adjustable via four threaded rods. The designed diameter is 3.8 m and the height is 3.65 m. The cylinder is sunk to a depth of 3.4 m. During the designing process, an all-stainless steel flocculation cylinder was discussed. The advantage is its longer lifetime; the disadvantage is its higher weight and higher investment costs. The next point for discussion was the roofing of the flocculation cylinder. The advantage of this measure is the partial reduction of smell; the disadvantage is the lack of the possibility to visually check the level in the cylinder.

3.2. Travelling Bridge

It is used to operate the ST, as well as acting a carrier for the floating sludge collector and for scraping the bottom of the tank.

It is a frame structure with a length of 7750 mm and a width of 1260 mm, mounted with one end on the inflow column and the other on the head of the tank wall. The rotational movement is enabled by a friction bearing connecting the bridge to the column; two driven travelling wheels above the tank wall; and one driving wheel suspended from the bottom side of the bridge, which is pressed against the inner vertical wall of the ST. The friction bearing is maintenance-free. The material used is polyethylene and the diameter of the bearing is 630 mm. The driven travelling wheels are 405 mm in diameter. A tank head scraper is mounted in front of these wheels. The driving wheel is suspended under the bridge. It rotates the travelling bridge by a rolling motion on the inner wall of the tank. The floor is made of stainless steel grating. The railings are 1100 mm high. They are divided in the middle of the railing height by a dividing handrail and in the lower part by a kick plate.

At the end of the bridge, near the middle of the ST, there is an opening gate. On the handrails, there are brackets for the floating sludge collector, bowls with a pump for the floating sludge, a winch for operating this equipment, and an electrical switchboard. A floating sludge pipe is routed under the bridge to allow the sludge to flow into a trough at the ST wall or into a flocculation cylinder. The bottom beam is fitted with draw rods used to scrape the bottom of the tank and sensors, which are used to start and stop the pumping of floating sludge into the sump. All devices are connected to the bridge via sleeves and are, therefore, adjustable. A set of brushes is suspended on the bridge above the outlet pipe to clean the upper part of the pipe.

Access to the bridge is possible by means of tipping steps. There are grab handles on the handrails. An emergency start/stop button is installed at the point of entry for the quick stopping of the bridge.

3.3. Bridge Drive Mechanism

It is used to set the travelling bridge, the floating sludge collection, and the bottom scraping into motion. It consists of a carrier plate, drive wheel, and drive. It is designed to have the power of 0.25 kW and 0.7 A. The starting current is 2.2 A. It operates non-stop as standard. In the event of a power failure, an automatic restart occurs. Manual control of the drive is enabled in the bridge switchboard. The drive wheel has Ø of 530 mm. To ensure correct operation, a pressure system made of flexible rubbers has been designed. The wheel tyre is fitted with a winter tread.

3.4. Floating Sludge Collection

The device is used to remove floating sludge from the ST surface and to discharge it outside the ST area.

The floating sludge collector system consists of two floating screw conveyors. The rotating motion of the screws moves the sludge into a hydraulically formed scraper container, which is connected to the screws.

Due to the floating arrangement of the screws and the fact that the floating sludge container is connected to the screws, the overflow height remains constant, even when the water level fluctuates.

In addition, the overflow height relative to the floating screws can be precisely adjusted and set using an adjustment wrench. The overflow height is, thus, used to set the desired amount of discharged sludge on one side and the desired mixing ratio of floating sludge, water, and air on the other side.

In this way, floating sludge layers of any consistency and thickness, even sludge bolsters and larger closed sludge covers, can be safely and quickly pumped out by a submersible pump fitted in the next part.

Water level fluctuations of up to 500 mm are possible, e.g., during flow changes and especially with submerged discharge devices or under wind loads, where the floating sludge scraper balances the sludge with millimetre precision using the weight-balanced bearings of the rotary screw conveyor arms.

Thanks to the floating screw arrangement and the connection of the scraping hopper to the floating screws, the SSR floating sludge scrapers ensure that the scraping hopper functions consistently without the need for operator intervention, even with these changes.

The modular design of the floating sludge removal system and standardized fasteners allow the simple and stable installation of the system in any separator, on any scraper design, and for any discharge system, with almost no restrictions.

The system is particularly suitable for retrofitting to existing circular scraping systems as it fully automatically compensates for any existing design inaccuracies caused by the floating arrangement.

The independent and precise adjustment of the floating screws to fluctuating water levels ensures the safe and simple fully automatic operation of the floating sludge collector.

With the aid of a time-controlled floating sludge pump, the mixture of sludge, water, and air is extracted at the required concentration and with a floating sludge volume of up to 50 m^3/day.

All materials used in the floating sludge collector are carefully selected for extremely long life and minimal maintenance.

Stainless steel is used for all wetted parts, UHMWPE for the maintenance-free bearings, UHMWPE and POM for the maintenance-free chain drive, and weatherproof elastomers resistant to atmospheric exposure for the seals. Commercially available materials are used only for the floating sludge collectors, submersible pump, and worm gear motor.

The drive unit is maintenance-free and is equipped with a lifetime lubricated gear motor of 0.18 kW and 0.59 A. The starting current is 2.14 A. The output shaft of the gear motor is in a special stainless steel design to prevent corrosion of the sealing surfaces and subsequent damage to the seal itself. Since the bearings are completely unloaded using the floating method, operation and maintenance work is limited to infrequent visual inspections, occasional re-setting of the slurry tank overflow height, and annual oil changes for the submersible motor pump.

For service work on the pump, it can be easily raised to operating level by means of a lifting device, sliding guide and special elastomer seals on the suction and pressure sides.

The new mechanical and technological equipment of the settling tank are presented in Figure 2.

Figure 2. New mechanical and technological equipment of the settling tank.

4. Conclusions

The design of the mechanical and technological equipment of the settling tank for the Boskovice wastewater treatment plant was carried out in accordance with the technological principles valid for this type of equipment.

A static assessment was carried out for the inflow column, which resulted in the conclusion that the given cross-section, under the worst-case condition of an empty tank and full load, is satisfactory.

A leakage test in accordance with CSN 75 0905 [9] was successfully performed on the reconstructed settling tank. The test has confirmed that the reconstructed settling tank is designed to be watertight.

The test operation confirmed that the requirement for an improved floating sludge removal capability was met. Thanks to the flexible bottom scraper wheels that follow the unevenness of the cambered concrete, the functional properties of the equipment have also improved in this respect.

Author Contributions: Conceptualization, P.B. and V.V.; methodology, P.B. and V.V.; formal analysis, P.B. and V.V.; investigation, P.B., V.V. and T.D.; resources, P.B. and V.V.; writing—original draft preparation, P.B., V.V. and T.D.; writing—review and editing, V.V. and T.D.; visualization, P.B., V.V. and T.D.; supervision, P.B. and V.V.; funding acquisition, T.D. All authors have read and agreed to the published version of the manuscript.

Funding: This research received no external funding.

Institutional Review Board Statement: Not applicable.

Informed Consent Statement: Not applicable.

Data Availability Statement: The authors declare that all data supporting the results of this research are available in this article.

Acknowledgments: This article has been produced with the financial support of the European Union under the REFRESH—Research Excellence For REgion Sustainability and High-tech Industries project number CZ.10.03.01/00/22_003/0000048 via the Operational Programme Just Transition.

Conflicts of Interest: The authors declare no conflict of interest.

References

1. Bürger, R.; Diehl, S.; Farås, S.; Nopens, I. On Reliable and Unreliable Numerical Methods for the Simulation of Secondary Settling Tanks in Wastewater Treatment. *Comput. Chem. Eng.* **2012**, *41*, 93–105. [CrossRef]
2. Abusam, A.; Keesman, K.J. Dynamic Modeling of Sludge Compaction and Consolidation Processes in Wastewater Secondary Settling Tanks. *Water Environ. Res.* **2009**, *81*, 51–56. [CrossRef] [PubMed]
3. Bürger, R.; Diehl, S.; Nopens, I. A Consistent Modelling Methodology for Secondary Settling Tanks in Wastewater Treatment. *Water Res.* **2011**, *45*, 2247–2260. [CrossRef] [PubMed]
4. Pilarski, G.; Kyncl, M.; Stegenta, S.; Piechota, G. Emission of Biogas from Sewage Sludge in Psychrophilic Conditions. *Waste Biomass Valor.* **2020**, *11*, 3579–3592. [CrossRef]
5. Kubonova, L.; Janakova, I.; Malikova, P.; Drabinova, S.; Dej, M.; Smelik, R.; Skalny, P.; Heviankova, S. Evaluation of Waste Blends with Sewage Sludge as a Potential Material Input for Pyrolysis. *Appl. Sci.* **2021**, *11*, 1610. [CrossRef]
6. CZECH REPUBLIC. Act No. 254/2001 Coll. on Water and Amendments to Certain Acts (Water Act). In *Collection of Laws*; Printing House of the Ministry of the Interior: Praha, Czech Republic, 2001; No. 98.
7. CZECH REPUBLIC. Government Decree No. 401/2015 Coll. on Indicators and Values of Permissible Pollution of Surface Water and Wastewater, Details of Permits for Discharge of Wastewater into Surface Water and Sewers and on Sensitive Areas. In *Collection of Laws*; Printing House of the Ministry of the Interior: Praha, Czech Republic, 2015; No. 166.
8. GRAU, P. *Nové Řešení Výtokových Otvorů z Rozdělovacího Hrnce Flokulačních Dosazováků*; AquaNova: Praha, Slovakia, 1999.
9. CSN 75 0905. *Water Suply and Sewerage Tanks—Testing of Water-Tightness*; ČSN: Praha, Czech Republic, 2014.

Disclaimer/Publisher's Note: The statements, opinions and data contained in all publications are solely those of the individual author(s) and contributor(s) and not of MDPI and/or the editor(s). MDPI and/or the editor(s) disclaim responsibility for any injury to people or property resulting from any ideas, methods, instructions or products referred to in the content.

Proceeding Paper

The Removal of Selected Pharmaceuticals from Water by Adsorption with Granular Activated Carbons [†]

Ján Ilavský * and Danka Barloková

Department of Sanitary and Environmental Engineering, Faculty of Civil Engineering, The Slovak University of Technology, Radlinského 11, 810 05 Bratislava, Slovakia; danka.barlokova@stuba.sk
* Correspondence: jan.ilasky@stuba.sk
[†] Presented at the 4th International Conference on Advances in Environmental Engineering, Ostrava, Czech Republic, 20–22 November 2023.

Abstract: The article deals with the removal of 20 selected drugs from drinking water by sorption on granulated activated carbon. Two different sorption materials were used, and the efficiency of removing these micropollutants was compared. Experiments were performed in laboratory conditions with two different values of pH of water (7.8 and 6.5, respectively), at laboratory temperatures and with an identical amount of added sorbent (Filtrasorb 400, WG12). Standard additions were made to drinking water, and the final concentration of 0.44–0.55 µg/L of pharmaceuticals was utilized in the experiments. Samples were taken after 30, 60, 120 and 240 min of contact of sorbent with water. The LC-MS method was used to analyze the selected pharmaceuticals, which was performed at the ALS laboratory in Prague, Czech Republic. The adsorption efficiency of the removal of the given pharmaceuticals from water, as well as the adsorption capacity of the granulated activated carbon for the given pharmaceuticals, depends on amount of time that the water and material were in contact. The adsorption efficiency for two different types of granulated activated carbon ranges from 13 to 90%. Water pH also influences the sorption abilities of granulated activated carbon.

Keywords: drinking water treatment; pharmaceuticals; adsorption; granulated activated carbon; static test; adsorption efficiency

1. Introduction

Over the previous decades, analytical methods have seen progress in determining micropollutants, leading to detection of trace concentrations of pharmaceuticals not only in surface and wastewater, but also in groundwater, and even in drinking water, in concentrations from nanograms to micrograms.

Pharmaceuticals are natural or synthetic substances that are included in the micropollutant group [1,2]. They can travel into surface or groundwater through municipal wastewater, and they can pose a risk to the environment even in low concentrations. Approximately 3000 various substances are used in human medicine in the European Union, such as analgesics, anti-inflammatory drugs, antibiotics, betablockers, lipid regulators, psychoactive drugs and many more. Many pharmaceuticals are also used in veterinary medicine, including antibiotics and anti-inflammatory drugs. After application, pharmaceutical formulae are ejected from the body in their original (unchanged) form, or in the form of metabolites, and enter water environments by various means [3]. From the environmental point of view, the biggest concern when using pharmaceutical products is their persistence and critical biological activity.

Residues of pharmaceutical products may enter the environment during production, usage, and liquidation. Medicinal products enter the environment mainly through runoff from municipal wastewater treatment plants, wastewater from manufacturing-industrial plants, wastewater from hospitals, application of sewage sludge to fields, veterinary medicine, application of animal manure, leachate from landfills, improper disposal

Citation: Ilavský, J.; Barloková, D. The Removal of Selected Pharmaceuticals from Water by Adsorption with Granular Activated Carbons. *Eng. Proc.* **2023**, *57*, 33. https://doi.org/10.3390/engproc2023057033

Academic Editors: Adriana Estokova, Natalia Junakova, Tomas Dvorsky, Vojtech Vaclavik and Magdalena Balintova

Published: 7 December 2023

Copyright: © 2023 by the authors. Licensee MDPI, Basel, Switzerland. This article is an open access article distributed under the terms and conditions of the Creative Commons Attribution (CC BY) license (https://creativecommons.org/licenses/by/4.0/).

of expired pharmaceuticals, improper disposal of unused medicines and contaminated waste [4].

Pharmaceuticals that are not removed in the wastewater treatment process are discharged into recipients, which leads to the contamination of rivers, lakes, and sometimes even underground and drinking water. According to studies published to date, the concentrations of pharmaceuticals in surface water and groundwater, which have been polluted by discharged wastewater from sewage treatment plants, are lower than 0.1 µg/L, and if they occur in drinking water, their concentrations are lower than 0.05 µg/L [5].

Figures 1 and 2 show the number of pharmaceuticals determined in surface, underground, tap and drinking water.

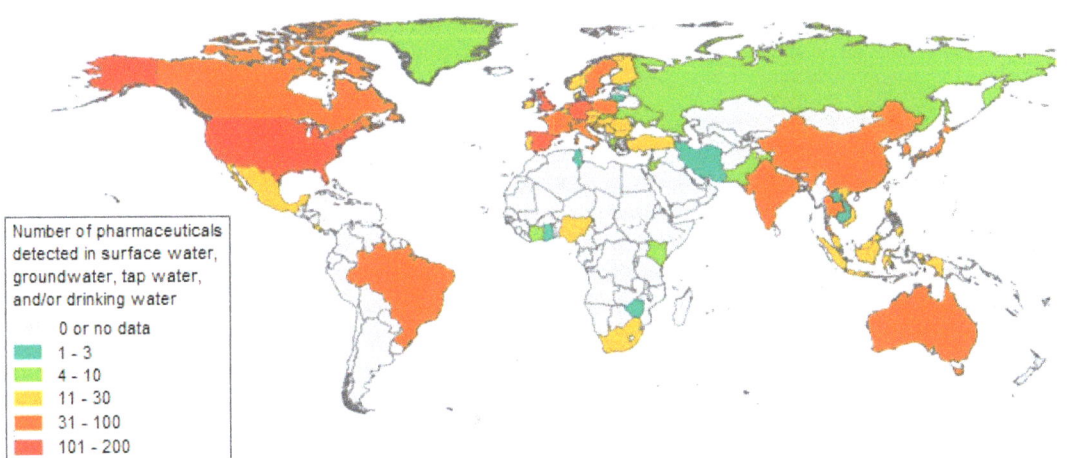

Figure 1. The number of pharmaceuticals determined in surface, underground, tap and drinking water [6].

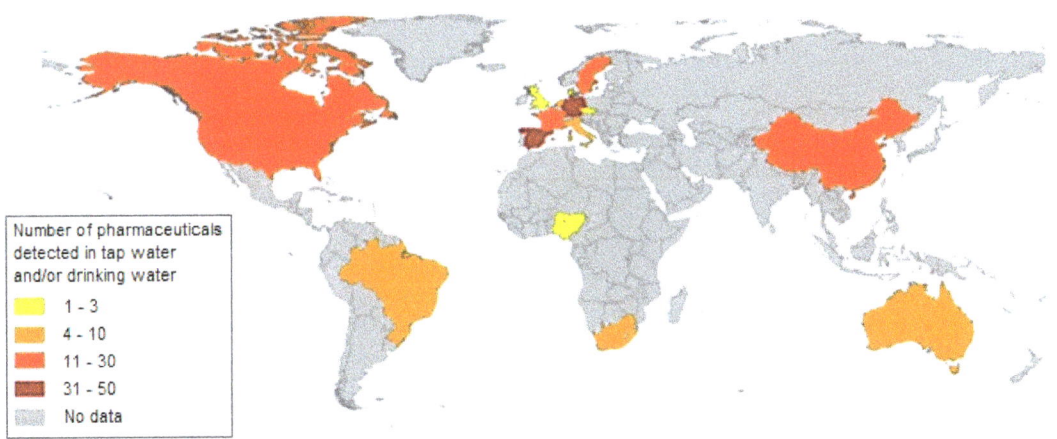

Figure 2. The number of pharmaceuticals determined in tap and drinking water [6].

On the basis of many published works, including monitoring and studies that dealt with the issue of drugs and their metabolites in water, it has been found that the pollution of water, whether surface, underground or, to a lesser extent, drinking water, by these micropollutants is a worldwide problem. Based on the study of the currently available literature, comprising more than 1000 works, a database was made, which showed that drugs or their transformation products were detected in the environment of 71 countries covering all continents [6].

Pharmaceutical products and their metabolites were also found in Slovak surface waters, and monitoring results indicated the frequent occurrence of the following pharmaceuticals: valsartan, venlafaxine, telmisartan, metoprolol, tramadol, clindamycin, erythromycin, carbamazepine and diclofenac. And some of them were even found in the waters of the High Tatras lakes [7].

1.1. Methods of Removal of Pharmaceuticals from Water

Pharmaceuticals are a diverse group of chemicals with various physical and chemical properties. The removal efficiency depends on their physical and chemical characteristics (e.g., hydrophobicity) and their reactivity to various removal processes, drug concentration, water temperature and hydraulic conditions. Therefore, treatment processes can only achieve a certain level of removal depending on the pharmaceutical product and used technology. These processes, such as reverse osmosis, ozonation and advanced oxidation technologies, can generally achieve higher drug removal rates compared to classical processes [8–10].

According to the available literature, the effectiveness of classical treatment of drinking water is low, usually up to 30% [5]. The literature [11] reports interesting results of the removal of some drugs (diclofenac, ibuprofen, bezafibrate, carbamazepine and sulfamethoxazole) from water by chemical coagulation. Based on the results of drug removal from water, which were achieved by coagulation, experiments were done with two types of water: MilliQ water and water reservoir, containing humic acids. Iron (pH 4,5) and aluminum (pH 6) salts were used as coagulants, and HPLC and UV detection were used to determine drugs in the monitored water samples. By coagulating MilliQ water with ferric sulfate, the removal efficiency reached less than 10% with the exception of diclofenac, which was removed with an efficiency of 66%. When using aluminum sulfate to coagulate water from the reservoir, the results were similar, i.e., less than 10% removal efficiency with the exception of diclofenac, which was removed by 30%.

In the presence of dissolved humic substances, diclofenac, as well as ibuprofen and bezafibrate, can be removed by coagulation with iron (III) sulfate with a maximum efficiency of 77% for diclofenac, 50% for ibuprofen and 36% for bezafibrate. A high amount of high molecular weight NOM improved the removal of ionizable drugs. The efficiency of removing non-ionizable compounds of carbamazepine and sulfamethoxazole by coagulation was very low. In the case where water had a high humic substance content, low pH values and iron coagulant, the effectiveness was higher, but basically, based on the results achieved with coagulation, removing drugs from water in this way is not effective.

Based on the research devoted to the removal of drugs from water, it follows that more suitable methods are the use of slow sand filtration and adsorption on granular activated carbon filters or the addition of powdered activated carbon to the water [5].

Adsorption on GAC is used in the water treatment processes, and it removes pharmaceuticals from water with higher efficiency than coagulation. Their efficiency is high, in some cases up to 98%, but this is not true in all situations. The efficiency depends on the nature of the organic substance being removed, the pH of the water and the presence of substances that affect sorption (e.g., humic substances, turbidity).

In the research carried out at Brno University of Technology (BUT), the effectiveness of removing ibuprofen (at a concentration of 1.02 µg/L) [12] and diclofenac (at a concentration of 1.28 µg/L) [13] from drinking water using the sorbents Filtrasorb F100, Bayoxide E33 and GEH was monitored. The measurements took place in a column with an internal diameter of 4.4 cm and a filling height of 70–80 cm, sampling 1 to 6 min after the start of the experiment. At the same time, the change in water quality influenced by the material used (pH, temperature and turbidity) was monitored. Ibuprofen removal efficiency ranged from 71 to 88% for F100, 85–92% for Bayoxide E33 and 81–85% for GEH in the first two minutes, then desorption occurred. In case of diclofenac, the GEH material achieved the best adsorption efficiency (99%), followed by Filtrasorb F100 (92–99%) and Bayoxide E33 (only 24–50%), depending on the sampling time (from 1 to 6 min).

Groundwater used for drinking purposes is water that is, in most cases, equal to the quality of drinking water. When treatment of such water is necessary, it is due to increased iron, manganese and heavy metal contents; the concentration of these materials must be reduced to values corresponding to drinking water. Often, it is sufficient to use disinfection. Disinfectants have demonstrated the ability to reduce drug concentrations in water; as stated in the WHO document [5], chlorine disinfectant can remove up to 50% of monitored drugs from water. The chlorine compound chloramine was also monitored, but here, the antibiotic removal efficiency was very low. Chlorine dioxide and ozone also show high removal efficiency [14–16].

Currently, some of the most effective ways to remove drugs from water are membrane processes. In the case of using membrane processes, the efficiency is influenced not only by the properties and nature of the drugs, but also by the properties of the membranes. The removal efficiency can reach up to 90% [17].

At present, pharmaceutical removal strategies can generally be divided into physical, chemical and biological methods. The advanced oxidation processes (AOPs), enzyme degradation, carbonaceous material Biochar, carbon nanotubes, nanoparticles, metals, biosorbents, granular activated carbons, resins, etc. may be an effective method for treating serious environmental pollution [18–25]. Of these methods, some are already being used in practical application to reduce the danger of the presence of drugs that threaten not only the aquatic environment, but also human health.

1.2. Pharmaceuticals Used in This Research

The supplied standard mixture contained the following types of pharmaceuticals: anti-inflammatory agents (diclofenac, indomethacin, ketoprofen, naproxen, paracetamol), antibiotics (metronidazole, sulfamethazine, sulfamethoxazole, trimethoprim), contrast agents (iomeprol, iopamidol, iopromide), circulatory system agents (valsartan, warfarin, hydrochlorothiazide) and others (bezafibrate, fluoxetine, ifosfamide, carbamazepine, caffeine).

For example, carbamazepine is an anticonvulsant drug used primarily to treat epilepsy and neuropathic pain. It is used to treat schizophrenia, along with other drugs, and is a second-line agent in bipolar disorder. Caffeine is a central nervous system stimulant from the methylxanthine group. It is the most widespread psychoactive drug in the world. Caffeine concentrations in surface waters were detected in the range of 0.1–6.9 µg/L.

Bezafibrate is a drug used as a lipid-lowering agent to treat hyperlipidemia. Fluoxetine is a drug used mainly in the treatment of depression. Ifosfamide is a chemotherapy drug used to treat many types of cancer. These include soft tissue sarcoma, osteosarcoma, bladder cancer, small cell lung cancer, cervical cancer and ovarian cancer.

Diclofenac is one of the most commonly prescribed medications, used to treat pain and inflammatory diseases. It is available in the form of a variety of medicinal, some of them are free of the counter without a prescription, but only at certain doses.

The occurrence of pharmaceuticals, their metabolites and transformation products in the environment has become the subject of research from the point of view of toxicity, monitoring and analytical determination. Various pharmaceuticals have been detected in many samples of environmental components around the world. Their occurrence has been recorded in runoff and sludge from wastewater treatment plants, surface water, seawater, groundwater, soil and sediments as well as in drinking water.

Therefore, the aim of our research was to verify the effectiveness of granular activated carbon in removing selected drugs from water to compare two different GACs, as well as the effect of water pH on their effectiveness.

2. Materials and Methods

The pharmaceutical standard was purchased from the ALS Czech Republic company in Prague, which also provided us with sample boxes and determined the drugs in the samples. The standard contained 20 different drugs. Granulated activated carbon WG12 (manufactured by Gryfskand Co., Hajnówka, Poland) was supplied by Envi-Pur, LtD. and Filtrasorb F400 (manufactured by Calgon Carbon Co., and delivered by Chemviron Carbon, Feluy, Belgium) by Jako LtD., both from the Czech Republic. The basic properties of the used granular activated carbon are summarized in Table 1.

Table 1. Properties of activated carbon WG12 and F400.

Parameter	WG12	F400
Form	pellet	granular
Iodine number [mg·g^{-1}]	min. 1000	min. 1050
Methylene blue [mg·g^{-1}]	min. 30	min. 260
Specific surface (BET) [m^2·g^{-1}]	min 1000	min 1050
Particle size [mm]	1.0–1.5	0.42–1.68
Operational density [g·cm^{-3}]	0.450 ± 30	0.425
Abrasion [-]	min. 85	min. 75
Hardness [-]	min. 95	min. 95
Uniformity coefficient	max. 1.3	max. 1.9
Moisture [wt.%]	max. 2	max. 2

A model solution with a drug content in the range from 0.44 µg/L to 0.55 µg/L was obtained by adding 50 mL of the drug mixture standard to 5 L of drinking water. The model water sample prepared in this way was divided into Erlenmayer flasks; 400 mL of model water was added to each sample, and 400 mg of granular activated carbon (Filtrasorb F400 and WG12) was added to the samples that were prepared in this way. The volume of model water together with activated carbon was mixed for 4 h using an OHAUS orbital shaker (OHAUS Europe GmbH, Nänikon, Switzerland) at 400 rpm. During the mixing, water samples were taken from the flasks at time intervals of 0, 30, 60, 120 and 240 min. The samples were collected in glass vials (40 mL) which contained a preservative (0.32 mL of 1% sodium thiosulfate).

The collected water samples (10 mL) after centrifugation and microfiltration were injected into a liquid chromatography device (Acquity UPLC I-Class, Waters Co., Milford, CT, USA) connected to a mass spectrometer (XEVO TQ-XS, Waters Co., Milford, CT, USA). The compounds were separated by a chromatographic column (InfinityLab Poroshell 10 EC-C18 (3.0 × 150 mm; 2.7 µm, Agilent Technologies, Santa Clara, CA, USA) mobile phase consisted of 0.01% HCOOH in water (A) and MeOH (B). The obtained data were evaluated with the Tar-getLynx quantification software. Quantification of analytes was conducted using the method of isotopically labeled internal standards addition.

3. Results and Discussion

To monitor the efficiency of the removal of drugs listed in Figures 3 and 4, two types of activated carbon, Filtrasorb F400 and WG12, were used in static tests. The efficiency of the sorbent materials was monitored at pH values of 7.8 and 6.5. The concentration of drugs in the water at the beginning of the tests was in the range of 0.44–0.55 µg/L, and the contact time of the model water with activated carbon varied in 30 min intervals up to four hours.

The adsorption efficiency (in %) and immediate adsorption capacity (in µg/g) of activated carbons WG12 and F400 were calculated for the individual pharmaceuticals depending on the water–material contact time on the base of the measured concentrations of the individual organic compounds [25]. The following formulas were used:

$$\eta = \frac{(c_0 - c_m) \cdot 100}{c_0} \ [\%] \qquad (1)$$

$$a_t = \frac{(c_0 - c_m) \cdot V}{m} \ [\mu g/g] \qquad (2)$$

where a_t is the immediate adsorption capacity in µg/g, η is the adsorption efficiency [%], c_o is the concentration of pesticides before the adsorption, c_m is the concentration of drugs after the adsorption at the time **t** [µg/L], **V** is the volume of water solution of 0.4 L and **m** is the weight of sorption material, 0.4 g.

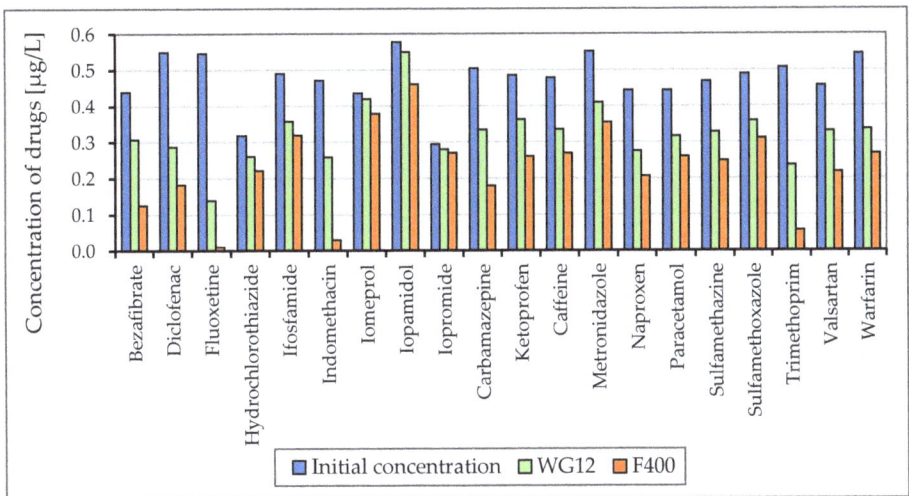

Figure 3. Concentration of drugs (µg/L) before and after 60 min adsorption with WG12 and F400 at the pH value 7.8.

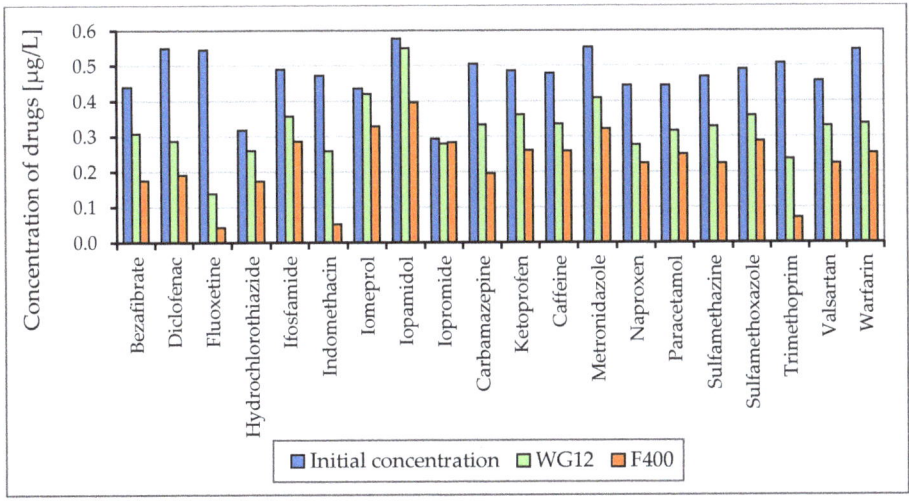

Figure 4. Concentration of drugs (µg/L) before and after 60 min adsorption with WG12 and F400 at the pH value 6.5.

Figures 5 and 6 show the efficiencies for each granular activated carbon calculated for 60 min of water contact time with the material used at the different pH of water.

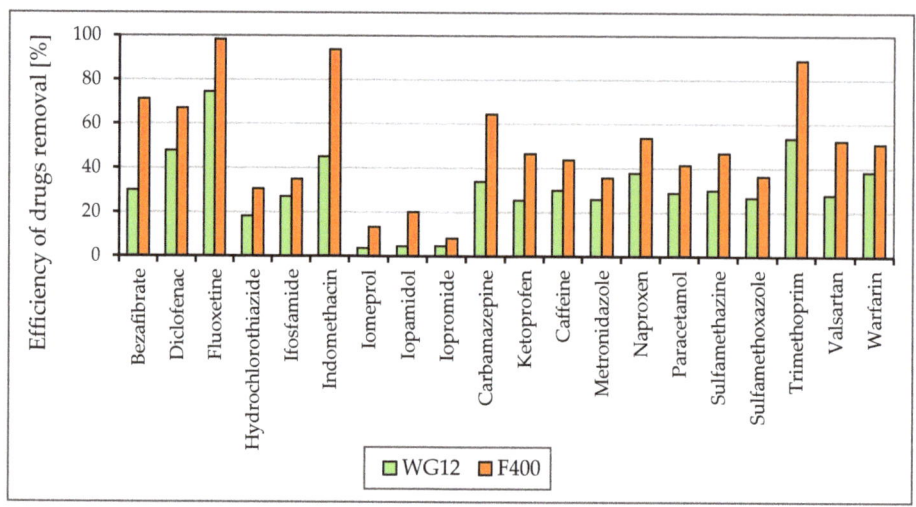

Figure 5. The efficiency of drug removal (in %) before and after 60 min adsorption with WG12 and F400 at the pH value 7.8.

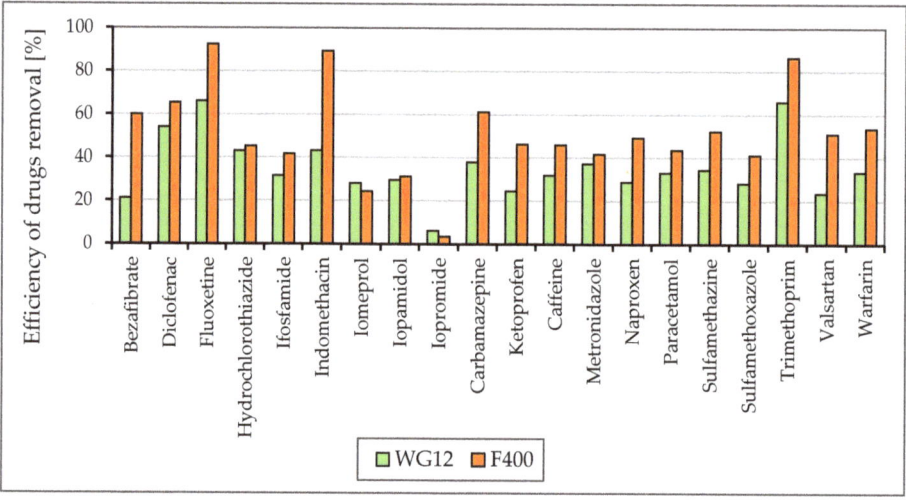

Figure 6. The efficiency of drug removal (in %) before and after 60 min adsorption with WG12 and F400 at the pH value 6.5.

As is obvious from Figures 3 and 4 the sorption materials F400 and WG12 have different adsorption efficiency for all pharmaceuticals used in this study. Adsorption efficiency ranges from 4 to 98%. There are few differences between pH value of water, but they are very significant. The pH value affects the efficiency of the removal of individual drugs.

Adsorption capacity for adsorption materials used was different. Adsorption capacity for F400 and contact time 60 min was between 0.023 to 214 µg/g, and for WG12, it ranged from 0.006 to 0.163 µg/g.

4. Conclusions

The results shows that Filtrasorb F400 achieved a higher drug removal efficiency from water than the WG12 material. At the same time, the results show the different effectiveness for individual drug. Therefore, it is necessary to verify the use of granular activated carbon directly for a specific contaminant and the quality of treated water. For some drugs, the pH of the water and the contact time of the water with granular activated carbon also play an important role. The higher the contact time, the greater the resulting effect of the method used. After 4 h of sorption, the removal efficiency of contaminants from water was more than 90% for each material.

Author Contributions: J.I. and D.B. worked out a concept and plan of experiments, ensured the installation of all equipment, assembly and verification of used technologies. D.B. and J.I. performed all experiments and water sampling. J.I. analyzed water samples (pharmaceuticals were analyzed in ALS laboratory of Prague, Czech Republic). D.B. and J.I. evaluated the obtained results from experiments. All authors have read and agreed to the published version of the manuscript.

Funding: The experiments were financially supported by Slovak Research and Development Agency of the Slovak Republic (Projects APVV-18-0205 and APVV-22-0610) and by the Ministry of Education, Science, Research and Sports of the Slovak Republic (Project VEGA 1/0825/21).

Institutional Review Board Statement: Not applicable.

Informed Consent Statement: Not applicable.

Data Availability Statement: Data will be available on request.

Acknowledgments: Thanks go to the workers at the ALS laboratory Prague, Czech Republic.

Conflicts of Interest: The authors declare no competing interests.

References

1. Rosenfeld, P.; Feng, L. *Risk of Hazardous Wastes*, 1st ed.; Elsevier: Amsterdam, The Netherlands, 2011; p. 472.
2. Yang, Y.; Ok, Y.S.; Kim, K.-H.; Kwon, E.E.; Tsang, Y.F. Occurrences and removal of pharmaceuticals and personal care products (PPCPs) in drinking water and water/sewage treatment plants: A review. *Sci. Total Environ.* **2017**, *596–597*, 303–320. [CrossRef] [PubMed]
3. Fent, K.; Weston, A.A.; Caminada, D. Ecotoxicology of human pharmaceuticals. *Aquat. Toxicol.* **2006**, *76*, 122–159. [CrossRef] [PubMed]
4. European Commission. Communication from the Commission to the European Parliament, the Council and the European Economic and Social Committee: European Union Strategic Approach to Pharmaceuticals in the Environment. Available online: https://eur-lex.europa.eu/legal-content/EN/TXT/HTML/?uri=CELEX:52019DC0128&from=SK (accessed on 2 July 2023).
5. WHO. *Pharmaceuticals in Drinking-Water*; WHO Press: Geneva, Switzerland, 2012.
6. Aus der Beek, T.; Weber, F.-A.; Bergmann, A.; Hickmann, S.; Ebert, I.; Hein, A.; Küster, A. Pharmaceuticals in the environment: Global occurrences and perspectives. *Environ. Toxicol. Chem.* **2016**, *35*, 823–835. [CrossRef] [PubMed]
7. Mackuľak, T.; Czölderová, M.; Grabic, R.; Bodík, I.; Vojs-Staňová, A.; Žabka, D.; Horáková, I. Výskyt liečiv a drog v povrchových vodách Slovenska. In Proceedings of the 38th International Scientific Symposium, Svit, Slovakia, 13–15 June 2018; pp. 91–105.
8. Baresel, C.; Ek, M.; Ejhed, H.; Allard, A.S.; Magnér, J.; Dahlgren, L.; Westling, K.; Wahlberg, C.; Fortkamp, U.; Söhr, S.; et al. Sustainable treatment systems for removal of pharmaceutical residues and other priority persistent substances. *Water Sci. Technol.* **2019**, *79*, 537–543. [CrossRef] [PubMed]
9. Wang, J.; Bai, Z. Fe-based catalysts for heterogeneous catalytic ozonation of emerging contaminants in water and wastewater. *Chem. Eng. J.* **2017**, *312*, 79–98. [CrossRef]
10. Cuerda-Correa, E.M.; Alexandre-Franco, M.F.; Fernández-González, C. Advanced Oxidation Processes for the Removal of Antibiotics from Water. An Overview. *Water* **2020**, *12*, 102. [CrossRef]
11. Vieno, N.; Tuhkanen, T.; Kronberg, L. Removal of Pharmaceuticals in Drinking Water Treatment: Effect of Chemical Coagulation. *Environ. Technol.* **2006**, *27*, 183–192. [CrossRef] [PubMed]
12. Kabelíková, E. Sledování Účinnosti Vybraných Adsorbentů na Odstraňování Mikropolutantů z Pitné Vody. Diploma Thesis, VUT Brno, Brno, Czech Republic, 2019.
13. Moravčíková, S. Sledování Účinnosti Odstraňování Léčiva z Vody Vybranými Adsorbenty. Diploma Thesis, VUT Brno, Brno, Czech Republic, 2020.
14. Vieno, N.M.; Härkki, H.; Tuhkanen, T.; Kronberg, L. Occurrence of pharmaceuticals in river water and their elimination in a pilot-scale drinking water treatment plant. *Environ. Sci. Technol.* **2007**, *41*, 5077–5084. [CrossRef] [PubMed]
15. Nam, S.W.; Choi, D.J.; Kim, S.K.; Her, N.; Zoh, K.D. Adsorption characteristics of selected hydrophilic and hydrophobic micropollutants in water using activated carbon. *J. Hazard. Mater.* **2014**, *270*, 144–152. [CrossRef] [PubMed]
16. Katsigiannis, A.; Noutsopoulos, C.; Mantziaras, J.; Gioldasi, M. Removal of emerging pollutants through Granular Activated Carbon. *Chem. Eng. J.* **2015**, *280*, 49–57. [CrossRef]

17. Licona, K.P.M.; Geaquinto, L.R.d.O.; Nicolini, J.V.; Figueiredo, N.G.; Chiapetta, S.C.; Habert, A.C.; Yokoyama, L. Assessing potential of nanofiltration and reverse osmosis for removal of toxic pharmaceuticals from water. *J. Water Process Eng.* **2018**, *25*, 195–204. [CrossRef]
18. Lu, Z.-Y.; Ma, Y.-L.; Zhang, J.-T.; Fan, N.-S.; Huang, B.-C.; Jin, R.-C. A critical review of antibiotic removal strategies: Performance and mechanisms. *J. Water Process Eng.* **2020**, *38*, 101681. [CrossRef]
19. Homem, V.; Santos, L. Degradation and removal methods of antibiotics from aqueous matrices—A review. *J. Environ. Manag.* **2011**, *92*, 2304–2347. [CrossRef] [PubMed]
20. Pertile, E.; Dvorský, T.; Václavík, V.; Heviánková, S. Use of Different Types of Biosorbents to Remove Cr (VI) from Aqueous Solution. *Life* **2021**, *11*, 240. [CrossRef] [PubMed]
21. Pertile, E.; Václavík, V.; Dvorský, T.; Heviánková, S. The Removal of Residual Concentration of Hazardous Metals in Wastewater from a Neutralization Station Using Biosorbent—A Case Study Company Gutra, Czech Republic. *Int. J. Environ. Res. Public Health* **2020**, *17*, 7225. [CrossRef] [PubMed]
22. Wang, X.; Yin, R.; Zeng, L.; Zhu, M. A review of graphene-based nanomaterials for removal of antibiotics from aqueous environments. *Environ. Pollut.* **2019**, *253*, 100–110. [CrossRef] [PubMed]
23. Zhang, A.; Li, X.; Xing, J.; Xu, G. Adsorption of potentially toxic elements in water by modified biochar: A review. *J. Environ. Chem. Eng.* **2020**, *8*, 104196. [CrossRef]
24. Kim, S.; Chu, K.-H.; Al-Hamadani, Y.A.J.; Park, C.-M.; Jang, M.; Kim, D.-H.; Yu, M.; Heo, J.; Yoon, Y. Removal of contaminants of emerging concern by membranes in water and wastewater: A review. *Chem. Eng. J.* **2018**, *335*, 896–914. [CrossRef]
25. Ilavský, J.; Barloková, D.; Marton, M. Removal of Specific Pharmaceuticals from Water using Activated Carbon. *IOP Conf. Ser. Earth Environ. Sci.* **2021**, *906*, 012065. [CrossRef]

Disclaimer/Publisher's Note: The statements, opinions and data contained in all publications are solely those of the individual author(s) and contributor(s) and not of MDPI and/or the editor(s). MDPI and/or the editor(s) disclaim responsibility for any injury to people or property resulting from any ideas, methods, instructions or products referred to in the content.

Proceeding Paper

Technology of the Biological Treatment of Mine Water at the Kohinoor II Mine [†]

Jaroslav Mudruňka, Kateřina Matunová Kavková, Radmila Kučerová *, Lucie Marcaliková, David Takač, Nikola Drahorádová, Martina Ujházy and Veronika Brašová

Department of Environmental Engineering, Faculty of Mining and Geology, VSB—Technical University of Ostrava, 17. Listopadu 2172/15, 708 00 Ostrava, Czech Republic; jaroslav.mudrunka@vsb.cz (J.M.); katerina.kavkova93@seznam.cz (K.M.K.); lucie.kucerova1@vsb.cz (L.M.); takac.dave@gmail.com (D.T.); nikola.drahoradova@vsb.cz (N.D.); martina.ujhazy@vsb.cz (M.U.); veronika.brasova@vsb.cz (V.B.)
* Correspondence: radmila.kucerova@vsb.cz
[†] Presented at the 4th International Conference on Advances in Environmental Engineering, Ostrava, Czech Republic, 20–22 November 2023.

Abstract: The aim of this work was to assess the effectiveness of the treatment process in the mine water treatment plant in Mariánské Radčice that pumps mine water from the MR1 pit and to evaluate whether this biotechnological unit is satisfactory in its treatment process with regard to the set limits for the discharge of treated mine water into watercourses, or whether this water can be discharged into Lake Most in the future, which is intended for recreation, and also with regard to the ecosystem that exists there.

Keywords: mine water; mine water pollution; biological treatment of mine water; root system; plant functions; elimination of ammonium ions

1. Introduction

The water around us is an essential natural resource from which we benefit on a daily basis. It is essential for the life of millions of species of plants and animals—from the largest inhabitants of the oceans to the smallest microorganisms. Water is a necessary need of the human body; it is also a resource that we use every day in households, industry, agriculture, transport and, last but not least, for recreation. Through our daily activities, we change the quality and availability of water, we threaten it with overuse, with various types of pollution, and we change its properties.

One of the sectors that fundamentally changes the quality and properties of surface and underground water is mining, whether it is coal mining or the mining of other natural resources. Mine water from surface coal mining is characterized by a low pH value, high hardness, high iron ion content, high concentrations of dissolved and suspended substances, and extremely low organic matter content. Such water must be purified.

Mine water can have quite a serious impact on the hydrological regime, and it is therefore necessary to address this problem and look for optimal solutions and possibilities for its reuse. Both conventional and alternative methods can be chosen for mine water treatment. Conventional methods include water treatment using wastewater treatment plants or sedimentation tanks, alternative methods include the use of wetland ecosystems or reed bed plants, with regard to the specifics of particular mine water sources [1].

2. Reed Bed Plants

Reed bed plants are passive methods of water treatment that use not only natural processes, but also natural materials. These treatment plants are built mainly from the natural building materials found in the surrounding area and the filtration areas are planted with

Citation: Mudruňka, J.; Matunová Kavková, K.; Kučerová, R.; Marcaliková, L.; Takač, D.; Drahorádová, N.; Ujházy, M.; Brašová, V. Technology of the Biological Treatment of Mine Water at the Kohinoor II Mine. *Eng. Proc.* **2023**, *57*, 34. https://doi.org/10.3390/engproc2023057034

Academic Editors: Adriana Estokova, Natalia Junakova, Tomas Dvorsky, Vojtech Vaclavik and Magdalena Balintova

Published: 7 December 2023

Copyright: © 2023 by the authors. Licensee MDPI, Basel, Switzerland. This article is an open access article distributed under the terms and conditions of the Creative Commons Attribution (CC BY) license (https://creativecommons.org/licenses/by/4.0/).

wetland vegetation. Biological stabilization tanks are most often used for water accumulation, water treatment, and additional water purification. Tanks may be aerobic or anaerobic in nature as well as aerated or non-aerated. Various final treatment ponds and aquaculture tanks are included. The most commonly used ones are aerobic biological tanks, where mechanical (sedimentation, adsorption), chemical (oxidation/reduction, decomposition of substances), and biological (decomposition, nutrient uptake, and metabolic processes) processes take place [2,3].

Passive systems for mine water treatment effectively treat even strongly diluted water and show a relatively high efficiency in removing pollution according to BOD_5, COD, suspended solids, and heavy metals. They cope well with the uneven inflows and even allow for intermittent operation. They are very simple in construction, require minimal maintenance, and are energy efficient. Also, the acquisition costs are usually lower than in the case of a conventional treatment plant. From an ecological point of view, they have a positive effect on the surrounding microclimate and have a certain landscaping function. Compared to the intensive systems, however, they are more demanding on land use, are partly dependent on climatic conditions, remove some nutrients, especially ammonia pollution, in a limited way, and have a low ability to regulate ongoing processes. In addition, the root systems of reed bed plants need to be permanently flooded [4–6].

Purified water passes through fields with wetland vegetation and lingers on them for some time. The main purpose of wetland vegetation is the filtration and then the insulation of the filter field, especially in the winter season. Filtration takes place in the soil and subsoil of the vegetation with the help of the root system. As the water passes through the vegetation, the impurities are collected directly in the natural filter, while the purified water flows further. It is necessary for vegetation stands to be designed with a sufficiently large area that takes into account the volume of treated water to effectively remove the required substances before the water returns to the watercourse. Passive settling lagoons and areas planted with wetland vegetation are the most environmentally friendly way of treating mine water. In addition to mine water treatment, wetlands built in this way are a rare habitat and are home to many different species of insects and birds [7,8].

Suitable wetland plants have an important role in the application of the passive method of water treatment. They tolerate constant and periodic flooding, the lack of oxygen, high salt content, and sudden changes in the pH values of the environment. Water rich in nutrients and waste products suits all plants that are appropriate for reed bed plants such as broadleaf and narrowleaf cattail (*Typha latifolia*, *Typha angustifolia*), great manna grass (*Glyceria maxima*), reed canary grass (*Phalaris arundinacea*), *Schoenoplectus lacustris*, lakeshore bulrush (*Sparganium erectum*), common reed (*Phragmites australis*) and common rush (*Juncus effuses*). Other plants include yellow flag (*Iris pseudocorus*), marsh-marigold (*Caltha palustris*), purple loosestrife (*Lythrum salicaria*), and meadowsweet (*Filipendula ulmaria*). It is recommended to mow the plants in the early spring (late February). It is possible to mow and harvest *Glyceria maxima* and *Phalaris arundinacea* even during the growing season and then use the harvested biomass for composting [9].

3. Mariánské Radčice Mine Water Treatment Plant

The treatment plant for mine water pumped from the MR1 pit of the Kohinoor II mine is located 0.8 km from the village of Mariánské Radčice. It is adjacent to the active Bílina mine, which is drained into the MR1 pit of the Kohinoor II mine. The Kohinoor II underground mine ceased its production in August 2002. In the Bílina mine, brown coal is still mined using the open-pit mining method [10,11].

The Mariánské Radčice mine water treatment plant, whose construction began in October 2018 and trial operation started in January 2021, is the largest biotechnological mine water treatment plant in Europe. Its area is 2.5 ha. It works on the principle of the biological treatment of mine water, similarly to natural reed bed plants. However, it is not a purely biological treatment plant; it has additional technological elements used especially for aeration and the pumping of water [12].

The entire biotechnological unit can be described as a system of wetlands; its aim is to ensure a stable method of mine water treatment using the physicochemical and biological processes that occur spontaneously in natural wetland systems. The system is designed so that its operation requires minimal service and maintenance, i.e., the lowest possible operating costs in the long term. The operation of the treatment plant is planned for the duration of pumping mine water from the MR1 pit, i.e., for the entire period of mining at the Bílina mine.

During 2021, a trial operation of the biotechnological mine water treatment plant in Mariánské Radčice took place. This process was closely monitored and evaluated in early 2022. The aim of the operation of this unit is to ensure such a quality of discharged water so that, with regard to the valid legislation of the Czech Republic, it is possible to discharge it into the recipient, which is the Radčický stream, or into Lake Most, if the standards for bathing water are reached after treatment [13].

The biotechnological system of this treatment plant consists of the following basic parts: a supply aeration object (pipe system, aeration cascade), seven tanks with an area of more than 2.5 ha, connecting troughs and pipes between individual objects, aeration piping system, blowers, pumps for pumping mine water, eventually a suction dredger for pumping sludge. The biotechnological treatment plant includes two electrical distribution stations, an outlet from the treatment plant to the recipient, and a sludge field. The site is divided into four height levels, the flow of mine water is ensured by gravity.

The whole system is populated with wetland plants (macrophytes). The main share consists of common reed (*Phragmites australis*), reed canary grass (*Phalaris arundinacea*), great manna grass (*Glyceria maxima*), cattail (*Typha*), and gradually associated self-seeding herbs from the surroundings. The surrounding areas were planted with shrubs and trees, partly fruit, and grassed. The entire mine water treatment plant is relatively well integrated into the surrounding landscape without a disturbing effect. During the construction, natural materials of local origin were used to the maximum extent (after previous mining activity).

An absolutely essential element of the biotechnological system, where the process of mine water treatment takes place, is a system of seven tanks. Six of them are settling and purifying tanks and are located in two lines next to each other. Tanks A1, B1, and C1 are located to the left of the entrance to the treatment plant. Parallel to them in the second line (to the right) there are tanks A2, B2, and C2. From a technological point of view, the A1 and A2 tanks differ; the others are identical in pairs (B1, 2 and C1, 2). The seventh tank is the tank marked D. Here the overall water treatment (Fe, Mn, and total N) and stabilization of the outflow parameters before discharge into the Radčický stream take place. Tank D has the function of final purification, mixing (incoming water from the previous 6 tanks), and stabilizing (Figure 1).

Mine water is pumped from the MR1 pit in the Kohinoor II mine site. Originally, it was a wind shaft of a mine into which mine water from the Bílina mine, where mining is still taking place, is drawn. The pit is 300 m deep. At present, the water is pumped from the pit by one or two 6 m long pumps. Depending on this, water flows to the water treatment plant at a speed of 60 to 120 L/s through the piping system. The pipe supply of pumped water has the parameter DN400, the gravity drainage pipe DN500, and a length of up to 605 m. The maximum design flow rate of the biotechnological system is 180 L/s.

On the inflow of mine water to the treatment plant there is a shaft made of concrete rings DN1000, a height of 3 m, with a steel lockable entrance. The concrete outlet has a capacity of 2.5 m^3 of water. From there, water is led through a DN400 pipe with a length of 40 m to the aeration cascade, which consists of a continuous staircase (1 pc). There is a considerable aeration of water and release of ammonium ions into the air on this cascade. The stairs are strongly coloured with sedimentary Fe particles (Figure 2). From there, mine water flows through concrete gutters to the first reservoirs A1 and A2.

Figure 1. Aerial view of the biotechnological unit of the water treatment plant. Tanks A1, B1, C1 in the left line, A2, B2, C2 in the right line, retention tank D in the upper right.

Figure 2. Cascade (staircase) behind the inflow of mine water to the treatment plant.

The vegetation sludge field (VSF) is part of the technological unit of the treatment plant. It serves to significantly reduce the volume of the sludge, which will later be disposed of as a waste. If necessary, the sludge will be pumped from the area of settling tanks using a suction excavator. The VSF is dimensioned so that the volume of the sludge is reduced to 1/3 of the original volume by dewatering and drying. This is done mainly by evaporation of water from the sludge and through evapotranspiration by wetland plants.

The biotechnological treatment plant has an active aeration system.

4. Evaluation of the Efficiency of the Purification Processes at the Treatment Plant

As part of ongoing research, samples of raw water were taken in the months of March–December 2021. The following parameters were evaluated: sulphates, iron, manganese, ammoniacal nitrogen, chemical oxygen demand by permanganate, pH values, undissolved substances, dissolved annealed substances, and benzo(a)pyrene. The limits for the discharge of treated mine water are set by Government Regulation No. 401/2015 Coll., on the indicators and values of permissible pollution of surface water and wastewater, the requirements for permits for the discharge of wastewater into surface waters and sewers, and on sensitive areas, as amended [14,15].

Based on the measurements and identified pollution parameters, respectively, the assessment of their values at the inflow and outflow at the biotechnological mine water treatment plant in Mariánské Radčice, the effectiveness of the treatment processes was evaluated with regard to the monitored parameters set by the legislation and the water management regulation of the Regional Authority of the Ústí nad Labem Region.

Based on the obtained measurements of water at the inflow to the treatment plant, it was found that mine water from the MR1 pumping pit does not show above-limit values of sulphates, manganese, COD_{Cr}, dissolved annealed substances, and undissolved substances, even at the input to the treatment process. These pollutants are therefore not a problem for these particular mine waters. On the other hand, the values of iron and ammoniacal nitrogen were unsatisfactory and high, the acidity of the waters was high and therefore the pH values were low, and the values of benzo(a)pyrene at the inflow were also exceeded. However, during the treatment process, it was effectively possible to reduce the values of all monitored substances to the values required for treated water currently flowing into the Radčický stream. The treated mine water fully complies with the legislative requirements imposed on it. The pollution values of sulphates were reduced by almost 30%, iron by 86%, manganese by almost 78%, and ammoniacal nitrogen by more than 89%. COD_{Cr} pollution increased slightly by less than 3% during the treatment process, which was negligible with respect to the satisfactory inflow values. The pH value increased by more than 22% to values between 7.85 and 8.26, thus the acidity was significantly reduced. The values of dissolved annealed substances decreased by almost 62% during the treatment, benzo(a)pyrene up to a hundredfold to the required below-limit values. Other monitored parameters were already satisfactory in the raw water.

The biotechnological unit for mine water treatment in Mariánské Radčice works very efficiently, and the treated water fully complies with the set limits with regard to its current use—outflow to the recipient. However, if the treated water should be drained into Lake Most, which is a locality for recreation and bathing, some of the permitted pollution in the treated water has been exceeded. The regulation in question sets stricter limits for bathing water, namely which would be exceeded for COD_{Cr} pollution and $N-NH_4^+$. For COD_{Cr} this is the permissible value of 26.2 mg/L, the annual limit average is 26 mg/L, but this limit was exceeded during 2021 only once. For $N-NH_4^+$, the permitted pollution level is 0.64 mg/L, but on an annual average this limit is only 0.23 mg/L. Compliance with this limit seems to be the biggest problem in the treatment process of the biotechnological unit of Mariánské Radčice.

If the treated mine water was to be drained into Lake Most, it would be necessary to monitor some other pollution parameters that are not currently under supervision. These include, namely: saturation of water with oxygen, biochemical oxygen demand (BOD), total organic carbon (TOC), total phosphorus, total nitrogen (nitrogen nitrate, nitrogen nitrite, nitrogen ammoniacal), total chlorides, and magnesium and calcium content. For bathing waters, the parameters of contamination by *Escherichia coli* bacteria, the intestinal enterococci, the thermotolerant coliform bacteria, radioactivity indicators, and possibly other values are also monitored. Since wastewater from human settlements is not treated at the treatment plant but only mine water, the biological indicators of bacterial pollution by *Escherichia coli* bacteria, the intestinal enterococci and the thermotolerant coliform bacteria do not appear to be problematic.

5. Conclusions

The biotechnological mine water treatment plant is at the very beginning of its existence, it has completed the first year of the trial operation, in the following years the improvement of the functioning of the entire system and its stabilization can be expected, as well as the improvement of the state of the still young vegetation, some parts of the fields are not covered by vegetation yet, and a fully developed, rich root system also cannot be expected.

During the one-year trial operation of the mine water treatment plant in Mariánské Radčice, it was found that purely biological natural processes should be supplemented with technological elements to the necessary extent, as these significantly affect the treatment process at the treatment plant. In particular, these are the elements of the active aeration of the tanks (root fields), blowers and oxygen distribution pipes. These elements were shut down in October 2021 for their inspection and maintenance, and immediately this fact had a negative impact on the quality of the treated water. The shutdown of the aeration system and the subsequently measured values constitute clear evidence that this element is desirable at the treatment plant and its inclusion is necessary.

The biotechnological treatment plant can undoubtedly be described as a functional unit in mine water treatment, which also takes into account the trend of using biological treatment plants where possible. At the same time, the nature and location of the mine water treatment plant fits very well into the surrounding landscape and is completely undisturbed in connection with the surrounding inhabited part of Mariánské Radčice. In addition, treated mine water could, if necessary, be used to irrigate agricultural areas in its immediate surroundings, especially in the event of the lack of natural moisture.

Author Contributions: Conceptualization, R.K. and K.M.K.; methodology, J.M.; software, D.T.; validation, R.K., K.M.K. and J.M.; formal analysis, V.B.; investigation, M.U.; resources, N.D.; data curation, K.M.K.; writing—original draft preparation, K.M.K.; writing—review and editing, L.M.; visualization, D.T.; supervision, R.K.; project administration, L.M.; funding acquisition, K.M.K. All authors have read and agreed to the published version of the manuscript.

Funding: This research received no external funding.

Institutional Review Board Statement: Not applicable.

Informed Consent Statement: Not applicable.

Data Availability Statement: Data are contained within the article: https://dspace.vsb.cz/bitstream/handle/10084/148482/KAV0034_HGF_N0788A290001_2022.pdf?sequence=1&isAllowed=y.

Conflicts of Interest: The authors declare no conflict of interest.

References

1. Pollert, J. Biologické Čištění—Alternativní Způsoby. Available online: http://kzei.fsv.cvut.cz/pdf/COV_pr_8.pdf (accessed on 7 September 2021).
2. Mlejnská, E.; Rozkošný, M.; Baudyšová, D.; Váňa, M.; Wanner, F.; Kučera, J. *Extenzivní Způsoby Čištění Odpadních Vod*; VÚV T. G. Masaryka: Praha, Czech Republic, 2009; ISBN 978-80-85900-92-7.
3. Vymazal, J. Constructed wetlands for wastewater treatment: Five decades of Experience. *Sci. Technol.* **2011**, *45*, 61–69. [CrossRef] [PubMed]
4. Vymazal, J.; Kröpfelová, L. Kořenové čistírny odpadních vod v České republice, jejich využití pro různé typy splaškových vod. In *Sborník Semináře Monitoring Těžkých Kovů a Vybraných Rizikových Prvků Při Čištění Odpadních Vod v Umělých Mokřadech*; ENKI: Třeboň, Czech Republic, 2008.
5. Greben, H.; Sigama, J.; Burke, L.; Venter, S. *Cellulose Fermentation Product Is an Energy Source for Biological Sulphate Reduction of Acid Mine Drainage Type Wastewater*; WRC Report No. 1728/1/08; Water Research Commission: Pretoria, South Africa, 2009; 128p, ISBN 978-1-77005-824-8.
6. Zajoncová, D. *Přírodní Čištění Vody*; ZO ČSOP Veronica ekologický Institut—Hostětín: Brno, Czech Republic, 2010; p. 16.
7. Lundquist, L. Novel Two-stage Biochemical Process for Hybrid Passive/Active Treatment of Mine-influenced Water. *Mine Water Environ.* **2021**, *41*, 14–155. [CrossRef]

8. The Coal Authority. Understanding Mine Water Treatment. Available online: https://assets.publishing.service.gov.uk/government/uploads/system/uploads/attachment_data/file/362236/Understanding_mine_water_treatment.pdf (accessed on 30 November 2021).
9. Korenovky. Kořenová Čistička—Funkce. Available online: https://www.korenova-cisticka.cz/o-korenovkach/fungovani/Korenova-cisticka%25E2%2580%2593korenova-cistirna%25E2%2580%2593funkce.html (accessed on 20 December 2021).
10. Mapy. Available online: https://mapy.cz/zakladni?x=13.6744749&y=50.5782755&z=13&base=ophoto (accessed on 22 August 2021).
11. Památkový Katalog—Národní Památkový Ústav. Důl Koh-i-Noor II. Část. Available online: https://www.pamatkovykatalog.cz/dul-koh-i-noor-ii-cast-1263911 (accessed on 12 October 2021).
12. VTEI—Vodohospodářské Technicko-Ekonomické Informace. Ročník 62, 4/2020. Available online: https://WWW.vtei.cz/wp-content/uploads/2020/08/6253-casopis-VTEI-4-20.pdf (accessed on 17 January 2022).
13. Seidl, M. Vedoucí střediska Kohinoor a báňský projektant [ústní sdělení]. Most, Czech Republic. 2021.
14. Nařízení Vlády č. 401/2015 Sb., o Ukazatelích a Hodnotách Přípustného Znečištění Povrchových Vod a Odpadních Vod, Náležitostech Povolení k Vypouštění Odpadních Vod do Vod Povrchových a do Kanalizací a o Citlivých Oblastech. Available online: https://aplikace.mvcr.cz/sbirka-zakonu/ViewFile.aspx?type=z&id=38506 (accessed on 22 June 2021).
15. Švec, J. *Zpráva o Průběhu Zkušebního Provozu Biotechnologického Celku*; Palivový kombinát Ústí: Chlumec, Česko, 2021; p. 7.

Disclaimer/Publisher's Note: The statements, opinions and data contained in all publications are solely those of the individual author(s) and contributor(s) and not of MDPI and/or the editor(s). MDPI and/or the editor(s) disclaim responsibility for any injury to people or property resulting from any ideas, methods, instructions or products referred to in the content.

Proceeding Paper

Biodegradation of the Personal Care Products [†]

Helena Hybská *, Martina Mordáčová and Mária Gregušová

Department of Environmental Engineering, Faculty of Ecology and Environmental Sciences, Technical University in Zvolen, 960 01 Zvolen, Slovakia; xlobotkova@tuzvo.sk (M.M.); xbenova@is.tuzvo.sk (M.G.)
* Correspondence: hybska@tuzvo.sk; Tel.: +421-455-206-488
[†] Presented at the 4th International Conference on Advances in Environmental Engineering, Ostrava, Czech Republic, 20–22 November 2023.

Abstract: Excessive consumption of cleaning and disinfecting agents, which constitute a distinct group of emergent pollutants known as PPCPs ("Pharmaceutical and Personal Care Products"), results in their accumulation in aquatic environments. Conventional wastewater treatment plants are unable to effectively remove the emergent pollutants that are present, including personal care product residues. This article focuses on the determination of surfactant substances in model samples prepared from selected personal care products and their biodegradability under laboratory-created aquatic ecosystem conditions. The conducted biodegradation processes, based on the monitored indicator (surfactants) in the model samples, confirm that the utilization of aquatic vegetation and gravel substrates can efficiently eliminate the present contaminants. Insights gained from researching the biodegradability of PPCP group products are applicable, including experiences with plant compositions used in aquatic environments, particularly in the construction of root-zone wastewater treatment systems.

Keywords: pharmaceutical and personal care products (PPCPs); surfactants; biodegradation; aquatic ecosystem

1. Introduction

Wastewater constitutes a high-risk mixture of pollutants which contaminates the environment and threatens human health. In most cases, wastewater treatment plants (WWTP) only partially reduce the content of the specific pollutants present in the treated water. Treated wastewater can still contain a complex mixture of toxic pollutants, which include, e.g., detergents, various disinfectants and cleaning agents, drug and pharmaceutical residues, or pesticides [1,2]. The wastewater treatment process is aimed at achieving the quality of discharged purified water in terms of selected physico-chemical indicators resulting from valid legislation, but not at monitoring the content of specific substances. The content of specific substances has been constantly increasing in recent years.

Specific pollutants can mix in wastewater causing a "cocktail effect" to occur. This effect represents a significant risk caused by the uncertain and unclear response of toxicants and their difficult to predict impact on the recipient and the aquatic environment [3]. The presence of specific pollutants in waters can result in a change in their sensory, organoleptic, and physical properties and affect chemical and biological processes [4,5]. Unstudied substances present in the aquatic environment can be called "contaminants of emerging concern" (CECs)—emerging pollutants. They can potentially or have even been confirmed to threaten human health and the environment [6]. These are substances that reach recipients for a long time, but their importance and impact on the quality of the aquatic environment has not been investigated.

Emerging pollutants have attracted worldwide attention due to their highly toxic effects, very low degradation, long-term action, and extensive distribution in the environment [7]. Determining the presence of emergent pollutants currently brings new challenges in the field of pollution control in all areas of the environment. The presence of emergent pollutants is beginning to be determined primarily in sewage wastewater, but also in surface and drinking water worldwide [6], with the aim of improving the level and quality of environmental analysis.

A special group of emergent pollutants consists of "Pharmaceutical and Personal Care Products" (PPCP's). These are new, modern polluting substances that have been entering the environment over the last several decades. These include, e.g., medicines, nutritional supplements, cleaning and disinfecting agents, cosmetics, and other products commonly used by society [8,9].

Their continuous use and release into the environment causes serious biological damage (persistence in the form of bioaccumulation in the environment, damage to fauna and flora, etc.) [10–12]. In addition to the abovementioned products, the consumption of washing powders, tablets, gels, capsules, etc. has also increased [13,14].

In article [15], we present the results of monitoring surfactants contained in PPCPs. The results of monitoring their content in surface waters prove the need to pay increased attention to them within the treatment process. This article deals with the biodegradation of surfactants contained in PPCPs in the aqueous environment in model samples.

2. Methods and Material

2.1. Model Samples

The model samples used in the research present the real use of the products in households, hotels, schools, etc. The model samples were prepared using products from the group of PPCPs (Tables 1 and 2). The selected products were dissolved in drinking water based on the dosage and method of use which is indicated on the labels of these products (the products in Table 1 have an MSDS—Material Safety Data Sheet, the products in Table 2 are health tested and harmless).

Table 1. Cleaning and disinfecting agents used for the preparation of model samples.

Standard Products	Ecological Products
Washing gels	
Savo chlorine free	Delizia Lavatrice Actilife
Persil Sensitive	Universe gel
Ariel Touch of Lenor	
Washing powders/tablets	
Persil	Universe tablets
Lovela Sensitive	
Products for the toilet	
Domestos Extended power	Toilet Supergel
	Frosch Eko WC
Means for cleaning and disinfecting surfaces, removing dirt—floors	
Alex Extra Protection	Lamino & Lino
Kitchen products	
Jar	Ecobalsam aquatix

Table 2. Cosmetic preparations used on model samples.

Ziaja—shower gel with goat's milk	Dove—hard soap for hands
Ziaja—shampoo with goat's milk	Sanytol—nourishing liquid soap
Balea—Creamy shower gel	Sensodyne—toothpaste
Dove—body and face gel for men	Elmex—toothpaste

2.2. Biodegradation

The model samples were poured into glass containers. In laboratory conditions, in the presence of selected aquatic plants, surfactant degradation processes took place. Their content was continuously checked. The surfactant indicator represents non-active substances that primarily ensure the cleaning effect in the tested products and are water soluble.

Detailed approach and optimization of conditions:
1. Description of the container (aquarium):
   - created to ensure a natural aquatic ecosystem;
   - container volume of 32 L, model sample volume of 25 L;
   - substrate: aquarium gravel substrate;
   - plant communities were formed by plants of the species: *Egeria, Limnophila, Cabomba, Anubias, Echinodorus, Ceratophyllum,* and *Lemna minor*;
   - aeration: aerating motors with pebbles;
   - addition of CO_2: the Neo CO_2 system based on the production of this oxide by the yeast Saccharomyces cerevisiae.
2. Evaluation of the dismantling process:
   - Determination of surfactants at the beginning, after 24 h, and after 48 h.
3. Optimization of surfactant degradation conditions in model samples:
   - the concentration of surfactants in the samples;
   - effect of aeration (addition of O_2) or addition of CO_2;
   - dismantling time.

The biodegradation process took place under laboratory conditions at a constant air-conditioned room temperature.

2.3. Determination of the Surfactant Content

The determination of surfactant content was based on the reaction of a water sample with methylene blue. Anionic surfactants in alkaline media form coloured ionic associates with methylene blue, which are extracted with chloroform. The absorbance of the samples at 650 nm is evaluated. The WTW CINTRA 20 spectrophotometer (GBS Scientific Equipment Pty. Ltd., Melbourne, Australia) was used (STN EN: 903; Water Quality) [16].

3. Results and Discussion

3.1. Process Optimization

The process of the biological degradation of surfactants, which was carried out in the model samples, required the optimization of the conditions.

We found that high surfactant contents in model samples are not suitable for degradation. The reason is the excessive load on aquatic plants and their related damage (destruction of phytomass), which caused the process of their regeneration to be long-term and in many cases their reuse was not possible. A special case were products that contained higher amounts of aggressive ingredients, such as hydrogen peroxides, hydroxides, and other biocidal components that caused an immediate lethal effect on aquatic plants.

Based on previous experience [17], we diluted the model samples so that the total surfactant content was in the range of 15–20 mg/L. We also based on the information that the total content of surfactants in sewage wastewater is lower than the determined content of individual personal care products (wastewater containing various cleaning and disinfecting agents is mixed and diluted in the sewage system).

During the biodegradation process, we used aeration motors to provide O_2. Based on the monitored concentrations of surfactants in the samples obtained during the degradation process, we found that the addition of oxygen significantly slowed down the process. A negative impact was also caused by the intensive formation of foam, which prevented the access of atmospheric oxygen to the aquatic ecosystem that was created.

CO_2 occurs naturally in aquariums (flora respiration and gas exchange above the water's surface). By using the Neo CO_2 system, we ensured the gradual release of natural CO_2 into the water. CO_2 is the basis for photosynthesis and its supply stimulates the growth of flora and maintains good health. The use of this system significantly slowed down the degradation of the surfactants that were present. The plants were in good condition. We conclude that as a source of nutrition, the plants used more easily available CO_2 (for the needs of photosynthesis) and therefore the content of the surfactants did not change significantly during the degradation process. Based on these findings, we continued the biodegradation process without the addition of O_2 and CO_2.

For the efficiency of the use of biodegradation processes on a larger scale in practice, it is necessary that the retention time and the time of the degradation of surfactants be as short as possible. If the concentration of surfactants in the model samples is kept to 20 mg/L, the degradation time is in the range of 24–48 h. The maximum permissible concentration of anionic surface substances in surface waters (1 mg/L) was also observed in accordance with the Regulation of the Government of the Slovak Republic, No. 269/2010 Coll., which specifies the requirements for achieving good water status.

3.2. Evaluation of the Biodegradation Process

The biodegradation process is described in Table 3. Ecological washing gels had a higher % degradation after only 24 h compared to "standard" washing gels. The average % degradation of "standard" washing gels after 48 h is 91.77%, i.e., 6.34% less than the average value for ecological gels (98.11%). In the Persil washing powder, the content of surfactants after 24 h was reduced by 51.21% and after 48 h, the surfactant concentration was only 0.98 mg/L. The Lovela Sensitive washing powder and Universe tablets had no significant differences during degradation and after 48 h > 98% was degraded. The biodegradation of Domestos WC gel ended after 0.5 h from the addition of the model sample to the aquatic ecosystem. This product contains biocidal ingredients that caused permanent damage and death to the aquatic plants. From the point of view of the protection of water ecosystems it is more appropriate to use ecological WC gels whose % degradation after 48 h was > 96%. Alex Extra Protection floor cleaner is 49.97% less biodegradable after 24 h and 9.34% less biodegradable after 48 h compared to the ecological product Lamino & Lino. The ecological product Ecobalsam is more biodegradable by 8.03% after 48 h compared to Jar.

Table 3. The biodegradation process.

Cleaning/Disinfectant Product	At the Beginning mg/L	Concentration of the Surfactants			
		After Biodegradation			
		24 h		48 h	
		mg/L	%	mg/L	%
Washing gels					
Savo chlorine free	10.32	4.22	59.11	0.83	91.96
Persil Sensitive	9.82	3.88	60.49	0.91	90.73
Ariel Touch of Lenor	10.57	4.15	60.74	0.78	92.62
Delizia Lavatrice Actilife	11.01	2.94	73.30	0.27	97.55
Universe gel	10.47	2.11	79.85	0.14	98.66
Washing powders/tablets					
Persil	12.85	6.27	51.21	0.98	92.37
Lovela Sensitive	10.38	4.15	60.02	0.13	98.75
Universe tablets	12.01	4.28	64.36	0.21	98.25

Table 3. Cont.

Cleaning/Disinfectant Product	Concentration of the Surfactants				
	At the Beginning mg/L	After Biodegradation			
		24 h		48 h	
		mg/L	%	mg/L	%
Products for the toilet					
Domestos Extended power	9.86	the process ended after 0.5 h. (biocide)			
Toilet Supergel	9.97	4.36	56.27	0.37	96.29
Frosch Eko WC	9.91	3.11	68.62	0.31	96.87
Means for cleaning and disinfecting surfaces. removing dirt—floor					
Alex Extra Protection	8.44	5.15	38.98	0.94	88.87
Lamino & Lino	9.15	2.02	77.925	0.18	98.03
Kitchen products					
Jar	11.39	5.3	53.47	0.96	91.57
Ecobalsam aquatix	12.87	1.47	88.54	0.06	99.57

Organic products biodegrade faster than "standard" products. Some of them (e.g., Lamino & Lino floor cleaner, Universe tablets and Universe washing gel) have a higher surfactant content than "standard" products from their categories but they still biodegrade faster.

4. Conclusions

Emergent pollutants, including PPCPs require attention due to their highly toxic effects, low degradation, long-term effects, and extensive distribution in the environment. Determining their presence in the environment brings new challenges in the field of pollution control and in finding ways to quickly remove them.

One of the methods of their removal is biological degradability using aquatic flora. By optimizing the conditions of the surfactant biodegradation process in model samples a suitable concentration of up to 20 mg/L was found. This is the concentration that plants can effectively break down. Excessive loading (at a higher concentration) of vegetation can result in permanent damage or the limitation of its functionality and thereby slows down the degradation process. We found that the artificial addition of O_2 to the biodegradation process is not suitable. It significantly slows down decomposition and increases foam production. The addition of CO_2 also slowed surfactant degradation as the plants began to use the more readily available CO_2 from the Neo system for their growth. For the effective use of biodegradation processes on a larger scale in practice, the retention time and the time of degradation must be as short as possible. At a concentration of up to 20 mg/L the degradation time is up to 48 h and the content of surfactants in the water meets the criterion from the Regulation of the Government of the Slovak Republic, No. 269/2010 Coll., which specifies the requirements for achieving good water status (<1 mg/L).

We found that products labelled as organic biodegrade faster than "standard" products. Through the implemented biodegradation processes under laboratory conditions, we managed to break down the surfactants in all samples made from cleaning and disinfecting agents with a break down time of up to 48 h without the addition of O_2 and CO_2.

PPCP products are widespread, and their assortment is very large, therefore, it is important to pay increased attention to the indicator of surfactants in wastewater as well as in surface water. Based on the principle of biodegradation which we implemented under laboratory conditions, root wastewater treatment plants work. The results of this research can be used precisely in the construction of root wastewater treatment plants (experience with plant compositions used in the aquatic environment). Another option for the use of root wastewater treatment plants is the addition of conventional municipal wastewater treatment plants where they would perform the function of purifying wastewater before discharging it to the recipient [18].

Author Contributions: Conceptualization, M.M. and H.H.; methodology, M.M. and H.H.; validation, H.H.; formal analysis, M.G.; resources, M.G.; data curation, M.M.; writing—original draft preparation, M.M.; writing—review and editing, H.H.; visualization, M.M. and M.G.; project administration, H.H. and M.M. All authors have read and agreed to the published version of the manuscript.

Funding: This work was supported by the Technical University in Zvolen under the project VEGA 1/0022/22—Evaluation methods of emergent pollutants by means of microcosms.

Institutional Review Board Statement: Not applicable.

Informed Consent Statement: Not applicable.

Data Availability Statement: Data are available in this manuscript.

Conflicts of Interest: The authors declare no conflict of interest.

References

1. Di Marcantonio, C.; Chiavola, A.; Paderi, S.; Gioia, V.; Mancini, M.; Calchetti, T.; Frugis, A.; Leoni, S.; Cecchini, G.; Spizzirri, M.; et al. Evaluation of Removal of Illicit Drugs, Pharmaceuticals and Caffeine in a Wastewater Reclamation Plant and Related Health Risk for Non-Potable Applications. *Process Saf. Environ. Prot.* **2021**, *152*, 391–403. [CrossRef]
2. Khan, M.T.; Shah, I.A.; Ihsanullah, I.; Naushad, M.; Ali, S.; Shah, S.H.A.; Mohammad, A.W. Hospital Wastewater as a Source of Environmental Contamination: An Overview of Management Practices, Environmental Risks, and Treatment Processes. *J. Water Process Eng.* **2021**, *41*, 101990. [CrossRef]
3. Gosset, A.; Polomé, P.; Perrodin, Y. Ecotoxicological Risk Assessment of Micropollutants from Treated Urban Wastewater Effluents for Watercourses at a Territorial Scale: Application and Comparison of Two Approaches. *Int. J. Hyg. Environ Health* **2020**, *224*, 113437. [CrossRef] [PubMed]
4. Aquatic Ecotoxicology—1st Edition. Available online: https://shop.elsevier.com/books/aquatic-ecotoxicology/amiard-triquet/978-0-12-800949-9 (accessed on 21 September 2023).
5. Gavrilescu, M.; Demnerová, K.; Aamand, J.; Agathos, S.; Fava, F. Emerging Pollutants in the Environment: Present and Future Challenges in Biomonitoring, Ecological Risks and Bioremediation. *New Biotechnol.* **2015**, *32*, 147–156. [CrossRef] [PubMed]
6. Liu, B.; Zhang, S.-G.; Chang, C.-C. Emerging Pollutants—Part II: Treatment. *Water Environ. Res. Res. Publ. Water Environ. Fed.* **2018**, *90*, 1792–1820. [CrossRef] [PubMed]
7. Gorito, A.M.; Ribeiro, A.R.; Almeida, C.M.R.; Silva, A.M.T. A Review on the Application of Constructed Wetlands for the Removal of Priority Substances and Contaminants of Emerging Concern Listed in Recently Launched EU Legislation. *Environ. Pollut.* **2017**, *227*, 428–443. [CrossRef] [PubMed]
8. Dey, S.; Bano, F.; Malik, A. Pharmaceuticals and Personal Care Product (PPCP) Contamination—A Global Discharge Inventory. In *Pharmaceuticals and Personal Care Products: Waste Management and Treatment Technology*; Butterworth-Heinemann: Oxford, UK, 2019; pp. 1–26. ISBN 978-0-12-816189-0.
9. Montes-Grajales, D.; Fennix-Agudelo, M.; Miranda-Castro, W. Occurrence of Personal Care Products as Emerging Chemicals of Concern in Water Resources: A Review. *Sci. Total Environ.* **2017**, *595*, 601–614. [CrossRef] [PubMed]
10. Chen, B.; Han, J.; Dai, H.; Jia, P. Biocide-Tolerance and Antibiotic-Resistance in Community Environments and Risk of Direct Transfers to Humans: Unintended Consequences of Community-Wide Surface Disinfecting during COVID-19? *Environ. Pollut.* **2021**, *283*, 117074. [CrossRef] [PubMed]
11. Ghafoor, D.; Khan, Z.; Khan, A.; Ualiyeva, D.; Zaman, N. Excessive Use of Disinfectants against COVID-19 Posing a Potential Threat to Living Beings. *Curr. Res. Toxicol.* **2021**, *2*, 159–168. [CrossRef] [PubMed]
12. Adams, J.; Bartram, J.; Chartier, Y. *Essential Environmental Health Standards for Health Care*; World Health Organization: Geneva, Switzerland, 2008; p. 57.
13. SanJuan-Reyes, S.; Gómez-Oliván, L.M.; Islas-Flores, H. COVID-19 in the Environment. *Chemosphere* **2021**, *263*, 127973. [CrossRef] [PubMed]
14. Yari, S.; Moshammer, H.; Asadi, A.F.; Jarrahi, A.M. Side Effects of Using Disinfectants to Fight COVID-19. *Asian Pac. J. Environ. Cancer* **2020**, *3*, 9–13. [CrossRef]
15. Lobotková, M.; Hybská, H.; Turčániová, E.; Salva, J.; Schwarz, M.; Hýrošová, T. Monitoring of the Surfactants in Surface Waters in Slovakia and the Possible Impact of COVID-19 Pandemic on Their Presence. *Sustainability* **2023**, *15*, 6867. [CrossRef]
16. STN EN 903 (75 7560) 1.3.1999 | Technická Norma | NORMSERVIS s.r.o. Available online: https://eshop.normservis.sk/norma/stnen-903-1.3.1999.html (accessed on 21 September 2023).

17. Lobotková, M.; Hybská, H.; Samešová, D.; Turčániová, E.; Barnová, J.; Rétfalvi, T.; Krakovský, A.; Bad'o, F. Study of the Applicability of the Root Wastewater Treatment Plants with the Possibility of the Water Recirculation in Terms of the Surfactant Content. *Water* **2022**, *14*, 2817. [CrossRef]
18. Vodohospodársky Spravodajca: Dvojmesačník pre Vodné Hospodárstvo a Životné Prostredie. Available online: https://sekarl.euba.sk/arl-eu/sk/detail-eu_un_cat-0239788-Vodohospodarsky-spravodajca/ (accessed on 21 September 2023).

Disclaimer/Publisher's Note: The statements, opinions and data contained in all publications are solely those of the individual author(s) and contributor(s) and not of MDPI and/or the editor(s). MDPI and/or the editor(s) disclaim responsibility for any injury to people or property resulting from any ideas, methods, instructions or products referred to in the content.

Proceeding Paper

Rapid Chloride Permeability Test of Mortar Samples with Various Admixtures [†]

Miriama Čambál Hološová [1,*], Adriana Eštoková [1] and Miloslav Lupták [2]

1. Institute for Sustainable and Circular Construction, Faculty of Civil Engineering, Technical University of Kosice, Vysokoskolska 4, 04200 Kosice, Slovakia; adriana.estokova@tuke.sk
2. Institute of Materials and Quality Engineering, Faculty of Materials, Metallurgy and Recycling, Technical University of Kosice, Vysokoskolska 4, 04200 Kosice, Slovakia; miloslav.luptak@tuke.sk
* Correspondence: miriama.cambal.holosova@tuke.sk
† Presented at the 4th International Conference on Advances in Environmental Engineering, Ostrava, Czech Republic, 20–22 November 2023.

Abstract: This paper is focused on the permeability of cement composite samples with various admixtures. Permeability was examined by the rapid chloride permeability test due to the simple and relatively quick performance among the many methods. Permeability is one of the durability parameters considering the pore system of the composite structure. Ion diffusion provides information about inner pore structure by passing a charge through the sample. In real-life conditions, not only chlorides but also other ions can penetrate into the structure and cause corrosion. Various cement supplements were used as admixtures. The reference sample consisted of cement, fine sand, and water, while the rest of the samples consisted of 20% blast furnace slag, bypass dust, eggshells or recycled glass instead of the cement. The results showed lower permeability in samples containing blast furnace slag and eggshells and a higher charge passage in samples with recycled glass and bypass dust than for the reference sample.

Keywords: RCP test; cement mortars; admixtures

Citation: Hološová, M.Č.; Eštoková, A.; Lupták, M. Rapid Chloride Permeability Test of Mortar Samples with Various Admixtures. *Eng. Proc.* 2023, *57*, 36. https://doi.org/10.3390/engproc2023057036

Academic Editors: Natalia Junakova, Tomas Dvorsky, Vojtech Vaclavik and Magdalena Balintova

Published: 8 December 2023

Copyright: © 2023 by the authors. Licensee MDPI, Basel, Switzerland. This article is an open access article distributed under the terms and conditions of the Creative Commons Attribution (CC BY) license (https:// creativecommons.org/licenses/by/ 4.0/).

1. Introduction

Concrete is exposed to different aggressive attacks due to various environmental conditions. The durability of concrete can be defined as the resistance to physical or chemical deterioration caused by environmental interaction. Coastal areas and materials in direct contact with water, acid rain, airborne ions, or ions found in de-icing salt are the main sources of aggressive ions that encounter the material, and the activity of aggressive ions causes the corrosion of the structure of the composite cement material. Aggressive ions penetration and transport through the concrete is a process that can take place through a number of mechanisms such as diffusion and convection, among others. When the concentration of ions on the outside of the concrete member is higher than on the inside, the diffusion process occurs [1–4]. Convection transport is only applicable to concrete structures that are in contact with liquid under a pressure head, such as liquid-retaining structures, and is expressed as the rate of flow of liquids through a porous material caused by a pressure head. Therefore, it depends on the pore structure and the viscosity of the liquids or gases involved. Dissolved ions and gases are therefore transported by convection with penetrating water into the concrete [5].

In the past, the durability of concrete used to be specified by maximum water–cement ratio and minimum cement content. The development of materials allowed for the use of various additives and admixtures, including secondary raw materials. This made it more difficult to determine the fundamentals of durability properties. Chemical deterioration is differentiated from physical deterioration by chemical interactions between constituents [6]. The case studies that Metha mentioned in [4] reaffirmed that the key to overall durability is the permeability of the concrete rather than its chemistry. The term permeability generally describes the durability of a material, specifically it describes the ability of liquid to move through the concrete.

To determine the permeability of the material, several methods can be used. Methods can be categorized into the three categories: diffusion test, migration test, and indirect test based on resistivity and conductivity. Among the useable methods are the following: European method EN 13396 (immersion test for repair products and system), Nordest methods NT BUILD 433 (immersion test) and NT BUILD 492 (rapid migration test), INSA steady-state migration test, multi-regime migration test, and resistivity test. The rapid chloride permeability test (RCPT) is method defined in two American standards, ASTM C 1202 [7] and AASHTO T 277 [8], widespread in the USA and Canada, and used in the present research. The methods used have been confirmed to be performance-based specifications, which also include measurements based on the principle of the transport of chloride ions, especially in conditions where the materials are exposed to the influence of the environment [4,9]. RCPT, also known as the Coulomb Test, was first developed by Whiting in 1981 [10]. The RCPT is often utilized for its easy performance and time efficiency.

It was found that several parameters, which were examined by appropriate test methods, were related to different mechanisms of chloride transport. These parameters included the following: chloride diffusion coefficient, capillary absorption, sorptivity, initial surface absorption, the volume of permeable voids, compressive strength, water permeability coefficients, chloride migration, water–cement ratio, etc. [4,11]. The rate of chloride ion ingress into concrete is primarily dependent on pore size, pore distribution, and the interconnectivity of the pore system. The pore structure depends on other factors such as mix design, the type of cement and other mix constituents, the use of supplementary cementitious materials, construction practices, concrete mix proportions, the degree of hydration, compaction, and curing conditions. Whenever the potential risk of chemical corrosion is impended, the concrete should be evaluated for chloride permeability [11–15].

This paper is focused on determining the permeability of various waste-based cement materials investigated by measuring the chloride ions penetration. Measured admixtures were selected to test traditional and non-traditional wastes. The test of ion transportation aimed to obtain more information about the material's internal structure and porosity system, which will provide information about the permeability and durability of the material, by simulating an accelerated model scenario of aggressive ions entering the material from the environment.

2. Materials and Methods

The fundamental tenet of the test is the passage of a charge through the sample during the 6 h test. According to [16] measured specimens with the diameter of 100 mm and thickness of 50 mm (obtained by sawing the central part of 100 × 200 mm concrete cylinders) should be used.

A constant voltage of 60 V was passing through the sealed sample. The sample was in contact with solutions on both sides; one side had a NaCl solution, and the other side had a NaOH solution according to standard [16]. To prevent sample from leaking, the sample was sealed with silicone putty. The samples were in contact with the solutions through a circular cross-section (as shown Figure 1) with sample dimensions of 44 mm circular diameter (on average) and 28 mm mean diameter. This RCP test has been modified to smaller dimensions compared to the guidelines. On the one hand, due to the difference in the composition of the mixtures, a mortar sample was used instead of a concrete sample, and on the other hand, due to the repeated current trend, the time frame was consequently shortened to 90 min.

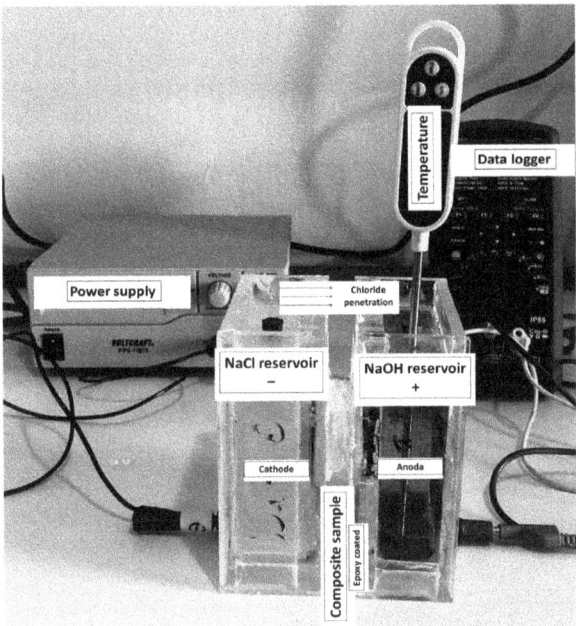

Figure 1. Apparatus for RCP test according to [12] (photo by author).

The amount of current passed through the sealed sample in mA at a constant voltage of 60 V ± 1 V was measured during a period of 90 min. Every minute, the highest and lowest value of the current in mA was recorded. To determine the permeability of mortar samples, the total charge ($Q_{(total)}$) passed through the mortars was calculated based on the maximum current detected within the 90 min measuring interval according to [16]:

$$Q_{(total)} = \int_0^t I dt \quad [C] \tag{1}$$

where:

Q—Charge (C);
I—Current (A);
t—Time (s).

The calculations of the total charge ($Q_{(total)}$) according to the standard [4], considering the 360-min duration of the test, was comparable to the results of other studies. $Q_{(total)}$ was established in accordance with the recurring current transition trend from the current charge achieved during 90 min.

The investigated preparations consisted of a reference sample (CEM) comprising a mix of Ordinary Portland Cement, fine sand, and water. The other samples had the cement component replaced by 20 wt.% of admixtures from secondary raw materials. The high chloride resistance of concrete depends on the type of cement and maximum water–cement ratio [17]. According to [4], concrete with a water-to-cement ratio of 0.38 reports a much lower permeability than the concrete with a water-to-cement ratio of 0.52. In the present research, the water coefficient was set to 0.5 for all mixtures, according to [18].

Blast furnace slag (BFS), bypass dust (BD), eggshells (ESs), and recycled glass (RG) were the secondary raw materials used in the cement composite samples as the 20 wt.% cement (CEM) supplement. Mixtures were taken from the bulk sample into round molds with dimensions of 44 × 28 mm and left for hydration for 28 days. After a period of 28 days, the samples were saturated with water for at least an hour before the RCP test to achieve a faster current increase.

3. Results and Discussion

The values of maximum passed current (I_{max}) through the individual samples measured during the 90 min experiment are reported in Table 1. These maximum values were used to calculate the charge transfer according to Equation (1).

Table 1. Measured maximum currents (I_{max}) per reference sample (CEM) and cement samples with supplements.

Sample	CEM	BFS	BD	ES	RG
Current I_{max} (mA) *	55.52	31.8	71.6	52.6	72.9

* Maximum current in milliamperes.

A comparison of the results of ion penetration between composite samples with supplementation and reference composite sample showed lower permeability for samples with BFS and ES, but higher permeability for samples with BD and RG by more than 30% (Table 1). However, this measurement provided only preliminary input data of the samples. Measured samples contain pozzolanic and latent hydraulic materials, which are subjected to continuous hydration processes, hardening of matrix, and changes in the inner structure.

Figure 2 represents the charge passed through each sample measured during a 6 h test according to the standard test method [7].

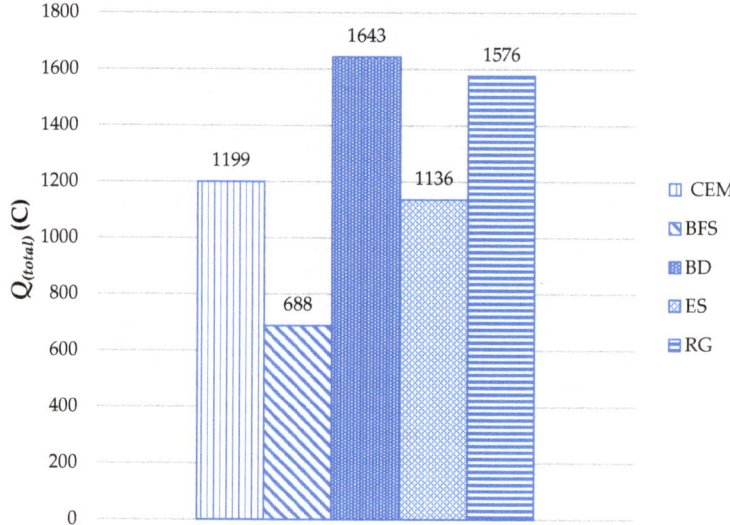

Figure 2. Total charge $Q_{(total)}$ passed through the reference composite sample and cement samples with supplements.

The sample with blast furnace slag (BFS) substitute and with eggshells (ESs) had a lower ion transition than the cement reference sample (CEM). The low permeability of the BFS samples corresponded with the results of the study [19]. Samples with bypass replacement (BD) and glass (RG) showed the highest ion permeability. The results for cement kiln dust (BD) from study [20] corresponded with the results obtained in the present research.

To express the differences in ion penetration and thus permeability among the samples with individual cement substitutes and reference sample, the charge index *(CHI)* was calculated by dividing the values of calculated total charges $Q_{(total)}$ of a sample with supplement and the reference sample:

$$CHI = \frac{Q_{(total)}(i)}{Q_{(total)}(ref)} \quad (2)$$

where:

$Q_{(total)}(i)$ is the total charge calculated per individual sample (C);
$Q_{(total)}(ref)$ is the total charge calculated for reference sample (C).

Figure 3 shows results of charge indexes of composites with supplementations in relation to the reference sample.

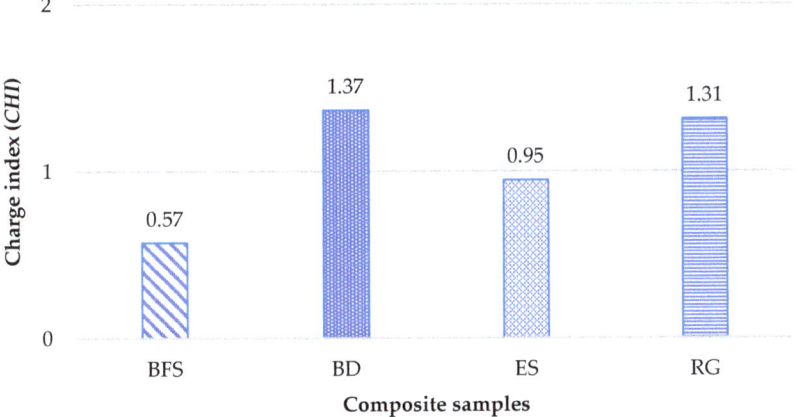

Figure 3. Charge indexes for the cement samples.

The charge transport does not depend only on the chlorides contained in the test solution, but also on the chlorides contained in the sample. Therefore, the sample with bypass dust content was optimized by interpolation calculation, and despite this, it turned out that the structure of the composite is the most permeable among investigated mixtures. This implies that the porous system of the structure is richer in pores filled with air or water, which allows for the passage of not only chloride ions in the pore solution of the composite but also the passage of other ions, for example, OH^-, SO_4^{2-}, Na^+, Ca^{2+}, etc., which participate in concrete corrosion [12].

4. Conclusions

The present work was focused on the estimation of the permeability of the cement composite material with different admixtures comprising secondary raw materials in comparison with a reference cement composite sample without admixture. The pore system of the material structure can lead to the corrosion of the material caused by ion penetration to the inner structure. To investigate the permeability of the material, the rapid chloride permeability test was used. RCPT is widely used test thanks to its relatively quick and simple performance. The results represent the amount of the passing charge transported through the sample under a constant voltage, which shows the permeability of the inner structure. The results showed a higher permeability for samples with bypass dust and recycled glass admixtures than the reference sample. On the other hand, a lower permeability than the reference sample was measured for blast-furnace-slag- and eggshell-based composites. The improvement in terms of decreasing the permeability was found for blast furnace slag samples by up to 30%.

Further investigations including the performance of X-ray fluorescence and diffraction analysis of the samples will provide more information about changes in the compound composition of the materials' structure.

Author Contributions: Conceptualization, A.E. and M.Č.H.; methodology, M.Č.H.; investigation, M.Č.H. and M.L.; writing—original draft preparation, M.Č.H. and A.E.; writing—review and editing, A.E. All authors have read and agreed to the published version of the manuscript.

Funding: This study was carried out as part of the project solution and has been supported by Scientific Grant Agency of the Ministry of Education, science, research and sport of the Slovak Republic and the Slovak Academy of Sciences, projects VEGA Grant No. 1/0230/21 and 2/0108/23.

Institutional Review Board Statement: Not applicable.

Informed Consent Statement: Not applicable.

Data Availability Statement: The data presented in this study are available upon request from the corresponding author.

Conflicts of Interest: The authors declare no conflict of interest. The funders had no role in the design of the study; in the collection, analyses, or interpretation of data; in the writing of the manuscript; or in the decision to publish the results.

References

1. Jau, W.C.; Tsay, D.S. A study of the basic engineering properties of slag cement concrete and its resistance to seawater corrosion. *Cem. Concr. Res.* **1998**, *28*, 1363–1371. [CrossRef]
2. Sanchez, T.; Conciatori, D.; Keserle, G.C. Influence of the type of the de-icing salt on its diffusion properties in cementitious materials at different temperatures. *Cem. Concr. Res.* **2022**, *128*, 104439. [CrossRef]
3. Joshi, P.; Chan, C. Rapid chloride permeability testing. *Concr. Constr.* **2002**, *47*, 37–43.
4. Concrete, C.; Australia, A. Chloride resistance of concrete. In *International Concrete Abstracts Portal*; CCAA: Sydney, Australia, 2009.
5. Wee, T.H.; Suryavanshi, A.K.; Tin, S.S. Evaluation of rapid chloride permeability test (RCPT) results for concrete containing mineral admixtures. *Mater. J.* **2000**, *97*, 221–232.
6. Iffat, S.; Emon, A.B.; Manzur, T.; Ahmad, S.I. An experiment on durability test (RCPT) of concrete according to ASTM standard method using low-cost equipments. *Adv. Mater. Res.* **2014**, *974*, 335–340. [CrossRef]
7. C1202-18; Standard Test Method for Electrical Indication of Concrete's Ability to Resist Chloride Ion Penetration. ASTM International: West Conshohocken, PA, USA, 2012.
8. AASHTO T 277; Standard Method of Test for Electrical Indication of Concrete's Ability to Resist Chloride Ion Penetration. American Association of State Highway and Transportation Officials: Washington, DC, USA, 2011.
9. Huang, K.S.; Yang, C.C. Using RCPT determine the migration coefficient to assess the durability of concrete. *Constr. Build. Mater.* **2018**, *167*, 822–830. [CrossRef]
10. Whiting, D. *Rapid Determination of the Chloride Permeability of Concrete*; Final Report Portland Cement Association: Skokie, IL, USA, 1981.
11. Chen, Y.; Ji, T.; Yang, Z.; Zhan, W.; Zhang, Y. Sustainable use of ferronickel slag in cementitious composites and the effect on chloride penetration resistance. *Constr. Build. Mater.* **2020**, *240*, 117969. [CrossRef]
12. Zych, T. Test methods of concrete resistance to chloride ingress. *Czas. Tech.* **2015**, *6-B*, 117–139.
13. Fenaux, M.; Reyes, E.; Gálvez, J.C.; Moragues, A. Modelling the transport of chloride and other ions in cement-based materials. *Cem. Concr. Compos.* **2019**, *97*, 33–42. [CrossRef]
14. Huang, K.-S.; Yang, C.-C. Using the Chloride Penetration Depth Obtained from Rcpt to Assess the Permeability of Concrete. *J. Environ. Sci. Technol.* **2020**, *28*, 2–5.
15. Abeywickrama, H.M.; Nanayakkara, S.M.A. Applicability of RCPT for performance-based durability design of reinforced concrete structures. Civil Engineering Research Symposium 2023. In Proceedings of the Civil Engineering Research Symposium 2023, Moratuwa, Sri Lanka, 27 September 2023.
16. ASTM C1202-12; Standard Test Method for Electrical Indication of Concrete's Ability to Resist Chloride ion Penetration. ASTM International: West Conshohocken, PA, USA, 2012; Volume 100, pp. 1–8.
17. Shi, C. Effect of mixing proportions of concrete on its electrical conductivity and the rapid chloride permeability test (ASTM C1202 or ASSHTO T277) results. *Cem. Concr. Res.* **2004**, *34*, 537–545. [CrossRef]
18. STN EN 206-1 (STN 73 2403); Concrete. Part 1: Specification, Performance, Production and Conformity. Publication No. 85349:72; Slovak Institute for Technical Normalization: Bratislava, Slovakia, 2002.

19. Sengul, O.; Tasdemir, M.A. Compressive strength and rapid chloride permeability of concretes with ground fly ash and slag. *J. Mater. Civ. Eng.* **2009**, *21*, 494–501. [CrossRef]
20. Kunal; Siddique, R.; Rajor, A.; Singh, M. Influence of bacterial-treated cement kiln dust on strength and permeability of concrete. *J. Mater. Civ. Eng.* **2016**, *28*, 04016088. [CrossRef]

Disclaimer/Publisher's Note: The statements, opinions and data contained in all publications are solely those of the individual author(s) and contributor(s) and not of MDPI and/or the editor(s). MDPI and/or the editor(s) disclaim responsibility for any injury to people or property resulting from any ideas, methods, instructions or products referred to in the content.

Proceeding Paper

Ecotoxicity of Wastewater in the Czech Republic [†]

Ida Antonie Bogáňová [1,*] and Petr Hluštík [2]

1 Vodárenská Akciová Společnost, a.s., Boskovice Division, 17. Listopadu 14, 680 19 Boskovice, Czech Republic
2 Faculty of Civil Engineering, Brno University of Technology, Veveří 331/95, 602 00 Brno, Czech Republic; hlustik.p@fce.vutbr.cz
* Correspondence: ida.antonie.boganova@vodarenska.cz
† Presented at the 4th International Conference on Advances in Environmental Engineering, Ostrava, Czech Republic, 20–22 November 2023.

Abstract: The following article is intended as an introduction to ecotoxicology/wastewater toxicology, as this has currently become a widespread and highly discussed topic with scarce information available to date. With the beginning of the COVID-19 pandemic, when disinfectants began to be used in larger quantities, problems also began at wastewater treatment plants. This situation was the reason for beginning to monitor the ecotoxicity/toxicity of the wastewater flowing into the studied wastewater treatment plant; the wastewater treatment plant entered a state of emergency and was unable to treat the inflowing wastewater. In this critical period, the following parameters were monitored at the inflow of the wastewater treatment plant—BOD5, CHSKCr, NL, N-NH4, N-inorg, N-total, P-total, RAS and pH. The effluent from the wastewater treatment plant was also monitored for BOD5, CHSKCr, NL, $N-NH_4$, $N-NO_3$, $N-NO_2$, N-inorg, N-total, P-total and RAS. Ecotoxicity (fish, barnacles, algae) and toxicity (*Vibrio fischeri*) were monitored at the inflow and outflow.

Keywords: ecotoxicity; water quality indicators; wastewater treatment plant; operation; maintenance; design example

1. Introduction

Ecotoxicity is a key factor regarding environmental protection in the European Union. In this regard, the European Union has strict standards and regulations to minimize the negative effects of chemical substances on the environment and human health.

The European Chemicals Agency (ECHA) [1] and the European Food Safety Authority (EFSA) [2] are responsible for evaluating and regulating chemicals for their ecotoxicity. These organizations carry out scientific evaluations of the effects of substances on aquatic organisms, soil, and other environments. The European Union has also introduced the registration, evaluation, authorization, and restriction of chemicals (REACH) system [3], which aims to improve the protection of human health and the environment from the risks associated with chemical substances. REACH requires the manufacturers and importers of chemicals to demonstrate their safety and carry out tests regarding their ecotoxicity. Another tool used by the European Union for ecotoxicity assessments is the system of classification, labeling, and packaging of chemical substances (CLP) [4]. This system establishes standardized criteria for classifying substances as hazardous to the environment and requires that such substances be appropriately labeled, and the information shared. The parameters of ecotoxicity and toxicity are not monitored, as a standard, in wastewater. However, in my view and in the opinion of some experts, these parameters should be monitored at least at the level of industrial producers, due to toxic wastewater discharges into public sewerage systems. Toxic wastewater can subsequently cause considerable damage to wastewater treatment plants, e.g., the mortality of nitrifying bacteria in activation reactors. These European standards are gradually being implemented by individual EU states.

The issue of toxicity/ecotoxicity is not legislated for operators of wastewater treatment plants in the countries of the European Union. Currently, there are valid European standards

Citation: Bogáňová, I.A.; Hluštík, P. Ecotoxicity of Wastewater in the Czech Republic. *Eng. Proc.* **2023**, *57*, 37. https://doi.org/10.3390/engproc2023057037

Academic Editors: Adriana Estokova, Natalia Junakova, Tomas Dvorsky, Vojtech Vaclavik and Magdalena Balintova

Published: 8 December 2023

Copyright: © 2023 by the authors. Licensee MDPI, Basel, Switzerland. This article is an open access article distributed under the terms and conditions of the Creative Commons Attribution (CC BY) license (https://creativecommons.org/licenses/by/4.0/).

and domestic national standards of individual member states that indicate the method of determining the effect of ecotoxicity on living microorganisms.

In the Czech Republic, a Methodological Instruction has been issued by the Ministry of the Environment of the Czech Republic—Waste Department [5], which defines ecotoxicity as a hazardous property of waste under code H14. In the Czech Republic, it is not mandatory to monitor the ecotoxicity/toxicity indicators for wastewater treatment plant operators. Some operators set this indicator themselves, due to the increased concentrations of toxic substances in wastewater. The reason for the increased levels of toxic substances was the disease COVID-19, which caused a higher concentration of disinfectants in wastewater and medicines. The increased concentration of toxic substances has a direct effect on the biological processes taking place in the activation tank, i.e., it can partially or completely stop the biological cleaning processes.

In the Slovak Republic, the monitoring and assessment of ecotoxicity is monitored according to Government Regulation No. 269/2010 Coll. of the Regulation of the Government of the Slovak Republic, which establishes specific requirements for achieving good water status [6]. This government regulation defines that the indicator of ecotoxicity on aquatic organisms has an indicative character, which is monitored regarding industrial and municipal wastewater discharged directly into surface waters.

Each member state of the European Union approaches the methods of determining ecotoxicity/toxicity differently in terms of legislation and standards. There is no standard for determining ecotoxicity in the Czech Republic. Ecotoxicity is determined according to the Slovak standard STN 83 8303 for testing the hazardous properties of waste, ecotoxicity, acute toxicity tests on aquatic organisms, and testing the inhibition of the growth of algae and higher cultivated plants using *Sinapis alba*, green algae, pearl oysters, and fish [7].

Wastewater toxicity affects living organisms, aquatic organisms, and soil biota. Wastewater discharged into the recipient waters affects the biochemical activity and growth of organisms (algae, bacteria, and protozoa). In the case of animal organisms, their immobilization or outright death occurs [8,9].

Mixed municipal wastewater flows into the sewage treatment plants. The operator should continuously analyze wastewater from important customers (from industrial sites, hospitals, car washes, etc.). Residual concentrations of pollutants in wastewater have a direct impact on the need to optimize wastewater treatment technology [10,11].

On the basis of these wastewater analyses, an evaluation of ecotoxicological risks and the global synergistic effects of ecotoxicity must be conducted, with an emphasis on the residual concentrations of pollutants. An example is hospital wastewater, which has the potential for ecotoxicity and antibiotic resistance in wastewater.

2. Test, Methods, and Evaluation of Ecotoxicological Bioassays

The basic terms employed in the field are ecotoxicology, ecotoxicity, the toxicity of substances, mortality, and immobilization. Their contextual meanings can be described as follows:

Ecotoxicology is a discipline established on the basis of findings from other scientific fields, namely, chemistry, applied ecology, biology, and toxicology. In other words, ecotoxicology can be defined as a discipline that combines knowledge from ecology (a science studying ecosystems) and toxicology (the study of the interaction of chemical substances with living organisms). In 1969, René Truhaut coined the first definition of ecotoxicology: "Study of the toxic effects of chemicals induced by natural and synthetic pollutants to protect natural species and societies" [12].

Ecotoxicity, designated as H14, is a dangerous property of substances with an instant or delayed negative, adverse effect on living organisms or on the environment [12].

Substance toxicity refers to the ability of substances to cause harm to a living organism. Toxicity depends on the physicochemical properties of the given substance, the way the substances enter the organism, their metabolism, and the frequency and dose of the supplied substances [12].

Mortality is a toxic effect that results in the death of the test organism [12].

Immobilization (inhibition of mobility) refers to short-term or long-term immobility, during which the test organism is unable to move in space [12].

Ecotoxicity/toxicity tests are divided into three basic groups—ecotoxicological bioassays, aquatic bioassays, and alternative bioassays.

2.1. Ecotoxicological Bioassays

These tests determine whether a certain substance will have a toxic effect on a given organism in the ecosystem, or whether a biotic bonds in the ecosystem will be disrupted. These tests can be divided according to the exposure time (acute, semi-acute, and chronic), target organism (soil, freshwater, and marine), complexity of the tested sample (natural samples, pure chemical substances, and mixtures of substances), method of sample preparation (direct tests of environmental matrices, the defined concentration of chemical substances, and the testing of leachates or extracts from natural samples), number of test organisms (single species, multi-species, and laboratory mixtures of species), advanced design of the test system (conventional tests with intact organisms, microbiotests, biosensors, biomarkers, and biofunds), degree of complexity of the detection system (enzymes, bioprobes, tissue and cell cultures, intact organisms, population, society), monitored responses (assessment of physiological activity, reproductive activity, and lethal or sub-lethal effects), tested matrix (air, water, soil, waste, sediments, and chemical substances) and trophic levels of test organisms (producers, consumers, and destroyers) [13].

2.2. Aquatic Bioassays

These are the most widespread tests used for determining toxicity. They are conducted with aquatic organisms, most often bacteria, algae, aquatic plants, crustaceans, and fish. They are used to evaluate chemical substances, preparations, and natural water. They are mainly applicable for substances that are soluble in water and for testing aqueous extracts in samples of solid substances. The objective of aquatic tests is to determine the effect of the tested substances on the aquatic ecosystem [13].

2.3. Alternative Bioassays

This type of test is used for several reasons, namely, to save laboratory space and chemicals, to save time, and due to the economic demands of breeding. The breeding of organisms intended for testing takes up a great deal of space in laboratories and is staff-, time-, and chemical consumption-intensive. Alternative bioassays use the resting phases of the organisms [13].

2.4. Determination Methods

Methods to determine ecotoxicity/toxicity in wastewater are described in several international standards, depending on the organism used. These organisms can be divided according to how they receive energy for their life processes (primary producers, primary consumers, secondary consumers, tertiary consumers, and destroyers).

Each of the standards describes the testing procedure using one of the organisms. To illustrate this point, we provide a brief overview:

- ČSN EN ISO 6341 Water quality—*Daphnia magna* Straus motility inhibition test (Cladocera, Crustacea)—Acute toxicity test [14];
- ČSN EN ISO 7346-1 Water quality—Determination of acute lethal toxicity of substances for the freshwater fish *Brachydanio rerio* Hamilton-Buchanan (Teleostei, Cyprinidae) Part 1: Static method [15];
- ČSN EN ISO 7346-2 Water quality—Determination of acute lethal toxicity of substances for the freshwater fish *Brachydanio rerio* Hamilton-Buchanan (Teleostei, Cyprinidae) Part 2: Semi-static method [16];

- ČSN EN ISO 7346-3 Water quality—Determination of acute lethal toxicity of substances for the freshwater fish *Brachydanio rerio* Hamilton-Buchanan (Teleostei, Cyprinidae) Part 3: Flow-through method [17];
- ČSN EN ISO 11348-1 Water quality—Determination of the inhibitory effect of water samples on the light emission of *Vibrio fischeri* (Test on luminescent bacteria)—Part 1: Method using freshly prepared bacteria [18];
- ČSN EN ISO 11348-2 Water quality—Determination of the inhibitory effect of water samples on the light emission of *Vibrio fischeri* (Test on luminescent bacteria)—Part 2: Method using liquid-dried bacteria [19];
- ČSN EN ISO 11348-3 Water quality—Determination of the inhibitory effect of water samples on the light emission of *Vibrio fischeri* (Test on luminescent bacteria)—Part 3: method using freeze-dried bacteria [20];
- ČSN ISO 20665 Water quality—Determination of chronic toxicity for *Ceriodaphnia Dubia* [21];
- ČSN ISO 20666 Water quality—Determination of chronic toxicity for *Brachionus calyciflorus* within 48 h [22];
- ČSN EN ISO 10712 Water quality—*Pseudomonas putida* growth inhibition test (*Pseudomonas* cell reproduction inhibition test) [23];
- ČSN EN ISO 8692 Water quality—Freshwater green algae growth inhibition test [24];
- ČSN EN ISO 20079 Determination of the toxic effect of water constituents and wastewater on duckweed (*Lemna minor*)—Duckweed growth inhibition test [25].

3. Course and Evaluation of Ecotoxicological Tests

The procedure for conducting ecotoxicological tests is shown in Figure 1. The preliminary test is performed first, followed by the verification test, the orientation test, and finally, the basic test.

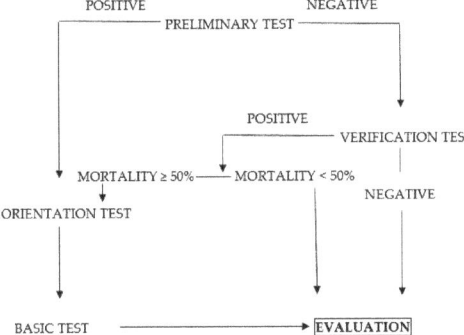

Figure 1. Course of ecotoxicological tests [26].

3.1. Course of Ecotoxicological Tests

Preliminary test: In a preliminary test, a sample of wastewater of unknown toxicity is tested with organisms. Two parallel determinations, including controls, are performed. If the result of this test reveals no mortality of the organism, it is marked as negative, and the verification test is conducted [26].

Verification test: After the verification test is marked as negative, six parallel determinations are performed. If there is no mortality in the wastewater sample at 10% higher than mortality in the control, the test result is marked as negative, and the test is terminated at this point. Otherwise, depending on the immobilization being demonstrated, the orientation test is conducted, namely, when the mortality is higher than 50% [26].

Orientation test: This test is used to determine the range in which a 50% effective concentration can be expected. This concentration is denoted as EC50 (the expressed concentration at which 50% of the tested organisms will die or become immobilized, or there is a change in growth or metabolic activity) [26].

Verification test. For the basic test, the concentration series is selected so that zero and 100% mortality are shown at the extreme concentrations, with the remaining concentrations kept within this range. The EC50 value is determined based on this test [26].

3.2. Evaluation of Ecotoxicological Tests

During these tests, the dependence of the substance concentration on the death of test organisms, and their reproductive capacity, growth, and immobilization are monitored. The so-called "ecotoxicological indices" express the nature of the effect of these substances on organisms (outputs of the toxicity tests). These indices can be used to compare the toxicity of various substances. The ecotoxicological indices describe the level of toxic effects, evaluate the acceptability of substances for the ecosystem, and then divide them into hazard classes. The ecotoxicological indices are divided into two groups—calculated and determined.

Calculated indices: These indices are defined by interpolating experimentally obtained data—EC50, IC50, and LC50. The EC abbreviation stands for effective concentration, LC denotes lethal concentration, IC stands for inhibitory concentration, LD stands for lethal dose, and ED stands for effective dose. The EC50 abbreviation represents the concentration at which 50% of the tested organisms will die or become immobilized, or when there is a change in their growth or metabolic activity.

Determined indices: This group of indices represents the concentrations of substances detected experimentally, the significance of which was verified by statistical comparison with the control. The most widely used indices are NOEC and LOEC. These values are determined by observation. The NOEC abbreviation denotes the highest tested concentration of a toxic substance at which there is no statistically significant effect on the organisms, compared to a control under the same conditions. LOEC stands for the lowest concentration of a toxic substance at which a statistically significant effect is observed compared to the control. NOAEL designates the toxicant dose at which the effect of the given substance that is observed is not considered undesirable.

After carrying out the above tests and evaluating them, the toxicity class can be determined. The individual toxicity classes are shown in Table 1 [26].

Table 1. Toxicity classes and concentration ranges of toxic substances [26].

Class	Concentration Range (mg·L^{-1})	Evaluation
0	$\geq 10^4$	non-toxic
1	10^3–10^4	very mildly toxic
2	10^2–10^3	mildly toxic
3	10^1–10^2	medium toxic
4	10–10^1	highly toxic
5	10^{-1}–10	very highly toxic
6	$\leq 10^{-1}$	extremely toxic

4. Ecotoxicity Monitoring at a Selected Wastewater Treatment Plant

Operating companies usually do not monitor toxicity/ecotoxicity in wastewater. The main reason for this is the legislation of the Czech Republic, which does not require these indicators to be monitored.

Ecotoxicity monitoring was carried out at the wastewater treatment plant operated by Vodárenská akciová společnost, a.s. The reason for taking a sample for ecotoxicity was an accident at the wastewater treatment plant in 2021, which caused the biological processes of wastewater treatment in the activation tank to stop functioning. In the case of the standard indicators of wastewater analyses, the cause of incapacity was not found; therefore, ecotoxicity and toxicity tests were carried out. Analyses carried out in the laboratory revealed the presence of ecotoxicity in the wastewater at the inflow to the wastewater treatment plant. This wastewater treatment plant is designed for 29,376 PE, with an average daily inflow of 4881.2 m^3·day^{-1}.

Since no substance was detected that would cause this cessation, the wastewater treatment plant mentioned below underwent repeated ecotoxicity monitoring. The monitoring took place at a mechanical–biological wastewater treatment plant with primary sedimentation, nitrification, denitrification, the chemical removal of phosphorus, anaerobic stabilization of sludge by digestion, sludge sanitation, sludge thickening, and dewatering.

4.1. Description of Technology Used at the Wastewater Treatment Plant

Wastewater from the service area flows to the pumping station at the wastewater treatment plant through combined sewerage. Water at flow rates reaching a certain value overflows into the river once the stormwater tank has been filled up. Downstream of the pumping station, the pre-treatment stage commences, consisting of mechanically scraped fine screens, a screening press, and a vortex sand trap. The pumps transport mechanically pre-treated wastewater to the primary sedimentation tank. After primary sedimentation, the wastewater is conveyed to the activation system. The biological treatment stage is designed as two independent lines, with two reactors for each of the lines. The activation process is based on controlled nitrification and denitrification, with the simultaneous chemical precipitation of phosphorus. The reactors are equipped with a fine-bubble aeration system and slow-moving horizontal stirrers. Subsequently, the mixed liquor is discharged over a degassing spillway in a split chamber and moves on to two circular secondary tanks. Following the sedimentation process, the water flows via a measuring channel into the river. Primary sludge is extracted from the primary tank into a primary sludge sump and is then fed into a gravity thickener. After settling, the thickened primary sludge is conveyed to a thickened mixed sludge sump. Surplus biological sludge from the secondary tanks is mechanically thickened using a strainer. The thickened mixed sludge is homogenized via a submersible stirrer and is then transported in a controlled manner to a stirred and heated digester. The digested sludge is pumped into sludge storage tank 1, where it is homogenized, and then passes through a pasteurization system as sanitized sludge into sludge storage tank 2, which stores the sludge before its mechanical dewatering using a decanting centrifuge. Once dewatered, the sludge is transported to the sludge management system storage area, from which it is subsequently removed for further processing [27].

4.2. Monitored Parameters

The monitoring took place in the period from June 2021 to March 2022. In the given period, the following parameters were monitored at the WWTP influent and effluent: BOD, COD, SS, N-NH4, N-inorg, N-total, P-total, RAS, and pH. Ecotoxicity and toxicity were monitored at the influent and effluent. Sampling was carried out on an irregular basis but occurred, on average, about 1–3 times a month.

BOD expresses biochemical oxygen demand, which is defined as the volume of oxygen consumed by microorganisms for the decomposition of organic matter under aerobic conditions. COD stands for chemical oxygen demand, which indicates the total oxygen consumption for organic matter oxidation in a wastewater sample. SS denotes suspended solids, where an exact volume of homogenized wastewater samples is filtered through a dried and weighed glass fiber filter under reduced pressure. The filter is then dried at 105 °C and the SS weight is determined by weighing [28]: N-NH$_4$—ammonia nitrogen, N-NO$_3$—nitrate nitrogen, N-NO$_2$—nitrogen dioxide, N-inorg—inorganic nitrogen, N-total nitrogen, P–total phosphorus, and RAS—dissolved inorganic salts. The test for the presence of toxicity in wastewater was performed on Vibrio Fischer in undiluted samples, 10-times-diluted, and 100-times-diluted samples. Ecotoxicity was tested on fish, pearl oysters, and algae in undiluted, 10-times-diluted, and 100-times-diluted samples.

4.3. Values of the Monitored Indicators

Tables 2 and 3 show the minimum and maximum values of the monitored influent and effluent parameters at the wastewater treatment plant.

Table 2. Values of selected parameters at the inflow to the wastewater treatment plant in mg/L [28].

Indicator	Minimum Value	Maximum Value
BOD	100	480
COD	220	950
SS	80	550
N-NH$_4$	15	85
N-inorg	15	85
N-total	25	88
P-total	2	10
RAS	720	720
pH	6.8	8.3

Table 3. Values of selected parameters at the effluent from the wastewater treatment plant in mg/L [28].

Indicator	Minimum Value	Maximum Value
BOD	1.5	8
COD	8	73
SS	1	15
N-NH$_4$	0	10
N-inorg	0.5	16
N-total	0	11
P-total	4	17
RAS	6	18
pH	0.7	3.6

Toxicity at the influent to the wastewater treatment plant, which was tested on Vibrio Fischer, ranged in the given period from 36 to 98% in the undiluted sample, 3–79% in the 10-times-diluted sample, and 0.2–10.9% in the 100-times-diluted sample.

Only one sample was taken at the effluent point and its toxicity was determined at 8% in the undiluted sample. As regards the ecotoxicity test at the influent to the wastewater treatment plant in an undiluted sample, fish mortality was 100%, water flea immobilization totaled 55–100%, and algae inhibition was at 20–100%. For the 10-times-diluted samples, fish mortality was 0–50%, water flea immobilization was 0–63%, and algae inhibition was 0–30%. Only one sample was taken at the effluent point and the values were only determined for the undiluted sample—fish mortality was at 0%, water flea immobilization at 7%, and algae inhibition at 19% [28].

5. Discussion

During the COVID-19 epidemic, the above-mentioned wastewater treatment plant was repeatedly in an emergency situation. The first such "signal" that something was happening at the wastewater treatment plant was the detection of higher nitrite values in the effluent and the impossibility of reducing the concentration of total phosphorus in the wastewater. Over the monitored period, the nitrite values were measured in tens of mg·L^{-1}. Such values are already very toxic for living organisms. As regards the problem of reducing the concentration of total phosphorus in wastewater, the doses of the precipitant, ferric sulfate, were increased. However, despite such increased doses, no reduction was achieved. Ferric sulfate dosing lowers the wastewater pH, which should not drop below 6.3. pH values of <6.3 cause the nitrification process to stop. During this period, it was necessary to bring activated sludge from another wastewater treatment plant, which was costly and time-consuming but offered an important solution to the given situation. The water in the secondary tank had a gray color. Another manifestation of the presence of ecotoxicity/toxicity in the wastewater was the formation and accumulation of froth at the effluent point from the wastewater treatment plant. The grass growing around the outlet was yellow-brown (considered to have been "burned" by the wastewater). The centrifuged sludge sample was more or less disinfected. To illustrate this conclusion, the concentrations of thermophilic coliform bacteria in a sample taken before the monitored event were at 57,000 KTJ. In the monitored period, this concentration dropped to a value of 1200 KTJ.

6. Conclusions

The topic that this article deals with has not yet been sufficiently explored in the context of the Czech Republic. In the Czech Republic, there is no valid legislation setting limiting values/boundaries for ecotoxicity/toxicity in discharged wastewater.

Legislation for the assessment of ecotoxicity/toxicity should be introduced in the Czech Republic. This legislation should define the methods of determining ecotoxicity, set limits for the discharge of wastewater into public sewers for significant customers, and set limits for discharges from wastewater treatment plants.

The operators of wastewater treatment plants and basin managers should be more concerned with the issue of ecotoxicity/wastewater toxicity due to potential accidents at wastewater treatment plants, which may result in the pollution of the recipient area. Each operator should continuously evaluate the ecotoxicological risks and global synergistic effects of ecotoxicity on water life.

Author Contributions: Conceptualization, I.A.B. and P.H.; methodology, I.A.B.; formal analysis, P.H. and I.A.B.; writing—original draft preparation, P.H.; writing—review and editing, I.A.B. and P.H.; funding acquisition, P.H. All authors have read and agreed to the published version of the manuscript.

Funding: This research received no external funding.

Institutional Review Board Statement: Not applicable.

Informed Consent Statement: Not applicable.

Data Availability Statement: The authors declare that all data supporting the results of this research are available in this article.

Conflicts of Interest: The authors declare no conflict of interest.

References

1. European Chemicals Agency (ECHA). Available online: https://european-union.europa.eu/institutions-law-budget/institutions-and-bodies/search-all-eu-institutions-and-bodies/european-chemicals-agency-echa_en (accessed on 10 October 2023).
2. European Food Safety Authority (EFSA). Available online: https://european-union.europa.eu/institutions-law-budget/institutions-and-bodies/search-all-eu-institutions-and-bodies/european-food-safety-authority-efsa_en (accessed on 12 October 2023).
3. Registration of Chemical Substances (REACH). Available online: https://europa.eu/youreurope/business/product-requirements/chemicals/registering-chemicals-reach/index_en.htm (accessed on 1 October 2023).
4. Regulation of the European Parliament and of the Council European Directive No. 1272/2008 on Classification, Labeling and Packaging of Substances and Mixtures (CLP Regulation). Available online: https://eur-lex.europa.eu/legal-content/EN/TXT/HTML/?uri=OJ:L:2008:353:FULL (accessed on 5 October 2023).
5. Waste Department, Ministry of the Environment of the Czech Republic. *Methodological Instruction of the Department of Waste to Determine the Ecotoxicity of Waste*; Ministry of the Environment of the Czech Republic: Prague, Czech Republic, 2007.
6. *Government Regulation No. 269/2010 Coll*, Ordinance of the Government of the Slovak Republic. Requirements for Achieving Good Water Status; Czech Standardization Institute: Praha, Czech Republic, 2010.
7. *STN 838303*; Testing of Hazardous Waste Properties, Ecotoxicity, Acute Toxicity Tests on Aquatic Organisms and Tests of Inhibition of the Growth of Algae and Higher Cultivated Plants. Czech Standardization Institute: Praha, Czech Republic, 1999.
8. Liwarska-Bizukojc, E. Evalution of Ecotoxicity of Wastewater from the Full-Scale Treatment Plants. *Water* **2022**, *14*, 3345. [CrossRef]
9. Hu, Y.; Lei, D.; Wu, D.; Xia, J.; Zhou, W.; Cui, C. Residual β-lactam antibiotics and ecotoxicity to *Vibrio fischeri*, *Daphnia magna* of pharmaceutical wastewater in the treatment process. *J. Hazard. Mater.* **2022**, *425*, 127840. [CrossRef] [PubMed]
10. Perrodin, Y.; Orias, F. Ecotoxicity of Hospital Wastewater. *Handb. Environ. Chem.* **2017**, *1590*, 60. [CrossRef]

11. Laquaz, M.; Dagot, C.; Bazin, C.; Bastide, T.; Gaschet, M.; Ploy, M.C.; Perrodin, Y. Ecotoxicity and antibiotic resistance of a mixture of hospital and urban sewage in a wastewater treatment plant. *Environ. Sci. Pollut. Res.* **2018**, *25*, 9243–9253. [CrossRef] [PubMed]
12. Dries, J.; Daens, D.; Geunes, L.; Blust, R. Evaluation of acute ecotoxicity removal from industrial wastewater using a battery of rapid bioassays. *Water Sci. Technol.* **2014**, *70*, 2056. [CrossRef] [PubMed]
13. Cvikýřová, Z. Ecotoxicity of Selected Musk Compounds. University of Technology in Brno. Faculty of Chemistry. Institute of Environmental Protection Chemistry and Technology. 2012. Available online: http://hdl.handle.net/11012/6379 (accessed on 1 October 2023).
14. ČSN EN ISO 6341; Water Quality—*Daphnia magna* Straus Motility Inhibition Test (Cladocera, Crustacea)—Acute Toxicity Test. Czech Standardization Institute: Praha, Czech Republic, 2023.
15. ČSN EN ISO 7346-1; Water Quality—Determination of Acute Lethal Toxicity of Sub-Stances for the Freshwater Fish *Brachydanio rerio* Hamilton-Buchanan (Teleostei, Cyprinidae) Part 1: Static Method. Czech Standardization Institute: Praha, Czech Republic, 1999.
16. ČSN EN ISO 7346-2; Water Quality—Determination of Acute Lethal Toxicity of Substances for the Freshwater Fish *Brachydanio rerio* Hamilton-Buchanan (Teleostei, Cyprinidae) Part 2: Semi-Static Method. Czech Standardization Institute: Praha, Czech Republic, 1999.
17. ČSN EN ISO 7346-3; Water Quality—Determination of Acute Lethal Toxicity of Sub-Stances for the Freshwater Fish *Brachydanio rerio* Hamilton-Buchanan (Teleostei, Cyprinidae) Part 3: Flow-Through Method. Czech Standardization Institute: Praha, Czech Republic, 1999.
18. ČSN EN ISO 11348-1; Water Quality—Determination of the Inhibitory Effect of Water Samples on the Light Emission of *Vibrio fischeri* (Test on Luminescent Bacteria)—Part 1: Method Using Freshly Prepared Bacteria. Czech Standardization Institute: Praha, Czech Republic, 2009.
19. ČSN EN ISO 11348-2; Water Quality—Determination of the Inhibitory Effect of Water Samples on the Light Emission of *Vibrio fischeri* (Test on Luminescent Bacteria)—Part 2: Method Using Liquid-Dried Bacteria. Czech Standardization Institute: Praha, Czech Republic, 1999.
20. ČSN EN ISO 11348-3; Water Quality—Determination of the Inhibitory Effect of Water Samples on the Light Emission of *Vibrio fischeri* (Test on Luminescent Bacteria)—Part 3: Method Using Freeze-Dried Bacteria. Czech Standardization Institute: Praha, Czech Republic, 2009.
21. ČSN ISO 20665; Water Quality—Determination of Chronic Toxicity for *Ceriodaphnia dubia*. Czech Standardization Institute: Praha, Czech Republic, 2010.
22. ČSN ISO 20666; Water Quality—Determination of Chronic Toxicity for *Brachionus calyciflorus* within 48 h. Czech Standardization Institute: Praha, Czech Republic, 2010.
23. ČSN EN ISO 10712; Water Quality—*Pseudomonas putida* Growth Inhibition Test (*Pseudomonas* Cell Reproduction Inhibition Test). Czech Standardization Institute: Praha, Czech Republic, 1997.
24. ČSN EN ISO 8692; Water Quality—Freshwater Green Algae Growth Inhibition Test. Czech Standardization Institute: Praha, Czech Republic, 2012.
25. ČSN EN ISO 20079; Determination of the Toxic Effect of Water Constituents and Wastewater on Duckweed (*Lemna minor*)—Duckweed Growth Inhibition Test. Czech Standardization Institute: Praha, Czech Republic, 2007.
26. Hamplová, V. Study of Semichronic Toxicity Test on *Sinapis alba* L. Seeds (White Mustard) for Selected Type of Aqueous Waste. College of Mining, Technical University of Ostrava. 2015. Available online: http://hdl.handle.net/10084/108197 (accessed on 5 October 2023).
27. Eyer, M.; Mikulášek, P.; Bláha, M. *Operating Regulations for the Permanent Operation of the Wastewater Treatment Plant at Company Vodárenská Akciová Společnost, a.s.*; Czech Standardization Institute: Praha, Czech Republic, 2008.
28. Vorálek, J. *Partial Evaluation of Progress, Monitoring and Implemented Water Management Measures at Wastewater Producers at Company Vodárenská Akciová Společnost, a.s.*; Czech Standardization Institute: Praha, Czech Republic, 2022.

Disclaimer/Publisher's Note: The statements, opinions and data contained in all publications are solely those of the individual author(s) and contributor(s) and not of MDPI and/or the editor(s). MDPI and/or the editor(s) disclaim responsibility for any injury to people or property resulting from any ideas, methods, instructions or products referred to in the content.

Proceeding Paper

Drought Risks Assessment Using Standardized Precipitation Index [†]

Martina Zeleňáková [1,*], Tatiana Soľáková [1], Mladen Milanović [2], Milan Gocić [2] and Hany F. Abd-Elhamid [3,4]

[1] Institute of Circular and Sustainable Construction, Faculty of Civil Engineering, Technical University of Kosice, 04200 Kosice, Slovakia; tatiana.solakova@tuke.sk
[2] Faculty of Civil Engineering and Architecture, University of Nis, 18000 Nis, Serbia; mmsmladen@gmail.com (M.M.); mgocic@yahoo.com (M.G.)
[3] Department of Water and Water Structures Engineering, Faculty of Engineering, Zagazig University, Zagazig 44519, Egypt; hany_farhat2003@yahoo.com
[4] Department of Environmental Engineering, Faculty of Engineering, Technical University Kosice, 04200 Kosice, Slovakia
* Correspondence: martina.zelenakova@tuke.sk
[†] Presented at the 4th International Conference on Advances in Environmental Engineering, Ostrava, Czech Republic, 20–22 November 2023.

Abstract: The paper explores the occurrence of minimal precipitation extremes at specific meteorological stations in the southeastern region of Serbia. With climate change leading to increased instances of droughts, these natural phenomena have garnered heightened interest due to their negative impacts on society, environment, and the economy. Employing the SPI-12 index in the southeastern part of Serbia from 1946 to 2021, the study sheds light on the vulnerability of this natural phenomenon in the observed stations. Understanding historical manifestation of these events helps water resource managers, farmers and policymakers manage these risks in the southeastern region of Serbia.

Keywords: drought; Serbia; vulnerability of drought

1. Introduction

Drought is a natural hazard whose likelihood increases with changing weather conditions [1]. The hydrologic cycle's parameters have been affected by climate, which led to changes in temperature and precipitation patterns that may increase the probability of floods and droughts [2,3]. The analysis of drought is a well-known topic in the scientific field. Starting from defining the main parameters of drought, to defining how to quantify it using different drought indices [4], the researchers have been occupied with how to describe drought risk assessment [5,6]. The SPI is used for drought analysis in different areas of the world [7]. This study aims to assess drought in Serbia from 1946 to 2021 using the Standardized Precipitation Index (SPI), which is based on measured monthly precipitation data at five synoptic stations located in the south-east of the country.

2. Materials and Methods

The Standardized Precipitation Index is a versatile and powerful tool for drought analysis [8]. It was developed to determine the precipitation change over a period of time. The SPI is calculated every n months at a different time scale, from 1 to 48 months depending on the time of interest. It can assist in diagnosing, defining, and monitoring drought that impacts a variety of human activities as well as ecosystems [9]. The SPI is used in this study to analyze drought at five meteorological stations (Nis, Leskovac, Dimitrovgrad, Vranje, Zajecar) located in the south-east part of Serbia. The monthly precipitation data were collected for the period from 1946 to 2021 and used for calculating the SPI for a 12-month time scale. The SPI is based on the cumulative probability of precipitation occurring at the observation station and the application of the Gamma function [8]. The analyzed region is characterized by a moderate precipitation regime with the average annual precipitation of 650 mm. The precipitation values are below average in Serbia.

3. Results and Concluding Remarks

The results of SPI for drought are presented and analyzed. The results of SPI-12 for five stations are shown in Figure 1 and extreme droughts are marked in green. Drought analysis using the SPI was done on a 12-month time scale for all selected stations for 75 years. The results of the SPI were used to identify the intensity of drought. The results for the selected five stations in south-east Serbia are presented in Table 1. The Nis station has the highest drought intensity (-3.809) and the Zajecar station has the highest wet intensity (3.231). According to results, the driest years for this region are 1993, 1949 and 1985, while the wettest years are 2010, 1955 and 2018, respectively. Based on the terrain topography and local hydro-climatic conditions, it is clear that drought periods and intensity are not the same at all stations. The Leskovac station stands out, from all stations, as the driest station with 142 dry months in observed period, while Vranje is the wettest station with 149 months. The analysis showed that extremely dry states are presented at Nis in 1947 (12 months) and in 1949 (4 months), Leskovac in 1959 and 1972 (5 months) and in 1964 (4 months), Dimitrovgrad in 1950 (6 months), in 1949 and 1993 (5 months) and in 2001 (4 months), Vranje in 1992 (5 months) and in 1990 and 2001 (4 months) and Zajecar in 2001 (4 months). It is clear that the driest stations are Zajecar and Dimitrovgrad (especially in the period 1946–1952) and the wettest is the Vranje station.

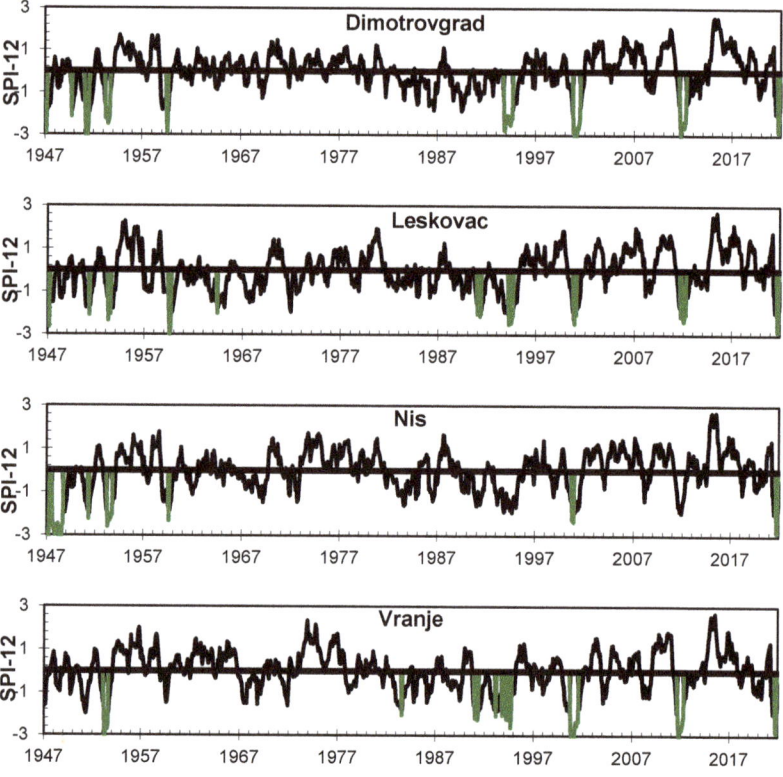

Figure 1. *Cont.*

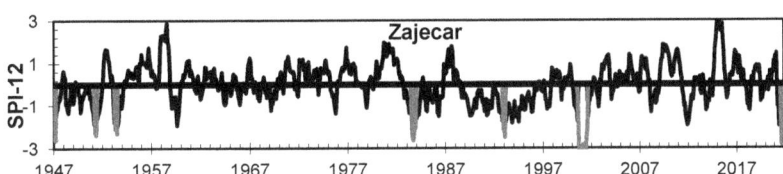

Figure 1. SPI-12—hydrographs for south-east part of Serbia during the period 1946–2021. Black are the SPI values; green are the extreme values of drought that occurred at this station.

Table 1. Identification of drought by SPI-12 for selected five stations.

Station	Maximum Drought Intensity	Driest Year	Maximum Wet Intensity	Wettest Year
Dimitrovgrad	−3.358	1950	2.771	2015, 2021
Leskovac	−3.177	1959, 1964	3.004	1955
Nis	−3.809	1947	3.110	2005
Vranje	−3.227	1993	2.814	2010
Zajecar	−3.337	1993	3.231	2010

The most frequent long-term minimum precipitation totals appear in the winter months. In the period from 1965 to 1983, no extremely long-term meteorological drought was recorded in the monitored stations. The Leskovac station is the most sensitive to the occurrence of extreme long-term precipitation deficits.

4. Conclusions

The analysis of drought using the SPI was calculated on a 12-month time scale for five stations in the south-east of Serbia for 75 years from 1946 to 2021. From the results, it is possible to see how extreme droughts often occurred during the period 1946–2021. Drought analysis could help decision-makers in water resources management. The results approved that SPI is a useful tool that could help decision-makers make efficient plans for drought risk management that could mitigate the impacts of such disasters. This analysis, on a local scale, can be easily applied in different locations around the world.

Author Contributions: Conceptualization, M.Z., H.F.A.-E. and M.G.; Data curation, T.S. and M.M.; Formal analysis, M.Z., H.F.A.-E. and M.G.; Investigation, T.S., M.M. and H.F.A.-E.; Methodology, M.Z., H.F.A.-E. and M.G.; Project administration, M.Z. and M.G.; Resources, M.Z. and M.G.; Software, T.S., M.M. and H.F.A.-E.; Supervision, H.F.A.-E. and M.G.; Validation, T.S., M.M. and H.F.A.-E.; Visualization, T.S. and M.M.; Writing—original draft, M.Z. and T.S.; Writing—review & editing, M.Z., M.G., T.S., M.M. and H.F.A.-E. All authors have read and agreed to the published version of the manuscript.

Funding: This research received no external funding.

Institutional Review Board Statement: Not applicable.

Informed Consent Statement: Not applicable.

Data Availability Statement: Data are contained within the article.

Acknowledgments: This work was supported by the Slovak Research and Development Agency under the Contract no. APVV-20-0281 and SK-SRB-21-0052.

Conflicts of Interest: The authors declare no conflict of interest.

References

1. Paulo, A.A.; Rosa, R.D.; Pereira, L.S. Climate trends and behaviour of drought indices based on precipitation and evapotranspiration in Portugal. *Nat. Hazards Earth Syst. Sci.* **2010**, *12*, 1481–1491. [CrossRef]
2. Chen, X.; Wang, S.; Hu, Z.; Zhou, Q.; Hu, Q. Spatiotemporal characteristics of seasonal precipitation and their relationships with ENSO in Central Asia during 1901–2013. *J. Geogr. Sci.* **2018**, *28*, 1341–1368. [CrossRef]
3. Arab Amiri, M.; Gocic, M. Analyzing the applicability of some precipitation concentration indices over Serbia. *Theor. Appl. Climatol.* **2021**, *146*, 645–656. [CrossRef]
4. Svoboda, M.; Fuchs, B. *Handbook of Drought Indicators and Indices*; Drought Mitigation Center Faculty Publications, University of Nebraska: Linkoln, NE, USA, 2016.
5. Wu, H.; Wilhite, D.A. An Operational Agricultural Drought Risk Assessment Model for Nebraska, USA. *Nat. Hazards* **2004**, *33*, 1–21. [CrossRef]
6. Yi, F.; Li, C.; Feng, Y. Two precautions of entropy-weighting model in drought-risk assessment. *Nat. Hazards* **2018**, *93*, 339–347. [CrossRef]
7. Zeleňáková, M.; Soľáková, T.; Purcz, P.; Abd-Elhamid, H.F. Analysis of drought in the eastern part of Slovakia using standardized precipitation index. In Proceedings of the International Conference on Water Resources Management and Sustainability: Solutions for Arid Regions, Dubai, United Arab Emirates, 22–24 March 2022.
8. McKee, T.B.; Doesken, N.J.; Kleist, J. Drought monitoring with multiple time scales. In Proceedings of the 9th Conference on Applied Climatology, American Meteorological Society, Dallas, TX, USA, 15–20 January 1995; pp. 233–236.
9. Mishra, A.K.; Singh, V.P. A review of drought concepts. *J. Hydrol.* **2010**, *354*, 202–216. [CrossRef]

Disclaimer/Publisher's Note: The statements, opinions and data contained in all publications are solely those of the individual author(s) and contributor(s) and not of MDPI and/or the editor(s). MDPI and/or the editor(s) disclaim responsibility for any injury to people or property resulting from any ideas, methods, instructions or products referred to in the content.

Proceeding Paper

Ecotoxicity Assessment of Substrates from a Thermally Active Coal Tailing Dump Using Tests for *Daphnia magna* †

Veronika Bilkova *, Bohdana Simackova, Oto Novak and Lukas Balcarik

Department of Environmental Engineering, Faculty of Mining and Geology, VŠB—Technical University of Ostrava, 17. Listopadu 2172/15, 708 00 Ostrava, Czech Republic
* Correspondence: veronika.polakova.st@vsb.cz
† Presented at the 4th International Conference on Advances in Environmental Engineering, Ostrava, Czech Republic, 20–22 November 2023.

Abstract: The aim of the study was to compare the ecotoxicity of waste materials formed from a mixture of construction materials with tailings obtained from the thermally active Ema coal tailing dump located in the city of Ostrava, Czech Republic. The ecotoxicity assessment was performed using acute lethality tests on the crustacean *Daphnia magna*. The test results are relevant for further possibilities of using technogenic substrates from tailings after mining activities and are an integral part of a comprehensive assessment of their biological effects on the environment.

Keywords: ecotoxicology; toxicity; *Daphnia magna*; coal mining; heaps; reclamation

1. Introduction

With the growing awareness of the negative impact of human activities on the planet, it is increasingly important to monitor and assess environmental stresses. One of the key tools in this area are ecotoxicity tests, which allow the identification of potentially hazardous substances and assess their impacts on ecosystems. These tests verify the reactions of living organisms to different concentrations and types of chemicals or mixtures of chemicals, thus providing relevant information on the risk effects of these substances and their direct effects on living organisms in the environment.

One of the areas that require close monitoring for ecotoxicity is anthropogenic environments following mining and industrial activities. The environment contaminated by mining and industry induces changes not only in the stability of soil ecosystems [1–4], but also leads to erosion and other changes in the structure and composition of soil substrates down to the level of their microstructures [5–10]. All of these changes then reduce the use of these substrates in the processes of revegetation and reclamation in these areas.

The tailings occurring in the Ostrava-Karviná district (Czech Republic) were never covered in any way and were therefore exposed to chemical weathering. Since the tailings also contained large amounts of fine-grained coal mass that was difficult to separate by conventional treatment methods, thermal processes caused by oxidation or combustion of the coal mass were common in the tailings [11]. The process of spontaneous combustion of the deposited tailings is mainly influenced by their chemical properties, especially the amount of water, ash, and combustible elements (e.g., pyritic sulphur, hydrogen). The ongoing endogenous combustion is closely related to the emission of gases, the possible resuspension of dust particles, and the presence of hazardous elements that may be present in the tailings material [12–15]. Another issue closely related to thermally active tailings is their acidification. Oxidation of pyrite produces sulfuric acid, which causes a decrease in pH [16]. Due to the low content of basic rocks in Ostrava-Karviná district tailings, acid neutralization occurs very slowly. Thus, considering the low pH value, the release of hazardous metals may occur. Thermally active tailings can thus represent a significant environmental burden, especially if the stored material burns through [17–19].

Today, coal tailings, which are mainly made up of rocks, are already being used, for example, in the construction industry and in the reclamation of landscapes after the end of deep mining. Within the construction industry, tailings are used as an available and cheap aggregate for earth embankments in transport structures [20]. However, for such use, it is necessary that the tailings have appropriate properties and do not endanger the construction environment with undesirable substances. Coal gangue has a low pH value, which can affect the mobility of toxic metals. However, its pH value can be adjusted with alkaline construction material, which is currently being used as a backfill material.

The aim of this study was to evaluate the ecotoxicity of tailings from the thermally active Ema tailings impoundment that were mixed with construction wastes for further use. The aquatic crustacean genus *Daphnia magna*, a species widely used in ecotoxicological studies, was selected as the test organism.

2. Materials and Methods

2.1. Description of the Origin of the Samples

Tailings samples were taken in September 2022 at the thermally active Ema coal tailing dump, Czech Republic. Sampling was carried out in places without vegetation from a depth of about 10 cm. Sampling points were chosen so that they were distributed evenly over the unvegetated area of the tailing dump. The samples of construction and demolition waste (CDW) came from renovation work in a prefab house. The waste mainly consisted of hollow burnt bricks, lime-cement mortar, plaster, including painting, and concrete. The construction waste was freed from the remains of foreign materials (plastics, paper, etc.). The tested samples were created by mixing tailings and CDW in a ratio of 9:1, 8:2, and 7:3. The proportions were chosen to preserve a greater amount of tailings. Another condition for the selection of parameters was maintaining the pH/H2O of the mixture around 7.0. Samples of the CDW mixture and tailings were mechanically treated by crushing on a Retsch jaw crusher type BB200 WC (Haan, Germany). After homogenization, they were fine-grained through Retsch sieves with a mesh size of 2 mm and subsequently dried to a constant weight in a vacuum dryer VO29 MEMMERT (Schwabach, Germany). The dried samples were kept for further analysis in a desiccator.

2.2. Methodologies Used for Determining Physico-Chemical Parameters and Bioavailability of Selected Metals

The active pH value (pH/H_2O) and the exchangeable pH value ($pH/CaCl_2$) were determined according to the standard ČSN EN ISO 10390 (836221) [21]. According to the standard ČSN ISO 11265 (836210) [22], the determination of soluble salts (conductivity, κ) in the samples was carried out. The obtained values were assessed according to APAL (2023).

To determine the bioavailability of selected risk metals (Co, Mn, Ni, Zn, and Fe), the BCR (Community Bureau of Reference) sequential extraction analysis according to Sutherland and Tack (2007) was chosen.

2.3. Analytical Methods

The F-AAS (flame atomic absorption spectrometry) method was applied to determine the concentration of selected hazardous metals in individual fractions using an AAS contrAA® 700 atomic absorption spectrometer from Analytik Jena GmbH company (Jena, Germany).

The element composition of the samples was determined using the X-ray powder diffraction (XRD) technique. XRD patterns were obtained using a Rigaku SmartLab diffractometer (Rigaku, Tokyo, Japan) with a D/teX Ultra 250 detector. The measured XRD patterns were evaluated using PDXL 2 software (version 2.4.2.0) and compared with the PDF-2 database, 2015 release (ICDD, Newton Square, PA, USA).

2.4. 48 h Acute Toxicity Test on Daphnia magna

Acute toxicity tests were performed using the Daphnotoxkit F test kit (MicroBiotests Inc., Ghent, Belgium). This test included pearl mussel epiphytes that are allowed to hatch 72 h prior to the start of tesus. This method allows simple age determinations of the individuals used in the test and ensures that only females are present. After prior activation, pearl-eyes were exposed to the test samples. For this test, aqueous leachates of a mixture of tailings from the Ema tailings impoundment and CDW were used at a ratio of 9:1, 8:2, and 7:3. A concentration series was then formed for each sample with 100%, 50%, 10%, and 1% aqueous leachate concentrations. Standardized water, prepared according to the kit manufacturer's instructions, was used to dilute the samples. Each sample test was performed in triplicate. Five individuals and 10 mL of the test solution were added to each test well. After 24 h, the immobility and mortality of the observed guinea fowl were calculated.

3. Results and Discussion

Active pH indicates the concentration of hydrogen ions. The addition of CDW increased both the active and exchange pH values. The exchange pH is determined by the concentration of hydrogen and aluminum cations, which can be exchanged for basic ions in the presence of a neutral salt solution. Mixing CDW with tailings caused an increase in conductivity, which indicates the salinity of the soil. However, even the mixture with the highest CDW ratio did not exceed the value of 2 dS m^{-1} and is therefore still evaluated as a low salinity sample. However, regarding the composition of CDW, it is mainly alkali metals such as calcium, etc. All the values found, including their evaluation, are shown in Table 1.

Table 1. Evaluation of the results of the tested mixtures of CDW (construction and demolition waste) and tailings.

	pH/H$_2$O	Interpretation	pH/CaCl$_2$	Interpretation	k dS m^{-1}	Interpretation
Tailing	5.54	Moderately acidic	5.78	Moderately acidic to slightly alkaline	0.125	Low salinity
CDW	11.33	Strongly alkaline	11.15	Moderately to strongly alkaline	0.772	Low salinity
Tailing/CDW, 9:1	6.01	Moderately acidic	7.17	Moderately acidic to slightly alkaline	0.274	Low salinity
Tailing/CDW, 8:2	6.44	Moderately acidic	7.37	Moderately acidic to slightly alkaline	0.308	Low salinity
Tailing/CDW, 7:3	6.86	Slightly acidic	7.91	Moderately to strongly alkaline	0.389	Low salinity

For an objective evaluation, an elemental analysis of tailings and CDW was also performed (see Table 2). Based on the results, it can be concluded that elements such as Si (27%), Fe (25%), and Al (11%) prevailed in the tailings. Ref. [19] states in his work that, considering the percentage of silicon, it can be assumed that acidification will not occur because of its loss but as a result of the lack of basic cations during the decomposition of aluminosilicates. The content of total sulphur contained in the anhydrous sample of OKR coal is generally low, which was also confirmed by elemental analysis. The total sulphur content was around 3.38%. In general, the content of heavy metals in OKR carbonaceous rocks is low and does not exceed the background of other industrial emissions from an ecological point of view.

Table 2. Chemical composition of construction and demolition waste and tailing [19].

%	CDW	Tailing	%	CDW	Tailing	%	CDW	Tailing
Ag	0.002	<0.0002	Ge	0.0002	0.002	S	0.53	3.38
Al	8.5	12.11	Hf	0.006	0.002	Sb	<0.0003	0.001
As	0.002	0.06	Hg	<0.0001	0.0005	Se	<0.00005	0.002
Ba	0.17	0.58	I	0.0002	<0.00030	Si	45	28.75
Bi	<0.0001	0.002	K	3.4	4.21	Sn	<0.0003	0.01
Br	0.0004	0.003	La	0.022	0.02	Sr	0.05	0.07
Ca	27	1.02	Mg	0.82	0.31	Ta	0.02	0.01
Cd	<0.0002	0.001	Mn	0.19	0.20	Th	0.003	0.01
Ce	0.02	0.03	Mo	0.001	0.002	Ti	1.09	1.09
Cl	<0.0002	0.02	Nb	0.006	0.01	Tl	0.0002	0.0006
Co	0.003	0.01	Nd	0.02	0.03	U	0.0002	0.003
Cr	0.10	0.04	Ni	0.02	0.02	V	0.03	0.05
Cs	0.01	0.01	P	0.14	0.38	W	0.0002	0,002
Cu	0.01	0.02	Pb	0.008	0.05	Y	0.008	0.01
Fe	8.0	31	Pr	0.0004	<0.00020	Zn	0.27	0.07
Ga	0.005	0.01	Rb	0.02	0.05	Zr	0.29	0.06

BCR sequential extraction analysis mimics the conditions to which the material may be exposed in natural conditions to determine the bioavailability of metals. For the analysis, several leaching agents are used, the strength of which gradually increases. According to the results of the BCR analysis (see Table 3), it cannot be unequivocally stated that the addition of CDW would reduce the metal content in the bioavailable fractions.

Table 3. Results of BCR (Community Bureau of Reference) analysis.

	Co mg kg^{-1}	Mn mg kg^{-1}	Ni mg kg^{-1}	Zn mg kg^{-1}	Fe mg kg^{-1}
			EXCHANGEBLE FRACTION		
Tailing/CDW, 9:1	17	300	8	75	30
Tailing/CDW, 8:2	2	100	9	90	30
Tailing/CDW, 7:3	14	60	9	140	25
			REDUCIBLE FRACTION		
Tailing/CDW, 9:1	2.08	4 691	1.04	89	906
Tailing/CDW, 8:2	2.08	4 820	3.65	99	1 078
Tailing/CDW, 7:3	0.52	4 871	3.13	89	1 042
			OXIDIZABLE FRACTION		
Tailing/CDW, 9:1	25	269	6.5	87	179
Tailing/CDW, 8:2	30	161	13.6	92	207
Tailing/CDW, 7:3	54	215	8.7	92	136

At last, ecotoxicity tests for *Daphnia magna* were carried out for the 9:1, 8:2, and 7:3 samples of the mixture of tailings from the Ema tailings and CDW. For the purpose of this test and for better orientation, the samples were renamed 1a (9:1 ratio), 2a (8:2 ratio), and 3a (7:3 ratio). The test results were read after 24 and 48 h. The mortality results after 24 h are graphically evaluated in Figure 1a, and the number of dead individuals after 48 h in Figure 1b. After 24 h of incubation, the highest mortality rates were found for the sample containing 70% tailings and 30% CDW. The highest lethality was observed for the 100% sample concentration, while the second highest lethality was observed for the sample treated at 1% concentration. There was no significant difference between samples that contained 10% CDW (sample 1a) and 20% CDW (sample 2a). After 48 h, we observed increased mortality in each sample. Again, the highest mortality was in the sample with the highest CDW content (sample 3a). The second highest mortality after 48 h was observed in sample 1a, which contained only 10% CDW. Considering the concentration of the sample, the highest mortality was observed at 100% concentration. Here again, there is no direct proportionality between concentration and mortality; the second highest mortality percentage was found at 1% concentration.

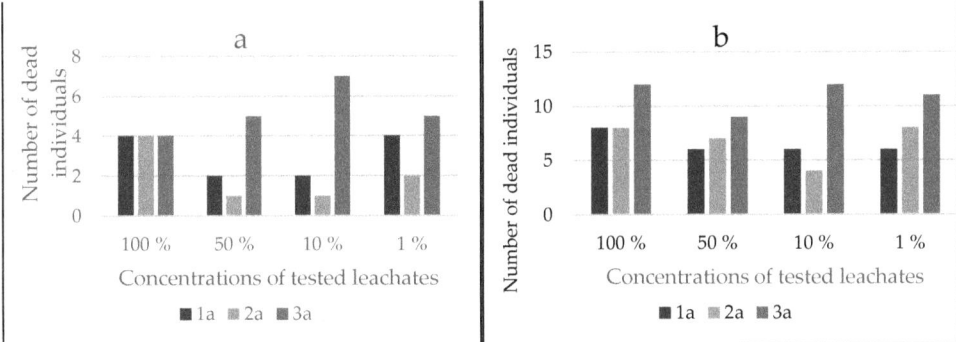

Figure 1. Graphical evaluation of mortality after 24 h (**a**: on the right) and 48 h (**b**: on the left).

The results show that the addition of CDW to the Ema tailings did not reduce toxicity to *Daphnia magna*. The increased mortality at 1% concentration was surprising. Samples could be retested at lower concentrations to determine the mechanism of action (not performed in this study due to small sample volumes).

4. Conclusions

The mixing of construction and demolition waste with the tailings of the selected tailings has been shown to influence the pH from acidic values towards neutral to slightly alkaline values, which could have a positive effect on the future development of the entire site. The increase in pH is demonstrably dependent on the amount of CDW added, so it can be said that increasing the amount will result in improved parameters.

The results of the sequential extraction analysis of BCR did not show a significant correlation between the increase in the amount of CDW added and the decrease in the content of hazardous metals in the bioavailable fractions.

Test results obtained in the acute toxicity test on *Daphnia magna* show increased negative effects in samples with higher CDW content. For a better evaluation, it is necessary to perform a larger number of tests with a larger volume of tested samples.

Author Contributions: Conceptualization, V.B. and B.S.; methodology, B.S. and V.B.; validation, V.B. and B.S.; formal analysis, L.B.; investigation, V.B. and O.N.; resources, B.S.; data curation, O.N.; writing—original draft preparation, V.B.; writing—review and editing, V.B.; visualization, V.B.; supervision, V.B.; project administration, V.B.; funding acquisition, V.B and B.S. All authors have read and agreed to the published version of the manuscript.

Funding: Research was funded by The Project for Specific University Research (SGS) No. SP2023/4 and SP2022/57 by the Faculty of Mining and Geology of VŠB—Technical University of Ostrava.

Institutional Review Board Statement: Not applicable.

Informed Consent Statement: Not applicable.

Data Availability Statement: Data are contained within the article.

Conflicts of Interest: The authors declare no conflict of interest.

References

1. Vojtková, H. New strains of copper-resistant pseudomonas bacteria isolated from anthropogenically polluted soils. In *International Multidisciplinary Scientific GeoConference Surveying Geology and Mining Ecology Management*; SGEM: Albena, Bulgaria, 2014; Volume 1, pp. 451–457. [CrossRef]
2. Šimonovičová, A.; Peťková, K.; Jurkovič, Ľ.; Ferianc, P.; Vojtková, H.; Remenár, M.; Kraková, L.; Pangallo, D.; Hiller, E.; Čerňanský, S. Autochthonous Microbiota in Arsenic-Bearing Technosols from Zemianske Kostoľany (Slovakia) and Its Potential for Bioleaching and Biovolatilization of Arsenic. *Water Air Soil Pollut.* **2016**, *227*, 336. [CrossRef]
3. Šimonovičová, A.; Ferianc, P.; Vojtková, H.; Pangallo, D.; Hanajík, P.; Kraková, L.; Feketeová, Z.; Čerňanský, S.; Okenicová, L.; Žemberyová, M.; et al. Alkaline Technosol contaminated by former mining activity and its culturable autochthonous microbiota. *Chemosphere* **2017**, *171*, 89–96. [CrossRef] [PubMed]
4. Nosalj, S.; Šimonovičová, A.; Vojtková, H. Enzyme production by soilborne fungal strains of *Aspergillus niger* isolated from different localities affected by mining. *IOP Conf. Ser. Earth Environ. Sci.* **2021**, *900*, 012027. [CrossRef]
5. Urík, M.; Polák, F.; Bujdoš, M.; Miglierini, M.B.; Milová-Žiaková, B.; Farkas, B.; Goneková, Z.; Vojtková, H.; Matúš, P. Antimony leaching from antimony-bearing ferric oxyhydroxides by filamentous fungi and biotransformation of ferric substrate. *Sci. Total Environ.* **2019**, *664*, 683–689. [CrossRef] [PubMed]
6. Farkas, B.; Vojtková, H.; Bujdoš, M.; Kolenčík, M.; Šebesta, M.; Matulová, M.; Duborská, E.; Danko, M.; Kim, H.; Kučová, K.; et al. Fungal mobilization of selenium in the presence of hausmannite and ferric oxyhydroxides. *J. Fungi* **2021**, *7*, 810. [CrossRef] [PubMed]
7. Matulová, M.; Bujdoš, M.; Miglierini, M.B.; Cesnek, M.; Duborská, E.; Mosnáčková, K.; Vojtková, H.; Kmječ, T.; Dekan, J.; Matúš, P.; et al. The effect of high selenite and selenate concentrations on ferric oxyhydroxides transformation under alkaline conditions. *Int. J. Mol. Sci.* **2021**, *22*, 9955. [CrossRef] [PubMed]
8. Balíková, K.; Vojtková, H.; Duborská, E.; Kim, H.; Matúš, P.; Urík, M. Role of Exopolysaccharides of *Pseudomonas* in heavy metal removal and other remediation strategies. *Polymers* **2022**, *14*, 4253. [CrossRef] [PubMed]
9. Šebesta, M.; Vojtková, H.; Cyprichová, V.; Ingle, A.P.; Urík, M.; Kolenčík, M. Mycosynthesis of metal-containing nanoparticles—Fungal metal resistance and mechanisms of synthesis. *Int. J. Mol. Sci.* **2022**, *23*, 14084. [CrossRef] [PubMed]
10. Šebesta, M.; Vojtková, H.; Cyprichová, V.; Ingle, A.P.; Urík, M.; Kolenčík, M. Mycosynthesis of metal-containing nanoparticles—Synthesis by *Ascomycetes* and *Basidiomycetes* and their application. *Int. J. Mol. Sci.* **2023**, *24*, 304. [CrossRef] [PubMed]
11. Martinec, P. *The Impact of Ending Deep Coal Mining on the Environment*; Anagram: Ostrava, Czech Republic, 2006.
12. Pertile, E.; Surovka, D.; Božoň, A. The study of occurrences of selected pahs adsorbed on PM10 particles in coal mine waste dumps Heřmanice and Hrabůvka (Czech Republic). In *International Multidisciplinary Scientific GeoConference Surveying Geology and Mining Ecology Management*; SGEM: Sofia, Bulgaria, 2016; Volume 3, pp. 161–168.
13. Pertile, E.; Surovka, D.; Sarčáková, E.; Božoň, A. Monitoring of Pollutants in an Active Mining Dump Ema, Czech Republic. *Inz. Miner.* **2017**, *2017*, 45–50.
14. Różański, Z. Fire Hazard in Coal Waste Dumps–Selected Aspects of the Environmental Impact. *IOP Conf. Ser. Earth Environ. Sci.* **2018**, *174*, 012013. [CrossRef]
15. Abramowicz, A.; Rahmonov, O.; Chybiorz, R. Environmental Management and Landscape Transformation on Self-Heating Coal-Waste Dumps in the Upper Silesian Coal Basin. *Land* **2021**, *10*, 23. [CrossRef]
16. Xu, J.; Zhao, H.; Yin, P.; Wu, L.; Li, G. Landscape Ecological Quality Assessment and Its Dynamic Change in Coal Mining Area: A Case Study of Peixian. *Environ. Earth Sci.* **2019**, *78*, 708. [CrossRef]
17. Surovka, D.; Pertile, E.; Dombek, V.; Vastyl, M.; Leher, V. Monitoring of Thermal and Gas Activities in Mining Dump Hedvika, Czech Republic. *IOP Conf. Ser. Earth Environ. Sci.* **2017**, *92*, 012060. [CrossRef]
18. Mi, J.; Yang, Y.; Zhang, S.; An, S.; Hou, H.; Hua, Y.; Chen, F. Tracking the Land Use/Land Cover Change in an Area with Underground Mining and Reforestation via Continuous Landsat Classification. *Remote Sens.* **2019**, *11*, 1719. [CrossRef]
19. Pertile, E.; Dvorský, T.; Václavík, V.; Syrová, L.; Charvát, J.; Máčalová, K.; Balcařík, L. The Use of Construction Waste to Remediate a Thermally Active Spoil Heap. *Appl. Sci.* **2023**, *13*, 7123. [CrossRef]
20. ČSN 73 6133; Navrhování a Provádění Zemního Tělesa Pozemních Komunikací. Czech Office for Standards, Metrology and Testing: Praha, Czech Republic, 2011.
21. ČSN EN ISO 10390 (836221); Půdy, Upravený Bioodpad a Kaly—Stanovení pH. Czech Office for Standards, Metrology and Testing: Praha, Czech Republic, 2022.
22. ČSN ISO 11265 (836210); Kvalita Půdy. Stanovení Elektrické Konduktivity. Czech Office for Standards, Metrology and Testing: Praha, Czech Republic, 1996.

Disclaimer/Publisher's Note: The statements, opinions and data contained in all publications are solely those of the individual author(s) and contributor(s) and not of MDPI and/or the editor(s). MDPI and/or the editor(s) disclaim responsibility for any injury to people or property resulting from any ideas, methods, instructions or products referred to in the content.

Proceeding Paper

A Comparison and Development of Municipal Waste Management in Three Countries, Slovakia, the Czech Republic and Poland, with an Emphasis on the Slovak Republic [†]

Lucia Domaracka [1,*], Simona Matuskova [2], Marcela Tausova [1], Barbara Kowal [3] and Katarina Culkova [1]

1. Institute of Earth Resources, Faculty of Mining, Ecology, Process Control and Geotechnologies, Technical University of Košice, 040 02 Košice, Slovakia; marcela.tausova@tuke.sk (M.T.); katarina.culkova@tuke.sk (K.C.)
2. Department of Environmental Science, Faculty of Mining and Geology, VŠB Technical University of Ostrava, 703 00 Ostrava, Czech Republic; simona.matuskova@vsb.cz
3. Faculty of Civil Engineering and Resource Management, AGH University of Krakow, 30-059 Kraków, Poland; bkowal@agh.edu.pl
* Correspondence: lucia.domaracka@tuke.sk
† Presented at the 4th International Conference on Advances in Environmental Engineering, Ostrava, Czech Republic, 20–22 November 2023.

Abstract: In this paper, we compare the development and possibilities for improvement of waste management in three countries: Slovakia, the Czech Republic, and Poland. Waste management is part of the circular economy. The circular economy is the basis for waste management today. This applies to municipal waste management, too. In this paper, we work with data available in the Eurostat database. We mainly deal with municipal waste in the three chosen countries. The output of this paper is an assessment of indicators such as recycling rates and waste production. The outputs are documented graphically.

Keywords: municipal waste; recycling rate; waste production

1. Introduction

Any country produces waste, some more than others. The volume of waste depends on the number of inhabitants. The more developed countries have a higher rate of recycled waste [1]. In northern and western European countries, waste stocks almost do not exist. In eastern and southern Europe, more than a half of the waste is stocking [2], not excluding Slovakia. The municipal waste volume increases every year. An EU goal to be achieved by 2050 concerns the ecologic limits of the planet and waste-to-energy plants, used to generate electricity [1,2].

With the aim to achieve long-term sustainable development in the world, the available sources have to be used effectively. The circular economy (CE) and waste treatments help to solve the problem. Waste treatment helps to avoid and decrease the volume of waste, as well as the negative influences of the waste on the living environment. The goal of waste treatment is to prevent waste levels rising, to increase recycling, and to decrease waste landfilling. The goal of the circular economy is, after products and goods life cycles end, to use them again to create something new. In this way, the rising levels of waste are decreasing.

The EU's long-term goal is to transform European societies, avoiding or decreasing the levels of waste rising and waste recycling into energy sources. This presents a key tool for providing effective energy use and sustainability in the EU.

The circular economy helps to provide healthy living environments without any waste or permanent use of the sources. Moreover, environment biodiversity is protected. Society produces a decreasing volume of greenhouse gases. According to EC goals in

2015, legislation started to be transmitted from the linear economy (LE) to a new one—the circular economy. This brought about other goals in circular and waste economies, such as recycling, stocking, waste avoiding, and increasing producers' responsibilities, repeating the use of sources, etc.

Municipal solid waste (MSW) nowadays presents as an inevitable factor of society and economic entities. With an aim to be ecologically sustainable, any country should have to be responsible for waste management, considering the aim of fulfilling of the mentioned goals [3].

Waste hierarchy presents one of the key factors of how to prevent waste rising in the EU [4], including recycling, the limitation of the waste's influence on the living environment, reducing resource consumption and decreasing costs [5]. Waste treatment has to be performed by solutions, considering the principles of living environment protection and material recovery [6].

Nowadays, waste management should be incorporated into the circular economy with increasing interest. The development of an efficient waste management system in the country presents a preparation period of the circular economy. The effectiveness of the circular economy is the theme of a number of studies in the literature [7]. For example, Martinho and Mourao reviewed scientific literature from the view of the CE concept in the EU, and found there was increasing importance and interest in the concept, as well as in a number of publications in chosen countries and organizations [8].

Waste treatment can be analyzed from different levels. From the level of municipalities, in almost all EU countries it is financed by local taxes or by a property tax. Radvan evaluated the mentioned conditions of the Czech Republic, Slovakia, and Poland, and found its revenue could be used for the waste management of municipalities [9]. In many municipalities, waste ends up in landfill in quite considerable volumes, with low levels of collection and recovery. This is the case in Dubrovnik's research [10]. From the levels of organizations' waste volumes, Sharma et al. pointed to the factors contributing to the transition of the prerequisites of the LE to the CE in SMEs [11]. The CE presents a good competitive advantage and sustainability of SMEs. Their study found low CE implementations in SMEs, suggesting tools for increasing CE implementations.

As for the territorial study of waste management, Pacurariu et al. analyzed the system of indicators, used for the monitoring of the transition to the CE, as being efficient and relevant in connection with the EU's sustainability [12]. Presently, countries also outside the EU, such as Russia, start to be orientated to circular and green economies when their companies transition to circular businesses as healthy and efficient business strategies. Fedotkina et al. conducted a study in Russia, trying to find the best waste management techniques [13]. One of the factors, helping to prevent waste levels from rising in the EU, was the application of a waste hierarchy [4], including recycling and reducing of sources of consumption [5]. Waste treatment should help to protect the environment and to contribute to material and energetic recovery [6].

Zorkociova and Paluskova measured the CE indexes in the EU with the aim to find out the trend of the indexes' developments, comparing them to the frame of V4 countries [14]. The results of the study show the need for continuous monitoring to provide greater efficiency and sustainability. Chovancova and Vavrek made a study of economic developments' dependence on energy consumption, also in V4 countries, finding that all analyzed countries should increase their speeds in implementing CE and waste treatment policies, regarding the need of energy [15]. Vokal and Stoch searched for waste management in V4 from the view of radioactive waste, and found differences in different countries [16]. Lacko et al. [7] found relations between waste treatment and CEs in V4. The results of the study show that V4 countries do not belong to the category of advanced countries in the sense of recycling and CE use [1,2].

Zaleski and Chawla discussed CE implementation in Poland to find out ways for improving the quality of municipal waste management before the COVID-19 pandemic [17]. The aim of this paper is to analyze the situation in post-pandemic countries, not only in

Poland, but also in the Czech Republic and Slovakia. This research provides data on V4 countries, contributing to overcoming the challenge of Martinho and Mourao by providing data according to individual countries [8]. Similarly, as was Pacurariu et al.'s, the paper's ambition is to analyze the CE and waste management in V4, as well as to discover whether the situation in 2023 has changed [12].

2. Methodology

The goal of the paper is to investigate the development of the presented situation. Research data are from Eurostat and analyzed in JMP 15 by © SAS Institute Inc. Analysis had been performed according to the following processes:

1. Analysis of production of the waste in municipalities in the Czech Republic, Slovak Republic and Poland in graphic expression-cartographer;
2. Graphic analysis of recycling rate-trend analyses;
3. Graphic summary of analysis of recycling rate of municipal waste with regard to trends of development;

In Figure 1, a flow chart about the development of research steps and the outcomes which were achieved can be seen.

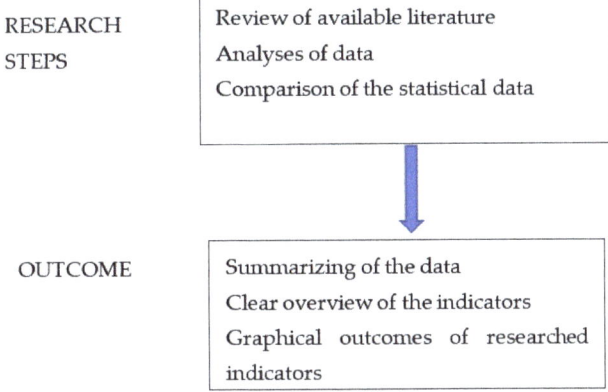

Figure 1. Flow chart of research steps and outcomes of the study, using our own processing.

3. Results

The EU goal is to protect the environment, support ecologic activities and stimulate ecological economy development. The long-term goal is for greater recycling in society, minimizing the rise of waste, and increasing the recycling of waste. In the research, the mentioned aspects have been analyzed in individual EU countries.

For the results, we evaluated the amount of municipal waste generated for the selected countries in kilograms per capita.

There is a demonstrable link between household consumption and municipal waste production in European economies, but also worldwide.

3.1. The Production of Municipial Waste in the Czech Republic, Slovak Republic and Poland

According to Figure 2, municipal waste production in Slovakia is growing the fastest of any EU country. In 2019, the average Slovak discarded of a record 435 kg of waste in waste bins. Meanwhile, the EU's reform package for the transition to a circular economy sets two targets for 2035: to recycle at least 65% of municipal waste and to increase landfill by no more than 10%. In 2019, according to data from the Slovak Statistical Office, Slovakia generated 2.37 million tons of municipal waste. Over the last ten years, there has been an increase of more than 30%. The production of municipal waste is growing fastest in the EU. And, it looks set to grow further. According to the analysis, municipal waste production goes hand in hand with consumption growth. Every purchased good and often service is sooner or later a source of waste. As consumption grows, so does the production of waste [18]. The total waste production in the Czech Republic in 2022 reached 38.5 million tons.

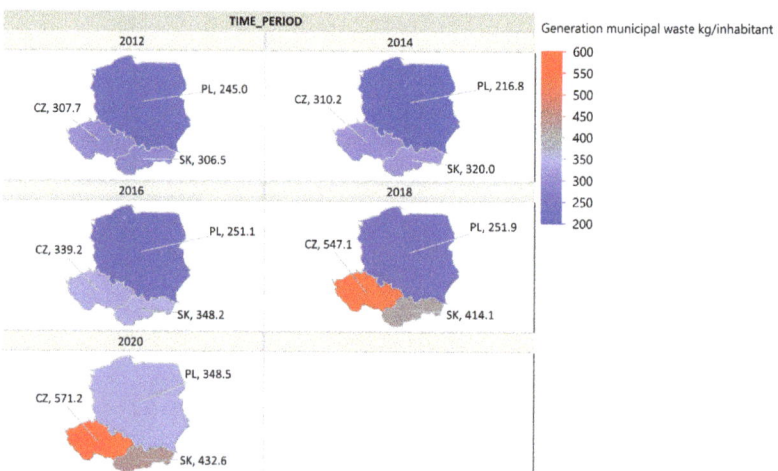

Figure 2. Municipal waste production in kilograms per inhabitant; source: our own processing.

Compared to the previous year, 2019 (37.4 million tons), this is an increase of approximately three percent. However, there has been a decline in municipal waste, i.e., millions of tons of waste were generated by households. The volume of municipal waste in the Czech Republic fell by 2.5% year-on-year to 5.7% (see Figure 2).

Per inhabitant of the Czech Republic, 536 kg of waste was generated. According to the most recent data from the Czech Statistical Office, the share of landfilled municipal waste was 48.4% in 2020. On the other hand, the recycling rate was reported at 43.7% and around 8% of municipal waste was recovered into electricity or heat in waste-to-energy plants (WEEE) last year. This is due not only to the sorting and subsequent recycling of waste, but also to the long-standing existence of waste-to-energy facilities [18,19].

Municipal Solid Waste in Poland; A Short Overview

MSW generation per capita increased in 2004–2005 to 319 kg. In 2005–2012, it remained at this level. Waste rising decreased to 272 kg in 2014; however, there was a recorded consequential increase to 336 kg in 2019. The lowest level occurred in 2014 (272 kg per capita), followed by decreasing due to a waste prevention program (WPP) in 2011. This trend can be influenced also by population and household expenses. The MSW rising in Poland of 336 kg per capita per year is under the EU average of 502 kg. The effect of WPP in 2011 cannot be objectively evaluated, since it reflects the global economic crisis, which is affecting waste increases as well [20,21].

From 2012, the recycling of paper, glass, metal and plastics was formally implemented in Poland (by Regulation of the Environment Ministry) through the preparation for re-use and recovery by other methods of waste fraction recycling (Journal of Laws of 2012, item 645). This triggered an increase in material recycling. This coincided with an energy recovery trend with high caloric waste fractions, mainly in the cement industry. This resulted in a waste landfilling decrease from 9194 Gg in 2004 to 5218 Gg in 2020. However, in 2017, the trend reversed—there was a recorded increase in waste landfilling in 2017–2019. Only a slight reduction had been recorded in 2020 [21,22].

3.2. A Comparison of the Three Chosen Countries: The Czech Republic, the Slovak Republic and Poland

Comparing the countries of the Czech Republic, Slovakia and Poland, Poland clearly produces the smallest amount of municipal waste. This is followed by the Slovak Republic, and the last country of the three is the Czech Republic.

If we want to see a development in municipal waste management, we need to look at the relevant indicator—the municipal waste recycling rate. The municipal waste recycling rate shows how waste from final consumers is used in the circular economy as a source of materials. Municipal waste rises mainly due to the final consumers, including households and other sources of waste, similar in origin and nature. It is recorded at the level of around 10% of total EU waste. However, its heterogeneous composition creates the challenges for municipal waste (MW) management. MW recycling rates present a good indicator of the MW management quality, measuring the recycled MW shares in the total rise of MW.

Material recycling, composting and anaerobic treatments are included in the recycling criteria. The ratio is shown in percentage. This indicator shows a trend that is positive over time, which means that it is increasing, and this is the case in all of the countries surveyed. Recycling rates are increasing in all of the selected countries. This may be due to the growing importance of the circular economy and to the fulfilment of EU commitments under existing agreements.

The EU average recycling rate is higher than in the three countries studied, but, looking at the trend over the last 20 years, the EU average recycling rate is half that of the Slovak Republic, the Czech Republic and Poland (see Figure 3).

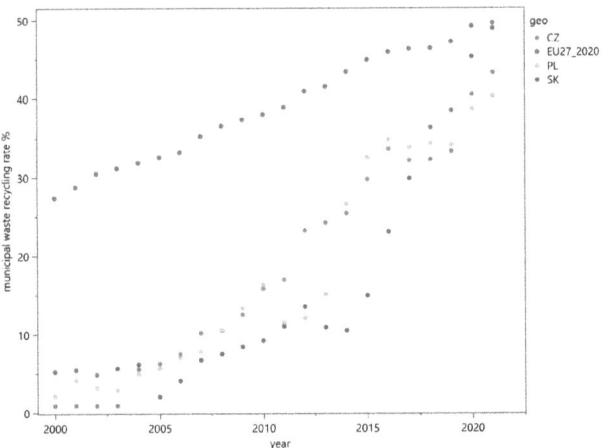

Figure 3. Development of waste recycling waste, using our own processing.

In Figure 4, we can see that the recycling rate is lowest in Poland, followed by the Czech Republic, and the highest recycling rate is in the Slovak Republic. Countries have to follow different protocols and documents. Slovakia must observe EU documents and recommendations to meet their environmental policy, such as the "Slovakian Waste Economy Program in 2021–2025" and the "Slovakian Program for waste avoiding in 2019–2025" [23]. The goal is to increase waste recovery with repeated use and recycling and support the minimizing of waste. The mentioned programs result from the decree of the European parliament and European council 2008/98/ES about waste and the EU Action Plan for a circular economy [24,25]. Part of the CE action plan is elaborated in the frame decree of waste; No 2018/851 (Year 2021) [26].

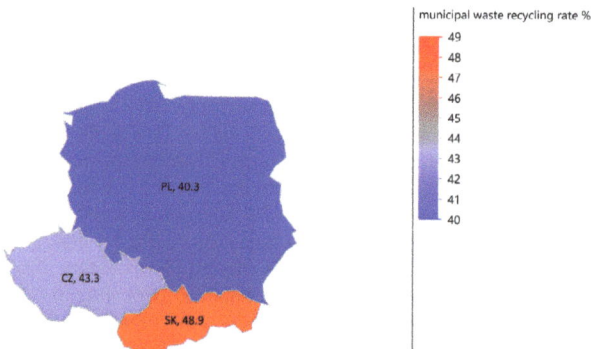

Figure 4. Cartograph of municipal waste recycling rate of chosen countries. Source: our own processing.

4. Conclusions

The separation of MW collection, especially of bio-degradable waste, still needs improvement and support. The highest investments in waste management were recorded in 2019. The trend followed the need to meet WM goals set by the EU. Investments aimed at the use of technological innovations to increase waste recovery rate, orientated toward reuse and recycling. Using the available database, this paper evaluated MW development and recycling in selected EU countries. The results show that, despite the increasing trend in MW, there is a positive development in waste recycling, having observed a positive trend in the analysed countries. In the assessment, the Czech Republic was the dominant country, generating the highest volume of municipal waste, but, on the other hand, it was the country with the second highest recycling rate among the countries compared. It should be highlighted in this comparison that Poland was the country that generated the least municipal waste and had the second highest municipal waste recycling rate of the countries compared [27,28]. Slovakia, with the second highest rate of waste production among the three monitored countries, was, however, in first place in terms of the recycling rate among our three monitored countries. The reasons for this situation are various, related to the standard of living and to the fact that MW production is directly dependent on social and economic indicators, especially GDP. Our evaluation of waste production and recycling can be used for the creation of governmental policies in the area of waste management, as well as for MW policies in individual municipalities. This paper presents partial results of the project, solving No ITMS 3131011T564 "Research of the RES implementation impacts to the processes of energetic management of the industries", and is limited to the number of individual states. Future orientations of research could be directed to the representative best cities from the analyzed states, and compared with the best EU results, as well as directed towards detail legislative support of the presented area.

Author Contributions: Conceptualization, L.D. and M.T.; methodology, B.K. and S.M.; software, M.T.; validation, K.C.; formal analysis, L.D. and M.T.; resources, K.C.; data curation, B.K.; writing—original draft preparation, L.D.; writing—review and editing S.M.; visualization, M.T.; supervision, B.K. and S.M.; project administration, K.C.; funding acquisition, L.D. All authors have read and agreed to the published version of the manuscript.

Funding: This research received no external funding.

Institutional Review Board Statement: Not applicable.

Informed Consent Statement: Not applicable.

Data Availability Statement: Data are contained within the article.

Conflicts of Interest: The authors declare no conflict of interest.

References

1. Taušová, M.; Mihaliková, E.; Čulková, K.; Stehlíková, B.; Tauš, P.; Kudelas, D.; Štrba, Ľ.; Domaracká, L. Analysis of municipal waste development and management in self-governing regions of Slovakia. *Sustainability* **2020**, *12*, 5818. [CrossRef]
2. Stehlíková, B.; Čulková, K.; Taušová, M.; Štrba, Ľ.; Mihaliková, E. Evaluation of communal waste in Slovakia from the view of chosen economic indicators. *Energies* **2021**, *14*, 5052. [CrossRef]
3. Tsoulfas, G.T.; Pappis, C.P. Environmental principles applicable to supply chains design and operation. *J. Clean. Prod.* **2006**, *14*, 1593–1602. [CrossRef]
4. Pomberger, R.; Sarc, R.; Lorber, K.E. Dynamic visualisation of municipal waste management performance in the EU using ternary diagram method. *Waste Manag.* **2017**, *61*, 558–571. [CrossRef] [PubMed]
5. Eriksson, O.; Carlsson Reich, M.; Frostell, B.; Björklund, A.; Assefa, G.; Sundqvist, J.O.; Granath, J.; Baky, A.; Thyselius, L. Municipal solid waste management from a systems perspective. *J. Clean. Prod.* **2005**, *13*, 241–252. [CrossRef]
6. Przydatek, G. Assessment of Changes in the Municipal Waste Accumulation in Poland. *Environ. Sci. Pollut. Res.* **2020**, *27*, 25766–25773. [CrossRef] [PubMed]
7. Lacko, R.; Hajduova, Z.; Zawada, M. The Efficiency of Circular Economies: A Comparison of Visegrad Group Countries. *Energies* **2021**, *14*, 1680. [CrossRef]
8. Martinho, V.D.; Mourao, P.R. Circular Economy and Economic Development in the European Union: A Review and Bibliometric Analysis. *Sustainability* **2020**, *12*, 7767. [CrossRef]
9. Radvan, M. Taxes on Communal Waste in the Czech Republic, Poland and Slovakia. *J. Local Self-Gov.* **2016**, *14*, 511–520. [CrossRef] [PubMed]
10. Smoljko, I.; Matic, M. Life-cycle assessment of municipal solid waste management systems: A case study of the city of Dubrovnik. *Hrčak* **2023**, *72*, 369–380. [CrossRef]
11. Sharma, N.K.; Govindan, K.; Lai, K.K.; Chen, W.K.; Kumar, V. The transition from linear economy to circular economy for sustainability among SMEs: A study on prospects, impediments and prerequisites. *Bus. Strategy Environ.* **2021**, *30*, 1803–1822. [CrossRef]
12. Pacurariu, R.L.; Vatca, S.D.; Lakatos, E.S.; Bacali, L.; Vlad, M. A Critical Review of EU Key Indicators for the Transition to the Circular Economy. *Int. J. Environ. Res. Public Health* **2021**, *18*, 8840. [CrossRef] [PubMed]
13. Fedotkina, O.; Gorbashko, E.; Vatolkina, N. Circular economy in Russia: Drives and barriers for waste management development. *Sustainability* **2019**, *11*, 5837. [CrossRef]
14. Zorkociova, O.; Paluskova, H. Waste management of EU countries related to circular economy issues. *Ad Alta* **2020**, *10*, 360–366.
15. Chovancova, J.; Vavrek, R. Decoupling Analysis of Energy Consumption and Economic Growth of V4 Countries. *Probl. Ekorozwoju* **2019**, *14*, 159–165.
16. Vokal, A.; Stoch, P. Czech Republic, Slovak Republic and Poland: Experience of radioactive waste (RAW) management and contaminated site clean-up. In *Processes, Technologies And International Experience*; Book Series Woodhead Publishing Series in Energy; Woodhead Publishing Limited: Cambridge, UK, 2013; Volume 48, pp. 415–437. [CrossRef]
17. Zaleski, P.; Chawla, Y. Circular Economy in Poland: Profitability Analysis for Two Methods of Waste Processing in Small Municipalities. *Energies* **2020**, *13*, 5166. [CrossRef]
18. Available online: https://www.odpady-portal.sk/Dokument/106395/komunalneho-odpadu-vlani-v-cesku-ubudlo-na-slovensku-to-bolo-naopak.aspx (accessed on 30 June 2023).
19. Eurostat. Environmental Protection Investments of General Government by Environmental Protection Activity–Waste Management. 2022. Available online: https://ec.europa.eu/eurostat/databrowser/view/env_ac_epigg/default/table?lang=en (accessed on 7 January 2022).
20. Overview of National Waste Prevention Programs in Europe-Poland. Available online: https://www.eea.europa.eu/en/advanced-search (accessed on 2 July 2023).
21. Country Profiles on Waste Prevention-2023. Available online: https://www.eea.europa.eu/themes/waste/waste-prevention/countries (accessed on 11 August 2023).
22. Poland. Available online: https://www.eea.europa.eu/en/countries/eea-member-countries/poland (accessed on 11 August 2023).
23. Statistical Office SR. Costs and Revenues in Living Environment Protection (2012–2019). Available online: https://slovak.statistics.sk/wps/portal/ext/themes/environment/environment/indicators (accessed on 15 January 2022).
24. Eurostat. Recycling Rate of Municipal Waste. 2022. Available online: https://ec.europa.eu/eurostat/databrowser/view/cei_wm011/default/table?lang=en (accessed on 7 January 2022).
25. Environmental Fund. Available online: http://www.envirofond.sk/sk/vyrocne-sprav (accessed on 22 January 2022).
26. European Green Agreement. Available online: https://www.enviroportal.sk/odpady/europska-zelena-dohoda-2019 (accessed on 10 January 2022).

27. Pawnuk, M.; Szulczyński, B.; Den Boer, E.; Sówka, I. Preliminary analysis of the state of municipal waste management technology in Poland along with the identification of waste treatment processes in terms of odor emissions. *Arch. Environ. Prot.* **2022**, *48*, 3–20. [CrossRef]
28. Brzeszczak, A.; Imiołczyk, J. Ratio analysis of Poland's sustainable development compared to the countries of the European Union. *Acta Oeconomica Univ. Selye* **2016**, *5*, 31–41.

Disclaimer/Publisher's Note: The statements, opinions and data contained in all publications are solely those of the individual author(s) and contributor(s) and not of MDPI and/or the editor(s). MDPI and/or the editor(s) disclaim responsibility for any injury to people or property resulting from any ideas, methods, instructions or products referred to in the content.

Proceeding Paper

Removal of Cadmium from Aqueous Solution Using Dried Biomass of *Euglena gracilis* var. *bacillaris* [†]

Martin Valica [1], Tomáš Lempochner [1], Linda Machalová [1], Vanda Adamcová [1], Patrícia Marková [2], Lenka Hutárová [2], Martin Pipíška [3], Juraj Krajčovič [2] and Miroslav Horník [1],*

[1] Institute of Chemistry and Environmental Sciences, Faculty of Natural Sciences, University of Ss. Cyril and Methodius in Trnava, Nám. J. Herdu 2, SK-917 01 Trnava, Slovakia; martin.valica@ucm.sk (M.V.); lempik.junior@gmail.com (T.L.); linda.machalova@gmail.com (L.M.); vanda.adamcova@ucm.sk (V.A.)
[2] Institute of Biology and Biotechnology, Faculty of Natural Sciences, University of Ss. Cyril and Methodius in Trnava, Nám. J. Herdu 2, SK-917 01 Trnava, Slovakia; patricia.markova9@gmail.com (P.M.); lenka.hutarova@ucm.sk (L.H.); juraj.krajcovic@ucm.sk (J.K.)
[3] Department of Chemistry, Faculty of Education, Trnava University in Trnava, Priemyselná 4, P.O. Box 9, SK-918 43 Trnava, Slovakia; martin.pipiska@truni.sk
* Correspondence: hornik@ucm.sk; Tel.: +421-33-55-65-392
[†] Presented at the 4th International Conference on Advances in Environmental Engineering, Ostrava, Czech Republic, 20–22 November 2023.

Abstract: The aim of this work was to evaluate the possibility of applying the dried biomass of *E. gracilis* var. *bacillaris* as a biosorbent for the removal of Cd. Experiments were carried out under conditions of batch systems involving aqueous solutions labelled with $^{109}CdCl_2$. From the kinetics of Cd biosorption, it can be assumed that the Cd removal was a rapid process that achieved the concentration equilibrium of $[Cd]_{biomass}:[Cd]_{solution}$ in the first minutes of the interaction. In individual experiments, the effect of solution pH, initial biosorbent, or Cd concentration was evaluated. According to MINEQL+ speciation modelling, it was found that the biosorption of Cd decreased linearly corresponding to a decrease in the proportion of Cd^{2+} in the solution. The biosorption data were well fitted to the Langmuir model of adsorption isotherm in comparison with the Freundlich model. The maximum biosorption capacity of the dried biomass of *E. gracilis* var. *bacillaris* for the removal of Cd was predicted and reached the value Q_{max} = 0.13 mmol/g or 14.1 mg/g (d.w.), respectively.

Keywords: Cd; biosorption; biosorbents; *Euglena gracilis*; adsorption isotherms; speciation

1. Introduction

Environmental pollution caused by heavy metals is one of the most pervasive problems in the world due to the fact that metals represent non-degradable and persistent contaminants that pose serious risks to both ecosystems and human beings [1]. The discharges of industrial wastewater containing the heavy metals are responsible for the contamination of rivers, lakes, and seas [2]. In the long term, the presence of heavy metals in the food chain [3] may lead to their bioaccumulation in humans, causing health problems [4].

Cadmium (Cd) represents a highly toxic, non-essential, and non-biodegradable heavy metal with a very long biological half-life, ca. 30 years [5]. Cd belongs to the most hazardous Class-I carcinogen and chronic nephrotoxin group [6]. The Agency for Toxic Substances and Disease Registry actually ranked Cd in seventh place within the priority list of hazardous substances based on their frequency of occurrence in the environment, toxicity, and adverse potential to affect human health [7]. Cd shows relatively high physico-chemical similarities with Zn, as an essential nutrient, leading also to significant interactions between these two heavy metals. Earth's crust is rich in Cd sources; however, its prevalence in huge amounts is due to anthropogenic activities, such as electroplating, mining, stabilizing plastics, manufacturing of alloys, cement, pigment, and batteries, fossil fuel combustion, high-phosphate fertilizers, and sewage sludge incineration [8,9]. The maximum contaminant level for cadmium in drinking water is 5 µg/L, according to the Environmental Protection Agency and the World Health Organization [10,11]. In wastewater and water systems, Cd occurs predominately in the form of Cd^{2+} cations, but it can also be present in the form of inorganic complexes, such as $CdCl^+$, $CdOH^+$, $CdHCO_3^+$, $CdCl_3^-$, $CdCl_4^{2-}$, $Cd(OH)_3^-$, and $Cd(OH)_4^{2-}$, or organic chelates. However, the most common valence state in the natural environment is Cd^{2+}, and the most important factors that control Cd mobility as well as chemical speciation are pH and redox potential [12].

The treatment of industrial wastewater containing heavy metals is a major economic and technical challenge because these contaminants cannot be removed by conventional wastewater treatment [13]. The conventional physico-chemical methods used in wastewater treatment systems and in the removal of heavy metals such Cd include chemical precipitation, adsorption, ion exchange, solvent extraction, and membrane filtration [14]. However, there exist limiting factors in the application of these technologies, such as inefficiency or the higher cost of techniques, energy consumption, incomplete removal of heavy metals, and the formation of potentially toxic by-products [15]. In general, adsorption is a suitable and feasible tool for the removal of metals in comparison to membrane technologies due to its low capital and operating costs and ease of operation, in addition to the potential application of a wide range of solid media for the removal of heavy metals [16].

The use of biomass as an adsorbent has drawn particular interest as microbial biomass can be produced at a low cost and can effectively concentrate heavy metals in the concentration range of 1–100 mg/L [17]. This process, known as biosorption, has the unique advantage that it can be performed in the absence of microbial metabolism, which allows the use of dead biomass, with the consequent economic savings [18]. In biosorption, the heavy metals attached within the cell membrane form a complex, leading to elimination from the environment [19]. Thus, this process is a passive remediation technique involving different mechanisms, such as physical adsorption, ion exchange, complexation, coordination, electrostatic interactions, or microprecipitation [20,21]. Several studies revealed that the presence of functional groups on biosorbent surfaces, such as the hydroxyl, thiol, carboxyl, and amino groups have a crucial role in metals biosorption [14].

A variety of biosorbents exist (e.g., yeast, fungi, bacteria, plants, microalgae) that show immense potential as effective biosorbents due to their high surface-to-area ratios, ubiquity, and ability to remove heavy metals at relatively dilute concentrations [22,23]. Biosorption of heavy metals, metalloids, or radionuclides using algae biomass has become a valuable technology for industrial wastewater treatment due to its lack of chemical requirements, high efficiency in detoxification of the diluted effluents, and low operation and maintenance costs [24]. In heavy metal biosorption, ion exchange is the principal mechanism in the case of algal biomass, which takes place on the surface wall [25].

The flagellate protist algae-like *Euglena gracilis* is a cosmopolitan organism able to grow under heterotrophic, photosynthetic, or photo-heterotrophic conditions [26]. *E. gracilis* has been found to tolerate and accumulate heavy metals [27]. Its metabolic flexibility [28,29], fast growth, and high biomass production under heterotrophic conditions [30,31] make this algal flagellate a suitable candidate for biosorption of heavy metals [32].

Our previous papers dealt with the biosorption of Cd using the dried biomass of freshwater moss *Vesicularia dubyana* [33], bacteria [34], terrestrial moss *Rhytidiadelphus squarrosus* [35], or activated sludge [36]. In this work, the dried biomass of *Euglena gracilis* var. *bacillaris* was applied to assess its ability to remove Cd from wastewater or contaminated solutions. In individual experiments, in addition to the effect of time, solution pH, biomass, and Cd concentration, or to the prediction of sorption parameters based on Langmuir and Freundlich adsorption isotherms models, the effect of Cd speciation in solution on Cd biosorption was also evaluated.

2. Materials and Methods

2.1. Biosorbent Preparation

The biomass of *Euglena gracilis* var. *bacillaris* was obtained by static cultivation of cells in Hutner's medium [37] for 21 d under permanent illumination (30 µmol photons/m²·s) and at 26 °C. The medium was sterilized through autoclavation at 121 °C for 20 min. After autoclaving, the pH (3.5) of the medium was adjusted using 1 M HCl. After 21 days of cultivation, the biomass yield was 3.4 g (w.w.) per 1 L of Hutner's medium. *E. gracilis* var. *bacillaris* cells were harvested by centrifugation at 4500 rpm for 5 min and washed 5 times with deionized water to remove any remaining medium. To obtain the defined biomass with a particle size of <300 µm, the biomass was dried at 60 °C for 48 h, mechanically homogenized using a mortar and pestle, and sieved through standardized sieves.

2.2. Batch Biosorption Experiments

All biosorption experiments were carried out in 50 cm³ Erlenmeyer flasks containing 20 cm³ of $CdCl_2$ solution of known concentration in deionized water radiolabeled with known initial activity of $^{109}CdCl_2$. Stock solution of $CdCl_2$ (1000 µmol/L) was prepared by dissolving a defined amount of $CdCl_2$ (analytical grade; CAS 10108-64-2; Sigma-Aldrich, St. Louis, MI, USA) into deionized water (0.05 µS/cm). For each $CdCl_2$ concentration, pH adjustment was realized using 1 M NaOH or 1 M HCl solution. Dried biomass of *E. gracilis* var. *bacillaris* (standard amount of biosorbent was 0.01 g/20 cm³ of $CdCl_2$ solution) with a particle size < 300 µm was suspended in the prepared $CdCl_2$ solutions and incubation was carried out on a shaker (250 rpm) at 25 °C. Evaporation of water from the surface of the solution was prevented by sealing the necks of the flasks with parafilm. At the end of the exposure, aliquots were taken from the flasks and the biomass was filtered through syringe filters (13 mm diameter; 0.45 µm permeability). The residual Cd concentration in the solution was calculated through the ^{109}Cd activity determined by scintillation gamma spectrometry. The percentage of Cd removal ($Q_\%$) and the biosorption capacity (Q_t) were evaluated with the following equations, Equations (1) and (2):

$$Q_\% = \frac{A_0 - A_S}{A_0} \times 100 \quad (1)$$

$$Q_t = (C_0 - C_t) \times \frac{V}{M} \quad (2)$$

where $Q_\%$ is the percentage of Cd biosorption (%), A_0 is the initial ^{109}Cd activity in the solution (Bq), A_S is the residual ^{109}Cd activity in the solution (Bq), Q_t is the amount of Cd sorbed on the biosorbent (µmol/g; d.w.), C_0 and C_t represent the initial concentration of Cd in solution and the concentration of Cd at the end of the experiment (µmol/L), respectively, and V and M are the volume of solution (L) and weight of the biosorbent (g; d.w.).

For the purpose of assessing the effect of Cd speciation on its biosorption by prepared biosorbent, the molar amounts of the disodium salt of ethylenediaminetetraacetic acid (EDTA-Na_2) were added to 20 cm³ of a solution containing 20 µmol/L of $CdCl_2$ to obtain the proportions of the Cd^{2+} form in the solution of 0%, 20%, 40%, 60%, 80%, and the maximum achievable amount of Cd^{2+} at pH 4.0. The calculation of the molar addition of EDTA-Na_2 was performed based on the prediction of Cd speciation in the solution using the modelling program MINEQL+ ver. 4.62.3.

2.3. Adsorption Isotherms Modelling

The sorption of sorbates shows an exothermic nature and is significantly affected by temperature changes. Therefore, temperature constancy is an essential requirement for obtaining reproducible results and for the analysis of experimental data by adsorption isotherms. The second requirement for the application of adsorption isotherms is reaching the equilibrium between the specific metal biosorption Q_{eq} (in µmol/g) and the residual metal concentration in the solution C_{eq} (in µmol/L).

The two-parametric Langmuir and Freundlich isotherm models were used to analyze the data of Cd biosorption by the dried biomass of E. gracilis var. bacillaris.

The Langmuir isotherm model, which operates on the assumptions that adsorption occurs in a monolayer on the solid, all sites are identical and may sorb only a single molecule, and are independent of adjacent site sorption [38], is given by Equation (3):

$$Q_{eq} = \frac{b \times Q_{max} \times C_{eq}}{1 + b \times C_{eq}} \quad (3)$$

where Q_{eq} is the equilibrium specific metal biosorption (μmol/g; d.w.); C_{eq} is the equilibrium concentration of metal in the solution (μmol/L); Q_{max} is the Langmuir constant characterizing the maximum biosorption capacity of the biosorbent (μmol/g; d.w.); b is the Langmuir constant describing the affinity between metal and biosorbent (L/μmol).

The Freundlich isotherm model is an empirical formulation which assumes heterogeneous surface adsorption [39] and is described by Equation (4):

$$Q_{eq} = K \times C_{eq}^{(\frac{1}{n})} \quad (4)$$

where Q_{eq} is the equilibrium specific metal biosorption (μmol/g; d.w.); C_{eq} is the equilibrium concentration of metal in the solution (μmol/L); K is the Freundlich constant related to the biosorption capacity of the biosorbent for metal (L/g; d.w.); n is the Freundlich constant characterizing the biosorption intensity (dimensionless constant).

2.4. Scintillation Gamma Spectrometry

For the determination of ^{109}Cd activity in the supernatant aliquots, a gamma-spectrometric scintillation detector, 76BP76/3 (Envinet, Třebíč, Czech Republic), with well-type crystal NaI(Tl) was used in combination with data processing software ScintiVision-32 (Ortec, Oak Ridge, TN, USA). A library including γ-photon peaks of radionuclides ^{109}Cd (E_γ = 88.04 keV), ^{137}Cs (E_γ = 661.66 keV), and ^{65}Zn (E_γ = 1115.52 keV) was built to calibrate the energies and efficiencies of the ^{109}Cd activity measurements. Gamma-spectrometric measurement with a counting time of 600 s allowed us to obtain the data with an error < 1.5%.

Standard and experimental solutions were prepared using a standardized ^{109}CdCl$_2$ solution (5.206 MBq/cm^3, 50 mg/L CdCl$_2$ in 3 g/L HCl; $T_{1/2}$ = 462.6 d) obtained from the Czech Metrological Institute (Prague, Czech Republic).

2.5. Cd speciation and Statistical Analysis of the Data

Prediction of Cd speciation in the model solutions containing different concentrations of EDTA-Na$_2$ was carried out using the speciation modelling program MINEQL+ ver. 4.62.3 (Environmental Research Software, Hallowell, ME, USA). This program, within the mathematical iterations, applies the USEPA database of thermodynamic constants and the distributions of individual chemical forms of metals in solution and makes predictions depending on the solution pH, temperature, ionic strength, concentration of cations or anions, and for the carbonate system naturally occurring in equilibrium with atmospheric CO$_2$ (pCO$_2$ = 38.5 Pa).

All analytical determinations were performed in triplicate. The graphical presentation of the results and the statistical analysis of the data obtained were performed using the program OriginPro ver. 8.5 (OriginLab Corporation, Northampton, MA, USA).

3. Results and Discussion

As shown in Figure 1a, the kinetics of Cd biosorption by the dried biomass of E. gracilis var. *bacillaris* from solutions containing 20 μmol/L CdCl$_2$ and at pH 4.0 was a rapid process that achieved the concentration equilibrium of [Cd]$_{biomass}$:[Cd]$_{solution}$ in the first minutes of the interaction. A Cd removal efficiency of 20% (8.2 μmol/g; d.w.) was reached after only 20 min of exposure and did not change after this time.

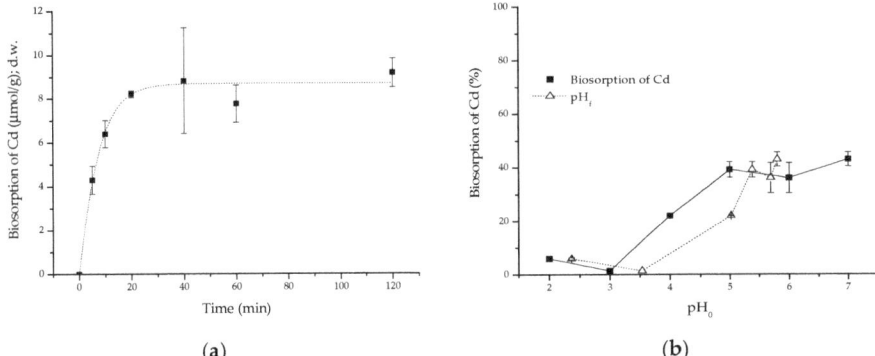

Figure 1. (a) Kinetics of Cd biosorption by dried biomass of *E. gracilis* var. *bacillaris* (biomass concentration C_B = 0.50 g/L) from solutions containing 20 µmol/L $CdCl_2$ (radiolabeled with 45.0 kBq/L $^{109}CdCl_2$) in deionized water, pH 4.0. (b) Effect of initial (pH_0) and final (pH_f) pH values on Cd biosorption by dried biomass of *E. gracilis* var. *bacillaris* (biomass concentration C_B = 0.50 g/L) from solutions containing 20 µmol/L $CdCl_2$ (radiolabeled with 44.5 kBq/L $^{109}CdCl_2$) in deionized water. Exposure of the biomass was carried out on a shaker (250 rpm) for 2 h (effect of pH) and at 25 °C. Error bars represent standard deviation of the arithmetic mean of the results of three independent experiments (±SD; n = 3).

The rapid kinetics demonstrated in this study are confirmed by the results of several other studies. The kinetics of Cd biosorption by the dried biomass of the microalga *Parachlorella* sp. was also studied by Dirbaz and Roosta [40], and they reached the concentration equilibrium after only 20 min. Those authors also found that increased agitation speed had a negative effect on the biosorption efficiency. Also, the time required to achieve the expected equilibrium condition decreased with the increase in the agitation speed. Gu and Lan [41] reported that the biosorption of Pb, Cd, and Zn by the dried biomass of green alga *N. oleoabundans* was rapid, with plateaus achieved within five minutes. At 30 min of contact time, the equilibrium biosorption was achieved for all metals studied.

It is generally known that biosorption kinetics is controlled by several mechanisms, which include (i) bulk diffusion; (ii) external diffusion through a liquid film layer around the biosorbent particles; (iii) intraparticle diffusion; and (iv) reaction rate [42]. It should be noted that by maintaining a well-mixed system with a sufficient agitation speed and avoiding the biosorbent's settling, the contribution to the bulk diffusion can be neglected [43].

In another series of experiments, the effect of the initial pH_0 on Cd biosorption was studied within a range from pH 2.0 to pH 7.0. The applied initial pH_0 value was related to the varying chemical speciation (chemical forms) of Cd as a function of pH. On the basis of speciation modelling realized using MINEQL+ ver. 4.62.3, it was found that at pH = 7 and $CdCl_2$ concentration C_0 = 20 µmol/L, Cd was present, practically, only in the form of free Cd^{2+} cations, and from this value the proportion of the Cd^{2+} form decreased significantly. For example, at pH = 9, the Cd^{2+} form proportion was 92% of the total Cd in solution in addition to the occurrence of $CdOH^+$, $CdCl^+$, or $Cd(OH)_2$ forms. It should be emphasized that the total concentration of Cd in solution also plays an important role in this respect, wherein a more significant decrease in the proportion of the Cd^{2+} form can be observed at higher Cd concentrations in solution as a function of pH.

Figure 1b depicts the relationship between the Cd biosorption by the dried biomass of *E. gracilis* var. *bacillaris* from solutions containing $CdCl_2$ C_0 = 20 µmol/L and the initial pH_0 value of these solutions, as well as the dependence between the pH_f of the solutions at the end of the experiment and the initial pH_0 value. As can be seen, the highest values of biosorption efficiency for Cd 43% (16.4 µmol/g; d.w.) were obtained within the range of pH_0 7.0 to pH_0 5.0 and the lowest Cd biosorption values occurred at initial pH_0 values of 2 and 3. A similar pattern of Cd biosorption was also described by Demey et al. [42], when they found that Cd biosorption by biosorbent prepared from the biomass of alga *Fucus vesiculosus* was essentially unchanged from pH 5.0 to pH 7.0. Depending on the pH, the functional groups, such as carboxyl and phosphoryl groups, can be protonated or unprotonated. At low pH, these functional groups are protonated, which results in the absence of binding sites for the removal of cationic metal ions, whereas more functional groups would be deprotonated with a pH increase and, thus, increase metal biosorption [44]. As was mentioned, the pH of the solution also has an effect on the metal speciation and zeta potential of the biosorbent as well.

Our results also showed that the initial value of $pH_0 = 6$–7 decreased towards the equilibrium value of $pH_f = 5.6$ during the experiments. On the other hand, for an initial $pH_0 = 4$, the pH value increased towards the equilibrium value of $pH_f = 5.6$. This phenomenon can be explained by the release of substances from the dried biomass of *E. gracilis* var. *bacillaris* that have the ability to buffer the surrounding environment.

Following the above findings, a series of experiments aimed at evaluating the effect of Cd speciation and the proportion of the Cd^{2+} chemical form on the biosorption of Cd by the dried biomass of *E. gracilis* var. *bacillaris* from solutions containing 20 µmol/L $CdCl_2$ without or with the addition of different amounts of EDTA-Na_2 was carried out. In a first step, the modelling of the occurrences of the different chemical forms of Cd without or in the presence of EDTA-Na_2 at pH 4 and at 25 °C using the speculation program MINEQL+ ver. 4.62.3 was performed. The concentrations of EDTA-Na_2 added to $CdCl_2$ solutions were chosen in such molar amounts, to 20 µmol/L $CdCl_2$, at which it was possible to achieve the proportion of the Cd^{2+} form in solutions of 0%, 20%, 40%, 60%, 80%, and the maximum achievable amount of Cd^{2+} at pH 4.0. This can be seen in Figure 2a, where a linear decrease in the Cd^{2+} form with increases in the concentrations of EDTA-Na_2 in the solution can be observed in favor of the formation of complexed forms $CdCl^+$, Cd-$EDTA^{2-}$, Cd-$HEDTA^-$, or Cd-H_2EDTA. The proportions of the mentioned Cd complexes in the solution decreased in the order Cd-$EDTA^{2-}$ >> Cd-$HEDTA^-$ > Cd-$H_2EDTA \geq CdCl^+$.

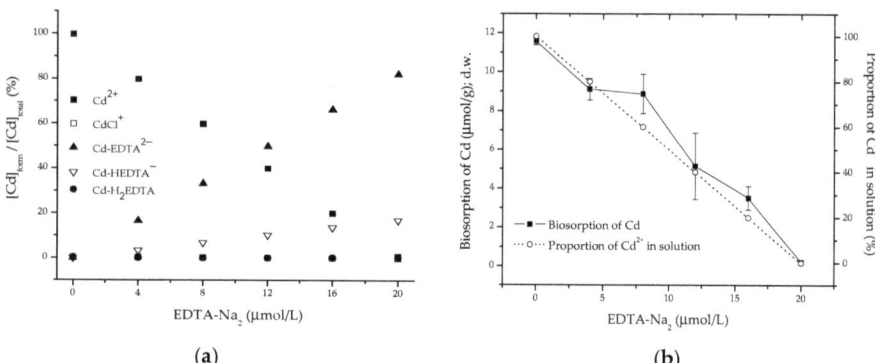

Figure 2. (a) Proportions of chemical forms of Cd in model solutions containing 20 µmol/L $CdCl_2$ without or with the addition of 4, 8, 12, 16, or 20 µmol/L EDTA-Na_2. (b) Effect of Cd speciation on Cd biosorption by dried biomass of *E. gracilis* var. *bacillaris* ($C_B = 0.5$ g/L) from model solutions containing 20 µmol/L $ZnCl_2$ (radiolabeled with 44.7 kBq/L $^{109}CdCl_2$) without or with the addition of 4, 8, 12, 16, or 20 µmol/L EDTA-Na_2, pH 4.0. Exposure of the biomass was carried out on a shaker (250 rpm) for 2 h and at 25 °C. Error bars represent standard deviation of the arithmetic mean of the results of three independent experiments ($\pm$SD; $n = 3$). Cd speciation was predicted using the modelling program MINEQL+ ver. 4.62.3 for pH = 4.0 and 25 °C.

It was found that the biosorption of Cd decreased linearly corresponding to the decrease in the proportion of Cd^{2+} in the solution given by the increasing concentration of

EDTA-Na$_2$ as a complexing agent in the solution. At an almost 100% initial proportion of the Cd^{2+} form in the solution, a maximum biosorption of 11.6 µmol/g (d.w.) was observed in this series. At the minimum proportion of Cd^{2+} (0.8%) in the solution, a more than 55-fold lower value of specific Cd biosorption (0.21 µmol/g; d.w.) was determined. Our previous work [33] also confirmed that with an increasing addition of NaCl to the solution, the initial presence of Cd^{2+} in the solution decreased in favor of CdCl$^+$ formation, which resulted in a decrease in the Cd biosorption by the dried biomass of the moss *Vesicularia dubyana*.

From this result, it can be concluded that Cd will be biosorbed mainly in the form of free Cd^{2+} cations. This finding is important from the point of view of the application of biosorbents in the removal of heavy metals from real wastewater or contaminated solutions, wherein, due to their heterogeneity in terms of the occurrence of inorganic and/or organic substances, it is necessary to take this fact into account as well.

Figure 3 describes the effect of biomass concentration, C_B, ranging from 0.25 to 2.5 g/L on Cd biosorption by the dried biomass of *E. gracilis* var. *bacillaris* from solutions with an initial Cd concentration of C_0 = 20 µmol/L. It was shown that the specific biosorption of Cd expressed in µmol/g (d.w.) decreased with an increasing biosorbent dose, while the removal efficiency (in %) increased slightly and linearly (R^2 = 0.888) with increasing biosorbent concentration. A similar observation has been reported by Cheng et al. [45] in the investigation of biosorption of Cd by the dead biomass of *Chlorella vulgaris*, when they confirmed that the removal efficiency of *C. vulgaris* increased from 89 to 95% as the adsorbent dosage was increased from 0.18 to 18 mg, and the biosorption capacity decreased with an increase in algal dose. This decrease can be due to the concentration gradient between the biosorbent and the sorbate–heavy metals; an increase in biosorbent concentration causes a decrease in the biosorption capacity or specific biosorption. Moreover, the increase in the biosorption of metals by increasing the biosorbent dose is due to an increase in the number of active sites and available surface area [46].

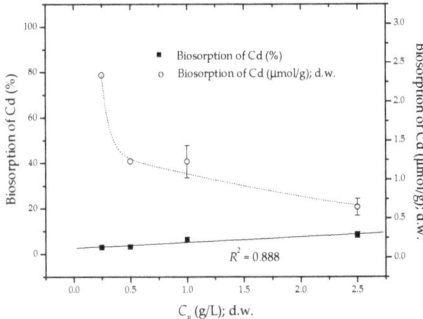

Figure 3. Effect of biomass concentration, C_B, on Cd biosorption by dried biomass of *E. gracilis* var. *bacillaris* from solutions containing 20 µmol/L CdCl$_2$ (radiolabeled with 44.9 kBq/L ^{109}CdCl$_2$) in deionized water, pH 4.0. Exposure of the biomass was carried out on a shaker (250 rpm) for 2 h and at 25 °C. Biosorption of Cd expressed in µmol/g (d.w.) and in % of total Cd in solution. Error bars represent standard deviation of the arithmetic mean of the results of three independent experiments (±SD; n = 3).

Similarly, the effect of an initial Cd concentration ranging from 1 to 80 µmol/L CdCl$_2$ on Cd biosorption by the dried biomass of *E. gracilis* var. *bacillaris* was also evaluated. As can be seen from Figure 4, the dependence of the Cd biosorption efficiency (in %) on the initial CdCl$_2$ concentration did not have a clear pattern. In contrast, however, the specific biosorption efficiency (in µmol/g; d.w.) increased linearly (R^2 = 0.975) with increasing initial Cd concentration in solution. Based on this fact, it can be argued that even at the highest concentration applied (80 µmol/L CdCl$_2$), saturation of the biosorbent was not observed, or active sites remained available on the surface of the biosorbent, respectively. Montazer-Rahmati et al. [46] characterized the biosorption capacity of marine algae *Cystoseira indica* and *Nizimuddinia zanardini* as a function of the initial concentrations of Cd, Ni, and Pb. They observed that the metal biosorption capacity increases with an increase in the initial metal ion concentration and explained it by the fact that increasing the initial metal concentration provides a larger driving force to overcome all mass transfer resistances between the solid and the aqueous phase, resulting in higher metal biosorption.

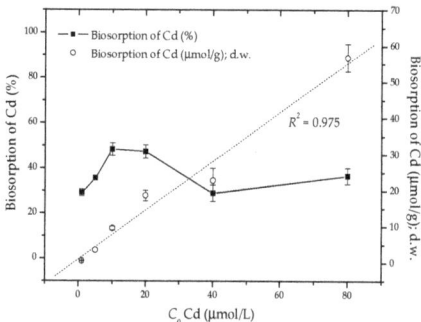

Figure 4. Effect of initial Cd concentration on Cd biosorption by dried biomass of *E. gracilis* var. *bacillaris* (biomass concentration C_B = 0.50 g/L) from solutions prepared in the form of $CdCl_2$ (radiolabeled with 47.5 kBq/L $^{109}CdCl_2$) in deionized water, pH 4.0. Exposure of the biomass was carried out on a shaker (250 rpm) for 2 h and at 25 °C. Biosorption of Cd expressed in µmol/g (d.w.) and in % of total Cd in solution. Error bars represent standard deviation of the arithmetic mean of the results of three independent experiments ($\pm$SD; n = 3).

Due to the fact that the biosorption of Cd by the dried biomass of *E. gracilis* var. *bacillaris* was a rapid process that achieved the concentration equilibrium of $[Cd]_{biomass}:[Cd]_{solution}$ within 20 min of exposure at constant temperature, it was possible to carry out a mathematical description of the data obtained in the study of the effect of the initial concentration of $CdCl_2$ in solution on the biosorption of Cd. For these purposes, the adsorption isotherms according to the Langmuir and Freundlich models in their nonlinear forms were applied. The description of the data obtained using a nonlinear regression method is shown in the plot of Figure 5. The relationship between the biosorption capacity, Q_{eq}, and Cd concentration in solution, C_{eq}, at equilibrium showed that the biosorption of Cd increased with the equilibrium concentration of Cd in solution, progressively, until the biosorbent's saturation. Experimental data were slightly better fitted by the Langmuir adsorption isotherm (R^2 = 0.999) than the Freundlich adsorption isotherm (R^2 = 0.997). The Langmuir isotherm model implies a monolayer adsorption of the metals on the surfaces of biosorbents. The parameters predicted from the Langmuir and Freundlich models of adsorption isotherm, such as the maximum biosorption capacity, Q_{max}, Langmuir affinity constant b, Freundlich constants K and n, are summarized in Table 1.

Table 1. The obtained parameters of adsorption isotherms according to Langmuir and Freundlich models. Details of the experiment are described in Figure 4.

Langmuir Adsorption Isotherm			Freundlich Adsorption Isotherm			
Q_{max} (µmol/g)	b (L/µmol)	R^2	K (L/g)	$1/n$	n	R^2
125 $\pm$ 8	0.0163 $\pm$ 0.0017	0.999	2.96 $\pm$ 0.49	0.753 $\pm$ 0.044	1.33 $\pm$ 0.08	0.997

To compare the ability of the dried biomass of *E. gracilis* var. *bacillaris* to remove Cd from model solutions, in Table 2 we summarized the Q_{max} values obtained by other authors when different biosorbents derived from macro- or microalgal biomasses were applied as biosorbents for Cd removal from single-component solutions. From the comparison, it is clear that for Cd biosorption, we did not obtain comparable data with those of other authors applying the different types of algal biomasses as biosorbents. These vastly different biosorption capacities for the same heavy metal by the same biosorbent or algal biomass are compatible with a commonly observed phenomenon. The biosorption capacity of algal biomass is dependent on many parameters, including the medium and condition of algal cultivation, the physiological state when algae were harvested, the condition of the drying process, as well as the biosorption conditions, such as the pH, metal concentration, incubation time, and biosorbent dose [47].

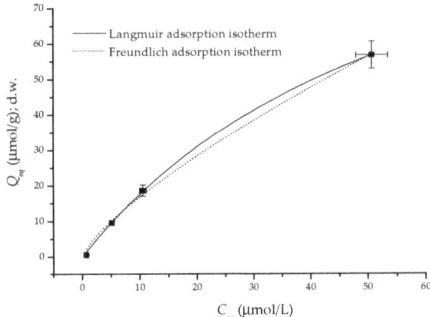

Figure 5. Description of Cd biosorption by dried biomass of *E. gracilis* var. *bacillaris* (biomass concentration C_B = 0.50 g/L) using adsorption isotherms according to Langmuir and Freundlich. Details of the experiments are described in Figure 4. Error bars represent standard deviation of the arithmetic mean of the results of three independent experiments ($\pm$SD; n = 3).

Table 2. Comparison of Q_{max} values quantifying Cd biosorption by biosorbents derived from macro- or microalgal biomasses.

Biosorbent	Maximum Biosorption Capacity Q_{max}		Reference
	(mmol/g); d.w.	(mg/g); d.w.	
Fucus vesiculosus (brown macroalgae)	1.27	143	Moreira et al. [48]
Scenedesmus quadricauda (green microalgae)	1.20	135	Mirghaffari et al. [49]
Sargassum polycystum (brown marine macroalgae)	0.93	105	Jayakumar et al. [50]
Fucus vesiculosus (brown macroalgae)	0.87	97.8	Demey et al. [42]
Parachlorella sp. (green microalgae)	0.76	86.1	Dirbaz and Roosta [40]
Sargassum filipendula (brown marine macroalgae)	0.42	47.2	Verma et al. [51]
Neochloris oleoabundans (green microalgae)	0.39	43.9	Gu and Lan [41]
Chlorella vulgaris (green microalgae)	0.19	21.4	Gu and Lan [44]
Cystoseira indica (brown marine macroalgae)	0.17	19.6	Montazer-Rahmati et al. [46]
Chlorella vulgaris (green microalgae)	0.15	16.7	Cheng et al. [45]
Euglena gracilis var. *bacillaris*	0.13	14.1	This work

4. Conclusions

In this work, the possibility of applying the dried biomass of E. gracilis var. bacillaris as a biosorbent for the removal of Cd from aqueous solutions was evaluated through batch biosorption experiments.

The kinetics of Cd biosorption by the dried biomass of E. gracilis var. bacillaris from solutions containing 20 μmol/L $CdCl_2$ and at pH 4.0 showed that it was a rapid process that achieved the concentration equilibrium of $[Cd]_{biomass}:[Cd]_{solution}$ in the first minutes of the interaction. The highest values of biosorption efficiency for Cd, 43% (16.4 μmol/g; d.w.), were obtained within the range of pH_0 7.0 to pH_0 5.0, and the lowest Cd biosorption values occurred at initial pH_0 values of 2 and 3. Attention has separately been focused on the effect of Cd speciation on its biosorption. According to MINEQL+ speciation modelling, a linear decrease in the Cd^{2+} form with increases in the concentration of EDTA-Na_2 in the solution was observed in favor of the formation of complexed forms $CdCl^+$, Cd-$EDTA^{2-}$, Cd-$HEDTA^-$, or Cd-H_2EDTA. It was found that the biosorption of Cd decreased linearly, corresponding to decreases in the proportion of Cd^{2+} in the solution given by the increasing concentration of EDTA-Na_2 as a complexing agent in the solution. At an almost 100% initial proportion of the Cd^{2+} form in the solution, a maximum biosorption of 11.6 μmol/g (d.w.) was observed in this series. At the minimum proportion of Cd^{2+} (0.8%) in the solution, a more than 55-fold lower value of specific Cd biosorption (0.21 μmol/g; d.w.) was determined. The specific biosorption of Cd, expressed in μmol/g (d.w.), decreased with a biosorbent dose increasing from 0.25 to 2.5 g/L, while the removal efficiency (in %) increased slightly and linearly ($R^2 = 0.888$) with an increasing biosorbent concentration. The obtained biosorption data were well fitted to the Langmuir model of adsorption isotherm in comparison with the Freundlich model. Also, from the Langmuir model of adsorption isotherm, the maximum biosorption capacity of the dried biomass of E. gracilis var. bacillaris for the removal of Cd was predicted and reached the value $Q_{max} = 0.13$ mmol/g or 14.1 mg/g (d.w.), respectively.

From the obtained results, it can be concluded that dried biomass of E. gracilis var. bacillaris could be used as a cost-effective, efficient, and environmentally friendly biosorbent for the removal of Cd from wastewater or contaminated water. However, limiting factors for the utilization of this type of biosorbent include their relatively low selectivity and stability, especially in terms of their regeneration or their reuse over long periods of time. Also, an important issue is their further treatment after use as hazardous waste. In particular, solidification or controlled incineration is expected. In some cases, controlled incineration could lead to the recovery of precious metals.

Author Contributions: Conceptualization, J.K., M.P. and M.H.; methodology, V.A., M.V., M.P. and M.H.; software, M.V.; validation, L.H., M.H. and M.V.; formal analysis, L.M. and V.A.; investigation, L.M., T.L. and P.M.; writing—original draft preparation, M.V.; writing—review and editing, M.P. and M.H.; supervision, J.K., M.P. and M.H. All authors have read and agreed to the published version of the manuscript.

Funding: This research was funded by the Slovak Research and Development Agency under contract no. APVV-17-0150.

Institutional Review Board Statement: Not applicable.

Informed Consent Statement: Not applicable.

Data Availability Statement: Data available on request due to privacy/ethical restrictions.

Conflicts of Interest: The authors declare no conflict of interest. The funders had no role in the design of the study; in the collection, analyses, or interpretation of data; in the writing of the manuscript; or in the decision to publish the results.

References

1. Wang, L.; Liu, J.; Filipiak, M.; Mungunkhuyag, K.; Jedynak, P.; Burczyk, J.; Fu, P.; Malec, P. Fast and efficient cadmium biosorption by *Chlorella vulgaris* K-01 strain: The role of cell walls in metal sequestration. *Algal Res.* **2021**, *60*, 102497. [CrossRef]
2. Jaafari, J.; Ghozikali, M.G.; Azari, A.; Delkhosh, M.B.; Javid, A.B.; Mohammadi, A.A.; Agarwal, S.; Gupta, V.K.; Sillanpää, M.; Tkachev, A.G. Adsorption of *p*-Cresol on Al_2O_3 coated multi-walled carbon nanotubes: Response surface methodology and isotherm study. *J. Ind. Eng. Chem.* **2018**, *57*, 396–404. [CrossRef]
3. Ali, H.; Khan, E.; Ilahi, I. Environmental chemistry and ecotoxicology of hazardous heavy metals: Environmental persistence, toxicity, and bioaccumulation. *J. Chem.* **2019**, *2019*, 6730305. [CrossRef]
4. Briffa, J.; Sinagra, E.; Blundell, R. Heavy metal pollution in the environment and their toxicological effects on humans. *Heliyon* **2020**, *6*, e04691. [CrossRef] [PubMed]
5. Rahoui, S.; Ben, C.; Chaoui, A.; Martinez, Y.; Yamchi, A.; Rickauer, M.; Gentzbittel, L.; El Ferjani, E. Oxidative injury and antioxidant genes regulation in cadmium-exposed radicles of six contrasted *Medicago truncatula* genotypes. *Environ. Sci. Pollut. Res.* **2014**, *21*, 8070–8083. [CrossRef] [PubMed]
6. Ma, H.; Wei, M.; Wang, Z.; Hou, S.; Li, X.; Xu, H. Bioremediation of cadmium polluted soil using a novel cadmium immobilizing plant growth promotion strain *Bacillus* sp. TZ5 loaded on biochar. *J. Hazard Mater.* **2020**, *388*, 122065. [CrossRef]
7. ATSDR. *Priority List of Hazardous Substances*; Agency for Toxic Substances and Disease Registry: Atlanta, GA, USA, 1997. Available online: www.atsdr.cdc.gov (accessed on 18 July 2023).
8. Cheng, K.; Tian, H.; Zhao, D.; Lu, L.; Wang, Y.; Chen, J.; Liu, X.; Jia, W.; Huang, Z. Atmospheric emission inventory of cadmium from anthropogenic sources. *Int. J. Environ. Sci. Technol.* **2014**, *11*, 605–616. [CrossRef]
9. Liu, Y.; Xiao, T.; Ning, Z.; Li, H.; Tang, J.; Zhou, G. High cadmium concentration in soil in the Three Gorges region: Geogenic source and potential bioavailability. *Appl. Geochem.* **2013**, *37*, 149–156. [CrossRef]
10. USA Environmental Protection Agency. *Aquatic Life, Ambient Water Quality Criteria, Cadmium-2016*; EPA 820-R-16-002; Office of Water, Office of Science and Technology, Health and Ecological Criteria Division: Washington, DC, USA, 2016.
11. World Health Organization (WHO). *Guidelines for Drinking-Water Quality*, 4th ed.; World Health Organization: Geneva, Switzerland, 2011; 327p.
12. Kabata-Pendias, A. *Trace Elements in Soils and Plants*, 4th ed.; CRC Press: Boca Raton, USA, 2011; 548p.
13. Jaafari, J.; Seyedsalehi, M.; Safari, G.; Arjestan, M.E.; Barzanouni, H.; Ghadimi, S.; Kamani, H.; Haratipour, P. Simultaneous biological organic matter and nutrient removal in an anaerobic/anoxic/oxic (A^2O) moving bed biofilm reactor (MBBR) integrated system. *Int. J. Environ. Sci. Technol.* **2017**, *14*, 291–304. [CrossRef]
14. Khan, Z.; Elahi, A.; Bukhari, D.A.; Rehman, A. Cadmium sources, toxicity, resistance and removal by microorganisms-A potential strategy for cadmium eradication. *J. Saudi Chem. Soc.* **2022**, *26*, 101569. [CrossRef]
15. Cardoso, S.L.; Costa, C.S.D.; Nishikawa, E.; da Silva, M.G.C.; Vieira, M.G.A. Biosorption of toxic metals using the alginate extraction residue from the brown algae *Sargassum filipendula* as a natural ion-exchanger. *J. Clean. Prod.* **2017**, *165*, 491–499. [CrossRef]
16. Albatrni, H.; Qiblawey, H.; El-Naas, M.H. Comparative study between adsorption and membrane technologies for the removal of mercury. *Sep. Purif. Technol.* **2022**, *257*, 117833. [CrossRef]
17. Ahluwalia, S.S.; Goyal, D. Microbial and plant derived biomass for removal of heavy metals from wastewater. *Bioresour. Technol.* **2007**, *98*, 2243–2257. [CrossRef]
18. Lezcano, J.M.; González, F.; Ballester, A.; Blázquez, M.L.; Muñoz, J.A.; García-Balboa, C. Sorption and desorption of Cd, Cu and Pb using biomass from an eutrophized habitat in monometallic and bimetallic systems. *J. Environ. Manag.* **2011**, *92*, 2666–2674. [CrossRef] [PubMed]
19. Bhatt, P.; Bhandari, G.; Turco, R.F.; Aminikhoei, Z.; Bhatt, K.; Simsek, H. Algae in wastewater treatment, mechanism, and application of biomass for production of value-added product. *Environ. Pollut.* **2022**, *309*, 119688. [CrossRef]
20. Martín-Lara, M.A.; Pagnanelli, F.; Mainelli, S.; Calero, M.; Toro, L. Chemical treatment of olive pomace: Effect on acid-basic properties and metal biosorption capacity. *J. Hazard Mater.* **2008**, *156*, 448–457. [CrossRef]
21. Ramesh, B.; Saravanan, A.; Senthil Kumar, P.; Yaashikaa, P.R.; Thamarai, P.; Shaji, A.; Rangasamy, G. A review on algae biosorption for the removal of hazardous pollutants from wastewater: Limiting factors, prospects and recommendations. *Environ. Pollut.* **2023**, *327*, 121572. [CrossRef]
22. Manikandan, A.; Suresh Babu, P.; Shyamalagowri, S.; Kamaraj, M.; Muthukumaran, P.; Aravind, J. Emerging role of microalgae in heavy metal bioremediation. *J. Basic Microbiol.* **2022**, *62*, 330–347. [CrossRef]
23. Monteiro, C.M.; Castro, P.M.L.; Malcata, F.X. Metal uptake by microalgae: Underlying mechanisms and practical applications. *Biotechnol. Prog.* **2012**, *28*, 299–311. [CrossRef]
24. Jaafari, J.; Yaghmaeian, K. Optimization of heavy metal biosorption onto freshwater algae (*Chlorella coloniales*) using response surface methodology (RSM). *Chemosphere* **2019**, *217*, 447–455. [CrossRef]
25. Michalak, I.; Chojnacka, K. Interactions of metal cations with anionic groups on the cell wall of the macroalga *Vaucheria* sp. *Eng. Life Sci.* **2010**, *10*, 209–217. [CrossRef]
26. Santiago-Martínez, M.G.; Lira-Silva, E.; Encalada, R.; Pineda, E.; Gallardo-Pérez, J.C.; Zepeda-Rodriguez, A.; Moreno-Sánchez, R.; Saavedra, E.; Jasso-Chávez, R. Cadmium removal by *Euglena gracilis* is enhanced under anaerobic growth conditions. *J. Hazard. Mater.* **2015**, *288*, 104–112. [CrossRef] [PubMed]

27. Rodriguez-Zavala, J.S.; Garcia-Garcia, J.D.; Ortiz-Cruz, M.A.; Moreno-Sanchez, R. Molecular mechanisms of resistance to heavy metals in the protist *Euglena gracilis*. *J. Environ. Sci. Health A* **2007**, *42*, 1365–1378. [CrossRef] [PubMed]
28. Gissibl, A.; Sun, A.; Care, A.; Nevalainen, H.; Sunna, A. Bioproducts from *Euglena gracilis*: Synthesis and applications. *Front. Bioeng. Biotechnol.* **2019**, *7*, 108. [CrossRef] [PubMed]
29. Inwongwan, S.; Kruger, N.J.; Ratcliffe, R.G.; O'Neill, E.C. *Euglena* central metabolic pathways and their subcellular locations. *Metabolites* **2019**, *9*, 115. [CrossRef]
30. Jasso-Chávez, R.; Campos-García, M.L.; Vega-Segura, A.; Pichardo-Ramos, G.; Silva-Flores, M.; Santiago-Martínez, M.G.; Feregrino-Mondragón, R.D.; Sánchez-Thomas, R.; García-Contreras, R.; Torres-Márquez, M.E.; et al. Microaerophilia enhances heavy metal biosorption and internal binding by polyphosphates in photosynthetic *Euglena gracilis*. *Algal Res.* **2021**, *5858*, 102384. [CrossRef]
31. Wang, Y.; Seppänen-Laakso, T.; Rischer, H.; Wiebe, M.G. *Euglena gracilis* growth and cell composition under different temperature, light and trophic conditions. *PLoS ONE* **2018**, *13*, e0195329. [CrossRef]
32. Lewis, A.; Guéguen, C. Using chemometric models to predict the biosorption of low levels of dysprosium by *Euglena gracilis*. *Environ. Sci. Pollut. Res.* **2022**, *29*, 58936–58949. [CrossRef]
33. Šuňovská, A.; Hasíková, V.; Horník, M.; Pipíška, M.; Hostin, S.; Lesný, J. Removal of Cd by dried biomass of freshwater moss *Vesicularia dubyana*: Batch and column studies. *Desalin. Water Treat.* **2016**, *57*, 2657–2668. [CrossRef]
34. Machalová, L.; Pipíška, M.; Trajteľová, Z.; Horník, M. Comparison of Cd^{2+} biosorption and bioaccumulation by bacteria—A radiometric study. *Nova Biotechnol. Chim.* **2015**, *14*, 158–175. [CrossRef]
35. Pipíška, M.; Horník, M.; Remenárová, L.; Augustín, J.; Lesný, J. Biosorption of cadmium, cobalt and zinc by moss *Rhytidiadelphus squarrosus* in the single and binary component systems. *Acta Chim. Slov.* **2010**, *57*, 163–172. [PubMed]
36. Remenárová, L.; Pipíška, M.; Horník, M.; Rozložník, M.; Augustín, J.; Lesný, J. Biosorption of cadmium and zinc by activated sludge from single and binary solutions: Mechanism, equilibrium and experimental design study. *J. Taiwan Inst. Chem. Eng.* **2012**, *13*, 433–443. [CrossRef]
37. Hutner, S.H.; Bach, M.K.; Ross, G.T. A sugar-containing basal medium for vitamin B12-assay with *Euglena*; application to body fluids. *J. Protozool.* **1956**, *3*, 101–112. [CrossRef]
38. Langmuir, I. The adsorption of gases on plane surfaces of glass, mica, and platinum. *J. Am. Chem. Soc.* **1918**, *40*, 1361–1403. [CrossRef]
39. Freundlich, H.M.F. Über die adsorption in lösungen. *Z. Phys. Chem.* **1906**, *57*, 385–470. [CrossRef]
40. Dirbaz, M.; Roosta, A. Adsorption, kinetic and thermodynamic studies for the biosorption of cadmium onto microalgae *Parachlorella* sp. *J. Environ. Chem. Eng.* **2018**, *6*, 2302–2309. [CrossRef]
41. Gu, S.; Lan, C.Q. Effects of culture pH on cell surface properties and biosorption of Pb(II), Cd(II), Zn(II) of green alga *Neochloris oleoabundans*. *Chem. Eng. J.* **2023**, *468*, 143579. [CrossRef]
42. Demey, H.; Vincent, T.; Guibal, E. A novel algal-based sorbent for heavy metal removal. *Chem. Eng. J.* **2018**, *332*, 582–595. [CrossRef]
43. Benettayeb, A.; Guibal, E.; Morsli, A.; Kessas, R. Chemical modification of alginate for enhanced sorption of Cd(II), Cu(II) and Pb(II). *Chem. Eng. J.* **2017**, *316*, 704–714. [CrossRef]
44. Gu, S.; Lan, C.Q. Lipid-extraction algal biomass for biosorption of bivalent lead and cadmium ions: Kinetics and isotherm. *Chem. Eng. Sci.* **2023**, *276*, 118778. [CrossRef]
45. Cheng, J.; Yin, W.; Chang, Z.; Lundholm, N.; Jiang, Z. Biosorption capacity and kinetics of cadmium(II) on live and dead *Chlorella vulgaris*. *J. Appl. Phycol.* **2017**, *29*, 211–221. [CrossRef]
46. Montazer-Rahmati, M.M.; Rabbani, P.; Abdolali, A.; Keshtkar, A.R. Kinetics and equilibrium studies on biosorption of cadmium, lead, and nickel ions from aqueous solutions by intact and chemically modified brown algae. *J. Hazard. Mater.* **2011**, *185*, 401–407. [CrossRef] [PubMed]
47. Yan, C.; Qu, Z.; Wang, J.; Cao, L.; Han, Q. Microalgal bioremediation of heavy metal pollution in water: Recent advances, challenges, and prospects. *Chemosphere* **2022**, *286*, 131870. [CrossRef] [PubMed]
48. Moreira, V.R.; Lebron, Y.A.R.; Lange, L.C.; Santos, L.V.S. Simultaneous biosorption of Cd(II), Ni(II) and Pb(II) onto a brown macroalgae *Fucus vesiculosus*: Mono- and multi-component isotherms, kinetics and thermodynamics. *J. Environ. Manag.* **2019**, *251*, 109587. [CrossRef]
49. Mirghaffari, N.; Moeini, E.; Farhadian, O. Biosorption of Cd and Pb ions from aqueous solutions by biomass of the green microalga, *Scenedesmus quadricauda*. *J. Appl. Phycol.* **2015**, *27*, 311–320. [CrossRef]
50. Jayakumar, V.; Govindaradjane, S.; Rajamohan, N.; Rajasimman, M. Biosorption potential of brown algae, *Sargassum polycystum*, for the removal of toxic metals, cadmium and zinc. *Environ. Sci. Pollut. Res.* **2022**, *29*, 41909–41922. [CrossRef]
51. Verma, A.; Kumar, S.; Kumar, S. Statistical modeling, equilibrium and kinetic studies of cadmium ions biosorption from aqueous solution using *S. filipendula*. *J. Environ. Chem. Eng.* **2017**, *5*, 2290–2304. [CrossRef]

Disclaimer/Publisher's Note: The statements, opinions and data contained in all publications are solely those of the individual author(s) and contributor(s) and not of MDPI and/or the editor(s). MDPI and/or the editor(s) disclaim responsibility for any injury to people or property resulting from any ideas, methods, instructions or products referred to in the content.

Proceeding Paper

Effects of Vermicompost Application on Plant Growth Stimulation in Technogenic Soils [†]

Marketa Dreslova

Department of Environmental Engineering, Faculty of Mining and Geology, VŠB–Technical University of Ostrava, 17. Listopadu 2172/15, 708 00 Ostrava, Czech Republic; marketa.dreslova.st@vsb.cz

[†] Presented at the 4th International Conference on Advances in Environmental Engineering, Ostrava, Czech Republic, 20–22 November 2023.

Abstract: The aim of this study was to support the use of waste materials formed from a mixture of technogenic soils for growing plants through adding vermicompost leachates. The effect on the growth of underground and above-ground biomass was evaluated on plants of the bush variety of tomato (Solanum lycopersicum). Three different types of biodegradable waste (apple pomace, matolin, and horse manure) were used in the experiments, from which individual vermicomposts were subsequently produced. The effect of the addition of vermicompost leachates to the soil was manifested in all the statistically evaluated parameters of the bush tomato plants. It was found that the highest values were achieved for the root weight (+2.91 g; $p < 0.01$) and for the stem (+1.92 g, $p < 0.01$). The lowest values were observed in the control plants without application of the vermicompost leachates.

Keywords: vermicompost; stimulation; biomass; technogenic soil; *Solanum lycopersicum*

Citation: Dreslova, M. Effects of Vermicompost Application on Plant Growth Stimulation in Technogenic Soils. *Eng. Proc.* **2023**, *57*, 42. https://doi.org/10.3390/engproc2023057042

Academic Editors: Adriana Estokova, Natalia Junakova, Tomas Dvorsky, Vojtech Vaclavik and Magdalena Balintova

Published: 18 December 2023

Copyright: © 2023 by the author. Licensee MDPI, Basel, Switzerland. This article is an open access article distributed under the terms and conditions of the Creative Commons Attribution (CC BY) license (https://creativecommons.org/licenses/by/4.0/).

1. Introduction

Nowadays, there is increasing pressure to recycle waste materials due to the efforts of maintaining the ecosystem and the use of energy and nutrients. Soils burdened with industrial activity, the so-called technosoils, are known for their reduced proportion of organic matter. A recently published meta-analysis [1] stated the average value of their organic carbon content is approximately 4.3%. In addition, these technogenic soils are also known for the content of toxic substances that cause changes in their biological activity, which are often manifested as the diversity of soil microorganisms [2–5], and also in their structure down to the level of soil microstructures [6–11]. All the above-mentioned changes reduce the functions of these technogenic soils and make their regeneration impossible. The enrichment of technosoils with vermicompost can contribute to their agricultural use.

Vermicomposting is a sustainable approach to waste management and it represents the process of the decomposition of organic waste through the cooperation of earthworms (genus *Eisenia*) and microorganisms [12]. It includes physical processes such as the fragmentation, aeration, and overturning of wastes, as well as biochemical processes such as enzymatic digestion, the transformation of waste materials, and their enrichment. Earthworms modify the physico-chemical and biological state of organic matter, reduce the C/N ratio, increase the surface area, expose more sites for microbial action and improve the decomposition process [13].

However, more generally, vermicompost can be described as, most often, a dark, weakly sticky mass of a solid state resembling humus or peat [14]. Factors affecting vermicompost quality include, for example, aeration, moisture [15], temperature [16], pH value [17,18], and the content of chemical substances, or elements, respectively [19]. A very important feature is the ability to supply crops with growth hormones such as auxin, gibberellin, or cytokinin, and the presence of fulvic acids and humic acids acting as plant growth regulators [20]. Aeration, mechanical mixing, additional nutrients (additives), extraction time, water quality, and temperature affect mainly the efficiency of the extraction of the substances mentioned above [21].

The aim of this work was to evaluate the effect of the application of vermicompost leachates extracts on the growth stimulation of tomato plants (*Solanum lycopersicum* L.) in a mixture with a technogenic soils.

2. Material and Methods

2.1. Preparation of Vermicomposts and Their Aqueous Solutions

Three different types of biodegradable waste (apple pomace, matolin, and horse manure) were used in experiments, from which individual vermicomposts were subsequently produced.

A specially designed glass laboratory extractor with a volume of 10 L was used for the preparation of aqueous extracts from vermicomposts. This extractor was equipped with a magnetic stirrer (850 rpm), an aeration device, and four probes (pH, EC, dissolved oxygen and temperature). During the production of aqueous extracts, a constant temperature of 30 °C and an air flow of 10 L/min were maintained. For leaching, one kilogram of fresh vermicompost was placed in a perforated basket (volume 1250 mL) and placed in a glass container containing 9 liters of demineralized water.

The ratio of vermicompost to water was 1:9. A dressing was then prepared from the leachates after 48 h of aeration, in a ratio of 1:10 with demineralized water. The dressing was applied in two variants, by spraying the leaf and dressing the substrate. One dose represented 100 mL of solution or demineralized water for control purposes. Pre-grown determinant varieties of *Solanum lycopersicum* plants were used as test plants.

2.2. Experimental Plants

The plants of Solanum lycopersicum were grown in special boxes (Figure 1) equipped with aluminum foil preventing temperature changes and the influence of results by external sources of heat or cold in the laboratory (radiators and windows). The temperature in boxes was 24 °C (day) and 18 °C (night) throughout the growing season. Sodium lamps of a power of approx. 10,500 Lx were used as a source of light and partly heat. Each plant was planted in 5L containers. Fertilizer was applied 5 times in total during the vegetation period, at intervals, in the 2nd, 6th, 8th, and 10th week. In the cases where irrigation with fertilizer was not applied, control demineralized water was used instead. The time interval since planting the plants to their harvest was, in total, 13 weeks for above-ground biomass and 14 weeks for underground biomass root. Fruits were harvested in a total of 4 phases, namely, in the 9th, 10th, and 13th week.

Figure 1. The growth of *Solanum lycopersicum* plants in boxes (author's photo).

2.3. Method of Plants Growth Assessment

At harvest, the plants were divided into fruits, other above-ground biomass (stem and leaf), and underground biomass. In the case of underground biomass, rough manual separation of the root system was performed after the substrate had dried (harvesting 1 week longer). The biomass collected in this way was then washed out 3 times in demineralized water and thereby freed of from the maximum of possible soil contamination. Each variant was performed in six repetitions. Fresh biomass weights and fruit weight and number were recorded during harvest.

2.4. Statistical Evaluation

The SAS 9.3 program was used for the plant yield evaluation. The MEANS and UNIVARIATE procedure was used to determine basic statistics. Correlations and regressions between the assessed variables were calculated using the CORR and REGG procedures. Subsequently, a detailed evaluation was performed using the GLM procedure, two-factor analysis of variance. The evaluated parameters were root weight (g), stem weight (g), fruit weight (g) and number of fruits (pcs). Several variants of the model were tested. Based on the Akaike information criterion [22], a model containing the effects of application form and dose was selected. The interaction effect was not statistically significant, so it was excluded from further evaluation. The Tukey–Kramer test [23] and statistical significance levels $p < 0.05$ were used to determine the significance between individual effect levels; $p < 0.01$; $p < 0.001$.

3. Results

3.1. Basic Statistics of the Evaluated Data Set

Table 1 shows the basic statistics of the evaluated data set. The average weight of the root system was 42.47 g, with a standard deviation of 6.04 g. The weight of the stem was in the range of 46.88–81.66 g. At harvest, an average of 22.27 pieces of fruit with an average weight of 6.49 g were obtained. The weight of the fruits of the bush varieties of tomatoes ranged from 4.41 g to 8.21 g with a standard deviation of 0.77 g.

Table 1. Basic statistics of the evaluated data set.

Variable	n	$\bar{x}$	s	min.	max.	s.e.	V (%)
root (g)	191	42.47	6.04	29.87	58.04	0.44	14.23
stem (g)	192	66.60	6.31	46.88	81.66	0.46	9.47
fruits (g)	192	6.49	0.77	4.41	8.21	0.06	11.82
fruits (pcs)	192	22.27	3.61	10	29	0.26	16.22

Notes: (**n**) number of measurements; ($\bar{x}$) arithmetic mean; (**s**) standard deviation; (**min**) minimum value; (**max**) maximum value; (**s.e.**) the mean error of the arithmetic mean; (**V**) coefficient of variance (%).

As part of the statistical analysis of the obtained data, a correlation analysis was also performed. Very strong and statistically significant correlations were observed mainly between the root weight and stem weight (g) (r = 0.705; $p < 0.01$) and the fruit weight (g) (r = 0.621; $p < 0.01$), respectively. A similarly strong relationship was found between the stem weight (g) and fruit weight (g) (r = 0.624; $p < 0.01$). Moderately strong correlations were also found between the root weight (g) and fruit number (pc) (r = 0.384; $p < 0.001$), stem weight (g) and fruit number (pc) (r = 0.341; $p < 0.01$), and between the fruit number and weight (g) (r = 0.369; $p < 0.01$), respectively.

3.2. Impact of Vermicompost Leachates on Plant Growth

The model equation for evaluating the effect of the application of vermicompost leachates on the amount of biomass of the bush tomato was statistically significant ($p < 0.01$) for all the monitored parameters and explained the variability of the evaluated biomass parameters from 19.4% to 82% (Table 2). The effect of the form of vermicompost application was statistically significant ($p < 0.01$) only for the root and stem weight. The interaction between the monitored effects was not statistically significant ($p > 0.05$) for any of the monitored parameters and was therefore not included in the evaluation.

Table 2. Basic statistics from the ANOVA procedure.

Rated Properties	Model		Form		Dose	
	r^2	p	F-Value	p	F-Value	p
root (g)	0.820	<0.001	59.66	<0.001	156.42	<0.001
stem (g)	0.51	<0.001	8.76	0.004	36.73	<0.001
fruits (g)	0.692	<0.001	1.04	0.308	82.79	<0.001
fruits (pcs)	0.194	<0.001	0.02	0.878	8.91	<0.001

The root weights reached statistically significantly higher values when vermicompost leachates were applied to the soil (+2.91 g; $p < 0.01$). Similar tendencies were also found for higher stem weights (g), which were also achieved when vermicompost leachates were applied to the soil (+1.92 g, $p < 0.01$). However, statistically significant application differences were not observed for fruit weights and fruit numbers.

4. Discussion

As is the evidence of previous research, the quality of vermicomposts varies depending on the input raw materials and production procedures [24,25]. A whole range of studies around the world looked at and confirmed the positive effect not only of vermicomposts, but also of their leachates, for example, on the growth of fruit [26] and vegetables [27]. Atiyeh et al. [28] confirm in their work, similar to my work, the increase in the biomass of bush tomatoes when using vermicompost based on food waste. Zeller [29] observed a greater increase in the root mass of bush tomatoes of about 10 g when vermicomposts were applied, which is also confirmed by the results presented by me. Moreover, the obtained results are consistent with the work of Edwards et al. [20]. Bachman and Metzger [30] observed a positive effect of vermicompost based on pig manure on the growth of the root systems, stems, and leaves of tomatoes. My results also align with the study by Joshi et al. [31], which describes the positive effect of vermicompost as an organic fertilizer and bioagent on plant growth. Furthermore, Pillai and Aswathy [32] observed, for example, an 18% higher yield of edible tomato berries when vermicomposts were applied to the leaf. Similarly, the positive effect of vermicompost application or vermicompost leachate was also observed in other field crops, such as (sown flax) in study [33]. Similarly to my study, Fritz [34] also evaluated the effect of leachates from vermicomposts on the yield of selected vegetable species (*Raphanus sativus*, *Rucola selvatica*, and *Pisum sativum*). In his work, this author, in contrast to my results, did not confirm the positive effect of the application of these extracts on the yield of the observed vegetable species. Contrary to the study by Fritz (2012), an even higher yield can be achieved by adding additives such as brewer's yeast.

In the case of this study, the specific kind of leachate was used as a result of the assumption that leachates prepared this way will improve the microflora in the rhizosphere and phyllosphere of technogenic soils, thereby increasing the biomass growth and plant yield.

5. Conclusions

This study shows the use of vermicompost for enriching the properties of technosoil suitable for the growth of cultural (agricultural) plants. Based upon the results achieved, in the case of *Solanum lycopersicum* plants, it can be inferred that the application of vermicompost in the form of leachates improves the overall health of soils and their productivity due to the improved intake of nutrients, the presence of humic substances and phytohormones, and that it also supports microbial activity. The effect of the addition of vermicompost leachates to the soil was manifested in all the statistically evaluated parameters of the bush tomato plants. The highest values were achieved for the root weight (+2.91 g; $p < 0.01$) and for the stem (+1.92 g, $p < 0.01$). The lowest values were observed in the control plants without the application of vermicompost leachate.

Funding: This research was funded by The Project for Specific University Research (SGS) No. SP2023/4 by the Faculty of Mining and Geology of VŠB—Technical University of Ostrava.

Institutional Review Board Statement: Not applicable.

Informed Consent Statement: Not applicable.

Data Availability Statement: Data are contained within the paper.

Conflicts of Interest: The author declares no conflict of interest.

References

1. Allory, V.; Séré, G.; Ouvrard, S. A meta-analysis of carbon content and stocks in Technosoils and identification of the main governing factors. *Eur. J. Soil Sci.* **2021**, *73*, e13141. [CrossRef]
2. Nosalj, S.; Šimonovičová, A.; Vojtková, H. Enzyme production by soilborne fungal strains of *Aspergillus niger* isolated from different localities affected by mining. *IOP Conf. Ser. Earth Environ. Sci.* **2021**, *900*, 012027. [CrossRef]
3. Šimonovičová, A.; Ferianc, P.; Vojtková, H.; Pangallo, D.; Hanajík, P.; Kraková, L.; Feketeová, Z.; Čerňanský, S.; Okenicová, L.; Žemberyová, M.; et al. Alkaline Technosol contaminated by former mining activity and its culturable autochthonous microbiota. *Chemosphere* **2017**, *171*, 89–96. [CrossRef]
4. Šimonovičová, A.; Vojtková, H.; Nosalj, S.; Piecková, E.; Švehláková, H.; Kraková, L.; Drahovská, H.; Stalmachová, B.; Kučová, K.; Pangallo, D. *Aspergillus niger* environmental isolates and their specific diversity through metabolite profiling. *Front. Microbiol.* **2021**, *12*, 658010. [CrossRef] [PubMed]
5. Vojtková, H. New strains of copper-resistant pseudomonas bacteria isolated from anthropogenically polluted soils. In Proceedings of the International Multidisciplinary Scientific GeoConference Surveying Geology and Mining Ecology Management, SGEM 2014, Albena, Bulgaria, 17–26 June 2014; Volume 1, pp. 451–457. [CrossRef]
6. Urík, M.; Polák, F.; Bujdoš, M.; Miglierini, M.B.; Milová-Žiaková, B.; Farkas, B.; Goneková, Z.; Vojtková, H.; Matúš, P. Antimony leaching from antimony-bearing ferric oxyhydroxides by filamentous fungi and biotransformation of ferric substrate. *Sci. Total Environ.* **2019**, *664*, 683–689. [CrossRef] [PubMed]
7. Farkas, B.; Vojtková, H.; Bujdoš, M.; Kolenčík, M.; Šebesta, M.; Matulová, M.; Duborská, E.; Danko, M.; Kim, H.; Kučová, K.; et al. Fungal mobilization of selenium in the presence of hausmannite and ferric oxyhydroxides. *J. Fungi* **2021**, *7*, 810. [CrossRef]
8. Matulová, M.; Bujdoš, M.; Miglierini, M.B.; Cesnek, M.; Duborská, E.; Mosnáčková, K.; Vojtková, H.; Kmječ, T.; Dekan, J.; Matúš, P.; et al. The effect of high selenite and selenate concentrations on ferric oxyhydroxides transformation under alkaline conditions. *Int. J. Mol. Sci.* **2021**, *22*, 9955. [CrossRef]
9. Balíková, K.; Vojtková, H.; Duborská, E.; Kim, H.; Matúš, P.; Urík, M. Role of Exopolysaccharides of *Pseudomonas* in heavy metal removal and other remediation strategies. *Polymers* **2022**, *14*, 4253. [CrossRef]
10. Šebesta, M.; Vojtková, H.; Cyprichová, V.; Ingle, A.P.; Urík, M.; Kolenčík, M. Mycosynthesis of metal-containing nanoparticles—Fungal metal resistance and mechanisms of synthesis. *Int. J. Mol. Sci.* **2022**, *23*, 14084. [CrossRef]
11. Šebesta, M.; Vojtková, H.; Cyprichová, V.; Ingle, A.P.; Urík, M.; Kolenčík, M. Mycosynthesis of metal-containing nanoparticles—Synthesis by *Ascomycetes* and *Basidiomycetes* and their application. *Int. J. Mol. Sci.* **2023**, *24*, 304. [CrossRef]
12. Singh, R.P.; Sing, P.; Araujo, A.S.F.; Ibrahim, M.H.; Sulaiman, O. Management of urban solid waste: Vermicomposting a sustainable option. *Resour. Conserv. Recycl.* **2011**, *55*, 719–729. [CrossRef]
13. Datta, S.; Singh, J.; Singh, S.; Singh, J. Earthworms, pesticides and sustainable agriculture: A review. *Environ. Sci. Pollut. Res.* **2016**, *23*, 8227–8243. [CrossRef] [PubMed]
14. Hanč, A.; Plíva, P. Vermikompostování—Perspektivní způsob nakládání s bioodpady. *Odpad. Forum* **2010**, *11*, 32.
15. Garg, V.K.; Gupta, R. Effect of temperature variations on vermicomposting of household solid waste and fecundity of *Eisenia fetida*. *Bioremediat. J.* **2011**, *15*, 165–172. [CrossRef]
16. Nagavallemma, K.P.; Wani, S.P.; Lacroix, S.; Padmaja, V.V.; Vineela, C.; Babu Rao, M.; Sahrawat, K.L. Vermicomposting: Recycling wastes into valuable organic fertilizer. *Glob. Theme Agrecosyst. Rep.* **2004**, *8*, 1–20.
17. Shak, K.P.Y.; Wu, T.Y.; Lim, S.L.; Lee, C.A. Sustainable reuse of rice residues as feedstocks in vermicomposting for organic fertilizer production. *Environ. Sci. Pollut. Res.* **2014**, *21*, 1349–1359. [CrossRef]
18. Garg, V.K.; Gupta, R. Vermicomposting of agro-industrial processing waste. In *Biotechnology for Agro-Industrial Residues Utilisation*; Springer: Dordrecht, The Netherlands, 2009; pp. 431–456. [CrossRef]
19. Pommeresche, R. *Žížaly a Jejich Význam pro Zlepšování Kvality Půdy*; Bioinstitut: Olomouc, Czech Republic, 2010.
20. Edwards, C.A.; Arancon, N.Q.; Sherman, R.L. *Vermiculture Technology: Earthworms, Organic Wastes, and Environmental Management*; CRC Press: Boca Raton, FL, USA, 2010. [CrossRef]
21. Salter, C.E.; Edwards, C.A. The production of vermicompost aqueous solutions or teas. In *Vermiculture Technology*; CRC Press: Boca Raton, FL, USA, 2010; pp. 153–163.
22. Akaike, H. Fitting autoregressive models for prediction. *Ann. Inst. Stat. Math.* **1969**, *21*, 243–247. [CrossRef]
23. Tukey, J.W. Comparing individual means in the analysis of variance. *Biometrics* **1949**, *5*, 99–114. [CrossRef]
24. Pant, A.P.; Radovich, T.J.K.; Hue, N.V.; Paull, R.E. Biochemical properties of compost tea associated with compost quality and effects on pak choi growth. *Sci. Hortic.* **2012**, *148*, 138–146. [CrossRef]

25. Hanč, A.; Bouček, J.; Švehla, P.; Drešlová, M.; Tlustoš, P. Properties of vermicompost aqueous extracts prepared under different conditions. *Environ. Technol.* **2017**, *38*, 1428–1434. [CrossRef]
26. Singh, R.; Gupta, R.K.; Patil, R.T.; Sharma, R.R.; Asrey, R.; Kumar, A.; Jangra, K.K. Sequential foliar application of vermicompost leachates improves marketable fruit yield and quality of strawberry (*Fragaria* x *ananassa* Duch.). *Sci. Hortic.* **2010**, *124*, 34–39. [CrossRef]
27. Ievinsh, G. Vermicompost treatment differentially affects seed germination, seedling growth and physiological status of vegetable crop species. *Plant Growth Regul.* **2011**, *65*, 169–181. [CrossRef]
28. Atiyeh, R.M.; Subler, S.; Edwards, C.A.; Bachman, G.; Metzger, J.D.; Shuster, W. Effects of vermicomposts and composts on plant growth in horticultural container media and soil. *Pedobiologia* **2000**, *44*, 579–590. [CrossRef]
29. Zaller, J.G. Foliar spraying of vermicornpost extracts: Effects on fruit quality and indications of late-blight suppression of field-grown tomatoes. *Biol. Agric. Hortic.* **2006**, *24*, 165–180. [CrossRef]
30. Bachman, G.R.; Metzger, J.D. Growth of bedding plants in commercial potting substrate amended with vermicompost. *Bioresour. Technol.* **2008**, *99*, 3155–3161. [CrossRef] [PubMed]
31. Joshi, R.; Singh, J.; Vig, A.P. Vermicompost as an effective organic fertilizer and biocontrol agent: Effect on growth, yield and quality of plants. *Rev. Environ. Sci. Bio/Technol.* **2015**, *14*, 137–159. [CrossRef]
32. Pillai, A.V.; Aswathy, K.K.; Preethy, T.T.; Renisha, M. Effect of organic liquid manures on crop growth and productivity. *Int. J. Curr. Res.* **2016**, *8*, 29023–29029.
33. Makkar, C.; Singh, J.; Parkash, C. Vermicompost and vermiwash as supplement to improve seedling, plant growth and yield in *Linum usitassimum* L. for organic agriculture. *Int. J. Recycl. Org. Waste Agric.* **2017**, *6*, 203–218. [CrossRef]
34. Fritz, J.I.; Franke-Whittle, I.H.; Haindl, S.; Insam, H.; Braun, R. Microbiological community analysis of vermicompost tea and its influence on the growth of vegetables and cereals. *Can. J. Microbiol.* **2012**, *58*, 836–847. [CrossRef]

Disclaimer/Publisher's Note: The statements, opinions and data contained in all publications are solely those of the individual author(s) and contributor(s) and not of MDPI and/or the editor(s). MDPI and/or the editor(s) disclaim responsibility for any injury to people or property resulting from any ideas, methods, instructions or products referred to in the content.

Proceeding Paper

End-of-Life Stage Analysis of Building Materials in Relation to Circular Construction [†]

Silvia Vilčeková [1,*], Peter Mésároš [2], Eva Krídlová Burdová [1] and Jana Budajová [1]

1. Institute of Sustainable and Circular Construction, Faculty of Civil Engineering, Technical University of Kosice, 04200 Kosice, Slovakia; eva.kridlova.burdova@tuke.sk (E.K.B.); jana.budajova@tuke.sk (J.B.)
2. Institute of Technology, Economics and Management in Construction, Faculty of Civil Engineering, Technical University of Kosice, 04200 Kosice, Slovakia; peter.mesaros@tuke.sk
* Correspondence: silvia.vilcekova@tuke.sk; Tel.: +421-55-602-4260
† Presented at the 4th International Conference on Advances in Environmental Engineering, Ostrava, Czech Republic, 20–22 November 2023.

Abstract: This article is focused on analyzing roof structure from environmental impact indicators and circularity point of view. The life cycle analysis of the roof structure includes the product phase, transport from the factory gate to the site, operational energy and operational water phase, and an end-of-life phase. Three end-of-life scenarios for built-in materials are designed to observe the reduction in environmental impacts throughout the life cycle of the structure. Scenario 1 mainly considers waste incineration, which accounts for almost 77% of the end-of-life phase. In addition, landfilling (15.4%) and recycling (7.7%) are considered. In scenario 2, landfilling accounts for 38.5% and incineration also accounts for 38.5%. Recycling (15.4%) and downcycling (7.6%) are also considered. In scenario 3, recycling and reuse represent 46.1% and 38.5%, respectively. Incineration (7.7%) and downcycling (7.7%) are also considered. The lifetime considered is 50 years and the functional unit is 1 m^2. One-Click LCA software was used for the analysis. Results for GWP-fossil are 415 kgCO$_{2eq}$, 381 kgCO$_{2qe}$ and 362 kgCO$_{2eq}$ for scenarios 1, 2 and 3. The circulation score of the roof composition for three scenario is determined to be 2%, 16% and 36%. It can be concluded that the end-of-life phase of the materials influenced these results to a large extent.

Keywords: roof structure; end-of-life; environmental impact; LCA; material circularity

1. Introduction

The end of life of buildings is being taken into account in improving waste management and in the environmental care construction sector [1]. Demolition waste makes up a significant proportion of the total waste generated and is intrinsically important both in terms of waste management and resource efficiency [2]. Due to the serious sustainability issues caused by the built environment, there are increasing demands to adopt circular economy principles in building design, such as flexibility and reversibility [3]. That study also shows that 14% of GHG emissions from a flexible building can be avoided if the foundation, supporting structure and ceiling elements are left in place for the next building. Such direct reuse leads to a significantly higher environmental value than recycling the same materials. Another study [4] found significant variations in lifetime carbon emissions between 145 properties ranging from 21 to 193 t CO$_{2eq}$, with lifetime carbon emission intensities ranging from 0.5 to 2.6 t CO$_{2eq}$/m^2. There is a strong correlation between lifetime carbon emissions and two factors: floor area and the number of inhabitants, followed by the number of bedrooms, the property type, the window frame material, the type of heating system, the age of the main occupant, the type of glazing and the thickness of the attic insulation. The authors of [5] concluded that carbonation of concrete in the post-use phase does not affect the validity of previous studies, showing that timber-framed buildings have significantly lower carbon emissions than concrete-framed buildings. The study found that carbonation of crushed concrete leads to significant CO_2 absorption. However, CO_2 emissions from fossil fuels used to crush concrete significantly reduce the carbon benefits gained from increased carbonation due to crushing. Long-term storage of crushed concrete will increase carbonation absorption but may not be practical due to space limitations. Overall, the effect of carbonation of concrete after use is small. According to [6], buildings could initiate smaller material flows and have improved environmental properties if they are intended for future disassembly and reuse. However, material flows in the life cycle of a building are complex maps, especially those initiated by material replacement and end of life. The results of study review [7] reveal the state of the art of the different biobased building products commercialized in the state of France, in which the energy recovery from bio-based insulation wastes expected in 2050 saves 4.1 million m^3 of land, 75,000 tons of fossil fuels and EUR 89 million while avoiding the rejection of 312,771 tCO_2eq. The results of another study [8] indicate that end-of-life recycling of biocomposite materials contributes to a reduction in environmental and economic costs in the construction industry. Strategic reuse of demounted concrete elements in new buildings may be one of the solutions that will support the transition to circular construction. In another study [9], a simple classification system for concrete quality proposed elements for reuse, where three main parameters were proposed, namely the calculation of the residual life, the extent of cracks and the target exposure class. The main goal of the research [10] was the development of a new lightweight construction material composed of gypsum, in which the conglomerate was partially replaced by dissolved expanded polystyrene (EPS) waste and the addition of textile fibers from end-of-life tires. The results obtained after the physico-chemical and mechanical characterization of the new gypsum composite show how to obtain a 31.3% lighter material with a 66.7% lower thermal conductivity and a 33.3% higher bending strength in boards compared to traditional gypsum material. This improvement in technical performance leads to a reduction in the consumption of natural resources and a large amount of waste recovered and reintroduced into the production process. Study [11] emphasizes that, due to the challenges of a roundabout, the circularity of buildings should be evaluated at the initial design stage to reduce the risks of circulation and environmental problems identified in the later stages of the project. Another study [12] points out that the implementation of circular building components can contribute to the transition to a circular economy. This is confirmed by a study [13] which says that bio-based circular building materials are "materials obtained in whole or in part from renewable biological origin or by-products and biological waste of plant and/or animal biomass that can be used as raw materials for building materials and decorative objects in construction, in their original forms or after being elaborated". Study [14] shows that simple green roof systems, without several artificial layer materials, are an environmentally responsible option. That study suggests leaving away rockwool, the egg carton-like plastic layers, and expanded clay when possible, or exploring for and developing alternative materials, in order to have minimal environmental impact.

Buildings are a significant contributor to climate change. This is why life-cycle assessments (LCA) are becoming increasingly popular for documenting environmental impacts during the detailed design stages of building projects. The LCA methodology calculates the potential environmental impacts caused by a product, such as a building [15]. The LCA methodology is implemented in this research work, whose main goal is to analyze the designed vegetation roof assembly from environmental impact categories during the whole life cycle and material circularity point of view. The three end-of-life scenarios are compared to determine the reduction in environmental impacts and investigate how the circularity score changes. The results of this research task are addressed to developers of new sustainable/green building materials, as well as to architects in the design of low-emission buildings considering building circularity. The limiting factor of this study is the number of compositions investigated.

2. Materials and Methods

2.1. Life Cycle Assessment Method

2.1.1. Goal and Scope

Roof structure designed for office buildings and placed in the city of Košice, Slovakia, is subjected to the analysis using the life cycle assessment (LCA) method. The goal and scope of the study is to determine environmental impact indicators and identify reductions in impacts based on end-of-life scenarios for a lifetime of 50 years. The analysis is performed for the "Cradle to Grave" system boundary and includes the following stages: A1–A3 (Product stage), B6 (Operational energy), B7 (Operational water) and C1–C4 (End-of-Life Stage). Functional unit (FU) is set to 1 m^2. One Click LCA software compliant with standards ISO 14040, ISO 14044, ISO 14025, EN 15804+A2 and EN 15978 is used for the analysis. The environmental impacts of the materials are based on EPDs and values representing average materials. CML is used as an impact assessment method. The core environmental impact indicators according to EN 15804+A2 are: Global Warming Potential-total (GWP-total), Global Warming Potential—fossil (GWP-fossil), Global Warming Potential—biogenic (GWP-biogenic), Global Warming Potential—LULUC (GWP-LULUC), Depletion Potential of the Stratospheric Ozone Layer (ODP), Acidification Potential—Accumulated Exceedance (AP-AE), Eutrophication Potential—Aquatic Freshwater (EP-AF), Eutrophication Potential—Aquatic Marine (EP-AM), Eutrophication Potential—Terrestrial (EP-T), Formation Potential of Tropospheric Ozone (POCP), Abiotic Depletion Potential of non-fossil resources (ADP-elements) and Abiotic Depletion Potential of fossil resources (ADP-fossil fuels), Water use (W) and other environmental impacts. These include impacts on human health, terrestrial toxicity, freshwater toxicity, seawater toxicity, air and water pollution, soil degradation and physical disturbances such as soil erosion and changes in landscape quality [16].

2.1.2. Life Cycle Inventory

Vegetation roof assembly is designed and investigated in terms of its environmental impact. The composition of the roof assembly is shown in Table 1. The roof structure is designed as a single-layer flat roof without an air gap and compliant current thermal and technical requirements. The composition meets current requirements set for thermophysical characteristics according to STN 73 0540-2+Z1+Z2. It is a diffusely closed structure using a vapor barrier and is designed as walkable roof structure. The thermal resistance of the roof is 10.172 m^2·K/W and exceeds the required value of 9.9 m^2·K/W.

Table 1. Materials of roof structure.

Material		Thickness [m]	Thermal Conductivity λ [W/m·K]	Density ρ [kg/m^3]	Area Density [kg/m^2]
Intensive substrate	-	0.400	-	1405	562.0
Filter geotextile	PP	0.0012	0.200	-	0.200
Drainage board	EPS	0.075	0.037	-	0.950
Waterproofing membrane	PES	0.0018	0.2	-	0.300
Waterproofing foil resistant to overgrowth of roots	PVC-P	0.0018	0.145	-	2.150
Separation geotextile	PE	0.0002	0.200	-	0.190
Mechanical anchor	PE/PP	-	-	-	0.136
Thermal insulation	XPS	0.140	0.035	35	4.200
Thermal insulation	XPS	0.140	0.035	35	4.200
Adhesive for thermal insulation		0.010	-	1120	0.200
Vapor barrier	PE	0.003	0.200	900	0.140
Reinforced concrete ceiling slab	Concrete + rebar	0.200	1.74	2500	500.0
Total thickness		0.970			1074.1

The environmental impacts for the product phase (A1–A3) are determined on the basis of the results presented in the Environmental product declarations (EPD) and the Ecoinvent database. Transportation (A4) methods and distances of the products from the factory gate to the site are presented in Table 2.

Table 2. Transportation processes.

Material		Distance [km]	Transportation Method
Intensive substrate	-	150	Dumper truck, 19-ton capacity
Filter geotextile	PP	520	Trailer combination, 40-ton capacity
Drainage board	EPS	520	Trailer combination, 40-ton capacity
Waterproofing membrane	PES	100	Trailer combination, 40-ton capacity
Waterproofing foil resistant to overgrowth of roots	PVC-P	1500	Trailer combination, 40-ton capacity
Separation geotextile	PE	520	Trailer combination, 40-ton capacity
Mechanical anchor	PE/PP	1700	Trailer combination, 40-ton capacity
Thermal insulation	XPS	430	Trailer combination, 40-ton capacity
Thermal insulation	XPS	430	Trailer combination, 40-ton capacity
Adhesive for thermal insulation		1500	Trailer combination, 40-ton capacity
Vapor barrier	PE	430	Trailer combination, 40-ton capacity
Ceiling slab	Concrete	60	Trailer combination, 40-ton capacity
Rebar	Steel	370	Trailer combination, 40-ton capacity

The operational energy (B6) includes the electricity consumption for automatic irrigation and represents a value of 13.5 kWh. The water consumption for irrigation of 0.3 m^3 represents the operational water (B7).

The scenarios for the end-of-life phase are presented in Table 3. Scenario 1 represents the market scenario that is most typical for that material in that market. End-of-life is considered in EPDs used in the analysis for Scenario 2. For products without EPDs, the end of life is set as the best practice. The purpose of Scenario 3 was to set the end-of-life phase to further reduce impacts through reuse, recycling and incineration of waste.

Table 3. End-of-life scenarios.

Material		Scenario 1	Scenario 2	Scenario 3
Intensive substrate	-	Landfilling	Recycling	Reuse
Filter geotextile	PP	Incineration	Landfilling	Recycling
Drainage board	EPS	Incineration	Incineration	Reuse
Waterproofing membrane	PES	Incineration	Landfilling	Recycling
Waterproofing foil resistant to overgrowth of roots	PVC-P	Incineration	Landfilling	Recycling
Separation geotextile	PE	Incineration	Incineration	Recycling
Mechanical anchor	PE/PP	Incineration	Landfilling	Recycling
Thermal insulation	XPS	Incineration	Incineration	Reuse
Thermal insulation	XPS	Incineration	Incineration	Reuse
Adhesive for thermal insulation		Incineration	Landfilling	Incineration
Vapor barrier	PE	Incineration	Incineration	Recycling
Ceiling slab	Concrete	Landfilling	Crushed to aggregate	Crushed to aggregate
Rebar	Steel	Recycling	Recycling	Reuse

2.2. Building Circularity

The Building Circularity Indicator (BCI) is a metric used to measure the circularity of a building's components, products, and materials. Circular economy principles aim to minimize linear flows, such as waste, and maximize restorative flows, such as recycling and reusing materials [17]. The BCI for the roof structure is calculated using the One-Click LCA software, specifically the Building Circularity tool. This tool likely considers various factors related to the life cycle of the materials used in the roof structure. Different weighting factors of individual materials were chosen for the evaluation of the materials. The calculation was performed for two parts, namely for materials recovered and for materials returned.

Virgin materials are new materials considered if they are found in the project. New materials typically have higher emissions compared to recycled, renewable, or reused materials. The multiplier for virgin materials is set to 0, indicating that the environmental impact of new materials is not favorable in this context. Renewable materials can be regrown. The default multiplier for renewable materials is set to 1, suggesting that they are considered environmentally friendly compared to new materials. Recovered recycled materials are materials obtained for the project that have been recycled. They generally have lower emissions compared to newly manufactured materials. The default multiplier for recovered recycled materials is set to 1, indicating that their environmental impact is comparable or better than that of new materials. If little or no processing is required for recovered recycled materials, they have almost no emissions. The default multiplier for cases where little or no processing is required is set to 1, suggesting that minimal processing is associated with minimal environmental impact.

Materials returned consist of materials that can be reused as materials with multiplier 1. It can be stated that reusing materials for other projects reduces emissions for the current project. This also includes materials that can be recycled with multiplier 1. Recycling produces some emissions, but the material can be structurally chained for reuse. Materials that can be downcycled have a multiplier of 0.5. Downcycled materials result in a product that is not as strong as the original and requires new materials to achieve similar strength. Materials that can be used as energy, i.e., combustible materials, have a multiplier of 0.5. It can be mentioned that combustion produces emissions, but recovered energy can be utilized (e.g., for heating, electricity, manufacturing). Materials that are disposed of (landfilled) have a multiplier of 0. Landfilled materials do not fully decay or decompose, leading to long-term environmental impact. These multipliers serve as a factor to adjust the environmental impact of a project or product based on the end-of-life fate of its materials. For instance, materials that can be reused or recycled are considered more environmentally friendly (multiplier of 1), while materials that end up in landfills have the highest environmental impact (multiplier of 0). Downcycled and combustible materials fall somewhere in between. Adjusting for these factors helps us assess the overall sustainability of a product or project.

3. Results and Discussion

Figures 1–3 present results for environmental impact indicators according to EN 15978 and EN 15804+A2. They present the percentage share of life cycle stages on the environmental impact indicators for three scenarios of EoL.

The operational energy phase (40.87%) and the product phase (40.65%) are the largest contributors to the GWP-fossil expressed as equivalent mass of CO_2 for scenario 1. The share of the other life cycle phases is negligible. The materials with the highest contribution to GWP-fossil are ready-mixed concrete (34.12%), rebar (26.43%, vegetation substrate (18.02%) and XPS insulation (12.49%). The share of the plastic waterproofing membrane and PVC root barrier is 1.76%. According to one study [18], low-density and high-density PE waterproofing membrane and the root barrier count for GWP 9.5%. Study [19] points out that the high thickness of the substrate (400 mm) required for adequate insulation increases the environmental impact of the systems, with values between 13.5% and 52.4% higher than the average roof systems. As a result of the change in the end-of-life scenario, the share of life-cycle phases per GWP-fossil has changed for scenarios 2 and 3. For scenario 2, the largest share is also for operational energy (44.53%) and product phase (44.29%). For scenario 3, the largest share is, again, for operational energy (46.86%) and the product phase (46.60%). The share of the other life cycle phases is negligible. Noticeably, as the share of recycling and reuse of materials increases, the share of impacts in the end-of-life phase decreases, and the share of impacts is assigned to the operational and product phases.

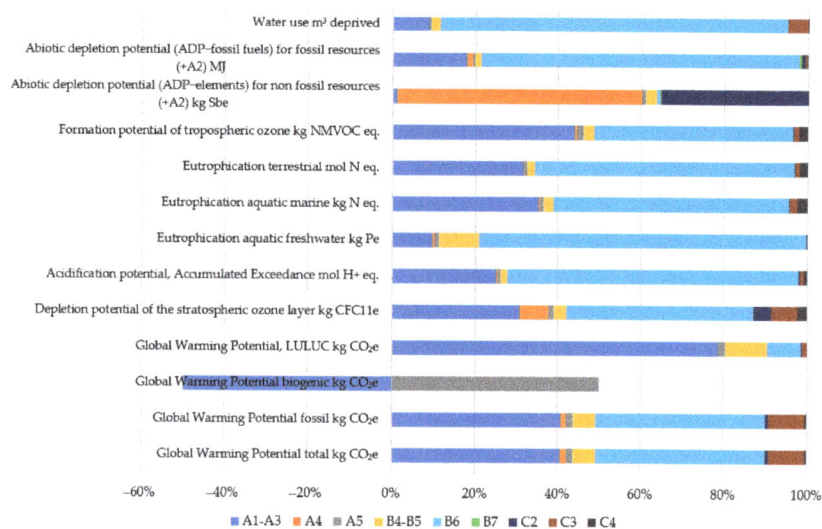

Figure 1. Environmental impact indicators—scenario 1.

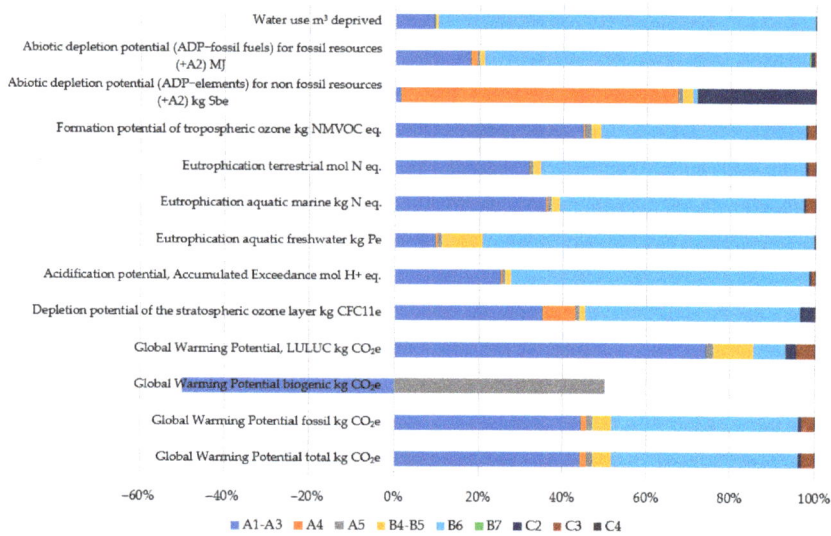

Figure 2. Environmental impact indicators—scenario 2.

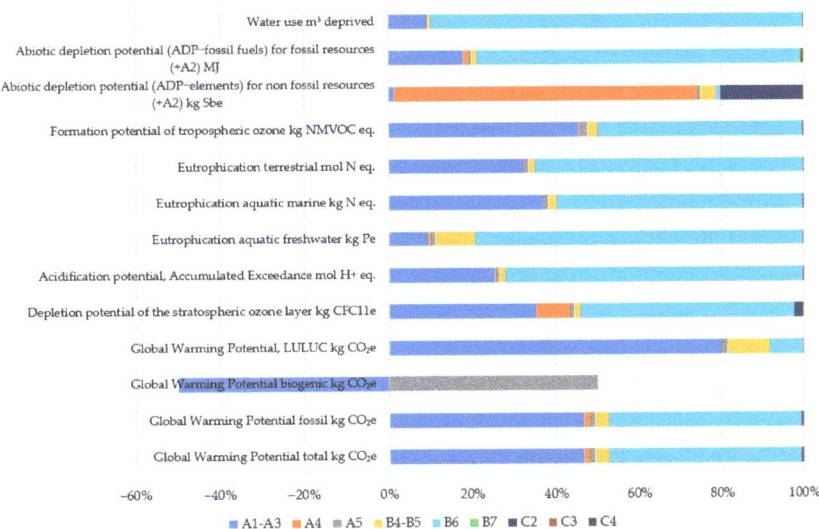

Figure 3. Environmental impact indicators—scenario 3.

Negative GWP-biogenic values in the product phase are attributed to waterproofing membranes. The largest contribution to ozone depletion potential expressed as equivalent emissions of CFC11 is from operational energy (45.07%), followed by the product phase (30.75%), transport from the factory gate to the site (6.89%) and the recovery process (6.36%) for scenario 1. As a result of the change in the end-of-life scenario, the contribution of the life cycle phases to ODP has also changed for scenarios 2 and 3. For Scenario 2, the operational energy (51.35%), the product phase (35.04%), the transport from the factory gate to the site (7.85%) and the transport to the disposal site (3.32%) also have the largest share. For scenario 3, the largest shares are again operational energy (51.87%), product phase (35.40%), transport from the factory gate to the site (7.93%) and transport to the disposal site (2.14%). The share of the other life cycle phases is negligible. It can be seen that, as the share of recycling and reuse of materials increases, the share of impacts in the end-of-life phase decreases, and the share of impacts is attributed to the operation phase, the product phase and also the transport from the factory gate to the recovery and disposal site, and the assessment process for Scenario 1.

Figure 4 shows the results for indicators where there are significant differences between the three scenarios in relation to the end-of-life phase.

These indicators are GWP-fossil, Abiotic depletion potential for fossil resources and Water use. As can be seen from the figure, the largest contributors to ADP-fossil fuels are transport to the disposal site, the recovery process and landfilling. For scenarios 2 and 3, landfilling is not considered, and thus the impacts are significantly lower compared to scenario 1, especially in the case of scenario 3, where there is a greater share of material reuse, as mentioned above.

In scenario 1, XPS isolation has the largest impact on the GWP-fossil in the C3 module with a contribution of 60.90% out of a total of 36.09 kg CO_{2eq}, considering combustion. In scenario 2, PE foil has the largest impact on GWP-fossil with a share of 35.93%, out of a total impact of 11.55 kg CO_{2eq}, considering incineration. In scenario 3, polyurethane foam has the largest contribution with 65.08% of the total impact of 0.63 kg CO_{2eq}.

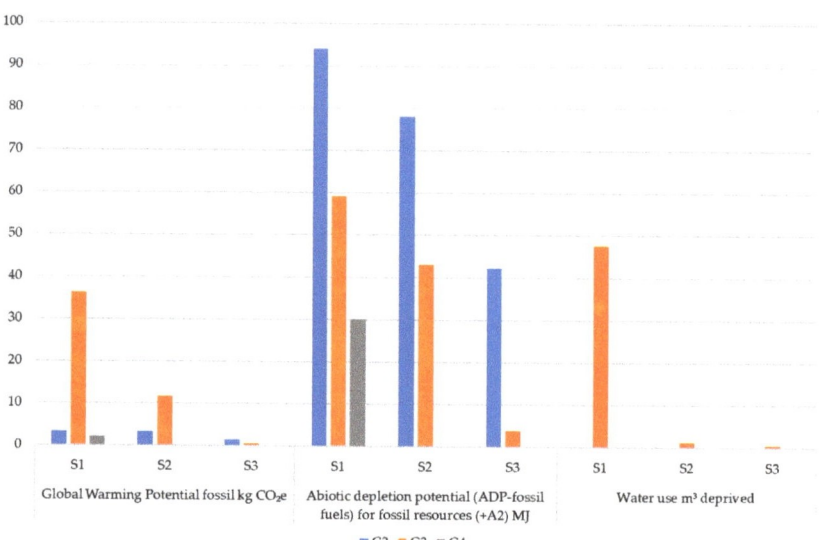

Figure 4. End-of-life scenarios for selected environmental impact indicators.

Figures 5–7 present the results of roof structure circularity for three scenarios of end-of-life. The highest circularity score is 36% for scenario 3, followed by scenario 2 with a percentage of 16%, and the lowest is 2% for scenario 1. Roof composition consists of virgin materials (100%). There are no renewable or reused materials in the composition. Scenarios 2 and 3 present scenarios according to which 56% of materials could be downcycled (56%) and recycled in percentages of 2.7% and 1.3%, respectively. According to scenario 1, 2.8% of materials could be used as energy and up to 94.5% put into landfill. Scenario 2 presents a scenario according to which 1.3% of materials could be used as energy and 40% put into landfill. And finally, according to scenario 3, 42.6% of materials could be reused and 0% landfilled.

Material Recovered	0 %	
Virgin	100 %	
Renewable	0 %	
Recycled content in A stage	0 %	
Reused	0 %	
Material Returned	4.1 %	2 %
Reuse as material	0 %	
Recycling	2.7 %	
Downcycling	0 %	
Use as energy	2.8 %	
Disposal	94.5 %	

Figure 5. Circularity for scenario 1.

Figure 6. Circularity for scenario 2.

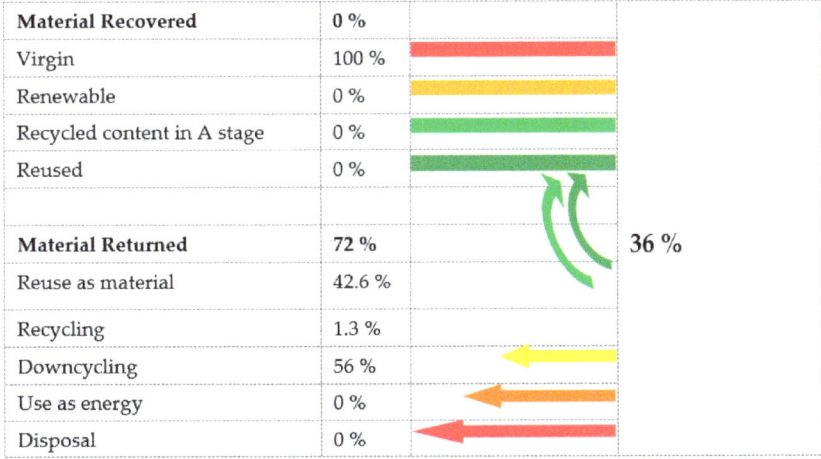

Figure 7. Circularity for scenario 3.

Another study contributes to the understanding of the application and decarbonization potential of circular strategies in the building industry by investigating real-life cases of new build, renovation, and demolition [20]. This study shows that circularity can be considered as a key strategy for mitigating carbon emissions in the building industry and that decarbonization potentials vary greatly between different building projects and applications of circular strategies, indicating that effective implementation of circular building strategies to capture potential environmental benefits is imperative.

4. Conclusions

In this study, the roof structure was evaluated in terms of environmental impact indicators and building circularity by using the LCA methodology during the whole lifespan. The three end-of-life scenarios were compared to determine the reduction in environmental impacts and investigate how the circularity score changed. The results show that the operational energy phase (40.87%) and the product phase (40.65%) are the largest contributors to the GWP-fossil expressed as CO_{2e} for scenario 1, in which waste incineration accounts for almost 77%, landfilling 15.4% and recycling 7.7%. In scenario 2, part of the load shifted to operational energy (44.53%) due to better waste utilization. The product phase contributed 44.29%. In this scenario, landfilling accounts for 38.5%, incineration also 38.5%. recycling 15.4% and downcycling 7.6%. For scenario 3, the largest share is again for operational energy (46.86%) and the product phase (46.60%). The share of the other life cycle phases is negligible. In this scenario 3, recycling and reuse represent 46.1% and 38.5%, respectively. Incineration is considered for 7.7% of materials and downcycling is also considered for 7.7%. It should be highlighted that as the share of recycling and reuse of materials increases, the share of impacts in the end-of-life phase decreases, and the share of impacts is assigned to the operational and product phases. The materials with the highest contribution to environmental impacts are ready-mixed concrete (34.12%), rebar (26.43%), vegetation substrate (18.02%) and XPS insulation (12.49%). The circularity of the roof composition for three scenario is determined to be 2%, 16% and 36%. The circularity score increases with the share of recycling and reuse of materials. Research focused on increasing the circularity scores of materials and whole buildings within the construction industry will be continued. The emphasis on reducing waste in manufacturing, construction, and demolition aligns with the principles of sustainability and resource efficiency. The concept of circularity involves designing and managing buildings in a way that minimizes waste generation, encourages reuse and recycling of materials, and promotes a closed-loop system. By incorporating circularity into construction practices, the industry can contribute to a more sustainable and resilient future. The potential benefits mentioned, such as creating more sustainable and resilient buildings and achieving cost savings, underscore the value of adopting circular practices. Circular economy principles help address environmental concerns by reducing waste and lead to economic advantages through resource optimization and improved efficiency. It is encouraging to see a focus on sustainability and circularity in the construction industry, as these efforts play a crucial role in advancing a more environmentally conscious and economically viable future. As the research progresses, the findings may provide valuable insights and contribute to the broader goal of enhancing sustainability in the built environment.

Author Contributions: Conceptualization, S.V. and P.M.; methodology, S.V.; software, S.V.; validation, E.K.B. and J.B.; formal analysis, J.B.; investigation, S.V.; resources, E.K.B.; data curation, S.V.; writing—original draft preparation, S.V. and E.K.B.; writing—review and editing, P.M.; visualization, S.V.; supervision, P.M.; project administration, S.V.; funding acquisition, S.V. All authors have read and agreed to the published version of the manuscript.

Funding: This research was funded by VEGA, grant number 1/0512/20 and by Erasmus +, grant number № 2021-1-SK01-KA220-HED-000023274 "Support of higher education system in a context of climate change mitigation through regional-level of carbon footprint caused by a product, building and organization".

Institutional Review Board Statement: Not applicable.

Informed Consent Statement: Not applicable.

Data Availability Statement: Data are contained within the article.

Acknowledgments: This research is the result of projects VEGA 1/0512/20 and Erasmus + 2021-1-SK01-KA220-HED-000023274.

Conflicts of Interest: The authors declare no conflicts of interest. The funders had no role in the design of the study; in the collection, analyses, or interpretation of data; in the writing of the manuscript; or in the decision to publish the results.

References

1. Quéheille, E.; Ventura, A.; Saiyouri, N.; Taillandier, F. A Life Cycle Assessment model of End-of-life scenarios for building deconstruction and waste management. *J. Clean. Prod.* **2022**, *339*, 130694. [CrossRef]
2. Mastrucci, A.; Marvuglia, A.; Popovici, E.; Leopold, U.; Benetto, E. Geospatial characterization of building material stocks for the life cycle assessment of end-of-life scenarios at the urban scale. *Resour. Conserv. Recycl.* **2017**, *123*, 54–66. [CrossRef]
3. Krohnert, H.; Itten, R.; Stucki, M. Comparing flexible and conventional monolithic building design: Life cycle environmental impact and potential for material circulation. *Build. Environ.* **2022**, *222*, 109409. [CrossRef]

4. Zheng, L.; Mueller, M.; Luo, C.; Menneer, M.; Yan, X. Variations in whole-life carbon emissions of similar buildings in proximity: An analysis of 145 residential properties in Cornwall, UK. *Energy Build.* **2023**, *296*, 113387. [CrossRef]
5. Dodoo, A.; Gustavsson, L.; Sathre, R. Carbon implications of end-of-life management of building materials. *Resour. Conserv. Recycl.* **2009**, *53*, 276–286. [CrossRef]
6. Vandervaeren, C.; Galle, W.; Stephan, A.; De Temmarman, N. More than the sum of its parts: Considering interdependencies in the life cycle material flow and environmental assessment of demountable buildings. *Resour. Conserv. Recycl.* **2022**, *177*, 106001. [CrossRef]
7. Rabbat, C.; Awad, S.; Villot, A.; Rollet, D.; Andres, Y. Sustainability of biomass-based insulation materials in buildings: Current status in France, end-of-life projections and energy recovery potentials. *Renew. Sustain. Energy Rev.* **2022**, *156*, 111962. [CrossRef]
8. Alshndah, Z.; Becquart, F.; Belayachi, N. Recycling of wheat straw aggregates of end-of-life vegetal concrete: Experimental investigation to develop a new building insulation material. *J. Build. Eng.* **2023**, *76*, 107199. [CrossRef]
9. Suchorzewski, J.; Santandrea, F.; Malaga, K. Reusing of concrete building elements—Assessment and quality assurance for service-life. *Mater. Today Proc.* **2023**, in press. [CrossRef]
10. Zaragoza-Benzal, A.; Ferrández, D.; Atanes-Sánchez, E.; Morón, C. New lightened plaster material with dissolved recycled expanded polystyrene and end-of-life tyres fibres for building prefabricated industry. *Case Stud. Constr. Mater.* **2023**, *18*, e02178. [CrossRef]
11. Van Der Zwaag, M.; Wang, T.; Bakker, H.; Van Nederveen, S.; Schuurman, A.C.B.; Bosna, D. Evaluating building circularity in the early design phase. *Autom. Constrction* **2023**, *152*, 104941. [CrossRef]
12. Towards implementation of circular building components: A longitudinal study on the stakeholder choices in the development of 8 circular building components. *J. Clean. Prod.* **2023**, *420*, 138287. [CrossRef]
13. Le, D.L.; Salomone, R.; Nguyen, Q.T. Circular bio-based building materials: A literature review of case studies and sustainability assessment methods. *Build. Environ.* **2023**, *24*, 110774. [CrossRef]
14. Chenani, S.B.; Lehvävirta, S.; Häkkinen, T. Life cycle assessment of layers of green roofs. *J. Clean. Prod.* **2015**, *90*, 153–162. [CrossRef]
15. Hansen, R.N.; Hoxha, E.; Rasmussen, F.N.; Ryberg, M.W.; Andersen, C.E.; Birgisdóttir, H. Enabling rapid prediction of quantities to accelerate LCA for decision support in the early building design. *J. Build. Eng.* **2023**, *76*, 106974. [CrossRef]
16. EN 15804:2012; Sustainability of construction works, Environmental product declarations, Core rules for the product category of construction products. CEN: Brussels, Belgium, 2012.
17. de Vries, B.B.; Chairman, T.U.; Kunen, I.T.T.; BV, B.G. Building Circularity Indicators. Ph.D. Thesis, Eindhoven University of Technology, Eindhoven, The Netherlands, 2016.
18. Gargari, C.; Bibbiani, C.; Fantozzi, F.; Campiotti, C.A. Environmental impact of Green roofing: The contribute of a green roof to the sustainable use of natural resources in a life cycle approach. *Agric. Agric. Sci. Procedia* **2016**, *8*, 646–656. [CrossRef]
19. Botejara-Antúnez, M.; Gonz'alez-Domínguez, J.; García-Sanz-Calcedo, J. Comparative analysis of flat roof systems using life cycle assessment methodology: Application to healthcare buildings. *Case Stud. Constr. Mater.* **2022**, *17*, e01212. [CrossRef]
20. Nußholz, J.; Çetin, S.; Eberhardt, L.; De Wolf, C.; Bocken, N. From circular strategies to actions: 65 European circular building cases and their decarbonisation potential. *Resour. Conserv. Recycl. Adv.* **2023**, *17*, 200130. [CrossRef] [PubMed]

Disclaimer/Publisher's Note: The statements, opinions and data contained in all publications are solely those of the individual author(s) and contributor(s) and not of MDPI and/or the editor(s). MDPI and/or the editor(s) disclaim responsibility for any injury to people or property resulting from any ideas, methods, instructions or products referred to in the content.

Proceeding Paper

Identification of the Hydrogeological Potential in Langos-San Andres, by Means of Electrical Resistivity Tomography Interpretation [†]

Benito Guillermo Mendoza Trujillo *, Leonardo Sebastián Cadena Rojas and Andrés Santiago Cisneros Barahona

Facultad de Ingeniería, Universidad Nacional de Chimborazo, Riobamba EC060150, Ecuador; lscadena.fia@unach.edu.ec (L.S.C.R.); ascisneros@unach.edu.ec (A.S.C.B.)
* Correspondence: benitomendoza@unach.edu.ec
[†] Presented at the 4th International Conference on Advances in Environmental Engineering, Ostrava, Czech Republic, 20–22 November 2023.

Abstract: This work aims to determine the hydrogeological potential in the community of Langos-San Andrés, belonging to the canton Guano. This was done by interpreting the regional geology–specific and subsurface stratigraphy. Stratigraphy was obtained by mathematical modeling of the electrical resistivity tomography results obtained in the study area. The interpretation of this information allowed us to know the structure of the soil and the levels at which the groundwater was located. Also, determining the optimal level of extraction of groundwater is a partition of 50 m depth onwards. Furthermore, the location of this water level will allow 1000 families in this area to obtain good quality water, reducing the water access gap that exists in the region and in Ecuador.

Keywords: stratigraphy; hydrogeology; Langos; geophysics; subsurface; resistivity electrical tomograph

1. Introduction

Water is of global interest mainly because in several places on the planet it is difficult to access [1–3]. The situation in Latin America is similar, with access to water being a problem in several countries, as it is not possible to provide drinking water to areas that are far from surface water sources [4,5]. In this sense, Ecuador has similar circumstances, even though it is one of the countries with many water resources because there are several sectors of the population that do not have access to water, especially for human use [6]. On the other hand, groundwater in Ecuador is the main source of water for human consumption because the existing springs are mostly located in areas with volcanic features, allowing water to be stored in subway strata [7].

However, the complexity of studying groundwater due to the cost of traditional methods, such as drilling, means that there is a shortage of hydrogeological data in the country [8]. In this context, geophysical methods are a solution to this problem because they are widely used worldwide and provide acceptable results, which allow an understanding of the behavior of groundwater in the subsurface strata [9–11].

In this context, according to data from the Decentralized Autonomous Government of Guano Canton (CAD Guano), the Langos–San Andres sector has a water distribution system for only a few families. This has changed due to population growth in the area since the construction of the new access road to Riobamba along Avenida de la República has led to the urbanization of the sector, which means that drinking water distribution does not supply the entire area, leaving more than 80% of the population in this sector without water [12]. In relation to this, as indicated by [7], this area is located within the perimeter of the Chambo aquifer, which has a large amount of groundwater and could serve to cover the needs of this sector.

Therefore, this work will allow the determination of stratigraphic characteristics in the Langos sector, using electrical resistivity tomography, to determine the groundwater level for water resource exploitation purposes.

2. Methodology

2.1. Study Area

The study area is located in the northwestern part of the Province of Chimborazo, in the Guano canton, San Andres parish, 8 km from the city of Riobamba. This zone was selected because of its physical and topographic characteristics. The San Andres parish has an average annual temperature of 11.19 °C and an altitudinal range of 2900 to 6310 msmn (Figure 1).

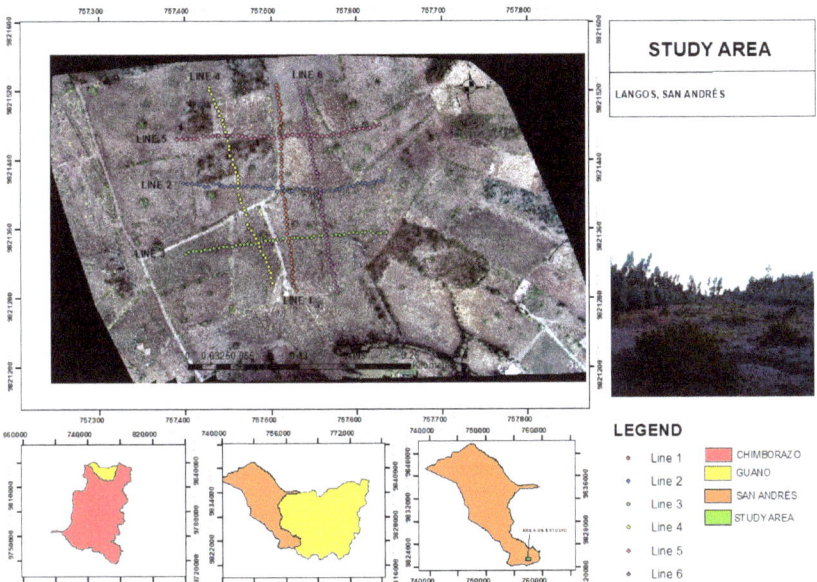

Figure 1. Location map of the tomography area in the Langos, Guano canton.

The methodology used was developed based on national and regional geological information [13] to determine the areas of hydrogeological interest in the Langos sector.

Then, by means of electrical resistivity topographies in the areas of interest, data were obtained for the reconstruction of the stratigraphy in two and three dimensions. The methods used to implement the proposed approach are described below:

2.2. Electrical Resistivity Tomography (ERT)

Electrical resistivity tomography allowed us to distinguish the distribution of the electrical resistivity of the subsoil by relating the solid constituents of the soil, water content, and temperature with respect to the electrical resistivity. This can be considered as a way to interpret the variability in the physical properties of the soil [10,14,15]. The results of this study are from a distance–depth stratigraphic section with a distribution of subsurface resistivity, which is easily understandable in geological terms, in which the electrical resistivity of the medium depends on the amount of water in the subsurface [16]. To determine the electrical resistivity of the structures, Equation (1) is used [17]:

$$\rho = \frac{a}{\phi^m S^n} \rho_w \qquad (1)$$

where ρ is the electrical resistivity of the structure, ρ_w is the electrical resistivity of the pore water, S is the degree of saturation, a and m are constants related to the saturation coefficient and the cementation factor, respectively, and n is a parameter related to the degree of saturation.

2.3. Acquisition of Data for Electrical Resistivity Tomography

To acquire data in the field, a place with easy access was determined, without interference from construction or trees, so that the cables used would not collide with these obstacles. The equipment used was the Syscal Pro, with two cables, 240 m long each, and 48 electrodes arranged at a distance of 5 m in such a way that it allows acquiring information up to an approximate depth of 60 m [15].

2.4. Electrical Resistivity Tomography Data Processing

The collected data were entered into the Res2dinvx64 ver. 4.07 software (licensed from the National University of Chimborazo) to perform the inversion in two dimensions, which is based on a smoothed least squares sequence with damping; the same sequence was refined with a trial–and–error comparison (Figure 2), using the finite difference method or the finite element method.

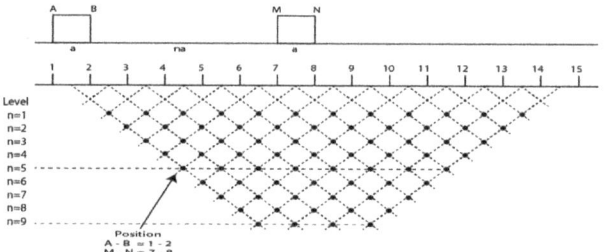

Figure 2. Rectangular block arrangement in 2D model [14].

From the data obtained in the inversion of the Res2dinv software, all the electrical topographies were projected in three dimensions using the Res3D software 3.13.50 (National University of Chimborazo license), which performs an inverse three-dimensional visualization of the resistivity data of the lithological materials that exist in the subsoil [18].

2.5. Electrical Resistivity Tomography Interpretation

To determine the groundwater level in the study area, the resistivity of rocks of water present in the found stratum was used (Table 1)

Table 1. Electrical resistivity of rocks or water [8].

Rocks or Water	Resistivity ($\Omega \cdot m$)
Seawater	0.2
Alluvial aquifer water	10–30
Faustian water	50–100
Dry sand and gravels	1000–10,000
Sands and gravels with fresh water	50–500
Sands and gravels with salt water	0.5–5
Clays	0.5–5
Marls	2–20
Limestones	2–20
Clayey sandstones	50–300
Quartzite sandstones	300–10,000
Volcanic tuffs, cinerites	20–100
Lavas	300–10,000
Graphitic Schists	0.5–5
Clayey or altered shales	100–300
Healthy shales	300–3000
Altered gneiss, granite	100–1000
Healthy gneiss, granite	1000–10,000

3. Results & Discussion

To understand the stratigraphy of the soil, it was found that the study area is located in the Riobamba Formation (PR) (Pleistocene), most of this formation is composed of rounded and angular gravels, which constitute the linarite volcanic phase of Chimborazo, and are also present in some stratification sectors in soil [7,8]. As described in the methodology, once the points where the ERTs should be performed according to the geological characteristics of the area under study were identified, 6 ERTs were performed that allowed to know the characteristics of the subsoil in this area, their results are presented below.

The results of tomography A (Figure 3A) show electrical resistivity ranges between 13.1 and 44.2 Ohm·m (blue color), revealing that groundwater could be found at approximately 50 m depth. On the other hand, resistivity ranges between 813 and 150 Ohm·m were observed in the green color range, and according to their resistivities, they were semi-permeable zones. In addition, in several parts of the tomography, the high resistivity ranges of 275 and 507 Ohm·m (sandy brown and red color) are characteristic of impermeable zones composed mostly of healthy lavas and granites.

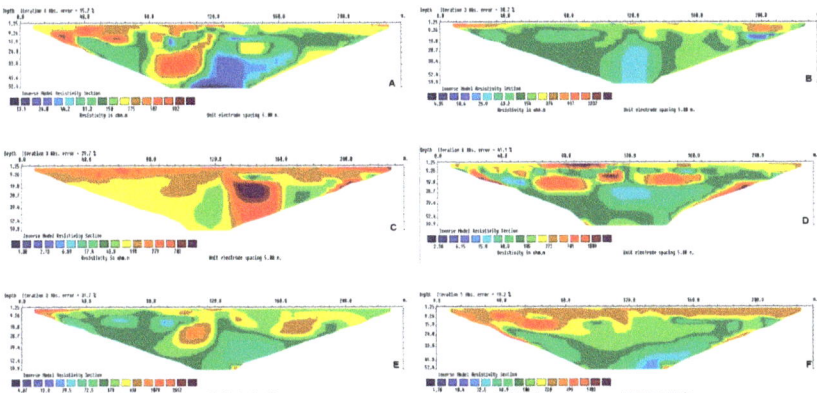

Figure 3. Electrical Resistivity Lines (**A–F**).

The second line (Figure 3B) has electrical resistivities of 25.9 Ohm·m in the light blue scale. These are semi-permeable zones in the lower center of the tomography at an approximate depth of 20 to 60 m. In the same way, resistivities between 632 and 154 Ohm·m are observed in the range of green colors that represent less permeable zones composed mostly of clayey sandstones and altered granite. There are also zones with resistivities between 376 and 917 Ohm·m, which were also observed in sandy brown and yellow colors at the corners of the upper limits, which represent impermeable zones of the subsoil.

The results of the third line (Figure 3C) show that there are resistivities of 43.8 Ohm·m, green color; this shows the presence of semi–permeable zones representing a good part of the tomography; there are also resistivities ranging between 110 and 702 Ohm·m (yellow, sandy brown, red and purple color) representing areas with low permeability, also the presence of large rocks that prevent the infiltration of rainwater to the subsoil, these lithological materials may be composed of healthy lavas and granites.

The fourth line (Figure 3D) shows electrical resistivities between 6.15 and 15.8 Ohm·m (light blue color), showing semi-permeable zones containing small pockets of water. In addition, there are resistivities between 40.9 and 105 Ohm·m in the green color range; this shows semi–permeable zones that allow water flow; likewise, we find resistivities ranging between 272 and 1809 Ohm·m in red and purple color, which indicates the presence of impermeable zones and rock intrusions.

The results of the fifth tomography (Figure 3E) represent electrical resistivities between 12 and 29.5 Ohm·m, in light blue color, which indicates the permeable zones of the subsoil, in the same way resistivities ranging between 72.5 and 178 Ohm·m are observed, which are observed in green color, this shows semi-permeable zones representing the majority of the tomography, resistivities between 438 and 1078 Ohm·m can also be observed, represented in sandy brown and yellow color, which reveals zones of low permeability.

The Sixth tomography (Figure 3F) shows electrical resistivities between 10.4 and 22.5 Ohm·m (light blue color), indicating the presence of permeable zones with a small water pocket that is surrounded by semi- permeable material since there are also resistivities that are between 48.9 and 106 Ohm·m, represented in the range of green colors that are the semi-permeable zones that are observed in most of the subsoil, in the same way we find resistivities ranging between 230 and 499 Ohm·m, graphically represented in sandy brown and yellow color that are found at shallow depths, these zones are of low permeability thus preventing the infiltration of rainwater to the subsoil.

From the data obtained in the topographies, the 3D reconstruction was performed (Figure 4), which revealed in a better way the presence of different lithological materials present in the study area. As shown in the figure, there are areas with low resistivity at depths of up to 50 m. There are also areas where the intrusions of rocks of different sizes have been observed. In addition, it was possible to detect the presence of a possible groundwater zone at a depth of 55 m.

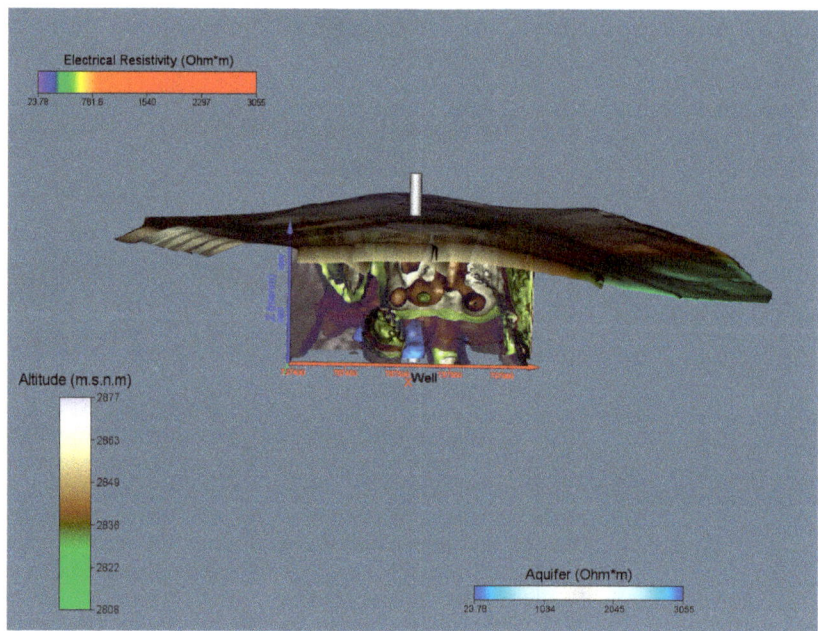

Figure 4. 3D model of ERT in Lagos San Antonio.

4. Conclusions

The sector selected for the topographies met the criteria in terms of accessibility and the absence of obstacles. The tomographic lines were performed in a high area of the Langos sector since in this sector there is a water supply tank for consumption in this area, so if in the future the exploitation of groundwater is carried out, it would save on the costs of construction of a storage tank, in addition to the fact that the land belongs to the GAD of Guano.

In this context, the ERTs and the interpretation of the obtained values of electrical resistivity determined that the stratigraphy of the area is of a heterogeneous type; that is, it does not have the same composition of rocky materials and soil because of traces of rock intrusions, poorly permeable zones and all this at a depth between 5 to 50 m were found, in this sense although the study area is located in an area with good hydrogeological power. It is not likely that there is a large volume of groundwater in this section of depth studied. It was observed that the low electrical resistivities were from 50 m depth, showing the possible presence of groundwater from here and towards greater depths. Therefore, if a water extraction project is to be conducted in this area, it is necessary to reach a drilling depth of approximately 120 m to ensure that a good flow of groundwater is obtained.

In this sense, it is advisable to continue with the geophysical study of the Langos area to determine more sectors that offer groundwater supply at less depth, implying a cost reduction in the drilling of possible groundwater production wells.

As shown in the site description, drilling a well in this area will allow the Langos community to access good–quality water, benefiting approximately 1000 families. This also demonstrates that these noninvasive methods are an acceptable solution for preliminary groundwater studies, reducing the cost of conducting test drillings that are not accessible to economically disadvantaged communities.

Author Contributions: Conceptualization and methodology: B.G.M.T. and A.S.C.B.; Software, validation, formal analysis, investigation, and data curation: B.G.M.T. and L.S.C.R.; Writing—original draft preparation, writing—review and editing: B.G.M.T., L.S.C.R. and A.S.C.B. All authors have read and agreed to the published version of the manuscript.

Funding: This research received no external funding.

Institutional Review Board Statement: Not applicable.

Informed Consent Statement: Not applicable.

Data Availability Statement: Data are contained within the article.

Conflicts of Interest: There are no conflicts of interest of any nature.

References

1. Coulibaly, H.; Santacruz de León, G. La Visión Africana Del Agua 2025 y La Realidad Sobre El Acceso Al Agua Para Consumo Humano En Mali, África. *Soc. Y Ambiente* **2019**, *20*, 29–51. [CrossRef]
2. Husic, A.; Fox, J.; Adams, E.; Pollock, E.; Ford, W.; Agouridis, C.; Backus, J. Quantification of Nitrate Fate in a Karst Conduit Using Stable Isotopes and Numerical Modeling. *Water Res.* **2020**, *170*, 115348. [CrossRef] [PubMed]
3. Liemberger, R.; Wyatt, A. Quantifying the Global Non-Revenue Water Problem. *Water Sci. Technol. Water Supply* **2019**, *19*, 831–837. [CrossRef]
4. Somma, N.M.; Bargsted, M.; Disi Pavlic, R.; Medel, R.M. No Water in the Oasis: The Chilean Spring of 2019–2020. *Soc. Mov. Stud.* **2021**, *20*, 495–502. [CrossRef]
5. Yager, K.; Valdivia, C.; Slayback, D.; Jimenez, E.; Meneses, R.I.; Palabral, A.; Bracho, M.; Romero, D.; Hubbard, A.; Pacheco, P.; et al. Socio-Ecological Dimensions of Andean Pastoral Landscape Change: Bridging Traditional Ecological Knowledge and Satellite Image Analysis in Sajama National Park, Bolivia. *Reg. Environ. Chang.* **2019**, *19*, 1353–1369. [CrossRef]
6. Wingfield, S.; Martínez-Moscoso, A.; Quiroga, D.; Ochoa-Herrera, V. Challenges to Water Management in Ecuador: Legal Authorization, Quality Parameters, and Socio-Political Responses. *Water* **2021**, *13*, 1017. [CrossRef]
7. Chidichimo, F.; Mendoza, B.T.; De Biase, M.; Catelan, P.; Straface, S.; Di Gregorio, S. Hydrogeological Modeling of the Groundwater Recharge Feeding the Chambo Aquifer, Ecuador. *AIP Conf. Proc.* **2018**, *2022*, 020003. [CrossRef]
8. Mendoza, B.T. Characterization of Real Aquifers Using Hydrogeophysical Measurements. An Application to the Chambo Aquifer (ecuador). Doctoral dissertation, Università Della Calabria, Rende, Italy, 2012.
9. Anomohanran, O. Hydrogeophysical Investigation of Aquifer Properties and Lithological Strata in Abraka, Nigeria. *J. Afr. Earth Sci.* **2015**, *102*, 247–253. [CrossRef]
10. Mahmoud, H.H.; Ghoubachi, S.Y. Geophysical and Hydrogeological Investigation to Study Groundwater Occurrences in the Taref Formation, South Mut Area–Dakhla Oasis-Egypt. *J. Afr. Earth Sci.* **2017**, *129*, 610–622. [CrossRef]
11. You, A.; Be, M.; In, I. Application of Geoelectric Method for Groundwater. *AIP Conf. Proc.* **2022**, *1977*, 020018.
12. Manya, P. *Valoración Ambiental Del Recurso Hídrico De La Parroquia San Andrés, Cantón Guano, Provincia De Chimborazo*; Escuela Superior Politécnica de Chimborazo: Riobamba, Ecuador, 2021.
13. Berrezueta, E.; Sánchez-Cortez, J.L.; Aguilar-Aguilar, M. Inventory and Characterization of Geosites in Ecuador: A Review. *Geoheritage* **2021**, *13*, 93. [CrossRef]
14. Holmes, J.; Chambers, J.; Meldrum, P.; Wilkinson, P.; Boyd, J.; Williamson, P.; Huntley, D.; Sattler, K.; Elwood, D.; Sivakumar, V.; et al. Four-Dimensional Electrical Resistivity Tomography for Continuous, near-Real-Time Monitoring of a Landslide Affecting Transport Infrastructure in British Columbia, Canada. *Near Surf. Geophys.* **2020**, *18*, 337–351. [CrossRef]

15. Mendoza, B.; Carretero Poblete, P.A.; Loaiza Peñafie, J.M.; Peñafiel Barros, G.O.; Tuaza Castro, L.A.; Osorio Rivera, M.A. Location of Archaeological Elements in the Puruha Necropolis of Payacucha (Rumicruz, Riobamba, Ecuador) by Means of Electrical Tomography Analysis | Localización de Elementos Arqueológicos En La Necrópolis Puruhá de Payacucha (Rumicruz, Riobamba, Ecuador). *Arqueol. Iberoam.* **2019**, *43*, 12–19.
16. Swileam, G.S.; Shahin, R.R.; Nasr, H.M.; Essa, K.S. Spatial Variability Assessment of Nile Alluvial Soils Using Electrical Resistivity Technique. *Eurasian J. Soil Sci.* **2019**, *8*, 110–117. [CrossRef]
17. Masi, M.; Ferdos, F.; Losito, G.; Solari, L. Monitoring of Internal Erosion Processes by Time-Lapse Electrical Resistivity Tomography. *J. Hydrol.* **2020**, *589*, 125340. [CrossRef]
18. Su, M.; Liu, Y.; Xue, Y.; Cheng, K.; Kong, F.; Guan, L. Detection Method for Boulders in Subway Shield Zones Based on Data Fusion Multi-Resistivity Three-Dimensional Tomography. *Bull. Eng. Geol. Environ.* **2021**, *80*, 8171–8187. [CrossRef]

Disclaimer/Publisher's Note: The statements, opinions and data contained in all publications are solely those of the individual author(s) and contributor(s) and not of MDPI and/or the editor(s). MDPI and/or the editor(s) disclaim responsibility for any injury to people or property resulting from any ideas, methods, instructions or products referred to in the content.

MDPI
St. Alban-Anlage 66
4052 Basel
Switzerland
www.mdpi.com

Engineering Proceedings Editorial Office
E-mail: engproc@mdpi.com
www.mdpi.com/journal/engproc

Disclaimer/Publisher's Note: The statements, opinions and data contained in all publications are solely those of the individual author(s) and contributor(s) and not of MDPI and/or the editor(s). MDPI and/or the editor(s) disclaim responsibility for any injury to people or property resulting from any ideas, methods, instructions or products referred to in the content.

www.ingramcontent.com/pod-product-compliance
Lightning Source LLC
LaVergne TN
LVHW070219100526
838202LV00015B/2062